D1571673

Northeast Lakeview College
33784 0001 1720 0

Soil and Water Conservation Engineering

Soil and Water Conservation Engineering

FIFTH EDITION

DELMAR D. FANGMEIER, P.E.
Professor Emeritus of Agricultural and Biosystems Engineering
The University of Arizona, Tucson, Arizona

WILLIAM J. ELLIOT, P.E.
Project Leader, Soil and Water Engineering Research Work Unit
U.S.D.A. Forest Service, Rocky Mountain Research Station, Moscow, Idaho

STEPHEN R. WORKMAN, P.E.
Associate Professor of Biosystems and Agricultural Engineering
University of Kentucky, Lexington, Kentucky

RODNEY L. HUFFMAN, P.E.
Associate Professor of Biological and Agricultural Engineering
North Carolina State University, Raleigh, North Carolina

GLENN O. SCHWAB, P.E.
Late Professor Emeritus of Agricultural Engineering
The Ohio State University, Columbus, Ohio

THOMSON
DELMAR LEARNING™

Australia Canada Mexico Singapore Spain United Kingdom United States

Soil and Water Conservation Engineering, 5th Edition
Delmar D. Fangmeier, William J. Elliot, Stephen R. Workman, Rodney L. Huffman, Glenn O. Schwab

Vice President, Career Education Strategic Business Unit:
Dawn Gerrain

Director of Editorial:
Sherry Gomoll

Acquisitions Editor:
David Rosenbaum

Developmental Editor:
Gerald O'Malley

Editorial Assistant:
Christina Gifford

Director of Production:
Wendy A. Troeger

Production Manager:
JP Henkel

Senior Production Editor:
Kathryn B. Kucharek

Director of Marketing:
Wendy Mapstone

Marketing Specialist:
Gerard McAvey

Cover Design:
Suzanne Nelson

Cover Images:
Getty Images

(side bar, top to bottom):
Tim McCabe, USDA
Natural Resources Conservation Service

R. L. Huffman, North Carolina
State University

USDA Natural Resources
Conservation Service

Gene Alexander, USDA
Natural Resources Conservation Service

Printed in the United States of America
1 2 3 4 5 XXX 09 08 07 06 05

For more information contact Delmar Learning,
5 Maxwell Drive, PO Box 8007, Clifton Park, NY 12065-2919.

Or you can visit our Internet site at
http://www.delmarlearning.com.

Library of Congress Cataloging-in-Publication Data
Soil and water conservation engineering /
Delmar D. Fangmeier ... [et al.].-- 5th ed.
p. cm.
ISBN 1-4018-9749-5
1. Soil conservation. 2. Water conservation. 3. Agricultural engineering. I. Fangmeier, D. D. (Del D.)
S623.S572 2005
631.4'5--dc22

2005022986

Dedication

We are dedicating this book to the memory of previous authors Glenn O. Schwab, Richard K. Frevert, Talcottt W. Edminster, and Kenneth K. Barnes, who developed and contributed much to previous editions. Dr. Glenn Schwab is especially recognized for his lifetime dedication to the preparation and publication of each edition of this book. He participated in the selection of new authors for this edition and the early planning and editing.

Contents

Preface

This updated edition continues to emphasize engineering design of soil and water conservation practices and their impact on the environment, primarily air and water quality. Furthermore, the production of food and fiber remains an important consideration because of increasing U.S. and world populations.

Many suggestions from instructors and colleagues have been included. The conversion from English to the International System of units (SI) is complete except in a few cases. Example problems, student problems at the end of chapters, and most illustrations are in SI units. In many cases, units are given in round numbers for ease of computation.

As in previous editions, the purpose of this book is to provide a professional text for undergraduate and graduate agricultural and biological engineering students and for others interested in soil and water conservation in rural and urban areas. Subject matter includes all the engineering phases of soil and water conservation for a one- or two-semester course. The coverage of water quality and urban applications has been increased compared to the previous edition. The first chapter covers general aspects with some worldwide applications; Chapter 2 is a new chapter discussing Water Quality; Chapter 3, "Precipitation" describes new sources and applications of intensity–frequency–duration data, Chapters 4, "Evapotranspiration," and 5, "Infiltration and Runoff" are revised; Chapter 6 contains an expansion and reorganization of Open Channel Flow topics and flow measurement; Chapters 7 through 9, on Erosion and its control, have been revised to include applications to the control of urban runoff; Chapter 10 is a new chapter on Channel Stabilization and Restoration; Chapter 11 is on Water Supply; Chapter 12 is a new chapter on Wetlands; Chapters 13 and 14, "Drainage Principles" and "Water Table Management," add new information on subirrigation; Chapters 15 through 18 cover Surface, Sprinkler, and Microirrigation and include current irrigation design methods ; Chapter 19 is on Pumps; and Chapter 20 is on Soil Erosion by Wind. All chapters include the best and most recent information available.

We have assumed that the student has a basic knowledge of calculus, surveying, mechanics, hydraulics, soils, and computers. We have attempted to emphasize the analytical approach, supplemented by sufficient field data to illustrate practical applications. The text material emphasizes engineering principles in the areas of erosion, drainage, irrigation and water resources. Sufficient tables, charts, and diagrams have been included to provide practicing engineers with readily usable information as well. Many examples and student problems have been included to emphasize the design principles and to facilitate an understanding of the subject matter. Computer models and software program sources have been described where applicable in the text, but detailed programs have not been included, because of space restrictions and the rapid changes in and obsolescence of software programs. Internet sources have been suggested that provide information to supplement the text and access to some computer programs and models. In many instances, students will find using a spreadsheet advantageous for reviewing example problems and solving homework problems.

Delmar D. Fangmeier, William J. Elliot, Stephen R. Workman, and Rodney L. Huffman

Acknowledgements

We are deeply indebted to many individuals and organizations for the use of material. We are especially grateful to The Ferguson Foundation, Detroit, Michigan, for making this text possible by defraying the cost of developing the first edition in 1955. Harold E. Pinches, formerly with the Foundation, was instrumental in promoting this project. We are grateful to Massey-Ferguson Ltd., Toronto, Canada, for providing funds to prepare the second edition. The following individuals have reviewed portions of the manuscript, provided information or made valuable suggestions for this edition: A. J. Clemmens, G. A. Clark, R. G. Evans, D. J. Hills, D. C. Kincaid, D. L. Martin, D. C. Slack, T. Strelkoff, R. G. Allen, A. D. Ward, F. H. Galehouse, G. D. Jennings, C. Agouridis, and T. Wiley. We also wish to recognize the contributions to this edition by Dr. Prasanta Kalita of the University of Illinois.

Special appreciation goes to many friends, colleagues and students at the University of Arizona, the U.S. Forest Service, the University of Kentucky, North Carolina State University, The Ohio State University, and the U.S. Department of Agriculture, who have contributed in many ways through frequent contacts. We also wish to express appreciation to our wives and families for their sympathetic understanding during the preparation of this edition.

Delmar D. Fangmeier, William J. Elliot, Stephen R. Workman, and Rodney L. Huffman

Internet Resources

In this edition we have added the addresses for Web pages at the end of each chapter. These sites were found to be sources of additional information to supplement the text, to provide technical data, and to access computer programs or models mentioned in the text. We realize that these addresses and or contents may change; however, most changes update the information.

An Internet site has been established for this book where the authors and others might add supplemental materials. The text site can be reached through http://www.agriculture.delmar.com.

Students and faculty are encouraged to conduct Internet searches for specific topics when more information is desired than is available in this text.

Abbreviations

Agr.	Agriculture
Agron.	Agronomy
ARS	Agricultural Research Service
ASAE	American Society of Agricultural Engineers
ASCE	American Society of Civil Engineers
ASTM	American Society for Testing Materials
AW	available water
BMP	Best Management Practices
BP	brake power
Bull.	Bulletin
Cons.	conservation
CPT	corrugated plastic tubing
CS	crop susceptibility
DC	drainage coefficient
dia.	diameter
DP	deep percolation
DU	distribution uniformity
DULQ	distribution uniformity low-quarter
EC	electrical conductivity
EPA	Environmental Protection Agency
ESSA	Environmental Science Service Administration
ET	evapotranspiration
EU	emission uniformity
Expt.	experiment
FC	field capacity
geophys.	geophysical
GIS	geographic information system
GPO	Government Printing Office
HDPE	high-density polyethylene
hp	horsepower
i.d.	inside diameter
kPa	kilopascals
kW	kilowatts
kWh	kilowatt hours
L.F.	load factor
mon.	month
NOAA	National Oceanic and Atmospheric Administration
NPSH	net positive suction head
NRCS	National Resource Conservation Service
o.d.	outside diameter
ppb	parts per billion
ppm	parts per million
ppt	parts per thousand
P.C.	point of curvature
PE	polyethylene
P.I.	point of intersection
P.T.	point of tangency
publ.	publication
PVC	polyvinyl chloride
PWP	permanent wilting point
RAW	readily available water
rep.	report
res.	research
rpm	revolutions per minute
RO	surface runoff
RUSLE	revised universal soil loss equation
SAR	sodium absorption ratio
SCS	Soil Conservation Service
SDI	stress day index
SI	International System of Units
Soc.	society
TMDL	Total Maximum Daily Load
UC	uniformity coefficient
USBR	U.S. Bureau of Reclamation
USDA	U.S. Department of Agriculture
USDC	U.S. Department of Commerce
USLE	universal soil loss equation
WEPP	water erosion prediction project
WP	water power

Symbols

a	cross-sectional area; constant; organic matter content
A	watershed area; annual soil loss; cross-sectional area; energy
b	constant; width; soil structure code; water table height
B	outside diameter; width of trench; transport coefficient; length (breadth) of sidewall
c	cut; chord length; profile permeability class
C	coefficient; cover management factor; energy; constant; climatic factor; cut; coefficient of variation; coefficient of skew; correction factor; Celsius; concentration
d	diameter; depth; distance; equivalent depth; effective surface roughness height; ridge spacing
D	diameter; depth; interrill erosion rate; runoff; degree of curvature; duration time; degree days; drainage coefficient; drainage area; deflection lag factor; length of water surface exposure
e	void ratio; distance; deflection angle; vapor pressure
E	efficiency; radiant energy; specific energy head; kinetic energy; degree of erosion factor; evaporation; wind erosion; elevation; modulus of elasticity
ET	evapotranspiration
f	infiltration rate; hydraulic friction; depth; fill; soil porosity
F	total infiltration; fertility factor; Froude number; force or load; dimensionless force; fill; dry soil fraction; safety factor
g	acceleration of gravity; gram
G	sensible heat; energy
h	head; wave height; height; hour; depth of channel; head loss
H	total head; height; head loss; suction head; specific energy head; riser pipe height
i	rainfall intensity; inflow rate; irrigation rate
I	total rainfall; irrigation depth; angle of intersection; soil erodibility index; rate of application; initial rainfall extraction; impact coefficient; moment of inertia
k	constant; permeability; time conversion factor; capillary conductivity; von Karman's constant
K	constant; hydraulic conductivity; bedding angle factor; soil erodibility; ridge roughness factor; ratio of tractive force; head loss coefficient; Scobey's coefficient of retardation; frequency factor; crop coefficient; sinuosity
l	slope-length
L	liter
L	length; slope length factor; slope length; leaching; wave length
m	meter; exponent; water content; water table height; rank order of events
M	watershed area; depth of snowmelt; particle size diameter
n	Manning roughness coefficient; drainable porosity; constant; number of values; hours of sunshine; pump specific speed
N	curve number; total number of events; revolutions per minute; hours of sunshine; Newton; normal concentration
o	outflow rate
p	atmospheric pressure; percent area shaded
P	probability; power; pressure; peak runoff rate; rainfall; conservation practice factor; water content; deep percolation; atmospheric pressure; wetted perimeter

q	seepage or flow rate; sprinkler discharge rate, emitter flow rate
Q	water volume; flow rate
r	radius; rate of application; reflectance coefficient; scale ratio (prototype to model)
R	hydraulic radius; radius; rainfall and erosivity index; ratio; residue cover; radiant energy; runoff; radius of influence
s	slope gradient or percent; rate of water storage; distance; standard deviation; second; well drawdown
S	slope gradient or percent; slope steepness factor; water storage; settlement; sprinkler or drain spacing; slope erosion factor; energy; maximum difference between rainfall and runoff; seepage loss; Siemens
t	time; temperature; thickness; width; Student's statistical level of significance
T	time; temperature; time conversion interval; time of concentration; time of lag; recession time; time of peak; time of advance or recession; return period; concentrated surface load; tangent distance; width; tractive force; transport capacity of runoff
u	volume conversion factor; wind velocity; settling velocity
v	velocity; rate of capillary movement; rate of soil water movement; threshold velocity
V	volume; vegetative cover factor
w	unit weight; flow conversion factor; wetted diameter
W	weight; width; watershed characteristics; water volume; furrow spacing; watt; average wind velocity; load on conduits; dry weight of soil; water depth
x	constant; variable; return period variate; distance; mean; water level; ratio
X	constant; variable; distance; horizontal coordinate
y	depth; duration variate; deviation of water depth
Y	constant; variable; minimum years of record; distance; vertical coordinate
z	sideslope ratio (horizontal to vertical); depth; height; soil roughness parameter; vertical coordinate; infiltrated depth
Z	vertical distance; infiltrated water volume or depth per unit area; depth
γ	psychometric constant; water density
Δ	slope of the saturated vapor pressure curve; length of upstream face
θ	sideslope angle; slope degrees; angle of derivation; soil water content
λ	latent heat of vaporization
ρ	density
μ	dynamic viscosity
ϕ	soil water potential (capillary)
Ψ	gravitational potential
Φ	potential
σ	Stefan-Boltzmann constant
τ	shear stress; hydraulic shear, infiltration opportunity time

Note: In the text various subscripts may be added to identify the specific definition for each symbol shown above.

CHAPTER 1

Conservation and the Environment

Soil and water conservation engineering is the application of engineering and biological principles to the solution of soil and water management problems. The conservation of natural resources implies *utilization without waste* while maintaining a continuous profitable level of crop production and while improving environmental quality. Engineers must develop economical systems that meet these requirements.

The engineering problems involved in soil and water conservation can be divided into the following topics: erosion control, drainage, irrigation, flood control, and water resource development and conservation. Although soil erosion takes place even under virgin conditions, the problems to be considered are caused principally by human exploitation of natural resources and the removal of the protective cover of natural vegetation. Urban–rural interface problems are even more serious because of high population density and increased runoff caused by severe changes in land use.

Drainage and irrigation involve water and its movement on the land surface or through the soil mass to provide optimum crop growth. To provide water at places and times when it is not naturally available, surface reservoirs or other storage facilities must be developed for irrigation and domestic use. Where available, ground water supplies can be developed and maintained by recharge techniques. Flood control consists of the prevention of overflow on low land and the reduction of flow in streams during and after heavy storms. In water-short regions, soil water should be conserved by modified tillage and crop management techniques, level terracing, contouring, pitting, reservoirs, water harvesting, and other physical means of retaining precipitation on the land and reducing evaporative losses from the soil surface.

Impact of Conservation Practices on the Environment

Some past civilizations have perished because they did not understand the impact of their agricultural practices or were unable to remedy their effects. For example, in some regions high soil salinity reduced productivity to the point where food supplies were insufficient for the population to survive. Currently some argue that we do not have the incentive or economically viable technology to maintain long-term food production. Thus it is important that we develop and apply the best conservation practices to protect the environment and conserve our soil and water resources for long-term sustainability.

Conservation practices can be beneficial or detrimental to the environment. For example, the terracing of rolling cultivated land can be beneficial by reducing erosion and sedimentation, or irrigation can be detrimental by increasing the salinity of downstream flow by the drainage water. Agricultural production of food and fiber is essential, but preserving and enhancing the environment is also necessary as population pressures increase. When drinking water becomes contaminated with nitrates, pesticides, sediment, and other materials, the public is and should be concerned. The reduction in wildlife and disappearance of some species on the endangered list causes suspicion that something is detrimentally affecting the environment. Changing land use will modify the quantity and quality of terrestrial and aquatic habitats. Agricultural and wildlife interests are increasingly being coordinated by local, state, and federal governments. Many conservation practices are being changed to improve the environment with little or no cost to agriculture. Environmental costs are paid either by taxes, when supported by the government, or by higher prices, when supported by the private sector. With the adverse economic conditions in agriculture in the 1980s, low-input, sustainable agriculture programs were promoted. One of the many ways proposed to improve farming, which could also enhance the environment, was to reduce fertilizers, pesticides, and other inputs. Others would argue that an increase in the use of fertilizers would increase production per hectare and require less land, so that erodible land could be taken out of cultivation and thus reduce erosion and pollution.

Erosion is one of the most serious problems, because it pollutes the water with sediment, nutrients, and pesticides, increasing the cost of water treatment. Considerable effort has been made since the 1930s to develop conservation practices and to encourage reduced tillage and similar practices to keep crop residues and cover on the land. Although reduced tillage practices have been effective in reducing cost of production, pesticide use is generally increased with minimum or no tillage.

In general, the impact of drainage on the environment has been beneficial. The drainage of the lakebed area in the Midwest was instigated by medical doctors, who realized that drainage reduced malaria long before the mosquito was known as the cause. Central Park in New York City, which was one of the earliest drainage projects, was also drained for health reasons. Obviously, subsurface drainage provides a desirable environment for the roots of agricultural crops and a stable base for roads and buildings. Without drainage much of our most productive cropland could not have been developed.

During the early development of the United States, agricultural interests prevailed and the drainage of swamps and other wetlands was done without question. In more recent years, drainage of these naturally wet areas was found to have some adverse effects on migrating terrestrial and aquatic life and other environmental aspects. As much as 50 percent of the wetlands have been drained, mostly in the Southeast, Midwest, and Pacific Coast states (USDA, 1989). Wetlands vary from permafrost in Alaska to the everglades in Florida to the desert wetlands in Arizona. In addition to providing habitat for fish and wildlife, wetlands enhance ground water recharge, reduce flooding, trap sediment and nutrients, and provide recreational areas.

Federal legislation known as the Swampbuster Provision of the 1985, 1990, and 1996 Farm Bills was passed to discourage wetland drainage.

Subsurface drainage water in humid areas and from irrigated lands generally contains higher concentrations of nitrate nitrogen, salts, and other chemicals than found in surface water or in the irrigation water. Wildlife interests often claim that open drainage ditches, when cleaned and straightened, reduce cover for wildlife. Because of this concern, some channel improvement projects have been redesigned to remove vegetation on only one side of the ditch and restrict cleanout and reshaping work to the opposite side. In large channels, low, small dams can be constructed to provide water pools for better fish habitat.

Irrigation in the central United States has been significantly changed by center-pivot irrigation. More land was converted to irrigation because of the center-pivot's ability to move over rolling terrain. One environmental benefit is that field corners have not been irrigated but planted to trees and other vegetation. These corner areas have become wildlife sanctuaries, and the wildlife population has greatly increased.

Irrigation projects, as in the U.S. West, have a major impact on the social and political structure of the entire community. Water storage and stream diversion facilities for irrigation, as well as stream flow and ground water use, influence minimum stream flow, depth of ground water, fish and wildlife populations, natural vegetation and crops grown, recreation activities, roads and public utilities, and other unique and site-specific factors. One example of a serious environmental problem is the chemical pollution (selenium and boron) from irrigation drainage water, such as occurred at the Kesterson National Wildlife Refuge in California (see Chapters 14 and 15). Blockage of the drainage system was mandated by the court. The Left Bank Outfall Drain in lower Pakistan, which is a constructed channel more than 250 km in length carrying only saline drainage water to the sea, illustrates the magnitude of the pollution problem. The Salton Sea in California is an example of an evaporation disposal site for drainage water. In the construction of the Aswan Dam in Egypt, many people were displaced and archaeological sites were covered with water. Similar impacts exist where large dams are built for flood control. Because of such problems as just described, environmental impact statements, which necessitate more overall planning effort, are now mandatory for most projects that impact soil or water resources. Details of various conservation practices and their environmental impacts are included in most chapters, and Chapter 2 has been devoted specifically to water quality.

1.1 Engineers in Soil and Water Conservation

Sound soil and water conservation is based on the full integration of engineering, atmospheric, plant, and soil sciences. Agricultural and biological systems engineers, because of their training in soils, plants, biology, and other basic agricultural subjects, in addition to their engineering background, are well suited to integrate these sciences. To develop and execute a conservation plan, engineers must have knowledge of the soil, including its physical and chemical characteristics, as well as a broad understanding of soil–plant–water–environment interactions. They have a unique role because their efforts are directed toward the creation of the proper environment for the optimum production of plants and animals. In addition to the agricultural application of their knowledge, they are playing an increasing role in the rural-urban sector, especially relating to air and water pollution control.

To be fully effective in applying technical training, engineers must be acquainted with the social and economic aspects that relate to soil and water conservation. They must have a full understanding of the various local, state, and federal government policies, laws, and regulations. They should also become familiar with ground and satellite mapping techniques, nationwide Geographic Information Systems (GISs), weather records and prediction systems, soil survey reports, and other physical data. To apply the vast amount of information available, engineers should be knowledgeable and able to use computers for solving job-related problems. Some of the presently available software will be referenced later.

1.2 Conservation Ethics

The increasing world population will dictate the necessity of conserving natural resources now and in future years. Fossil fuels, soils, minerals, timber, and many other materials are being exhausted at a rapid rate. The average 70-year-old American in a lifetime will use 1 000 000 times his or her weight in water, 10 000 times in fossil fuels and construction materials, and 3000 times in metal, wood, and other manufactured products (Wolman, 1990). Recycling of paper, glass, metals, and other items is being practiced, partly because of the increasing costs of waste disposal and partly because of public support and appreciation for conservation. The decreasing population of wildlife and the disappearance of many species are evidence that much of the problem is related to air and water pollution as well as loss of habitat. In agriculture, soil erosion from farmland is not only one of the major causes of water pollution, but the loss of the land itself reduces the production of food and fiber. Government incentive programs since the 1930s have been helpful, and research efforts have developed many useful agronomic, tillage, and mechanical practices. Exploitation of the soil and other natural resources for economic benefit has been practiced since the early pioneers developed this country. Private enterprise systems encourage this concept, but it is not in the best interest of society. Natural resources should be passed on to future generations in as good or better condition than previous generations have left them. Conversion of prime farmland to urban development and other nonfarm uses without regard to future food needs continues because farmland cannot compete with other, higher-value uses. Political solutions to these problems are not likely, but we can continue to promote and teach appropriate conservation ethics.

1.3 Environmental Regulations

Agricultural production has increased through more efficient farming systems and the use of fertilizers and pesticides. These on-site practices have had off-site impacts such as increasing the nitrogen content of ground and surface water supplies, as well as increasing the chemical content of streams and lakes from pesticides in runoff. Because most of the population is in urban areas, this has reduced the availability or quality of their already limited water supplies. To reduce the impact of various agricultural practices, environmental regulations have been established. Each state is authorized/required to establish Best Management Practices (BMPs). These practices establish guidelines based on Best Available Technology (BAT) that will minimize the pollution of our soil and water resources. For example, guidelines recommend the timing and amount of fertilizer to be applied to various crops that will just meet crop needs and thus reduce the nitrogen content of runoff and percolation.

As further criteria, Total Maximum Daily Loads (TMDLs) have been established for sediment and various chemicals in streams and rivers.

1.4 Hydrologic Cycle

Engineering for soil and water conservation and environmental control is based on understanding the processes of the hydrologic cycle. The hydrologic cycle (Figure 1-1) is composed of the processes of precipitation, evaporation, infiltration, runoff, transpiration, percolation to ground water, and interflow. Precipitation is the primary source of our usable water supply. The measurement or prediction of precipitation is required for most of the designs for soil and water conservation. Evaporation is the major depletor of water once it falls as precipitation. Infiltration fills the soil profile and adds to ground water storage. Runoff must be measured or predicted to estimate and control soil erosion, to predict flood events, and to estimate the availability of surface water supplies. Transpiration is difficult to measure and, combined with evaporation, is called evapotranspiration. Evapotranspiration is important for irrigation and drainage design. Ground water supplies are replenished by percolation. Interflow is the lateral flow of water to streams or lakes. More detailed descriptions of the major processes in the hydrologic cycle are discussed in Chapters 3, 4, and 5.

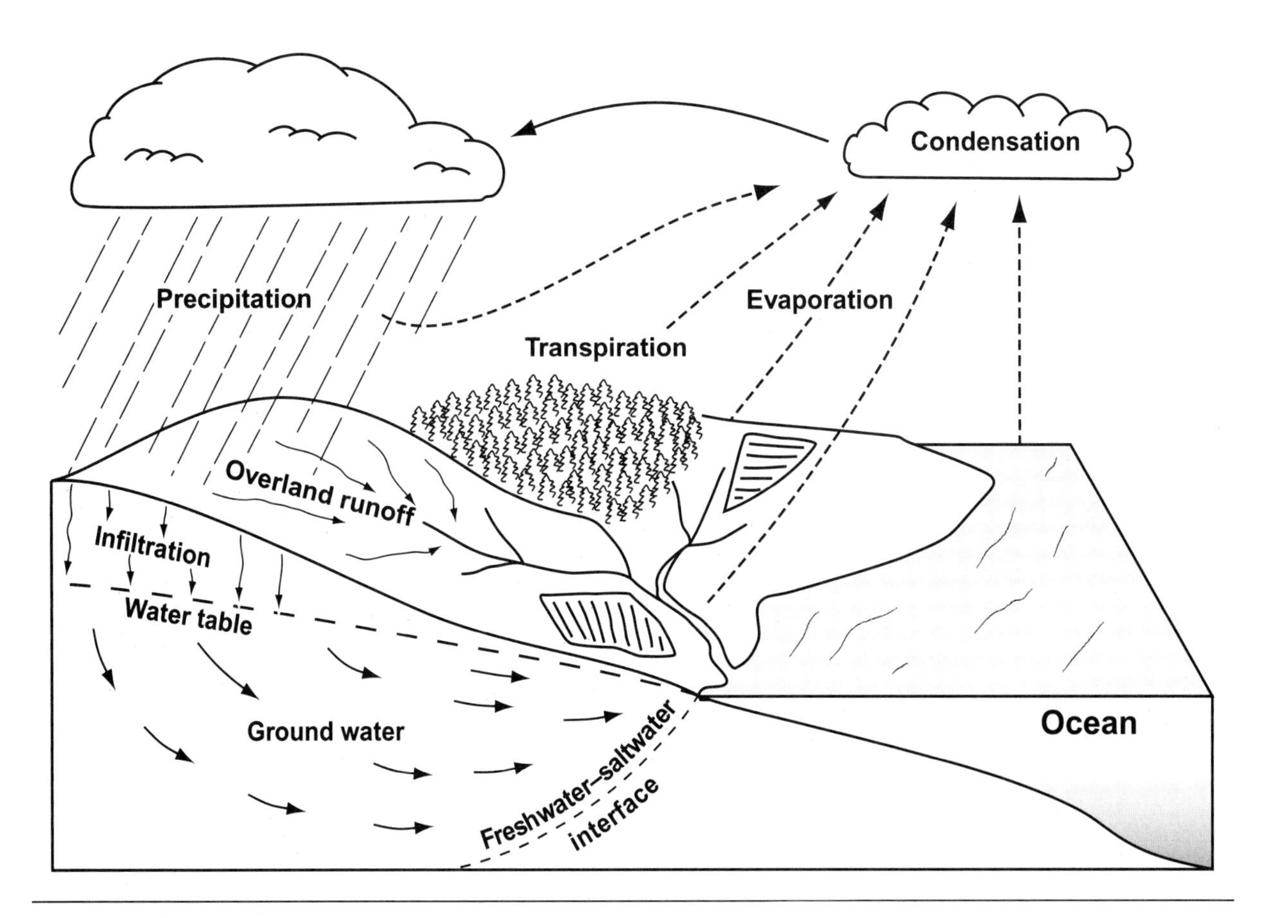

Figure 1–1 The hydrologic cycle (Adapted from Heath, 1980)

Major Conservation Practices

This text considers conservation practices for erosion control, drainage, irrigation, and flood control and water supply.

1.5 Soil Erosion

Erosion is a serious problem worldwide. The control of soil erosion by water and wind is essential to maintain crop productivity, reduce sedimentation, and reduce pollution in streams and lakes. The amount of soil eroded by water from fields and pastures in the United States is estimated to be equivalent to about 1 m from 200 000 hectares (ha), of farmland each year. Additional losses are caused by wind erosion. Not only is soil lost by erosion, but a proportionally higher percentage of plant nutrients, organic matter, pesticides, and fine soil particles is lost than in the original soil. One of the most dramatic wind erosion events was the dust storm of 1934, which swept across the United States from the Great Plains to the Atlantic Coast. In the United States, federal legislation has provided technical assistance and financial incentives for controlling erosion and reducing pollution.

1.6 Drainage

Drainage practices in the world date back thousands of years. Some notable examples of large drainage projects are the polders in Holland and the fens in England, which are lowlands reclaimed from the sea. Drainage has long been recognized as essential for permanent crop production in low flatlands, in humid areas, and in irrigated agriculture. In the humid U.S. Midwest many drainage systems have been installed to develop and improve crop production. These systems may provide both surface and subsurface drainage. An estimated one third of the irrigated land in the world is affected by sodicity and salinity. Subsurface drainage systems are required to reclaim salt-affected soils and to prevent salinity problems by maintaining a low water table. In areas that experience periods of both excessive water and water shortages during a growing season, subsurface pipe drains are also used to deliver water to the soil profile.

1.7 Irrigation

Irrigation expanded worldwide after 1950 because of the increased need for more food and fiber and because of the development of new and more efficient methods. Much of the expansion began in arid regions where few crops can be grown without irrigation. Later expansion was in more humid areas to supplement rainfall for maximum crop production. Many large dams were constructed for water storage. More efficient well-drilling methods and better pumps also significantly contributed to the water supply and pressure needs of sprinkler and microirrigation systems. Areas with productive soils, adequate drainage, and a reliable water supply were irrigated in arid, semiarid, and subhumid climatic regions. Relatively large quantities of water are required for crop production in arid areas, with irrigation using up to 80 percent of the available water supply.

1.8 Flood Control and Water Supply

Flooding is one of the most significant natural phenomena, causing loss of life, crops, and property as well as creating health hazards, water pollution, and interruption of services, such as transportation, utilities, police, fire protection, and emergency operations. Flood damage in upstream areas occurs primarily on agricultural lands, whereas downstream floods cause major damage in metropolitan areas. This text considers conservation designs that reduce erosion and flooding in upstream areas. Many practices that reduce flooding add to the water supply by allowing more water to infiltrate, which may increase crop production as well, or reduce the peak flow by water storage. Stored water may also be used for on-farm irrigation.

Internet Resources

NRCS photos, publications, specialty centers, and technical information
http://www.nrcs.usda.gov

NRCS Water and Climate Center publication "Conservation and the Water Cycle"
http://www.wcc.nrcs.usda.gov/factpub/aib326.html

"Major Uses of Land in the United States, 1997"
http://usda.mannlib.cornell.edu

References and Bibliography

Heath, R. C. (1980). *Basic Elements of Ground-Water Hydrology with Reference to Conditions in North Carolina.* U.S. Geological Survey Water Resources Investigations Open-File Report 80–44. USGS Open-File Services Section, Denver, CO 80225.

Postel, S. (1989). *Water for Agriculture Facing Limits.* Paper 93. Washington, DC: World Watch Institute.

U.S. Department of Agriculture (USDA). (1989). *The Second RCA Appraisal. Soil, Water and Related Resources on Nonfederal Land in the United States.* Washington, DC: U.S. Government Printing Office.

Vesterby, M., & K. S. Krupa. (1997). *Major Uses of Land in the United States.* Statistical Bulletin No. 973. Resource Economics Division, Economic Research Service, U.S. Department of Agriculture. Washington, DC.

Wolman, M. G. (1990). The impact of man. *EOS, Transactions of the American Geophysical Union, 71*(52), 1884–1886.

CHAPTER

2 Water Quality

Water is a primary component of the biosphere. The ability of the biosphere to support life as well as the health and enjoyment of that life depends on water quality. Adequate supplies of clean water are vital for agriculture, domestic use, recreation, wildlife, and thousands of manufacturing and mining processes. As competition for water grows, wise management and protection of that resource becomes increasingly important.

Historically, the primary objective of agricultural development has been production, i.e., maximizing arability and yields. Where production was the primary objective, environmental quality often suffered. Declines in natural habitat have spurred major changes in engineering and production practices. Environmental quality objectives now play a major role in development of designs and management strategies. Agriculture, forestry, construction, and other types of development are subject to federal and local regulations regarding water use, flow management, and associated transport of contaminants.

Many dramatic examples can be found throughout the world where production-oriented practices have resulted in environmental threats or actual degradation. The following are a few examples.

- In the San Joaquin Valley of California, selenium concentrations in the Kesterson Reservoir rose to levels that caused serious malformations in nesting birds. Naturally occurring selenium was leached from irrigated areas and carried by the San Luis Drain into the reservoir. Since the reservoir did not have an outlet, it acted as an evaporation basin, concentrating the solutes until toxic concentrations were reached. The drain and reservoir were ordered closed. The Department of the Interior eliminated discharges from the drain in 1986. Drainage for the San Luis Unit of the Central Valley is still unresolved.
- Intensive livestock production in some areas of the United States has created a nutrient imbalance where nutrients are imported in feed (and subsequently excreted as waste) in greater quantities than can be effectively utilized by local crops. Excess application of animal wastes to cropland has resulted in contamination of ground and surface waters. Feed additives such as copper and zinc can

accumulate in the soil to the point of toxicity for some crops. Concerns are also growing about antibiotics and hormones added to feeds.

- A major irrigation project in the former Soviet Union diverted much of the water from two rivers (Amu Darya and Syr Darya) that feed the Aral Sea on the Kazakhstan–Uzbekistan border. Evaporative losses now exceed inflows. The surface area of the sea has been reduced by nearly half and the volume by about three quarters since the early 1960s, with the shoreline having receded tens of kilometers. A sea that had been a productive fishery was reduced to concentrated brine where very little lives.
- Heavy withdrawals of water from the Colorado River in the western United States and Mexico are changing the ecology of the lower reaches of the Colorado and its delta on the Gulf of California. Withdrawals are for both irrigation and urban use. With these extractions, the Colorado River now reaches its outlet on the Gulf of California only in very wet years.
- Large feedlots in the central United States concentrate thousands of cattle on relatively small areas. Runoff from those feedlots is high in nutrients and can cause severe impacts if allowed to reach surface waters. Runoff collection and treatment facilities can greatly reduce the export of nutrients from feedlot operations.
- Many commonly used herbicides have been found in water supplies—sometimes at concentrations exceeding health advisory levels. Although normal usage is sometimes the source, careless handling of chemicals and equipment washwater is often to blame. Concerns have led to major changes in the types of chemicals permitted for use. Approved practices for mixing and rinsing have been developed to prevent direct contamination events.
- Many municipal waste treatment plants land-apply wastewater and sludge that contains trace amounts of heavy metals. The pH of the soils on the application sites must be carefully managed for many years so that the metals stay in insoluble, immobile forms. Failure to do so would allow the metals to leach into the ground water.

These examples illustrate that well-intentioned practices can have unintended effects—often far beyond the area where the practices are implemented. The impacts of a particular practice on a small area may be insignificant, but the cumulative effects of regional implementation of that same practice can be devastating. Various agencies with responsibility for ecosystem management are now given authority to regulate practices on local, state, regional, and river basin scales.

Human influence has changed many ecosystems, and in all likelihood change will continue. The pressures of growing populations and desires for improved standards of living will continue to drive development. Engineers are in the middle of all this, charged with providing for society's needs while protecting society's interests in the quality of the environment and the long-term health of these resources. To meet these challenges, engineers must be aware of the effects of various practices and be able to select those that will maximize utility with minimal environmental impacts. They must be aware of the regulations that restrict, prohibit, or require certain practices. Competent engineers can be a voice of reason in the shaping of public policies and regulations.

Water Quality Issues

Direct measures of water quality are the concentrations of biological, chemical, and physical contaminants. Quality of aquatic habitat involves additional parameters such as temperature, channel characteristics, turbidity, and dissolved oxygen.

Water quality standards vary with the intended use. Standards and regulations have been established for drinking water, discharge from industrial or municipal treatment plants, discharge or runoff from agricultural operations, runoff from forestry operations, and land application of wastewaters, to name a few. The primary regulatory agency in the United States is the Environmental Protection Agency (EPA), but various state and local authorities impose standards as well. Most agencies publish current standards via the Internet.

Many factors may be involved in any water quality situation: pH, alkalinity, temperature, dissolved oxygen, turbidity, sediment, macronutrients, other inorganic species, hardness, organic matter, salinity, pesticides, nonaqueous-phase liquids, or other contaminants resulting from human activity such as solvents, PCBs, and dioxin. In addressing any situation, consider, as appropriate:

- Cause or source. Natural, anthropogenic, or both? Magnitude or intensity of the source(s)?
- Concentrations that are (1) desirable, (2) tolerable, (3) toxic with chronic exposure, (4) toxic with acute exposure.
- Effects on soils, plants, or aquatic habitats.
- Solubility, volatility, density. Dominant physical/chemical forms.
- Transport mechanisms. Advection and/or diffusion.
- Reactions, biological or chemical, in ground water or surface water. Rates.
- Partitioning of solutes between water and solids. Affinity for soil or organic matter.
- Persistence in the environment. Half-life. Daughter products. Modes of breakdown.
- Assessment methods. Can it be quantified? How is it measured? Standard analytical methods.
- Treatment/removal. Available methods. Efficacy. Cost. Practicality.
- Regulatory controls and requirements. Which agencies have authority?

Water quality issues are seldom easy. As this list suggests, many factors—physical, chemical, biological, economic, social, and political—may be important. As an introduction to some of the issues in water quality, a number of major topic areas will be discussed briefly.

2.1 Trophic States

Waters may be classified according to trophic conditions (related to the availability of nutrients) or productivity. Trophic classification uses a continuous scale ranging from oligotrophic (relatively poor in nutrients and having low productivity) to mesotrophic (moderate nutrient availability and moderate productivity) to eutrophic (rich in nutrients and having high productivity). High productivity is not necessarily good. Many game fish do best in oligotrophic waters, which tend to be

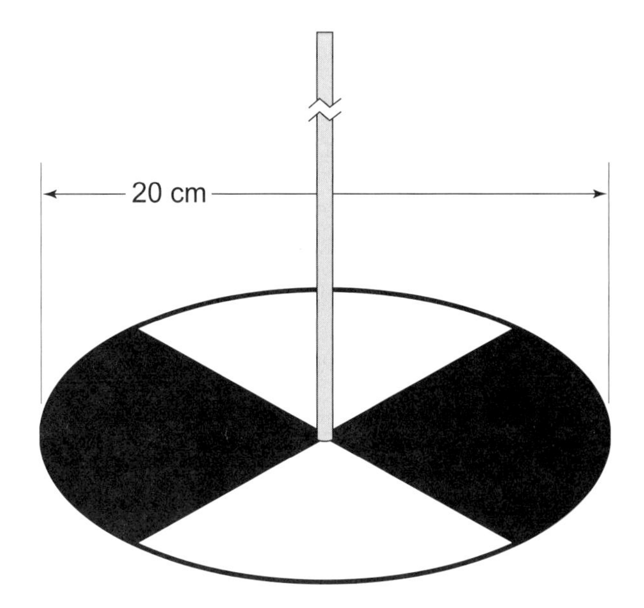

Figure 2–1
Secchi disk. The disk is lowered into a body of water until it is no longer visible. The depth is an index of the transparency of the water.

clear (Secchi disk depths averaging about 3 meters—see Figure 2-1). Eutrophic (and hypereutrophic) waters tend to be murky (Secchi disk depths averaging less than 1 meter). The high organic matter content of eutrophic systems (Figure 2-2) make them more prone to becoming anoxic. Table 2-1 provides standard definitions of the various trophic states.

Figure 2–2 Algae in a lake in Iowa. (Photo by Lynn Betts, NRCS)

TABLE 2–1 Definitions of Trophic States

Trophic State	*Definition*
Oligotrophic	Clear waters with little organic matter or sediment and minimum biological activity.
Mesotrophic	Waters with more nutrients, and therefore, more biological productivity.
Eutrophic	Waters extremely rich in nutrients, with high biological productivity. Some species may be choked out.
Hypereutrophic	Murky, highly productive waters, closest to the wetland status. Many clear-water species cannot survive.
Dystrophic	Low in nutrients, highly colored with dissolved humic organic material. (Not necessarily a part of the natural trophic progression.)

Source: Environmental Protection Agency (2004).

In the course of natural aging, conditions in lakes change from oligotrophic to eutrophic as sediments and organic matter accumulate. This process normally takes thousands of years. Human influence can radically accelerate the process, but may be reversed somewhat through improved practices of nutrient management and erosion control (Chapters 7–10, 12–14).

The degree of eutrophication of a body of water can be quantified using a trophic state index that relates various combinations of Secchi disk depths and concentrations of total nitrogen, total phosphorus, and chlorophyll *a* (e.g., Carlson, 1977). A number of such indices have been developed for different types of water bodies and situations.

2.2 Dissolved Oxygen

The quality of natural waters as habitat for aquatic species is strongly related to the amount of oxygen available in those waters. With a free surface in contact with the atmosphere, the exchange of gases between air and water is usually adequate to support a wide range of species. Under some conditions, however, the gas exchange may not take place fast enough to supply all demands. Those organisms that can survive with lower oxygen concentrations will survive, whereas those that require higher concentrations suffocate. Fish kills are the result of such competition.

The amount of dissolved oxygen (DO) naturally varies with depth in the water column. In swift or shallow streams, vertical mixing carries oxygen to the bottom. In slow streams and swamps, vertical mixing is weaker. In waters where oxygen demand is higher than the rate of supply by mixing and diffusion, the waters near the bottom may be anoxic.

2.3 Contaminant Sources

Any substance or organism that is present in an amount or concentration that is objectionable or harmful may be considered to be a contaminant. In some cases, contaminants are naturally occurring, but their release may be influenced by human activity, such as the problem with selenium in the Kesterson Reservoir cited earlier. More often, contaminants are directly produced and distributed by human activity.

Countless agricultural, municipal, and industrial sources have contributed contaminants to both ground and surface waters.

Contaminant sources that are particularly important in rural areas include fertilizers, pesticides, septic tank effluent, animal wastes, and agricultural and municipal sludges. Particulate and gaseous emissions, such as smokestack or vehicle discharges, may be transported long distances through the atmosphere before eventual deposition in the landscape.

Biological Contaminants

Microorganisms in drinking water are a problem throughout the world. They include viruses, bacteria, algae, and protozoa. Although most microorganisms are harmless or beneficial, many are pathogenic. The very young, very old, and immune-compromised are most susceptible to water-borne pathogens. The most common sources of these pathogens are human and animal wastes. Figure 2-3 shows the sizes of the biological contaminants discussed below.

2.4 Protozoa

Protozoan cysts range from 2 to 15 microns. The most common include *Giardia lamblia*, *Entamoeba histolytica*, and *Cryptosporidium*. These cause diarrhea and gastroenteritis. *Schistosomiasis* is found in streambanks in the tropics and is a major cause of disease in those regions (McCutcheon et al., 1992).

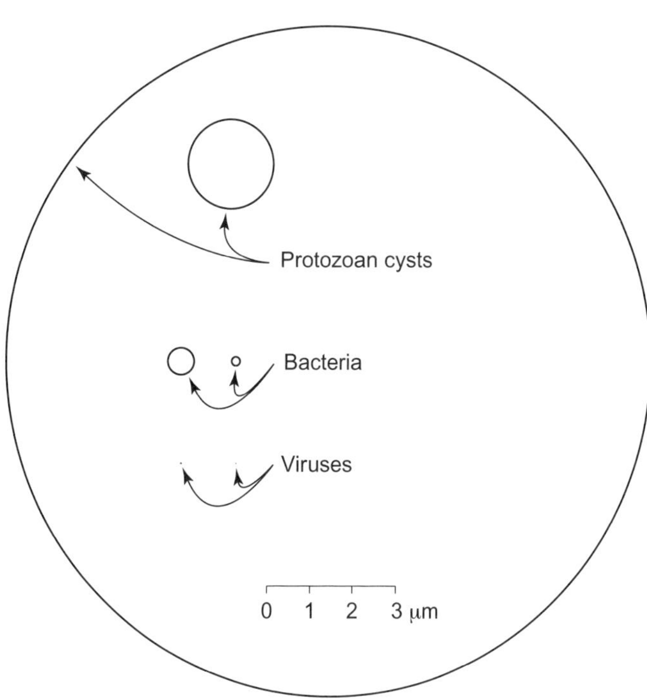

Figure 2–3

Relative sizes of biological contaminants.

2.5 Bacteria

Bacteria range in size from 0.2 to 0.6 microns. They can cause cholera (*Vibrio cholerae*), typhoid fever (*Salmonella* serogroup Typhi), and epidemic dysentery (*Shigella dysenteriae* type 1), among others. Water-borne bacterial infections usually are associated with poor sanitation and hygiene. *Escherichia coli* includes many coliform bacteria, mostly benign, that inhabit the intestines of warm-blooded animals. A test for fecal coliform is commonly used as an indicator of fecal contamination of a water supply. In recent decades, several toxic strains have been discovered that can cause diarrhea and even kidney failure (Mead & Griffin, 1998). Infection is usually via contaminated, uncooked, or undercooked foods or via direct contact with infected individuals.

2.6 Viruses

Viruses are the smallest microorganisms, ranging from 0.01 to 0.03 microns. Enteric viruses infect the gastrointestinal tract of mammals and are excreted in feces. Where water supplies are contaminated by feces, there is the potential for transmission. Water-borne viruses of particular concern include hepatitis A, Norwalk-type viruses, rotaviruses, adenoviruses, enteroviruses, and reoviruses (AWWA, 1990), most of which infect the intestine and/or the upper respiratory tract. Viruses can also cause aseptic meningitis, encephalitis, poliomyelitis, and myocarditis. The most effective means of prevention are good sanitation and hygiene.

Chemical Contaminants

2.7 Concentration Units

There are several methods for expressing the amount of a solute in a given mass or volume of solution. Each has advantages for particular applications.

Molal concentration (molality) is the number of moles of solute per kilogram of solvent.

Molar concentration (molarity) is the number of moles of solute per liter of solution.

Normal concentration (normality) is the number of gram equivalent weights of solute per liter of solution. The gram-equivalent weight is the gram molecular weight divided by the hydrogen equivalent (i.e., valence) of the substance.

Mass concentration is the mass of solute per unit volume of solution. Most water quality standards and laboratory reports express concentrations in mg/L or µg/L.

Equivalents per liter is the number of moles of solute multiplied by the hydrogen equivalent of the solute per liter of solution.

Parts per million (ppm) is the number of grams of solute per million grams of solution. Very small concentrations are often reported in parts per billion (ppb) or parts per trillion.

For a water solution, if the total solute concentration is less than 10 000 ppm and the solution temperature is near 4°C, molarity and molality may be considered equal and 1 ppm equals 1mg/L. Likewise, 1 ppb equals 1 µg/L. For most practical purposes, these equivalencies may be used for total solutes less than 10 000 ppm and temperatures lower than 100°C. For higher concentrations, density corrections will be necessary (Freeze & Cherry, 1979).

Where concentrations reported in mg/L (or ppm) must be used in chemical computations, they can be converted to molarity or molality.

$$\text{Molarity} = \frac{\text{mg/L}}{1000 \times \text{formula weight}} \tag{2.1}$$

Example 2.1

Potassium bromide (KBr) is to be used in a tracer experiment. The KBr is to be applied using a sprayer calibrated to apply 200 L of water per hectare. To apply 8 g/ha of Br, how much 0.1N solution of KBr should be mixed with water to yield 200 L of solution for application?

Solution. A 0.1N solution contains 0.1 gram-equivalent weight of solute per liter of solution. One gram-equivalent weight of bromide is 79.9 g, so the solution contains $0.1 \times 79.9 = 7.99$ g Br per liter. To get 8 g Br, mix 1 L of 0.1N KBr with 199 L of water.

2.8 Macronutrients—Nitrogen and Phosphorus

Nitrogen (N) and phosphorus (P) are the primary macronutrients of concern with water quality. Both N and P are present in natural waters. Where significant N or P is added, algal blooms may occur. Algae compete very strongly for aquatic nutrients and respond very quickly to changes in trophic conditions. An abnormal abundance of nutrients permits rapid growth of algae (a bloom), which may cause offensive tastes or odors. As the algae die and decompose, dissolved oxygen can be depleted, killing fish and other aquatic fauna.

In freshwater systems, cyanobacteria (blue-green algae) are able to fix nitrogen from the atmosphere, so P is usually the limiting nutrient. In estuarine waters, N is usually limiting. Concentrations as low as 0.1 mg/L nitrate–nitrogen (NO_3–N) can be sufficient to trigger algal blooms in estuarine systems (Mallin, 1994).

Enrichment of rivers, lakes, estuaries, and coastal oceans with macronutrients from agricultural and urban sources has been clearly linked to eutrophication (Carpenter et al., 1998) and suggested as a cause of increases in populations of toxic microorganisms such as *Pfisteria piscicida* (Glasgow & Burkholder, 2000). More frequent occurrences of harmful algae blooms (often misnamed "red tides") have been correlated with increases in pollution levels in many locations around the world, but there is still disagreement whether there is a causal link.

Nitrogen accounts for about 78% of the atmosphere, but the dominant gaseous form (N_2) cannot be used directly by most plants. To enhance plant growth, nitrogen is applied in chemical fertilizers (such as urea $(NH_2)_2CO$, ammonium nitrate NH_4NO_3, or anhydrous ammonia NH_3) or in organic forms in sludges or animal and vegetable wastes. The inorganic forms become available as they dissolve in the soil water. The organic forms must first be mineralized (converted to inorganic forms). As the organic matter decomposes, ammonia(um) is released. Under anaerobic conditions, ammonia(um) may remain in solution or adsorbed to soil indefinitely. In aerobic conditions, with sufficient moisture and carbon, ammonia(um) can be quickly converted to nitrite and then nitrate, an anionic form that is highly soluble and very mobile. N as nitrate is readily leached from the soil and can be a problem in ground water. Nitrate can migrate for considerable distances in the ground water without appreciable reduction in concentration. Riparian buffers can be effective for removing modest concentrations of nitrate from discharging ground water (Gilliam et al., 1997). Water table management and controlled drainage (Chapter 14) can reduce nitrogen export in drainage waters from agricultural lands. Figure 2-4 depicts the nitrogen cycle.

Nitrate is the form of N most associated with health risks. High concentrations of nitrate in drinking water contribute to methemoglobenemia (blue baby syndrome), which primarily affects infants and has been associated with spontaneous abortions (both human and animal). The maximum contaminant level (MCL) established by the EPA is 10 mg/L nitrate–N. The MCL for nitrite–N is 1 mg/L, but high nitrite concentrations are very rare in nature because microbes in the soil rapidly convert nitrite to nitrate.

Nitrogen can also reach surface waters in airborne forms. Gaseous ammonia is released into the atmosphere from animal wastes. Oxides of nitrogen are produced by burning fossil fuels in power plants, waste incineration, and internal combustion engines. Atmospheric deposition is the return of these nitrogen compounds to land and water either by dry deposition or with precipitation. Nitrogen oxides and sulfur dioxide are the primary causes of acid rain.

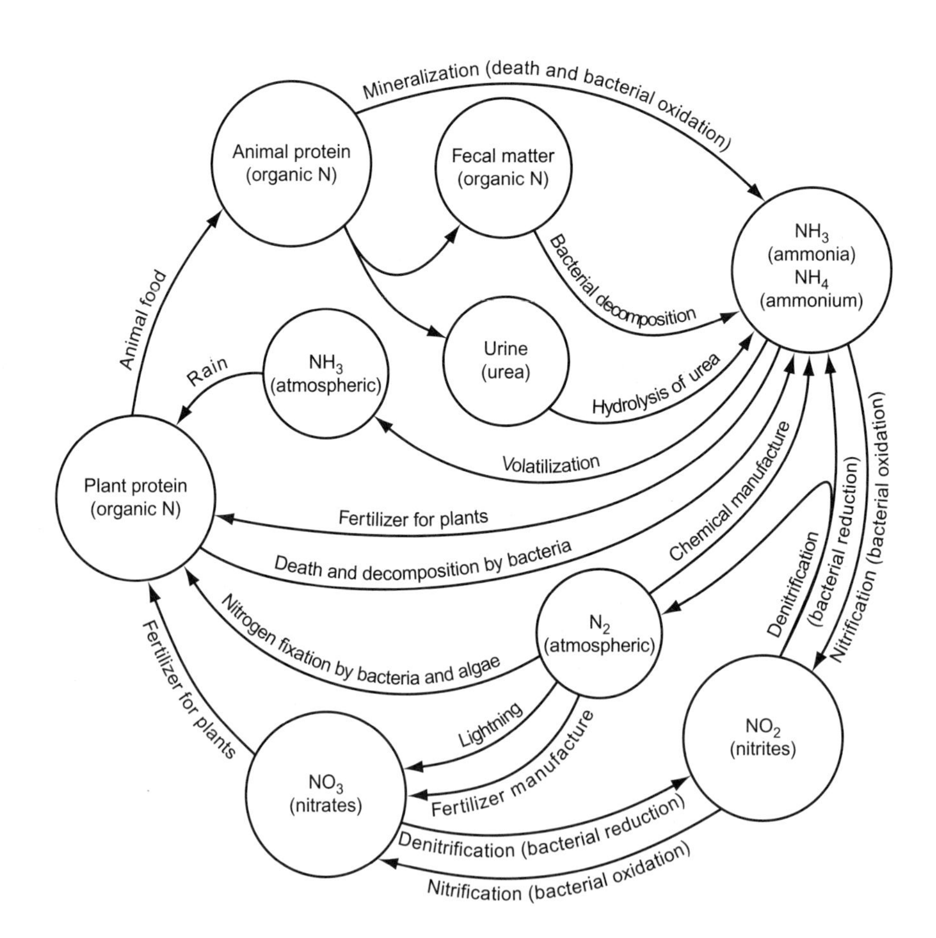

Figure 2–4

The nitrogen cycle. (Redrawn from SCS, 1992)

Phosphorus occurs naturally in many mineral forms and is gradually released by chemical weathering. Phosphate-rich deposits are mined to produce fertilizers and other phosphate products. Phosphorus is also found in organic matter. Phosphorus is applied to the land in both organic and chemical fertilizers (e.g., triple superphosphate $Ca(H_2PO_4)_2 \bullet H_2O$). Figure 2-5 presents an abbreviated phosphorus cycle specific to a waste application site. Soluble phosphate readily adsorbs onto soil solids or combines with iron or aluminum to form low-solubility compounds. Unless loading to the soil is so high that it overwhelms the immobilizing capacity of the soil, P will not migrate with soil water. The primary mode of transport is erosion by water, which carries P adsorbed to soil particles into surface waters. Erosion control practices (Chapters 7–10) and wetlands (Chapter 12) are very effective for limiting P in runoff.

If phosphorus-containing wastes are applied to the soil surface without incorporation, they may be easily detached and transported by water. Soluble P can enter surface waters this way.

There is no MCL for phosphorus in drinking water. Excess phosphorus is readily excreted and does not normally pose a health risk.

Potassium is usually grouped with N and P as a macronutrient, but it is not associated with health or water quality problems. There is no MCL for potassium in drinking water.

2.9 Inorganic Chemicals

There are thousands of chemicals that can be found in water. Solubility and toxicity vary widely. A few of the most common that are subject to federal regulation include arsenic, barium, beryllium, cadmium, chromium, copper, cyanide, fluoride, lead, mercury, selenium, and thallium. Sources include discharge from metal and petroleum refineries, industrial discharges, and decaying piping systems. Natural

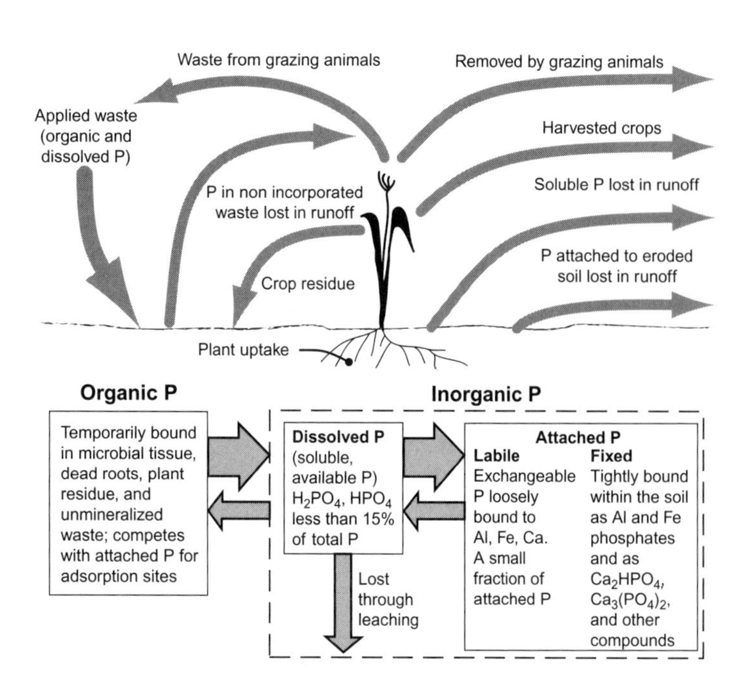

Figure 2–5

Phosphorus inputs and losses at a waste application site with transformations within the soil (abbreviated phosphorus cycle). (Redrawn from SCS, 1992)

mineral deposits may contribute to locally high concentrations. Some are present in fertilizers or are added to animal feeds.

The term **heavy metals** is variously used to refer to antimony, arsenic, cadmium, chromium, copper, lead, mercury, and zinc, among others. It has been defined to include all elements with atomic weights from that of copper to that of mercury (Kennish, 1992) or having a specific gravity greater than 4.0 (Connell & Miller, 1984). Many are essential trace nutrients. Non-essential elements may interfere with organisms by substituting for chemically similar elements. Several (e.g., arsenic, lead, and mercury) have been used in pesticides, but that practice has been largely discontinued because of the persistence of those elements and their concentration in the food chain.

Heavy metals typically adsorb to soil minerals or organic matter, or form complexes or chelates. Dissolved concentrations are usually low. Increases in salinity, increases in oxygen availability, or decreases in pH tend to release metal ions into solution, where they are more mobile (Connell & Miller, 1984).

Inorganics also include compounds containing common ions such as sodium, calcium, magnesium, and chloride. Water hardness is often defined as the total concentration of the metallic cations (the most common are calcium and magnesium) in solution that react with sodium soaps to produce solids or scummy residue and that react with anions. Hardness is expressed as milligrams per liter equivalent $CaCO_3$. Table 2-2 shows concentration ranges corresponding to hardness classes used by the U.S. Geological Survey.

Hardness is a common problem in ground water in areas underlain by carbonate rock such as limestone. Natural rainfall is slightly acidic and gradually dissolves these minerals as it percolates through them. Excessive hardness interferes with detergents and can cause scale deposits in heating systems or irrigation equipment. Excessive sodium (relative to calcium and magnesium) in soils tends to disperse clays, which restricts the movement of air and water through the soil profile. Chloride in excess of 250 mg/L, in the presence of sodium, gives water a salty taste. Crops have varying tolerances to salt concentrations in the soil water (Chapter 15).

Current MCLs for selected inorganics are presented in Table 2-3. Consult the EPA and regulatory agencies in your area for complete lists.

2.10 Organic Chemicals

Organic chemicals include thousands of compounds, both synthetic and natural. Those of greatest concern for water quality include pesticides (e.g., alachlor,

TABLE 2–2 Water Hardness Classes

Description	*Equivalent $CaCO_3$ (mg/L)*
Soft	0–60
Moderately hard	60–120
Hard	120–180
Very hard	> 180

Source: U.S. Geological Survey (2004).

atrazine, carbofuran, lindane) and industrial chemicals, particularly solvents (e.g., benzene, toluene, xylene, carbon tetrachloride). Many such organics are confirmed or suspected carcinogens.

The trend in pesticide development has been toward high specificity and short half-lives in the environment. This is a great improvement over persistent pesticides based on arsenic, mercury, and chlorinated hydrocarbons (e.g., DDT). Current MCLs for selected organics are presented in Table 2-3. Consult the EPA and regulatory agencies in your area for complete lists.

Many ground water contamination incidents could be traced to operators who dumped unused pesticide mixes and equipment rinse water directly on the ground—often next to the well that supplied the water. Current best management practice requires a concrete pad with a sump—located some distance from the well—where chemical mixing and equipment cleaning are performed. Proper handling of the water collected in the sump prevents direct contamination of the soil and ground water.

The great variety and complexity of organic materials found in water makes discrete analytical description impractical for many purposes. Organic materials can be quantified with collective parameters, such as chemical oxygen demand (COD), biological oxygen demand (BOD), dissolved organic carbon (DOC), and total organic carbon (TOC). COD is the equivalent amount of an oxidizing agent, such as permanganate, required for oxidation of the organic constituents; usually expressed as oxygen equivalents. BOD measures the amount of oxygen consumed in microbially mediated oxidation of the organic matter. TOC is a measure of the CO_2 produced in the oxidation or combustion of a water sample from which carbonate has been removed. These measures, especially BOD, are often used as indicators of the amount of oxygen that may be depleted from waters containing organics. The dependence of microbial activity on many other factors makes these measures difficult to apply in nature (Stumm & Morgan, 1996).

Physical Contaminants

2.11 Sediment

Particulate solids have various origins and effects on water quality. The most common contaminant and by far the largest (by mass) is sediment, i.e., soil particles that have become suspended either through erosion from upland areas (agricultural and other disturbed lands, Chapter 7), streambank erosion (Chapter 10), or detachment of streambed particles. The size and amount of sediment that can be transported depend on the energy of the flow—fast-flowing streams have higher capacities for transport than do slow-flowing streams (Chapters 6, 7, and 10). Vigorous flows can move boulders, whereas the quiet flows in lakes can transport only the finest silts and clays. Deposition of sediment can occur wherever flowing water slows (Chapters 7–9, and 12).

Sediment also includes carbonaceous materials, whether derived from geologic formations (such as coal) or modern plant detritus. Organic materials provide a food source for microbes and may harbor pathogens.

Carbonate and silica biocrystals are a minor source of sediment. Biocrystals are formed by organisms such as mollusks, foraminifers, sponges, and diatoms.

TABLE 2–3 Selected Drinking Water Quality Standards

Microorganisms	*MCLG[1] (mg/L)[2]*	*MCL or TT[1] (mg/L)[2]*	*Potential Health Effects from Ingestion of Water*	*Sources of Contaminant in Drinking Water*
Cryptosporidium	zero	TT[3]	Gastrointestinal illness (e.g., diarrhea, vomiting, cramps)	Human and animal fecal waste
Giardia lamblia	zero	TT[3]	Gastrointestinal illness (e.g., diarrhea, vomiting, cramps)	Human and animal fecal waste
Legionella	zero	TT[3]	Legionnaire's disease, commonly known as pneumonia	Found naturally in water; multiplies in heating systems
Total coliforms (including fecal coliform and *E. coli*)	zero	5.0%[4]	Used as an indicator that other potentially harmful bacteria may be present[5]	Coliforms are naturally present in the environment; fecal coliforms, and *E. coli* come from human and animal fecal waste.
Viruses (enteric)	zero	TT[3]	Gastrointestinal illness (e.g., diarrhea, vomiting, cramps)	Human and animal fecal waste

Disinfectants and Disinfection Byproducts	*MCLG[1] (mg/L)[2]*	*MCL or TT[1] (mg/L)[2]*	*Potential Health Effects from Ingestion of Water*	*Sources of Contaminant in Drinking Water*
Chloramines (as Cl_2)	MRDLG = 4[1]	MRDL = 4.0[1]	Eye/nose irritation; stomach discomfort, anemia	Water additive used to control microbes
Total Trihalomethanes (TTHMs)	n/a[6]	0.080	Liver, kidney, or central nervous system problems; increased risk of cancer	By-product of drinking water disinfection

Inorganic Chemicals	*MCLG[1] (mg/L)[2]*	*MCL or TT[1] (mg/L)[2]*	*Potential Health Effects from Ingestion of Water*	*Sources of Contaminant in Drinking Water*
Asbestos (fiber >10 micrometers)	7 million fibers per liter	7 MFL	Increased risk of developing benign intestinal polyps	Decay of asbestos cement in water mains; erosion of natural deposits
Nitrate (measured as nitrogen)	10	10	Infants below the age of six months who drink water containing nitrate in excess of the MCL could become seriously ill and, if untreated, may die. Symptoms include shortness of breath and blue-babysyndrome.	Runoff from fertilizer use; leaching from septic tanks, sewage; erosion of natural deposits
Nitrite (measured as nitrogen)	1	1	Infants below the age of six months who drink water containing nitrite in excess of the MCL could become seriously ill and, if untreated, may die. Symptoms include shortness of breath and blue-baby syndrome.	Runoff from fertilizer use; leaching from septic tanks, sewage; erosion of natural deposits
Selenium	0.05	0.05	Hair or fingernail loss; numbness in fingers or toes; circulatory problems	Discharge from petroleum refineries; erosion of natural deposits; discharge from mines

Table 2.3 continued

Organic Chemicals	*MCLG*[1] *(mg/L)*[2]	*MCL or TT*[1] *(mg/L)*[2]	*Potential Health Effects from Ingestion of Water*	*Sources of Contaminant in Drinking Water*
Alachlor	zero	0.002	Eye, liver, kidney, or spleen problems; anemia; increased risk of cancer	Runoff from herbicide used on row crops
Atrazine	0.003	0.003	Cardiovascular system problems; reproductive difficulties	Runoff from herbicide used on row crops
Carbofuran	0.04	0.04	Problems with blood or nervous system; reproductive diffiulties	Leaching of soil fumigant used on rice and alfalfa
Chlordane	zero	0.002	Liver or nervous system problems; increased risk of cancer	Residue of banned termiticide
2,4-D	0.07	0.07	Kidney, liver, or adrenal gland problems	Runoff from herbicide used on row crops
Dinoseb	0.007	0.007	Reproductive difficulties	Runoff from herbicide used on soybeans and vegetables
Diquat	0.02	0.02	Cataracts	Runoff from herbicide use
Endothall	0.1	0.1	Stomach and intestinal problems	Runoff from herbicide use
Glyphosate	0.7	0.7	Kidney problems; reproductive difficulties	Runoff from herbicide use
Lindane	0.0002	0.0002	Liver or kidney problems	Runoff/leaching from insecticide used on cattle, lumber, gardens
Methoxychlor	0.04	0.04	Reproductive difficulties	Runoff/leaching from insecticide used on fruits, vegetables, alfalfa, livestock
Picloram	0.5	0.5	Liver problems	Herbicide runoff
Polychlorinated biphenyls (PCBs)	zero	0.0005	Skin changes; thymus gland problems; immune deficiencies; reproductive or nervous system difficulties; increased risk of cancer	Runoff from landfills; discharge of waste chemicals
Simazine	0.004	0.004	Problems with blood	Herbicide runoff

Source: Excerpted from Environmental Protection Agency (2003).

[1] Definitions:

Maximum Contaminant Level (MCL). The highest level of a contaminant that is allowed in drinking water. MCLs are set as close to MCLGs as feasible using the best available treatment technology and taking cost into consideration. MCLs are enforceable standards.

Maximum Contaminant Level Goal (MCLG). The level of a contaminant in drinking water below which there is no known or expected risk to health. MCLGs allow for a margin of safety and are nonenforceable public health goals.

Maximum Residual Disinfectant Level (MRDL). The highest level of a disinfectant allowed in drinking water. There is convincing evidence that addition of a disinfectant is necessary for control of microbial contaminants.

Maximum Residual Disinfectant Level Goal (MRDLG). The level of a drinking water disinfectant below which there is no known or expected risk to health. MRDLGs do not reflect the benefits of the use of disinfectants to control microbial contaminants.

Treatment Technique (TT). A required process intended to reduce the level of a contaminant in drinking water.

[2] Units are in milligrams per liter (mg/L) unless otherwise noted. Milligrams per liter are equivalent to parts per million (ppm).

[3] EPA's surface water treatment rules require systems using surface water or ground water under the direct influence of surface water to (1) disinfect their water, and (2) filter their water or meet criteria for avoiding filtration so that the following contaminants are controlled at the following levels:

- *Cryptosporidium:* (as of 1/1/02 for systems serving >10 000 and 1/14/05 for systems serving <10 000) 99% removal
- *Giardia lamblia:* 99.9% removal/inactivation
- Viruses: 99.99% removal/inactivation
- *Legionella:* No limit, but EPA believes that if *Giardia* and viruses are removed/inactivated, *Legionella* will also be controlled.
- Turbidity: At no time can turbidity (cloudiness of water) go above 5 nephelolometric turbidity units (NTU); systems that filter must ensure that the turbidity go no higher than 1 NTU (0.5 NTU for conventional or direct filtration) in at least 95% of the daily samples in any month. As of January 1, 2002, for systems servicing >10 000, and January 14, 2005, for systems servicing <10 000, turbidity may never exceed 1 NTU, and must not exceed 0.3 NTU in 95% of daily samples in any month.

[4] No more than 5.0% samples total coliform-positive in a month. (For water systems that collect fewer than 40 routine samples per month, no more than one sample can be total coliform-positive per month.) Every sample that has total coliform must be analyzed for either fecal coliforms or *E. coli* if two consecutive TC-positive samples, and one is also positive for *E. coli* fecal coliforms, system has an acute MCL violation.

[5] Fecal coliform and *E. coli* are bacteria whose presence indicates that the water may be contaminated with human or animal wastes. Disease-causing microbes (pathogens) in these wastes can cause diarrhea, cramps, nausea, headaches, or other symptoms. These pathogens may pose a special health risk for infants, young children, and people with severely compromised immune systems.

[6] Although there is no collective MCLG for this contaminant group, there are individual MCLGs for some of the individual contaminants:

- *Haloacetic acids:* dichloroacetic acid (zero); trichloroacetic acid (0.3 mg/L)
- *Trihalomethanes:* bromodichloromethane (zero); bromoform (zero); dibromochloromethane (0.06 mg/L)

Excessive sediment can degrade aquatic habitat by restricting penetration of light, which affects photosynthetic activity and predator–prey relationships, and by altering benthic habitat (Wilber, 1983). Deposition of sediment in impoundments and channels also reduces capacity, increasing the frequency of flooding. Sediment from agricultural lands may carry nutrients or pesticides. Sediment can also change the benthic environment by filling in the spaces between rocks and gravel.

2.12 Turbidity

The term *turbidity* refers to the murkiness of the water. It is quantified by measuring the degree to which light is scattered by suspended particulates (sediment and organic matter) in the water. The common unit is the Nephelometric Turbidity Unit

(NTU). A number of sensors are commercially available for laboratory or field use. The Secchi disk (Figure 2-1) provides a method for quick evaluation of turbidity, although its visibility is influenced by color of the water in addition to particulates.

Turbidity is not a health threat in itself. Organic matter that contributes to turbidity can harbor pathogens and tends to deplete dissolved oxygen. Suspended solids increase treatment requirements.

Turbidity affects growth of phytoplankton, algae, and aquatic plants, generally favoring those organisms closest to the surface. Elevated turbidity places fish under stress, reducing feeding success, growth, and hatching rates. Benthic invertebrates may suffer from reduced oxygen levels. Particulates settling to the bottom can suffocate eggs and larvae.

Water Quality Regulations

In the United States, major water quality legislation began with the Federal Water Pollution Control Act of 1948, which authorized preparation of comprehensive programs for eliminating or reducing pollution of interstate waters and tributaries and improving sanitary conditions of surface and underground waters. This act has been amended many times. The Reorganization Plan No. 3 of 1970 abolished the Federal Water Quality Administration in the Department of Interior and transferred its functions to the newly created EPA. The 1972 amendments included national objectives for restoration and maintenance of the chemical, physical, and biological integrity of the nation's waters. Limitations for point source discharges were to be determined.

The 1977 amendments renamed the legislation as the Clean Water Act (CWA) and authorized development of best management practices (BMPs) and procedures for assumption of the regulatory programs by the states. The CWA was again amended by the Water Quality Act of 1987 that required states to develop strategies for toxics cleanup from waters where best available technology (BAT) discharge standards are inadequate. The act also authorized a $400 million program for states to develop watershed-scale nonpoint source management and control programs under EPA oversight, and required EPA to study and monitor water quality effects attributable to dammed impoundments. Section 303(d) of the CWA requires states to identify pollutant-impaired waters and develop Total Maximum Daily Loads (TMDLs), which specify the amounts of various pollutants that can be tolerated from various sources in a watershed (Chapter 5). Section 319 established the Nonpoint Source Management Program, which provides grants to support demonstration projects, technical and financial assistance, education, training, and monitoring. Section 402 established the National Pollutant Discharge Elimination System (NPDES) to authorize EPA issuance of discharge permits. Section 404 gave the U.S. Army Corps of Engineers (COE) authority over discharge of dredged or fill materials into navigable waters.

The Safe Drinking Water Act (SDWA) of 1974 and its amendments in 1986 and 1996 were designed to protect public health by regulating drinking water supplies—rivers, lakes, reservoirs, springs, and wells (except private wells serving fewer than 25 people). The SDWA authorized EPA to set health-based standards for both natural and man-made contaminants in drinking water. The original act focused on treatment, whereas the 1996 amendments added powers to protect sources and require operator training. To set a standard, the EPA first determines which contaminants may affect public health, then establishes maximum contaminant level goals (MCLGs) below which there is no known or expected health hazard, and finally sets maximum contaminant levels (MCLs) as close to the MCLGs as considered feasible. An MCL is enforceable, whereas an MCLG is not.

The Wetlands Reserve Program (WRP) was established under the Food Security Act (FSA) of 1985 (as amended by the 1990 and 1996 Farm Bills). The WRP, a voluntary program that encourages restoration and protection of wetlands, is administered by the U.S. Department of Agriculture (USDA) Natural Resources Conservation Service (NRCS) in consultation with the Farm Service Agency. Landowners retain control of eligible lands, which may be used or leased for hunting, fishing, or other undeveloped recreational activities. Other wetlands damage legislation includes Sections 402 and 404 of the CWA. A 1990 Memorandum of Agreement (MOA) between the COE and EPA established the foundation for wetlands damage mitigation practices, including restoration of degraded wetlands or creation of new wetlands to compensate for unavoidable damage to existing wetlands. A 1994 multiagency MOA established the NRCS as the lead agency for wetlands delineation on agricultural lands under both the CWA and FSA. (Chapter 12 discusses wetland processes and design.)

New legislation is introduced in each session of Congress. Current summaries of these acts and related regulations are available on the Internet from the U.S. EPA, the U.S. Army Corps of Engineers, the USDA Natural Resources Conservation Service, the U.S. Fish and Wildlife Service, and the U.S. Bureau of Reclamation.

Water Treatment

Whenever the quality of a water supply does not meet the standards for its intended use, some type of treatment will be necessary. Treatment processes vary in sophistication, but all add to the cost of the water. Application of best management practices can markedly improve the quality of the water supply, reducing treatment costs as well as providing environmental benefits.

This section briefly discusses major water treatment practices, their applicability, and efficacy. Treatment methods may include physical, chemical, and biological processes, either alone or in combination. A treatment method is selected to best fit the type of water quality problem, the intended use of the treated water, and the economics of the situation.

2.13 Clarification

Settling is often the first step in water treatment. Large or heavy particulates will fall to the bottom quickly in still water. Very fine suspended particulates can be made to settle more quickly by addition of flocculating agents, such as gypsum or polyacrylamide. Adjustment of pH may be needed. Flocs that do not settle may be removed by skimming, if they float, or by subsequent filtration.

2.14 Filtration

Filtration can be effective for removal of particulate contaminants. The size and amount of particulates present determines the type of filter that is appropriate.

Filters are needed with microirrigation systems (Chapter 18) to remove suspended particulates, which may include fine sands, silt, clay, or algae. They are also common in public and private water treatment systems. Sand filters have varying capacities and efficacies, depending on the size of the filter and the media used, and are easily regenerated by backwashing. Sand filters can remove most suspended materials but are not effective with very fine particulates or bacteria.

Cartridge filters can be constructed with finer pores than sand filters. Materials include paper fiber, fiberglass, ceramics, and precision-etched polycarbonates. The finer pore structure requires more cross-sectional area than for a sand filter for equivalent capacities. Cartridge filters are most effective where the concentration of particulates is low, so that cleaning or replacement is required less frequently. Cartridge filters can remove bacteria but should be regularly maintained in drinking water systems, as the cartridge itself could become a problem if bacteria accumulate and multiply. Some cartridge filters contain carbon, which can remove chlorine and some organic compounds.

For very demanding purposes, micro- (0.1–2 μm), ultra- (0.001–0.1μm), and nanofiltration are available. Nanofiltration can remove organic compounds having molecular weights of 300 to 1000 and reject some salts.

2.15 Ion Exchange

The most familiar ion exchange treatment process is water softening. Sodium ions are stored in a filter bed of zeolite (an aluminosilicate resin). As hard water flows through the bed, sodium ions are exchanged for calcium and magnesium ions. The zeolite bed is periodically regenerated (charged with Na^+ and Cl^- ions) by flushing with brine.

Deionized water is produced in a similar process. Ions in the water are exchanged for H^+ and OH^-, which can then combine to form water. Deionization is used where extremely pure water is needed. Deionization will not generally remove nonionic compounds, organics, or pathogens.

Many other ion exchange processes have been developed. The reader should consult other texts or suppliers for additional information.

2.16 Disinfection

Chlorination is the most common method of disinfection. Chlorine gas is added to the water after pretreatments (clarification, filtration, etc.) to kill microbes. An excess of chlorine is added to provide a residual concentration (usually about 5 ppm) to control regrowth throughout the distribution system. If other organic compounds are present, chlorine tends to react with them first, forming chloramines or chlorinated hydrocarbons that may be carcinogens. This reaction also increases the amount of chlorine needed to achieve disinfection. Sodium hypochlorite and calcium hypochlorite are alternative sources of chloride that are less dangerous to handle than chlorine gas. These are commonly used in small or private water systems or as algicides in microirrigation systems.

Ozonation uses ozone (O_3), a powerful oxidant, to kill pathogens. Unlike chlorine, ozone will not leave a lasting residual, so there is the possibility of subsequent regrowth of pathogens. Because of this, ozonation should not be used alone where water may reside in the system for extended periods before use.

Ionizing radiation, such as ultraviolet light, can kill many pathogens. Ultraviolet light can also break down low-level organics. Like ozonation, it has no residual effect, which limits its application.

2.17 Reverse Osmosis

In reverse osmosis, a pressure difference on the order of 10–70 bars drives water across special membranes, rejecting nearly all organics and 90–99 percent of all ions. Over 99.9 percent of viruses and bacteria are also removed. A fraction of the source water, the concentrate or reject water, carries the rejected ions and other constituents to disposal. The fraction that passes through the membrane, the permeate, is the desired product. Many types of membranes are commercially available to meet different permeate requirements.

2.18 Distillation

Distillation is the collection of condensed steam that is produced by boiling water. This can remove all types of impurities, although organics with boiling points near that of water require very close control and perhaps multiple distillations. Distillation is energy intensive, but can produce water with impurity concentrations as low as 10 parts per trillion.

Water Quality Modeling

With the advent of inexpensive and powerful computers, modeling of hydrologic processes and associated water quality parameters has become commonplace. Researchers and regulators make frequent use of well-tested models to study and predict behavior of systems from field to river basin scales. Models, however, can easily be misapplied, and results can be misinterpreted or extrapolated beyond reason. Model users should have a thorough understanding of the physical processes of the system in question, the assumptions inherent in the conceptual model, and the limitations imposed by sparse or uncertain inputs.

Water quality modeling is, by nature, a problem with spatial aspects. Geographic Information Systems (GISs) are often used to manage the spatially distributed inputs and to store, manipulate, and display the model outputs.

Several good summaries of current models are available. The reader should consult sources such as the EPA, USGS, COE, NRCS, or Parsons et al. (2001). Many consulting groups provide programs, documentation, support, and training for water quality modeling.

Engineering in Water Quality

What is the engineer's role in protecting and improving water quality? What tools are needed?

First, the engineer must thoroughly understand the physical processes involved (rainfall, runoff, infiltration, erosion, overland and channelized flow, etc.). Second, he or she must be able to adequately quantify those processes. Only then can the most appropriate management strategies, structures, and systems be selected to maximize the objectives of utility and environmental quality.

Water is absolutely essential to individuals, society, and wildlife. Conflicts over use and allocation will continue as opposing interests clash over a limited resource. Engineers, in cooperation with the physical and life sciences, must continually strive to discover and implement the most practical and economical solutions to these ever-changing challenges.

Internet Resources

U.S. Environmental Protection Agency (EPA)
http://www.epa.gov
http://www.epa.gov/ost/wqm
http://www.epa.gov/storet

U.S. Geological Survey (USGS)
http://water.usgs.gov/software/water_quality.html

USDA Natural Resources Conservation Service (NRCS)
http://www.nrcs.usda.gov
http://www.wcc.nrcs.usda.gov/water/quality/common/h2oqual.html

U.S. Bureau of Reclamation (USBR)
http://www.usbr.gov

U.S. Army Corps of Engineers (COE)
http://www.usace.army.mil

U.S. Fish and Wildlife Service (FWS)
http://www.fws.gov

References

American Water Works Association (AWWA). (1990). *Water Quality and Treatment: A Handbook of Community Water Supplies*, 4th ed. New York: McGraw-Hill.

Carlson, R. E. (1977). A trophic state index for lakes. *Limnology and Oceanography, 22,* 361–369.

Carpenter, S. R., N. F. Caraco, D. L. Correll, R. W. Howarth, A. N. Sharpley, & V. H. Smith. (1998). Nonpoint pollution of surface water with phosphorus and nitrogen. *Ecological Applications, 8*(3), 559–568.

Connell, D. W., & G. J. Miller. (1984). *Chemistry and Ecotoxicology of Pollution*. New York: Wiley.

Freeze, R. A., & J. A. Cherry. (1979). *Groundwater*. Upper Saddle River, NJ: Prentice Hall.

Gilliam, J. W., J. E. Parsons, & R. L. Mikkelsen. (1997). Nitrogen dynamics and buffer zones. In N. E. Haycock, T. P. Burt, K. W. T. Goulding, & G. Piney (eds), *Buffer Zones: Their Processes and Potential and Water Protection* (pp. 54–61). Harpenden Herts, UK: Quest Environmental.

Glasgow, Jr., H. B., & J. M. Burkholder. (2000). Water quality trends and management implications from a five-year study of a eutrophic estuary. *Ecological Applications, 10*(4), 1024–1046.

Kennish, M. J. (1992). *Ecology of Estuaries: Anthropogenic Effects*. Boca Raton, FL: CRC Press.

Mallin, M. A. (1994). Phytoplankton ecology in North Carolina estuaries. *Estuaries*, 17, 561–574.

McCutcheon, S. C., J. L. Martin, & T. O. Barnwell, Jr. (1992). Water quality. In D. R. Maidment (ed), *Handbook of Hydrology*. New York: McGraw-Hill.

Mead, P. S., & P. M. Griffin. (1998). *Escherichia coli* O157:H7. *Lancet, 352*(9135), 1207–1212.

Parsons, J. E., D. L. Thomas, & R. L. Huffman (eds). (2001). *Agricultural Non-point Source Water Quality Models: Their Use and Application*. Southern Coop. Series Bull. 398. Online at www3.bae.ncsu.edu/Regional-Bulletins/. Accessed 10 January 2004.

Stumm, W., & J. J. Morgan. (1996). *Aquatic Chemistry: Chemical Equilibria and Rates in Natural Waters*, 3rd ed. New York: Wiley.

USDA Soil Conservation Service (SCS). (1992). *National Engineering Handbook Part 651—Agricultural Waste Management Field Handbook*. Online at www.info.usda.gov/CED/. Accessed 7 October 2004.

U.S. Environmental Protection Agency (EPA). (2003). *National Primary Drinking Water Standards*, EPA 816-F-03-016. Online at www. epa.gov/safewater/mc/.html.

———.(2004). *Eutrophication*. Online at www.epa.gov/maia/html/eutroph.html. Accessed 10 January 2004.

U.S. Geological Survey (USGS). (2004). *Explanation of hardness*. Online at water.usgs.gov/owq/Explanation.html. Accessed 10 January 2004.

Wilber, C. G. (1983). *Turbidity in the Aquatic Environment: An Environmental Factor in Fresh and Oceanic Waters*. Springfield, IL: Charles C Thomas.

Problems

2.1 Research a current water problem in your area. Examine the history of the problem. How did it develop? What is being done about it?

2.2 Shock chlorination is a one-time method for disinfection of wells. Household bleach (5.25 percent NaOCl, by weight, specific gravity 1.08) is added to provide a chloride concentration of 100 ppm. Calculate the volume of bleach that must be added, per meter of water standing in the well, to 10-, 15-, and 25-cm-diameter wells.

2.3 Choose a local water system (e.g., a municipal system), and determine the source of its water and identify of the treatment processes it uses. Are the treatment processes dependent on the season of the year? If so, why and how?

2.4 Estimate the time required for sand (0.05–1 mm), silt (0.002–0.05 mm), and clay (<0.002 mm) particles to settle to the bottom of a 2-m-deep settling basin. Use Stokes's law (Equation 9.14), assuming spherical particles having the density of quartz (2.65 g/cm^3).

2.5 Calculate the energy required per liter to distill water, assuming the source water is at 20°C and there is no energy recovery. Assuming an energy cost of $0.10 per kilowatt-hour, estimate the daily cost for a household of four if it had to distill 1000 L/day for all its domestic needs.

2.6 Search the EPA STORET database to find water quality data for your area. Report the location, types of data available, and periods of record.

CHAPTER 3 Precipitation

Precipitation includes all forms of liquid and solid water that falls from the atmosphere and reaches the earth's surface. Familiar examples such as rain, snow, sleet, and hail, make precipitation one of the most recognized portions of the hydrologic cycle. Identification and or prediction of the intensity, duration, and frequency of precipitation events are important aspects of most environmental engineering applications.

Description

3.1 Formation of Precipitation

Condensation is the conversion of water from a vapor into a liquid. For condensation to occur, air must be cooled to the dewpoint temperature. The primary means of cooling for precipitation events is vertical uplift, which includes a drop in temperature and pressure. As air rises, it is cooled at the dry adiabatic lapse rate of 1 °C/100 m. Once condensation begins, latent heat is produced and the temperature will decrease at the moist adiabatic lapse rate, which varies with temperature, vapor pressure, and elevation but is approximately one half of the dry rate (0.5–0.7 °C/100 m). The cooling can also occur by mixing with a cooler body of air, conduction to a cool surface, or a lowering of atmospheric pressure.

Typically, warm moist air rises in the atmosphere and is cooled. As the air is cooled, its capacity to hold water vapor decreases (see Chapter 4 concerning saturation vapor pressure). After sufficient cooling, the air mass becomes saturated, but raindrops may not form. In order for raindrops to form, water molecules must be attracted to and condense onto particles called cloud condensation nuclei, or CCN (Dingman, 1994). Natural sources of CCN are windblown clay and silt particles, smoke from fires, volcanic materials, and sea salt. Cloud seeding is sometimes used during a drought to add CCN to the atmosphere to promote condensation.

The final component of the precipitation event is a lateral supply of moist air to the storm center. If all of the moisture in the earth's atmosphere were to condense and fall, a depth of about 25 mm of water would be added to the earth's surface (Dingman, 1994). For storms to produce significant amounts of precipitation, it is necessary for lateral flow to occur, which supplies much of the water for a rainfall event. One of the primary limitations to cloud seeding is the lack of a lateral source of moist air to sustain the precipitation event.

Warm air can rise as a result of convection, convergence, and topography. When the soil surface warms up very rapidly, the temperature of the adjacent air layer is increased and the density is decreased making the air lighter. The lighter air can rise to great heights to form billowing thunder clouds. Precipitation from clouds formed by this process is called convective precipitation. Low- to very-high-intensity rainfall characterizes this type of precipitation.

Uneven heating or cooling of the earth's surface causes air masses to move. The movement is usually associated with the mixing of low-pressure and high-pressure air masses (Figure 3-1). The convergence of these air masses causes significant mixing of the air fronts (boundaries between air masses). The convergence can occur either with cold fronts, warm fronts, or stationary fronts. A cold front (cooler and denser air) lifts the warm moist air ahead of it. When the air rises, it condenses and forms showers and thunderstorms. When a warm front (warm and less dense air) encounters colder air, it moves up and over the colder air ahead of the front. Since the warm front movement is generally slower than a cold front, when condensation takes place, precipitation is typically steady and widespread. Sometimes a stationary front can form with little movement of the air masses. Instabilities near the front cause vertical uplift, cooling, and precipitation. Precipitation events caused by the convergence of fronts is called cyclonic or frontal precipitation.

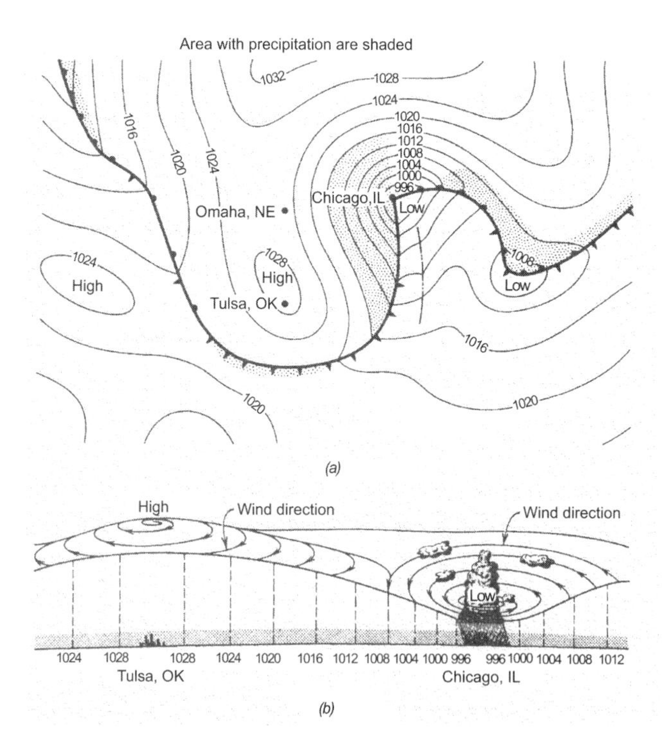

Figure 3–1

(a) Portion of a weather map in April showing cloudy weather in the East, rain in Chicago, and clear skies in the Southwest. (b) Wind circulation around the high-pressure center (clockwise) at Tulsa and a low-pressure center (counterclockwise) at Chicago.

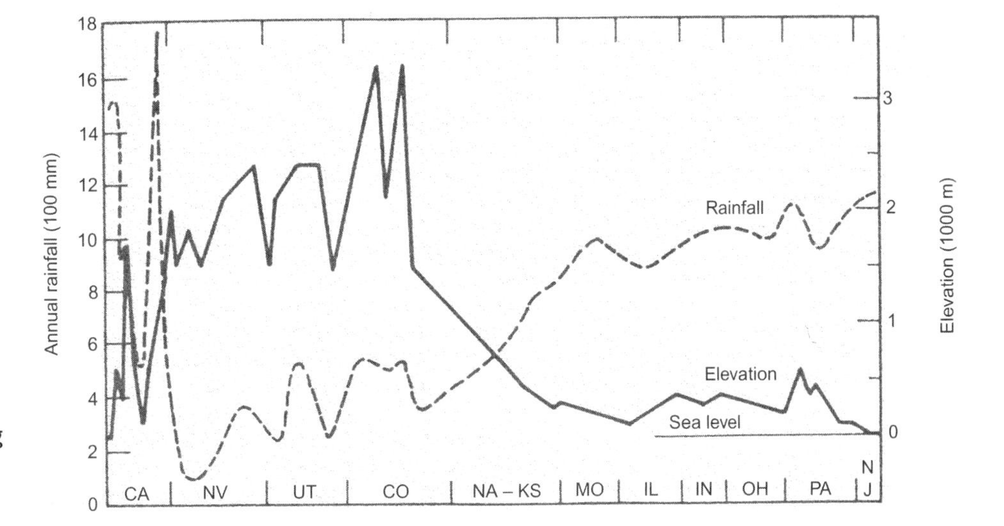

Figure 3–2

Average annual rainfall and elevation across the United States along the 40th parallel of latitude.

Warm air may also be lifted up when it encounters and moves over high topographic surfaces such as mountain ranges. The air is lifted and cooled as it moves. In the United States, this process contributes to precipitation mostly on the western side of the mountain ranges (Figure 3-2), because the prevailing wind direction is generally from west to east. Precipitation caused by the uplift over topographic surfaces is called orographic precipitation. Once the air mass has moved over the mountains, it will be very dry, causing arid conditions to prevail on the downwind side.

3.2 Characteristics of Precipitation

The largest portion of precipitation occurs as rain, and because rainfall directly affects soil erosion, the characteristics of raindrops are of interest. Raindrops include water particles as large as 7 mm in diameter. The size distribution in any one storm covers a considerable range and varies with the rainfall intensity. Storms with higher rainfall intensity produce large-diameter raindrops and a wider range of raindrop diameters.

Raindrops are not necessarily spherical. Falling raindrops can be deformed from a spherical shape because of pressure difference and air resistance. Large raindrops (more than 5 mm in diameter) are generally unstable and split in the air.

The velocity of fall for raindrops depends on the size of the particle with larger drops falling more rapidly. As the height of fall increases, the velocity increases only to a height of about 11 m; the drops then approach a terminal velocity, which varies from about 5 m/s for a 1-mm drop to about 9 m/s for a 5-mm drop.

Precipitation may also occur as frozen water particles including snow, sleet, and hail. Snow is a grouping of ice crystals. Sleet forms when raindrops are falling through air having a temperature below freezing. A hailstone is an accumulation of many thin layers of ice over a snow pellet. Of the forms of precipitation, rain and snow make the greatest contribution to our water supply.

Direct condensation from the atmosphere near the earth's surface, commonly referred to as dew, can also contribute to water at the soil surface. Studies of dew formation show that about 30 mm per year condenses on bare soil, 25 mm on a grass cover, 15 mm on corn leaves during the summer, and 33 mm on soybean leaves. Although dew is normally evaporated by noon, it helps to reduce the rate of soil water depletion. An important contribution to the hydrologic budget of western forests is fog drip. Fog condenses on the leaves and branches of trees overnight and drips to the ground. In some places, such as the coastal forests of California, more than 50 percent of the annual water budget is produced from fog drip. Orographic cooling can also result in fog drip as the clouds contact the trees in higher elevations.

3.3 Time Distribution

The time of a day in which precipitation may be expected to occur depends on the type of precipitation. Frontal and orographic storms are not influenced by diurnal effects. Because storms of the convective type are caused by surface heating, these storms are more likely to occur in the afternoon and early evening.

Rainfall distribution at any location varies greatly with the season. A considerable difference in the seasonal distribution of precipitation throughout the United

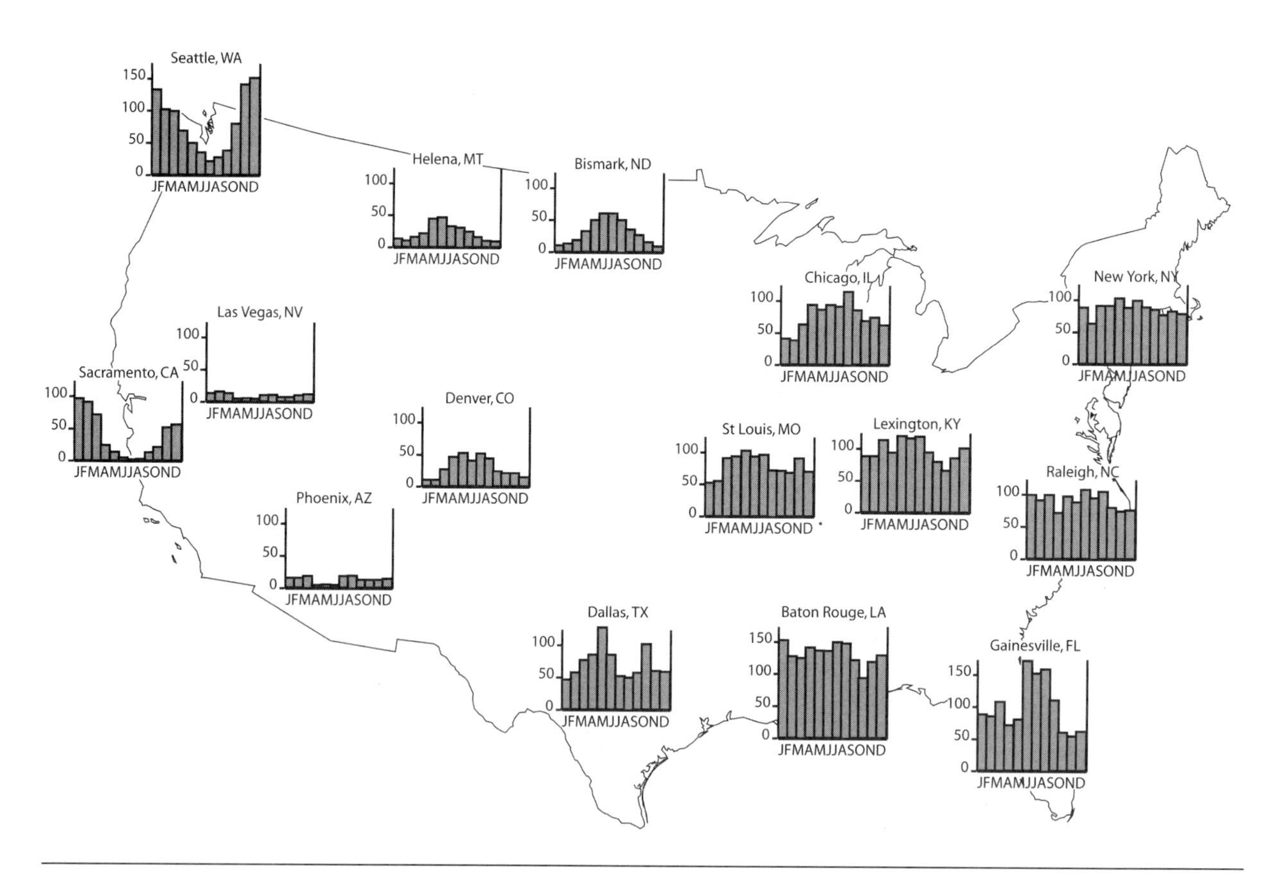

Figure 3–3 Mean monthly precipitation (mm) for selected locations in the United States (1971–2000).

States is shown in Figure 3-3. On the West Coast, where annual precipitation is high, summertime precipitation is generally very low, making irrigation necessary. In the Midwest and the South, the monthly precipitation in summer is generally somewhat higher than in the other seasons. On the East Coast, there is little difference between summer and winter precipitation.

The annual rainfall distribution over the United States is shown in Figure 3-4. Annual rainfall amounts vary from less than 100 mm in some southwestern areas to over 2500 mm in some mountainous areas. Annual precipitation is not in itself a good index of the amount of water available for plant growth, because evaporation, seasonal distribution, and water-holding capacity of the soil vary with geographical location. The Spatial Climate Analysis Center (SPAC) at Oregon State University provides access to a wide range of climate parameters.

There are some indications that precipitation occurs in cycles; however, relationships between such cycles and other natural phenomena are yet to be established. The effects of natural phenomena such as the El Niño Southern Oscillation and potential global warming on precipitation cycles are being investigated. Some evidence exists that sunspot activity is related to summer temperature and severe droughts. Thompson (1973) showed that average July–August temperatures in the Corn Belt of the United States since 1900 follow about a 20-year cycle of sunspot numbers. Similar observations have been made in other countries at the same latitudes.

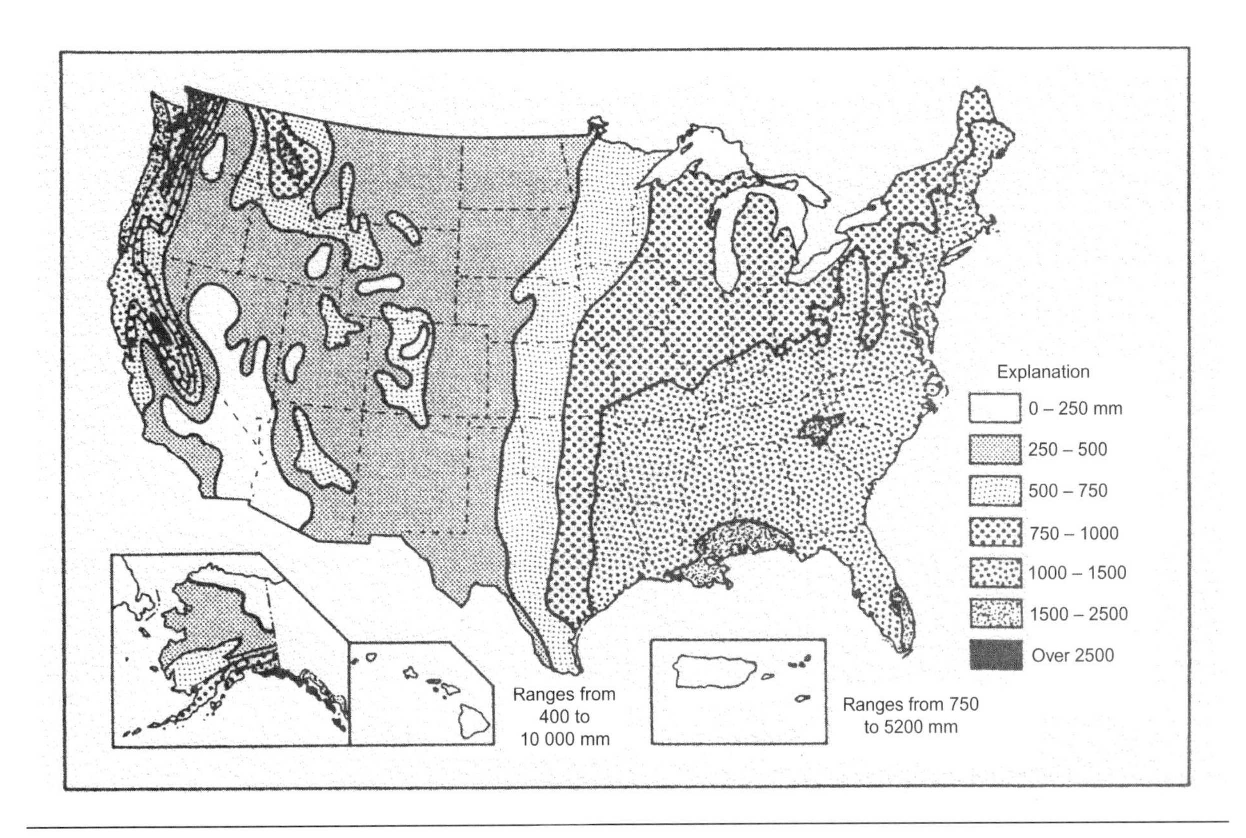

Figure 3–4 Average annual precipitation in the United States in mm. (Redrawn from USDA, 1989)

3.4 Geographical Distribution

The geographical distribution of rainfall is largely influenced by the location of large bodies of water, movement of the major air masses, and changes in elevation. Figure 3-2 illustrates the effects of elevation and air mass movement on annual rainfall across the United States along the 40th parallel of latitude. Moving from west to east in Figure 3-2, one can notice that the highest rainfall occurs as the air initially rises over the western mountains (orographic precipitation). The dry air provides some additional precipitation as the air masses rise to higher elevations near the Rocky Mountains. As the air moves down the mountain slopes, lower annual rainfall is generally observed. The rainfall does not increase until the effects of the maritime tropical air moving up from the Gulf of Mexico become apparent. Then the rainfall gradually increases toward the eastern boundary of the United States because of the orographic effects of the Appalachian Mountains and the increased moisture along the Atlantic Ocean.

Measurement of Precipitation

Because most estimates of runoff rates are based on precipitation data, information regarding the timing and quantity of precipitation is of great importance.

3.5 Measuring Rainfall

Rain gauges are used to measure the depth and intensity of rain falling on a flat surface. Rain gauges generally used in the United States are vertical, cylindrical containers with top openings 203 mm in diameter. A funnel-shaped hood is inserted to minimize evaporation losses. Problems associated with rainfall measurements with gauges include effects of topography and nearby vegetation, buildings, and other structures.

Rain gauges may be classified as recording or nonrecording. Nonrecording rain gauges are economical, requiring service only after a rainy day, and are relatively free of maintenance. The gauges are generally read on a daily basis to obtain daily rainfall; however, they do not provide rainfall intensity data or the exact time of rainfall.

Recording rain gauges may be of several types. The weighing type accumulates water in a container placed above the recording mechanism. The weight of the water creates tension on the spring. The amount of displacement is recorded electronically or through an appropriate linkage to a chart placed on a clock-driven drum. Another commonly used recording type of rain gauge is the tipping bucket instrument. In this rain gauge, precipitation is funneled to a tipping bucket assembly. The assembly has two rain collection compartments balanced on a fulcrum, each of which can hold a known increment of precipitation. Precipitation is directed to one of the compartments. When that compartment is filled to capacity, it tips to the opposite position, bringing the other compartment in place to collect the precipitation. The movement momentarily activates a magnetic switch and sends a signal to a data logger.

Radar measurement of cloud density and rainfall rate is now common. NEXRAD (Next Generation Radar) or Doppler radar works by sending out a pulse of energy and measuring the portion of emitted energy that is reflected back to the radar after striking the raindrops in its path. Computers analyze the strength of the returned pulse and the phase shift of the pulse to determine the direction and magnitude of the storm. A network of ground-based radars and weather surveillance satellites covers most parts of the continental United States. The National Weather Service (NWS) makes these data available to the public in both image (weather maps) and original coded forms. Weather maps and other information on satellite rainfall data can be obtained from the NWS via the Internet.

3.6 Measuring Snowfall

Because the water content of freshly fallen snow varies from less than 40 mm to over 400 mm of water per meter of snow, snowfall is much more difficult to measure than rainfall. Although this wide variation in density makes it difficult to indicate the amount of snow by simple depth measurements, a water-equivalent depth of 10 percent of snow depth is commonly accepted. Water content of compacted snow, however, is often 30 to 50 percent of snow depth.

Snowfall measurements are often made with regular rain gauges, the evaporation hood having been removed. A measured quantity of some noncorrosive, nonevaporative, environmentally benign antifreeze material is generally placed in the rain gauge to cause the snow to melt on entrance. Errors caused by wind are more serious in measuring snowfall than in measuring rain. Snow may also be measured by sampling the depth on a level surface with a metal sampling tube or with the top of the rain gauge.

Another method of measuring snowfall is by determining the depth of snow by a snow survey. Such surveys are particularly useful in mountainous areas. These snow courses consist of ranges that are sampled at specified intervals. The sampling equipment consists of specially designed tubes that take a sample of the complete depth of the snow pack. The sample is then weighed and the equivalent depth of water recorded.

By measuring the snow courses for a period of years and comparing the equivalent water depth with the observed runoff from the snowfield, watershed managers can make predictions of the amount of runoff. Aerial snow surveys can be made by photographing depth gauges or by picking up radio-transmitted signals from depth-measuring equipment. Such devices are set up on snow ranges at suitable locations. These predictions are of particular value in planning for irrigation needs and forecasting the probability of spring floods.

3.7 Errors in Rain Gauge Measurement

Many errors in precipitation measurement result from carelessness in handling the equipment and in analyzing data. Errors characteristic of the nonrecording rain gauges include water creeping up on the measuring stick, evaporation, leaks in the funnel or can, and denting of the cans. The volume of water displaced by the measuring stick is about 2 percent and may be taken as the correction for evaporation.

Another type of error is caused by obstructions such as trees, buildings, and uneven topography. These errors can be minimized by proper location of the rain gauge. The gauges are normally placed with the opening about 760 mm above the soil surface. The location should have minimum turbulence of the wind passing across the gauge. A typical rule is to have a clearance of 45 degrees from the vertical centerline through the gauge, but a safer rule is to be sure that the distance from the obstruction to the gauge is at least twice the height of the obstruction.

The wind velocity also affects the amount of water caught in a rain gauge. A wind velocity of 16 km/h would cause a deficit catch of about 17 percent, but at 48 km/h the deficit is increased to about 60 percent. Whenever possible, the gauge should be located on level ground, because the upward or downward wind movement may easily affect the amount of precipitation caught.

3.8 The Gauging Network in the United States

Precipitation records have been kept in the United States ever since it was settled; however, the recording rain gauges that provide the intensity of precipitation have been used since about 1890. Rain gauges have steadily increased in number since then; the gauging network in the United States now consists of about 11 000 nonrecording and 3500 recording instruments. Volunteers service many of the nonrecording gauges. Most of the recording equipment is connected with local, state, or federal installations. The results of these extensive gauging activities are published by the National Oceanic and Atmospheric Administration (NOAA) via the Internet.

Rainfall Depth

Many activities require the rainfall depth over a large watershed to be computed. The watershed may include several rain gauges that are unevenly distributed over the area. A simple arithmetic mean of the precipitation totals can be used to represent the precipitation for the watershed, but methods that account for the area represented by each rain gauge will generally provide a better estimate of the precipitation. Two of the most widely accepted techniques to compute average precipitation depth are the Thiessen Polygon method and the Isohyetal method.

3.9 The Thiessen Polygon Method

The Thiessen method is illustrated in Figure 3-5. When using this method, the locations of the rain gauges are plotted on a map of the watershed and connected with straight lines. Perpendicular bisectors are then drawn on each of the straight lines and extended in such a way that the bisectors enclose areas referred to as Thiessen Polygons. All points within one polygon will be closer to its rain gauge than to any of the others. The rain recorded from the rain gauge within a polygon is considered to represent the precipitation within the polygon area.

Some difficulty may be encountered in determining which connecting lines to construct in forming the sides of the polygon. Because only one set of Thiessen Polygons generally needs to be drawn for a given watershed and a set of rain gauge locations, this procedure does not present a serious limitation. The mean precipitation over a watershed can be determined by using the equation

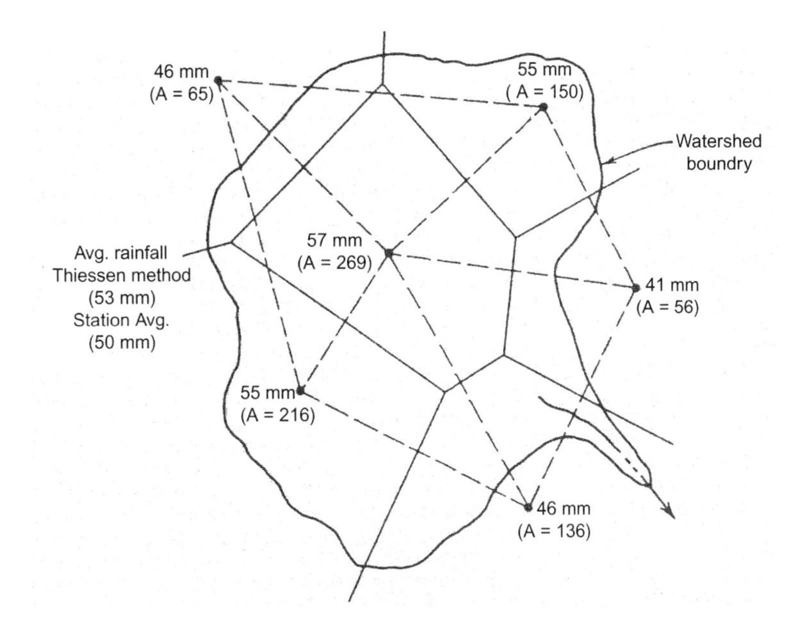

Figure 3–5
Thiessen network for computing average rainfall depth over a watershed.

$$P = \frac{A_1P_1 + A_2P_2 + \ldots + A_nP_n}{A} \tag{3.1}$$

where P represents the average depth of rainfall in a watershed of area A and $P_1, P_2, \ldots, P_n$ represent the rainfall depths from rain gauges within the polygon having areas $A_1, A_2, \ldots, A_n$ within the watershed. The areas $A_1, A_2, \ldots, A_n$ can be determined by using a digitizer or a planimeter.

Example 3.1

A storm on the watershed illustrated in Figure 3-5 produces rainfall at the various gauge locations as indicated. Compare the mean precipitation as determined by the arithmetic mean and the Thiessen Polygon methods.

Solution. By the arithmetic mean method, the mean precipitation depth is

$$\frac{(46 + 55 + 57 + 41 + 55 + 46)}{6} = 50\,\text{mm}$$

By the Thiessen Polygon method, the areas represented by the polygons surrounding each rain gauge are determined with a planimeter and substituted in Equation 3.1:

$$P = \frac{(65 \times 46)+(150 \times 55)+(269 \times 57)+(216 \times 55)+(56 \times 41)+(136 \times 46)}{892}$$
$$= 52.7\,\text{mm}$$

3.10 The Isohyetal Method

The Isohyetal method consists of recording the depths of rainfall at the locations of the various rain gauges and plotting isohyets (lines of equal rainfall) by the same methods used for locating contour lines on topographic maps. The area between isohyets may then be measured with a digitizer or a planimeter, and the average rainfall determined by using Equation 3.1. The choice of analysis method depends partly on the area of the watershed, the number of rain gauges, the distribution of the rain gauges, and in some situations, the characteristic of the rainstorm.

Design Storms

Historical precipitation data can be used to estimate the likelihood of future storm events. The rainfall intensity i (L/T) or rainfall depth are characteristics of the rainfall event that are generally required for design purposes. Storms of high intensity generally have fairly short durations and cover small areas. Storms covering large areas are seldom of high intensity but may last several days. The infrequent combination of relatively high intensity and long duration storms produces large total amounts of rainfall. These rainfalls are generally associated with a slowly moving warm front or the development of a stationary front and can cause significant erosion damage and devastating floods.

Statistical methods are used to analyze the rainfall records to determine the magnitude of storm events for specific return periods (Haan et al., 1994). The return period is sometimes called recurrence interval or frequency of the rainfall event. The relationship between return period T in years and probability of occurrence P can be expressed by

$$T = 100/P \quad \textbf{3.2}$$

The probability of occurrence is the probability in percentage points that an event equaling or exceeding a given event will occur in a given year.

3.11 Frequency Analysis

If long-term precipitation data are available, a frequency analysis can be conducted to quantify precipitation uncertainty. A rainfall record of at least 20 years is recommended for a statistically representative sample (Serrano, 1997). For best results, the length of record should be greater than the design life of the structure (Serrano, 1997).

Data for the hydrologic frequency analysis can be selected by two methods: the annual series and the partial-duration series method. In the annual series method, only the largest single storm event for each year is selected for the analysis. Thus, for 20 years of record only 20 values would be analyzed, regardless of whether multiple large events occurred during a particular year. With the partial-duration series method, all values above a given base value are chosen regardless of the number of events within a given period or the largest N events in N years are selected. The partial-duration series is applicable when the second largest event value of the year would affect the design. An example is the design of drainage channels where damage may result from flooding caused largely by flows lower than the annual peak flow rate. The annual and partial duration series methods give essentially identical results for return periods greater than 10 to 25 years. For shorter return periods, the partial duration series method gives a larger value.

Regardless of the method of selecting the data, the values must satisfy two important criteria: (1) that each event is independent of a previous or subsequent event, and (2) that the data for the period of record for analysis is representative of the long-term record. The first criterion is necessary for statistical analysis, and the second criterion ensures that the predicted values will reflect the pattern of occurrences from the period of record. When the data are obtained from the highest annual values, the number of observations during the year should be large. Selection of a water year, such as October 1 to September 30, rather than the calendar year has proved beneficial for some types of applications.

Although it is possible to extend the analysis to predict return periods greater than the length of record, the selection of an appropriate probability distribution function becomes important (Haan et al., 1994). One of the most versatile probability distributions is the Weibull distribution because it can be used to approximate exponential, normal, or skewed distributions (Weibull, 1951).

The Weibull cumulative distribution function $P(x)$ is

$$P(x) = e^{-(x/\alpha)^{\beta}} \qquad 3.3$$

where $P(x)$ = Weibull cumulative distribution function
x = rainfall depth (L),
α = characteristic depth (L),
β = shape parameter.

The parameters α and β can be determined by linear regression. Equation 3.3 can be algebraically manipulated to

$$\log_e\left[\log_e\left(\frac{1}{P(x)}\right)\right] = \beta\log_e x - \beta\log_e\alpha \qquad 3.4$$

Although it may not be obvious, Equation 3.4 has the familiar form of a line ($y = mx + b$) where the large term on the left side of the equation is y and natural logarithm of rainfall depth is x. A regression of these will allow the Weibull parameters to be determined.

The first step in the analysis is to rearrange the data in decreasing order of magnitude. The data are assigned a rank m from 1 to N where N is the number of observations. From the ranking, a plotting position (frequency) can be determined by

$$P = \frac{m - a}{N + 1 - 2a} \qquad 3.5$$

where a is a parameter that depends on the distribution (Bedient & Huber, 2002). The parameter a varies from 0 for the original Weibull formula to 0.375 for normal or lognormal distributions, 0.3 for median ranks, and 0.44 for the Gumbel distribution. Bedient and Huber (2002) recommend 0.4 for a when the distribution is unknown. Once the plotting position P has been determined, the left side of Equation 3.4 can be computed. A linear regression of the left side and the natural logarithm of precipitation produces the slope, which is equivalent to β. The characteristic depth α can be computed from the intercept b of the regression line and slope β as

$$\alpha = e^{(-b/\beta)} \tag{3.6}$$

Example 3.2

Determine the Weibull cumulative distribution function for the 20 years of 24-hr rainfall data presented below for Lexington, Keutucky. Estimate the 24-h 10-yr, 25-yr, and 100-yr storms using the distribution.

Year	*24-h Rainfall (mm)*	*Year*	*24-h Rainfall (mm)*
1984	49.28	1994	60.20
1985	45.47	1995	107.95
1986	38.10	1996	61.72
1987	59.69	1997	141.22
1988	85.85	1998	128.02
1989	78.74	1999	44.20
1990	67.06	2000	55.37
1991	84.33	2001	55.37
1992	126.75	2002	57.15
1993	66.04	2003	52.58

Solution. The rainfall data are rearranged in decreasing order and assigned a rank m in the table below. The rank $m = 1$ is assigned to the largest 24-h rainfall over the 20-year period. Rank $m = 2$ is given to the second highest rainfall, etc. This proceeds until the last rainfall total is given the rank $m = N$, where N is the number of years of record.

The return period is estimated by the plotting position of the ranked series. For $a = 0.4$ in Equation 3.5, the plotting position P is

$$P = \frac{m - 0.4}{N + 0.2}$$

I (mm)	*Rank m*	*Plotting Position P*	*1/P*	*$\log_e(\log_e[1/P])$*	*$\log_e(I)$*
141.22	1	0.030	33.667	1.257	4.950
128.02	2	0.079	12.625	0.930	4.852
126.75	3	0.129	7.769	0.718	4.842
107.95	4	0.178	5.611	0.545	4.682

(Continued)

I (mm)	Rank *m*	Plotting Position *P*	*1/P*	$\log_e(\log_e[1/P])$	$\log_e(I)$
85.85	5	0.228	4.391	0.392	4.453
84.33	6	0.277	3.607	0.249	4.435
78.74	7	0.327	3.061	0.112	4.366
67.06	8	0.376	2.658	−0.022	4.206
66.04	9	0.426	2.349	−0.158	4.190
61.72	10	0.475	2.104	−0.296	4.123
60.20	11	0.525	1.905	−0.439	4.098
59.69	12	0.574	1.741	−0.589	4.089
57.15	13	0.624	1.603	−0.751	4.046
55.37	14	0.673	1.485	−0.927	4.014
55.37	15	0.723	1.384	−1.125	4.014
52.58	16	0.772	1.295	−1.353	3.962
49.28	17	0.822	1.217	−1.628	3.897
45.47	18	0.871	1.148	−1.982	3.817
44.20	19	0.921	1.086	−2.495	3.789
38.10	20	0.970	1.031	−3.501	3.640

A linear regression of the last two columns gives estimates for the slope (2.961) and intercept (−13.058). The slope is β or 2.961, and α can be computed from Equation 3.6 as

$$\alpha = e^{(-b/\beta)} = e^{(-(-13.058)/2.961)} = 82.27$$

Equation 3.3 can be used to determine rainfall values for various return intervals. The frequency for a 10-yr storm can be computed from Equation 3.2:

$$P = \frac{1}{T} = \frac{1}{10} = 0.1$$

Similarly, the frequency for 25- and 100-yr storms are 0.04 and 0.01, respectively. Solving Equation 3.3 for x and substitution of $P = 0.1$ gives the 24-h 10-yr storm depth:

$$x = \alpha[-\log_e(P)]^{1/\beta} = 82.27[-\log_e(0.1)]^{1/2.961} = 109 \text{ mm}$$

The 24-h 25-yr rainfall is 122 mm, and the 24-hr 100-yr rainfall is 138 mm

3.12 Intensity–Duration–Frequency (IDF) Curves

The design of most hydrologic structures requires knowledge of how frequently storms of specific intensities and durations occur at a particular location. Return periods or frequencies are usually reported as storm events that are expected to occur on average once in 2, 5, 10, 25, 50, or 100 years. Storm duration can be as small as 5, 10, or 15 minutes or for longer periods such as 24 hours, 2 days, or more. A general expression for the relationship between rainfall intensity i for a given duration t and return period T is given by

$$i = \frac{KT^x}{(t + b)^n} \tag{3.7}$$

where K, x, b, and n are constants for a given geographic location, and can be determined statistically from rainfall data analyses. Plots of rainfall intensity and storm durations are often constructed for various return periods and are called intensity–duration–frequency (IDF) curves (Figure 3-6). Since rainfall depth is simply the intensity times the duration, depth–duration–frequency (DDF) curves are sometimes plotted. Current estimates of IDF and DDF curves indicate they are not the smooth functions that Equation 3.7 might indicate.

The National Weather Service Hydrometeorological Design Studies Center has developed an interactive Internet site based on the NOAA Atlas 14 publications for producing IDF and DDF data (Bonnin et al., 2003). The site is called the Precipitation Frequency Data Server (PFDS) and allows the user to select a location in the United States for retrieval of either IDF or DDF data. Once a site is selected (from the map, by station, or coordinates), a table of IDF or DDF data is produced (Table 3-1). In addition to the table of data, tables are produced indicating the upper and lower 90 percent confidence limits of the data. The data are displayed graphically (Figure 3-6).

Prior to the Atlas 14 publications, Hershfield (1961) completed an analysis of rainfall frequency data and provided isohyetal maps (maps showing lines of equal rainfall depths) for the United States (Figure 3-7). These maps were often referred to as TP-40 maps (available at the PFDS Internet site). Weiss (1962) developed a procedure for using the TP-40 maps to obtain IDF data for any location.

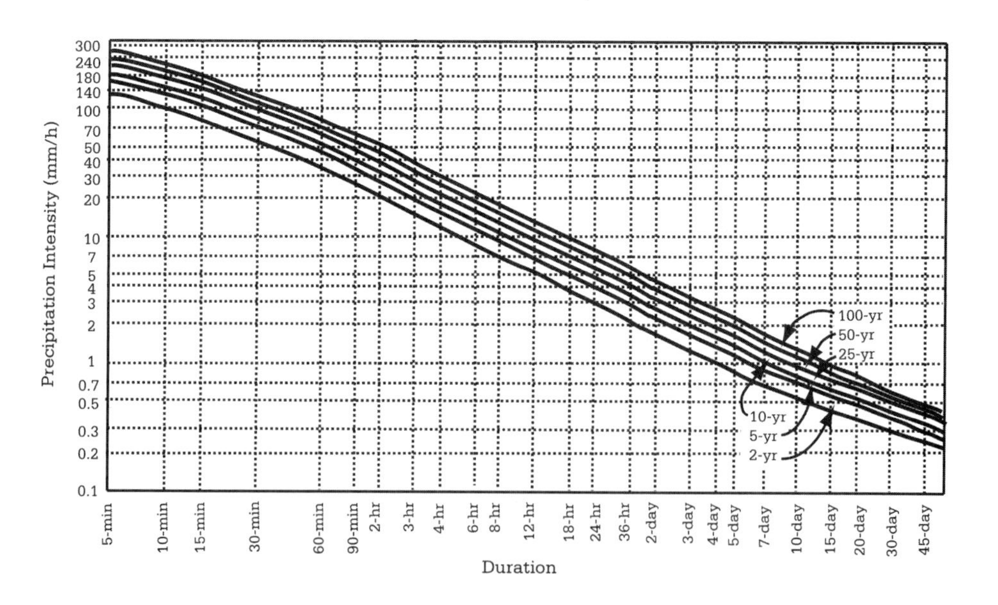

Figure 3–6

Rainfall intensity–duration–frequency data near St. Louis, Missouri.

TABLE 3–1 Typical Intensity Duration Frequency Table Produced at the Atlas 14 Interactive Internet Site for an Area Near St. Louis, Missouri.

	Precipitation Frequency Estimates (mm)																
Return Period (yr)	***5 min***	***10 min***	***15 min***	***30 min***	***60 min***	***120 min***	***3 hr***	***6 hr***	***12 hr***	***24 hr***	***48 hr***	***4 day***	***7 day***	***10 day***	***20 day***	***30 day***	***45 day***	***60 day***
2	11.04	17.26	21.10	28.23	34.64	40.58	43.20	51.36	61.07	72.18	83.45	95.21	111.82	126.60	172.56	210.99	262.95	306.92
5	13.95	21.68	26.62	36.45	45.73	53.85	57.35	68.01	80.54	94.75	109.79	125.21	145.22	163.92	219.19	264.80	327.98	380.12
10	15.94	24.60	30.25	42.02	53.47	63.63	68.03	80.55	95.11	112.36	130.17	148.18	169.73	191.23	251.56	301.08	371.44	428.66
25	18.39	28.13	34.76	49.08	63.65	77.72	83.59	99.12	116.35	139.37	161.38	182.93	205.65	230.82	296.42	350.94	429.95	493.82
50	20.29	30.76	38.07	54.42	71.70	89.83	97.15	115.18	134.87	163.78	189.46	214.13	236.77	265.03	333.79	391.39	477.23	546.27
100	22.21	33.44	41.54	59.98	80.20	103.55	112.60	133.68	155.86	193.03	222.80	250.73	272.29	303.81	374.80	435.13	527.80	602.25
200	24.31	36.28	45.13	65.88	89.38	119.23	130.44	155.03	180.20	227.91	262.42	294.17	313.27	348.16	420.16	483.10	582.50	662.67
500	27.30	40.15	50.09	74.23	102.66	143.80	158.55	188.83	218.43	285.19	327.18	364.12	377.30	417.25	487.98	553.45	662.01	750.07
1000	29.73	43.26	54.09	81.06	113.84	165.91	184.07	219.54	253.09	338.72	387.57	428.76	436.25	478.77	546.14	612.68	728.27	822.60

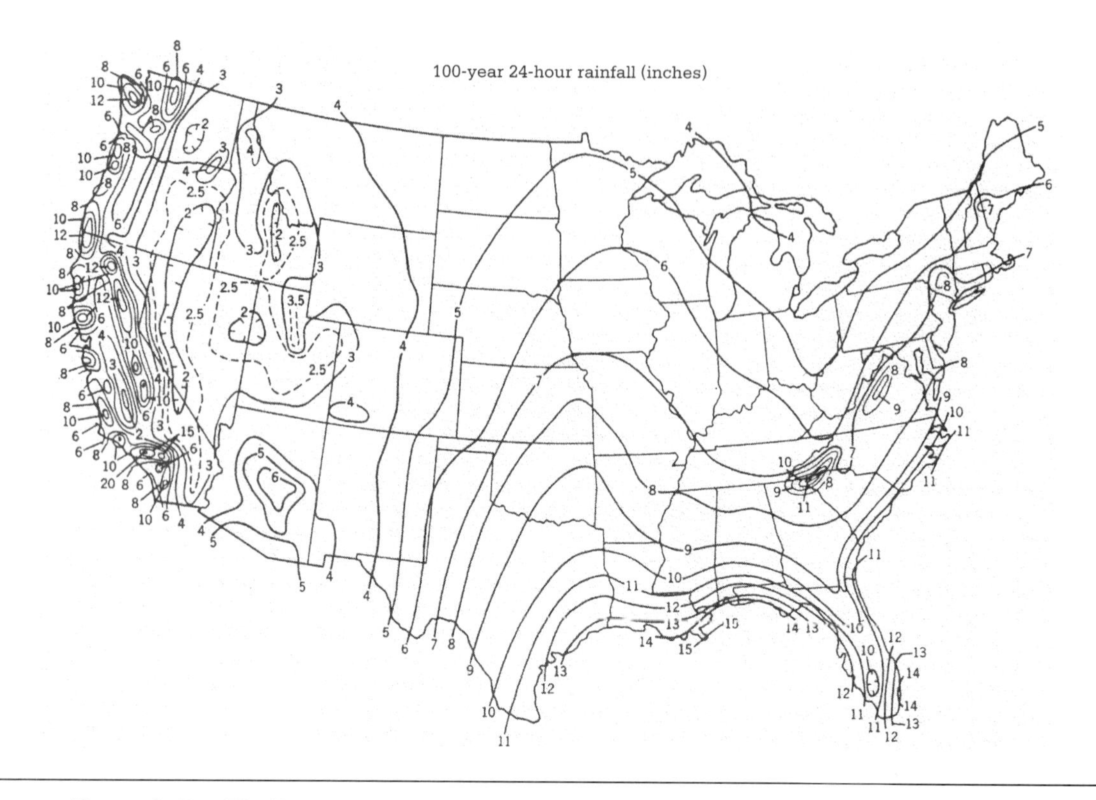

Figure 3–7 TP-40 data for 24-h 100-yr return period storm. (Herschfield, 1961)

Example 3.3

Determine the rainfall intensities for 10-min and 6-h storms that might be expected to occur once in 5 years and 50 years at O'Hare International Airport near Chicago, Illinois.

Solution. Access the Precipitation Frequency Data Server and select the state of Illinois. By either selecting a rainfall station near Chicago or by identifying Chicago in the pull-down list, obtain the estimates from the NOAA Atlas 14. The 10-min and 6-h storms can be determined for both return periods from the table produced at the PFDS.

	10-min	*6-h*
5-yr	21.7	71.6
50-yr	30.4	122.3

3.13 Average Depth of Precipitation over an Area

Precipitation gauges and DDF data represent information at a point in the watershed. If the point data are to represent a much larger area, corrections are required to adjust (reduce) the point data and account for uneven distribution of the storm over the watershed. Figure 3-8 provides adjustment factors for converting point rainfall over an area for storms of various durations. The design point rainfall may

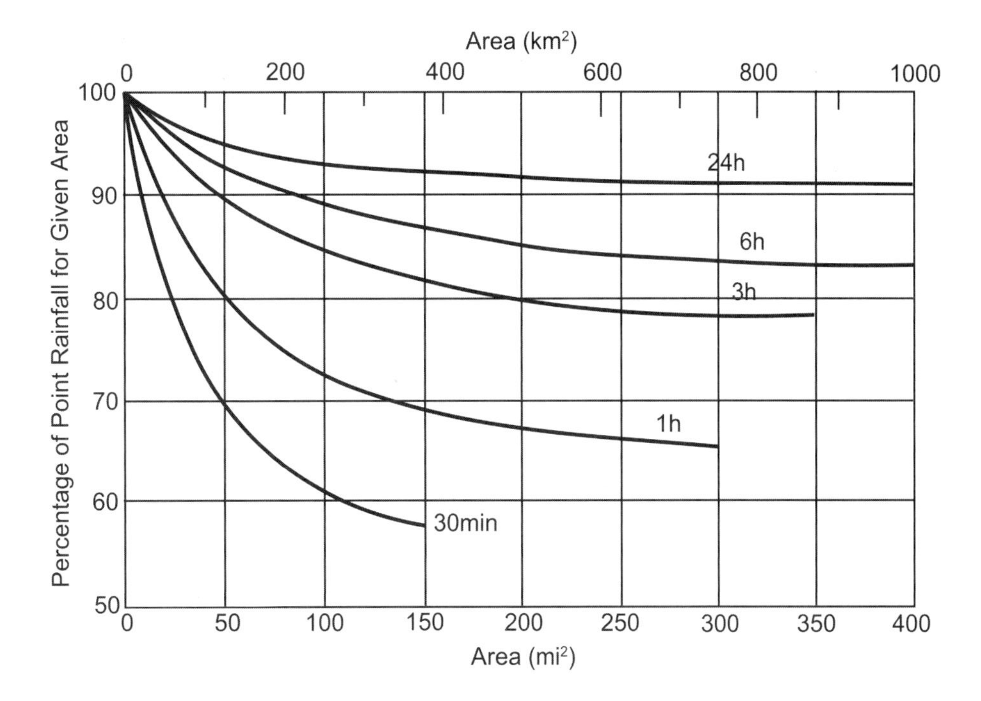

Figure 3–8
Area–depth curves in relation to point rainfall. (Redrawn from Hershfield, 1961)

be considered as the maximum for the storm, and thus the average rainfall over a watershed will be less than that of the maximum. As part of the development of the NOAA Atlas 14 project the depth area reduction factors are being statistically analyzed and updated. These data will be available on the PFDS.

Example 3.4

Determine the rainfall amount and intensity for a 6-h storm that will occur once in 50 years over an area of 400 km^2 in Chicago, Illinois.

Solution. The 6-h 50-yr rainfall in Chicago (from Example 3.3) was 122.3 mm. From Figure 3-8, for a 6-h storm over a 400-km^2 area, read the percentage of point rainfall as 87 percent. Therefore, the 6-h 50-yr rainfall over an area of 400 km^2 in Chicago can be approximated by $0.87 \times 122.3 = 106.4$ mm.

3.14 Storm Synthesis

Most storm events represent a combination of rainfall intensity patterns (Figure 3-9). As the storm event moves across the area, the rainfall intensity fluctuates. The event may include short-duration, high-intensity precipitation in combination with low-intensity precipitation or short periods of no precipitation. To determine maximum rainfall intensity for a particular duration, the time period must be selected from the most intense portion of the storm. The return period of any rainfall intensity for a particular duration can be obtained from the IDF curves for any specific location, similar to Figure 3-6 for St. Louis, Missouri.

Example 3.5

Determine the return period for the maximum rainfall intensity occurring for any 20-min period and for the first 140 min (2.33 h) during the storm shown below for St. Louis, Missouri.

Time (A.M.)	*Time Interval (min)*	*Cumulative Time (min)*	*Rainfall during Interval* (mm)*	*Cumulative Rainfall (mm)*	*Rainfall Intensity for Interval (mm/h)*
6:50					
7:00	10	10	1	1	6
7:10	10	20	10	11	60
7:15	5	25	11	22	132
7:35	20	45	46	68	138
7:45	10	55	19	87	114
8:25	40	95	31	118	47
9:10	45	140	6	124	8
10:50	100	240	6	130	4

*Corrected rainfall based on nonrecording gauge depth.

Solution. The maximum rainfall intensity for 20 min is 138 mm/h. Interpolating from Figure 3-6, read a return period of 25 years. For the first 140-min storm, the average rainfall intensity is 124/2.33 = 53 mm/h, for which Figure 3-6 indicates a return period of over 100 years.

The rainfall intensity pattern with time is of particular interest in hydrologic studies. The storm patterns can be used as input to hydrologic models to predict runoff. In some cases, historical storm patterns that have caused significant damage are selected and used as input. In most cases, a synthetic storm is derived.

During the development of the PFDS, an intensive statistical analysis was performed on the precipitation records to produce temporal estimates of rainfall during a storm event. Figure 3-10 illustrates the temporal distribution of 6-, 12-, 24-, and 96-h storms in the Ohio River basin. Each of these temporal distributions

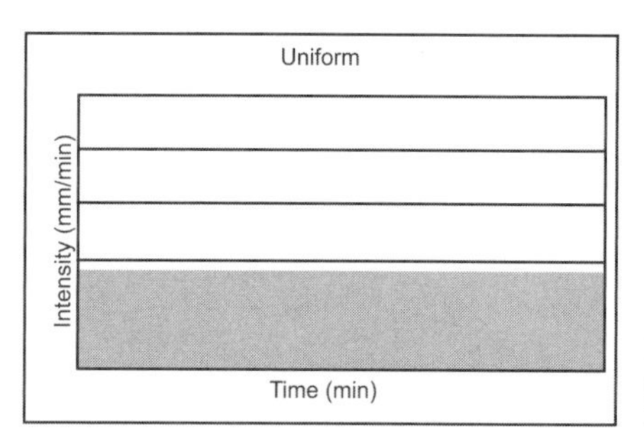

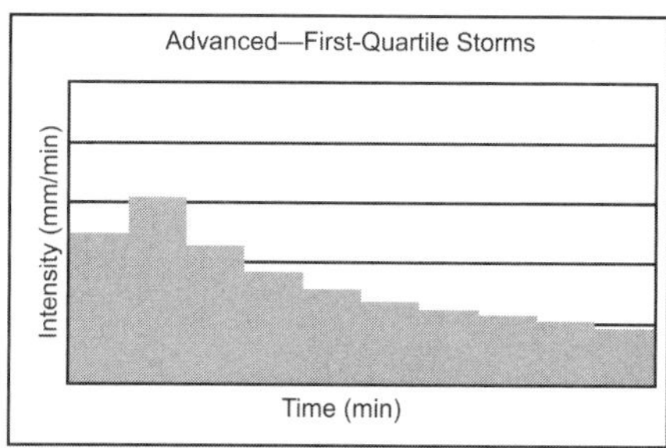

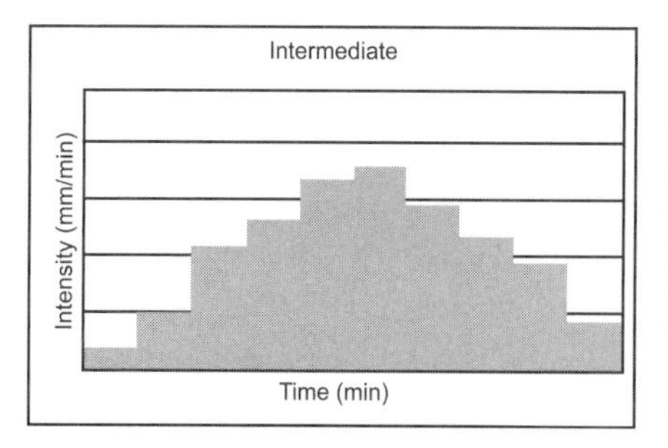

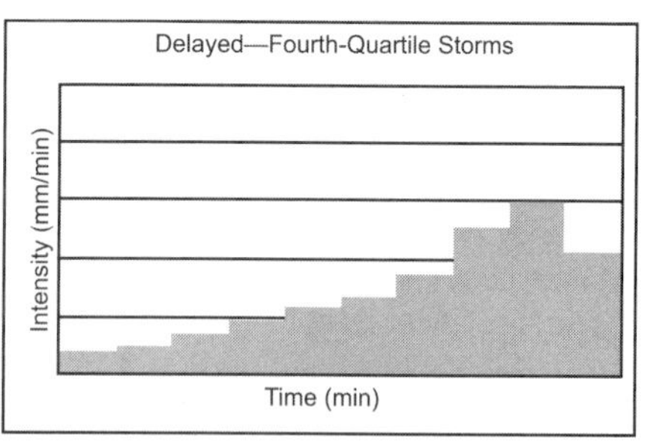

Figure 3–9
Rainfall intensity patterns.

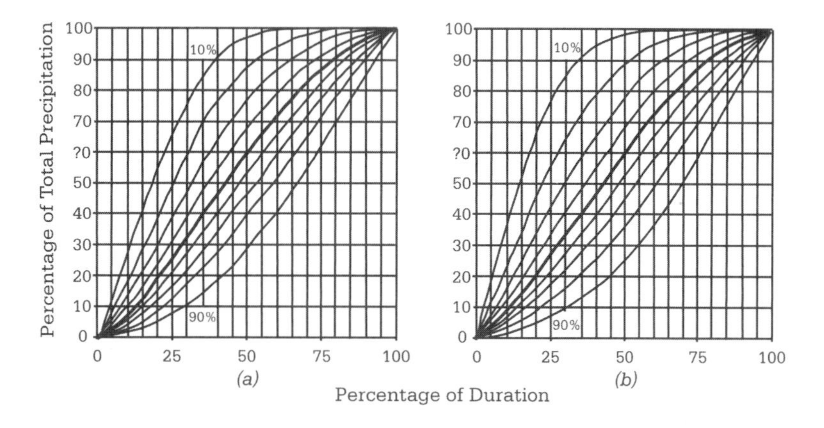

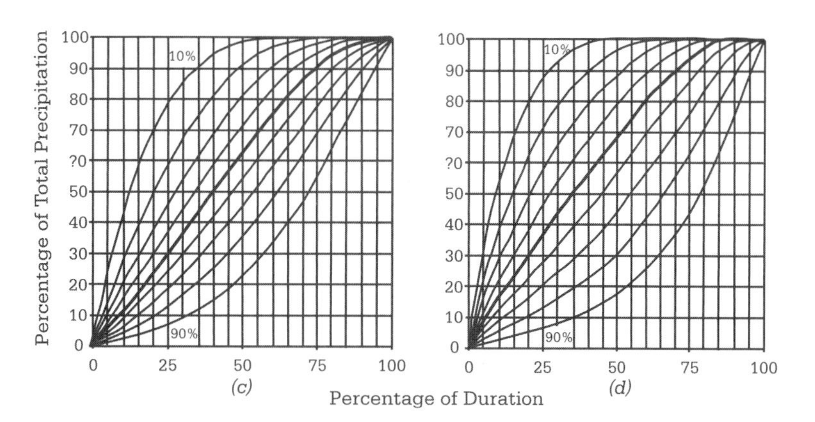

Figure 3–10

Temporal distribution of rainfall for the Ohio River basin and surrounding states for (a) 6-hr duration, (b) 12-hr duration, (c) 24-hr duration, and (d) 96-hr duration.

can be further divided into storms where the bulk of the precipitation occurred in the first, second, third, or fourth quartile of the storm duration (Figure 3-11 for 6-h storms). Of the 17 000 storms that occurred in the Ohio River basin lasting up to 6 hours, most were either first or second quartile storms, meaning that most of the rainfall fell during the early stages of the event. Example 3.6 illustrates the use of the temporal storm distributions and the PFDS to develop a synthetic storm.

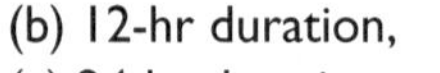

Example 3.6

Develop a 6-h 25-yr synthetic storm for Lexington, Kentucky, that represents lower and upper 90 percent confidence limits of the storm total. The design storm is classified as a second-quartile storm, indicating that the rainfall is temporally distributed such that the bulk of the rainfall occurs during the second quarter of the event.

Solution. From the PFDS site, determine that the precipitation estimate is 93.1 mm for a 6-h 25-yr storm in Lexington, Kentucky. The lower-bound confidence interval is 85.05 mm and the upper-bound confidence interval for the estimate is 101.0 mm. Construct a table to calculate the rainfall depths with time according to Figure 3-11b. Obtain the storm duration and total precipitation percentages from Figure 3-11b. The accumulated storm depth is the design rainfall depth (93.1 mm) times the total precipitation percentage (Figure 3-12). The depth increment is the difference in storm depth at successive time increments. The storm depth and depth increment can be computed for the upper and lower confidence intervals.

Storm Duration			*Design Rainfall*		*Confidence Interval*			
					Upper 90%		*Lower 90%*	
(%)	*(min)*	*Total Precipitation (%)*	*Storm Depth (mm)*	*Depth Increment (mm)*	*Storm Depth (mm)*	*Depth Increment (mm)*	*Storm Depth (mm)*	*Depth Increment (mm)*
0	0	0	0	0	0	0	0	0
10	36	5	4.65	4.65	5.05	5.05	4.25	4.25
20	72	15	13.96	9.31	15.15	10.10	12.76	8.51
30	108	32	29.79	15.83	32.32	17.17	27.22	14.46
40	144	50	46.55	16.76	50.50	18.18	42.53	15.31
50	180	67	62.38	15.83	67.67	17.17	56.98	14.46
60	216	80	74.48	12.10	80.80	13.13	68.04	11.06
70	252	88	81.93	7.45	88.88	8.08	74.84	6.80
80	288	93	86.58	4.66	93.93	5.05	79.10	4.25
90	324	97	90.31	3.72	97.97	4.04	82.50	3.40
100	360	100	93.10	2.79	101.00	3.03	85.05	2.55

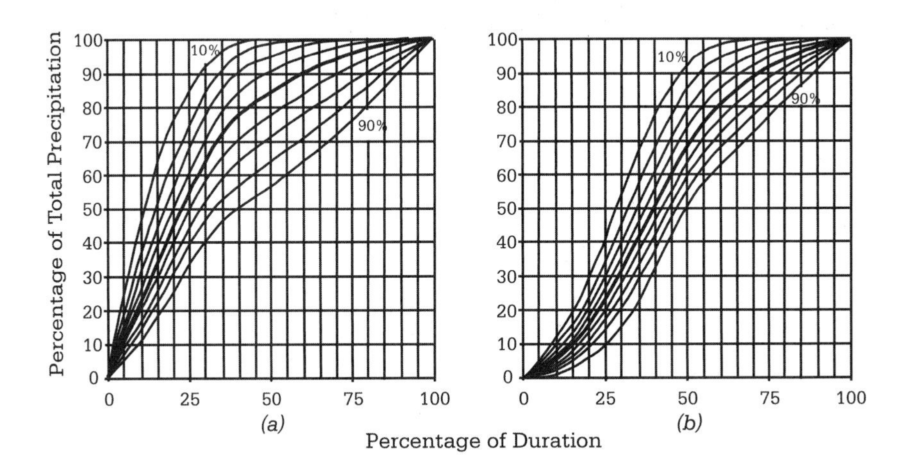

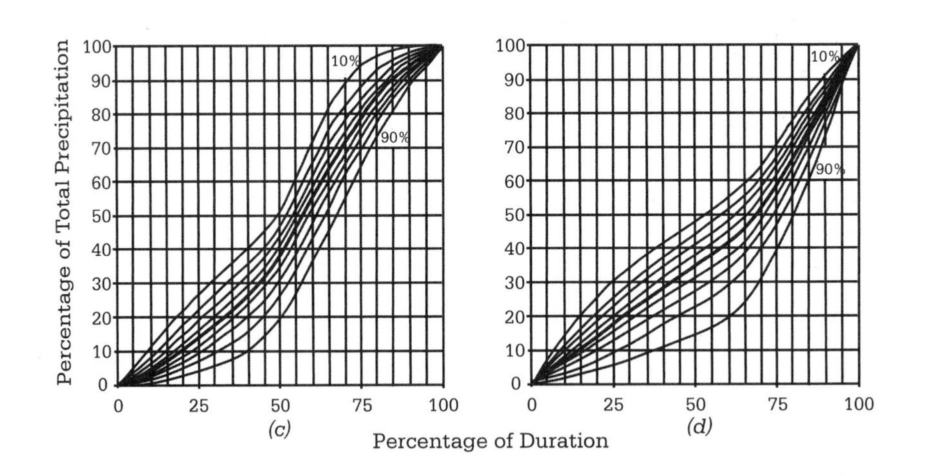

Figure 3–11

Temporal distribution of 6-h duration storms for the Ohio River basin and surrounding states with (a) first-quartile storms, (b) second-quartile storms, (c) third-quartile storms, and (d) fourth-quartile storms.

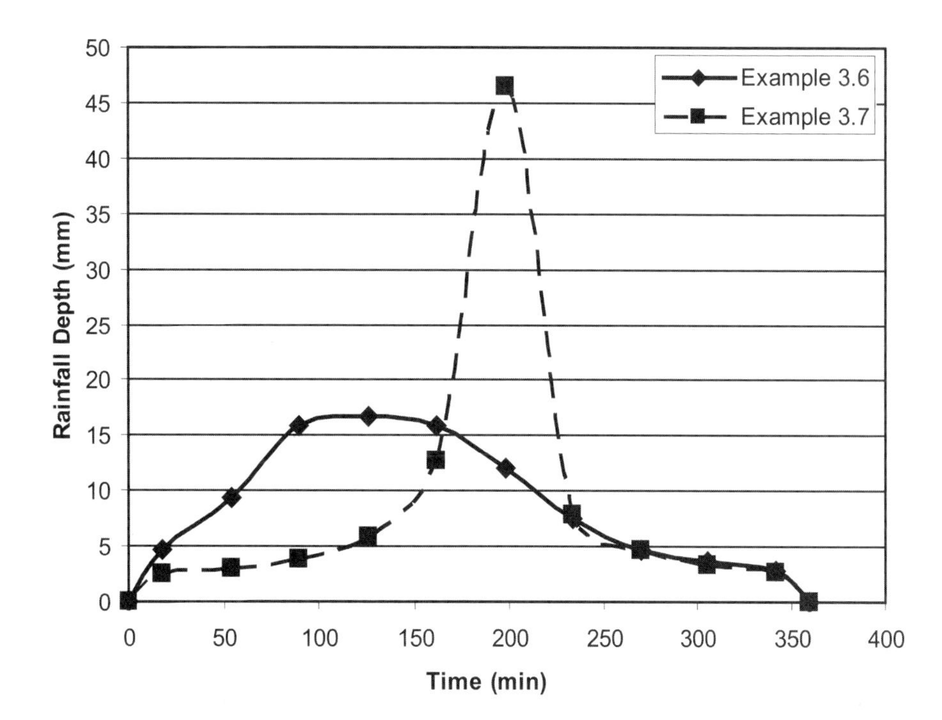

Figure 3–12
Time distribution of rainfall predicted for Examples 3.6 and 3.7.

The SCS developed a method of storm synthesis based on rainfall patterns called type curves for the United States (SCS, 1986). The 24-hr rainfall depth for a desired storm frequency can be distributed over a 24-hr period based on the type curve. Type II and type III curves represent all but the extreme western portions of the contiguous United States. Although a rainfall depth is distributed over a 24-h period in the SCS method, approximately 50 percent of the rainfall is projected to fall during the one-hour period from 11:30 A.M. to 12:30 P.M.

Haan et al. (1994) described a method of using the IDF curve to develop a synthetic storm that produces similar results to the SCS method. A design storm of a specific duration and frequency is chosen. The storm duration is broken up into equal portions (e.g., 15-minute increments). Equation 3.3 or an IDF curve is used to determine the average rainfall intensity for each duration increment of the storm. For example, if a 3-h 25-yr storm was to be developed with 15-minute increments, the first increment would represent a 15-minute 25-yr storm, the second increment would represent a 30-minute 25-yr storm, and proceed until the final increment, which would represent a 3-h 25-yr storm. Rainfall depth is determined by multiplying the duration times the intensity. The incremental depth is the difference between the computed rainfall depths at each time increment. The incremental depths are then rearranged into the storm pattern of choice. If the depths are arranged in a symmetrical (intermediate) pattern, the resulting storm is very similar to the SCS Type II storm (Haan et al., 1994).

Example 3.7

Construct a 6-h 25-yr storm for Lexington, Kentucky. The parameters for Equation 3.3 are $K = 8.21$, $x = 1.593$, $n = 0.76$, and $b = 8.44$. Use a time increment of 36 minutes.

Solution. Construct a table of time increments, and determine the associated rainfall intensity and depth for each increment.

- The rainfall intensity for the 36-min increment is
$$i = \frac{8.21(25)^{1.593}}{(36 + 8.44)^{0.76}} = 77.4\,\text{mm/h}.$$
- The rainfall intensity for a 72-min increment is
$$i = \frac{8.21(25)^{1.593}}{(72 + 8.44)^{0.76}} = 49.3\,\text{mm/h}.$$
- The rainfall depth that occurs over the first 36 minutes is
$77.4(36/60) = 46.5$ mm.
- The rainfall depth that occurs over the first 72 minutes is
$49.3(72/60) = 59.2$ mm.
- The incremental depth is
$59.2 - 46.5 = 12.7$ mm.

Once the incremental depths are determined, the synthetic storm can be assembled (Figure 3-12). An intermediate storm is shown where the maximum intensity occurred between 180 and 216 minutes (46.5-mm rainfall), the second greatest intensity preceded it from 144 to 180 minutes (12.7 mm), and the third largest intensity occurred from 216 to 252 minutes (7.8 mm), etc. The intermediate storm pattern is similar to that produced by the SCS Type II curve. An advanced or delayed storm could have been developed by arranging the 36-minute incremental depths differently. Note that the total storm depth is the same as the 6-h 25-yr storm.

Duration (min)	Intensity (mm/h)	Depth (mm)	Incremental Depth (mm)	Synthetic Storm (mm)
0	0	0	0	0
36	77.4	46.5	46.5	2.5
72	49.3	59.2	12.7	3.0
108	37.2	67.0	7.8	3.9
144	30.3	72.8	5.8	5.87
180	25.8	77.5	4.7	12.7
216	22.6	81.4	3.9	46.5
252	20.2	84.8	3.4	7.8
288	18.3	87.9	3.0	4.6
324	16.8	90.6	2.7	3.4
360	15.5	93.1	2.5	2.7
Total			93.1	93.1

3.15 Probable Maximum Precipitation

In cases where failure of a hydraulic structure has the potential to cause massive economic damage or loss of lives, an analysis based on the probable maximum precipitation (PMP) is required. The PMP is the "theoretical greatest depth of precipitation for a given duration that is physically possible over a given size storm area at a particular geographical location at a certain time of year" (Dingman, 1994). The storm area is taken as the area of the drainage basin of interest, and the duration is the time of concentration of the basin (Chapter 5).

In application, the PMP value is input into a hydrologic model to predict the largest flood that might be expected to ever occur in the drainage basin, the probable maximum flood (PMF). The flood value is used in design of the hydraulic structures such as channels (Chapter 6) and emergency spillways (Chapter 9).

Internet Resources

Precipitation data server:
http://hdsc.nws.noaa.gov/hdsc/pfds/

Internet weather source:
http://weather.noaa.gov

References

Bedient, P. B., & W. C. Huber. (2002). *Hydrology and Floodplain Analysis*. Upper Saddle River, NJ: Prentice Hall.

Bonnin, G. M., D. Todd, B. Lin, T. Parzybok, M. Yekta, & D. Riley. (2003). *Precipitation Frequency Atlas of the United States "NOAA Atlas 14 Volume I, Version 2."* Silver Spring, MD: NOAA, National Weather Service.

Dingman, S. L. (1994). *Physical Hydrology*. Englewood Cliffs, NJ: Prentice Hall.

Haan, C. T., B. J. Barfield, & J. C. Hayes. (1994). *Design Hydrology and Sedimentology for Small Catchments*. San Diego: Academic Press.

Hershfield, D. N. (1961). *Rainfall Frequency Atlas of the United States*. May. Washington, DC: U.S. Government Printing Office.

Serrano, S. E. (1997). *Hydrology for Engineers, Geologists and Environmental Professionals*. Lexington, KY: HydroScience.

Soil Conservation Service (SCS). (1986). *Urban Hydrology for Small Watersheds*, Technical Release No. 55. Washington, DC: Soil Conservation Service, U.S. Department of Agriculture.

Thompson, L. M. (1973). Cyclical weather patterns in the middle latitudes. *Journal of Soil and Water Conservation, 29*, 87–89.

U.S Department of Agriculture (USDA). (1989). *The Second Appraisal. Soil, Water, and Resources on Nonfederal land in the United States*. Washington, DC: Author.

Weibull, W. (1951). A statistical distribution function of wide applicability. *Journal of Applied Mechanics, 73*, 293–297.

Weiss, L. L. (1962). A general relation between frequency and duration of precipitation. *Monthly Weather Review, 90*, 87–88.

Problems

3.1 Determine the total rainfall to be expected once in 5, 25, and 100 years for a 60-min storm at your present location.

3.2 Determine the maximum rainfall intensity to be expected once in 10 years for storms having durations of 20, 30, 120, and 360 min, respectively, at your present location.

3.3 Determine the parameters for Equation 3.3 for your present location.

3.4 Compute the average rainfall for a given watershed by the Thiessen Polygon method from the following data. How do the weighted average and the gauge average compare?

Rain Gauge	*Area (ha)*	*Rainfall (mm)*
A	14.0	58
B	4.5	41
C	5.3	51
D	4.9	43

3.5 During a 60-min storm the following amounts of rain fell during successive 15-min intervals: 33 mm, 23 mm, 15 mm, and 5 mm. What are the maximum intensity for 15 min and the average intensity? If the storm had occurred in Lexington, Kentucky, how often would you expect such a 60-min storm to occur? What type of storm pattern was it?

3.6 For your present location, compute a synthetic storm with a duration of 5 hours and a 50-yr return period using a time increment of 30 minutes.

CHAPTER 4

Evaporation and Evapotranspiration

Two phases of the hydrologic cycle of particular interest in agriculture are evaporation and transpiration. About three fourths of the total precipitation received on land areas of the world returns directly to the atmosphere by evaporation or transpiration. Most of the balance returns to the ocean as surface or subsurface flow.

Evaporation is the transfer of liquid surface water into vapor in the atmosphere. The water molecules, both in the air and in the water, are in rapid motion. Evaporation occurs when the number of moving molecules that break from the water surface and escape into the air as vapor is larger than the number that re-enter the water surface from the air and become entrapped in the liquid. Evaporation, which may occur from water surfaces, wet leaf surfaces, or from water on soil particles, is important in water management and conservation.

The vapor pressure of water is the partial pressure exerted by the water molecules in their gaseous form. For example, if liquid water is introduced into a closed container, water will evaporate from the surface until there is a balance or equilibrium between the molecules leaving the water and those re-entering the water. A pressure gauge attached to the container will indicate an increase in pressure. This increase is the vapor pressure of the water. The saturation vapor pressure is related to temperature and can be determined by the following equation (ASCE, 2005)

$$e_s(T) = 0.6108 \exp\left[\frac{17.27\,T}{T + 237.3}\right] \tag{4.1}$$

where e_s is the saturation vapor pressure in kPa and T is the air temperature in °C.

Wind increases the rate of evaporation, particularly as it disperses the vapor layer found directly over the evaporating water surface under stagnant conditions. Because of this mixing, the characteristics of the atmosphere above the surface are of interest. As might be expected from the decreased concentration of water molecules, evaporation increases with decreased vapor pressure. Also, the rate of evaporation decreases slightly with increases in the salt content of the water.

Transpiration is the process through which water vapor passes into the atmosphere through the tissues of living plants. The amount of water that passes through plants by the transpiration process is often a substantial portion of the total water available during the growing season and, besides energy availability, is governed by total leaf area and plant stomatal control. It can vary from near zero to as much as 2000 mm per year, depending largely on the water available, type of plant, density of plant growth, amount of sunshine, climatic dryness, and soil fertility and structure. Less than 1 percent of water uptake is actually retained by the plant. The rate of evaporation or transpiration increases with a rise in temperature of the surface because saturation vapor pressure at the surface increases with increases in surface temperature.

In areas with growing plants, water passes into the atmosphere by evaporation from soil surfaces and by transpiration from plants. Because evaporation and transpiration are difficult to separate, they are frequently considered together and called evapotranspiration. Estimated evapotranspiration is needed for determining irrigation requirements for crops as well as water storage in ponds and reservoirs. High evapotranspiration from such crops as grass may be beneficial for the removal of soil water. Methods for predicting evaporation from water surfaces or evapotranspiration can be grouped into three categories. The newest methods use combinations of these.

Mass Transfer. This approach recognizes that water moves away from evaporating and transpiring surfaces in response to the combined phenomena of turbulent mixing of the air and the vapor pressure gradient. Thornthwaite and Holzman (1942) proposed such a method. Application of methods based on mass transfer principles are often combined with other methods.

Energy Balance. Energy is required for evaporation of water, so if there is no change in water temperature, the net radiation or heat supplied can be related to evaporation. Most methods include an energy component.

Empirical Methods. Several such methods, developed from experience and field research, are based primarily on the assumption that energy available for evaporation or evapotranspiration is proportional to the temperature. Blaney and Criddle (1950), Thornthwaite (1948), and many others have proposed equations of this type.

4.1 Evaporation from Water Surfaces

Dalton's Law. Dalton's law for evaporation from free water surfaces is

$$E = C(e_s - e_a) \tag{4.2}$$

where E = rate of evaporation (mm/day),
C = a constant (mm day^{-1} kPa^{-1}),
e_s = saturation vapor pressure at the temperature of the water surface (kPa),
e_a = actual vapor pressure of the air (e_s of the air times relative humidity) (kPa).

Rohwer (1931) evaluated the constant C in Equation 4.1 as (in SI units, mm/day)

$$C = (3.30 + 1.973\, U_{0.15})(1.465 - 0.00548P) \tag{4.3}$$

Where $U_{0.15}$ = average water surface wind velocity (estimated to be at a height of 0.15 m) (m/s),
P = atmospheric pressure (kPa).

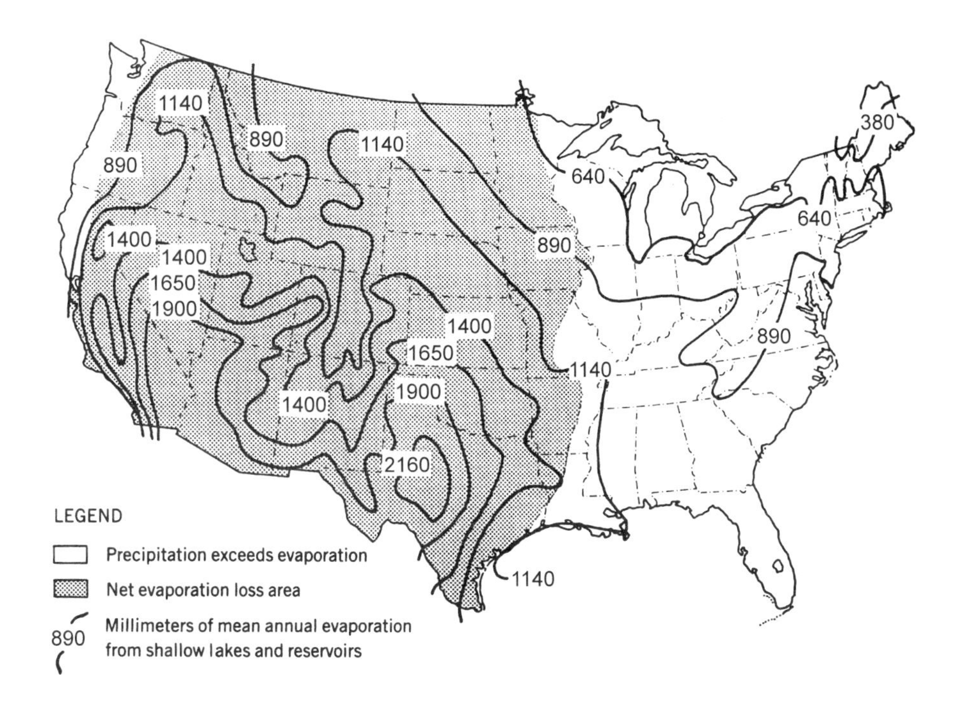

Figure 4–1

Average annual evaporation from shallow lakes and net evaporative loss area. (Revised from USDA, 1981)

Based on measured evaporation from large and small water surface areas, Rohwer (1931) determined that the evaporation from the small water surface areas can be multiplied by 0.77 to estimate the evaporation from large water surface areas. Equation 4.20 or 4.21 can be used to calculate wind speeds for the desired height from wind speed measured at other heights.

Meyer (1942) evaluated the constant C for pans and shallow ponds (in SI units, mm/month),

$$C = 112.5 + 25.1U_{7.6} \tag{4.4}$$

and for small lakes and reservoirs,

$$C = 82.6 + 18.5U_{7.6} \tag{4.5}$$

where $U_{7.6}$ = average wind velocity for the period in m/s at a height of 7.6 m. The vapor pressure e_a should be measured at 7.6 m height in application of Equations 4.4 and 4.5, and the air temperature is the average of the daily minimum and maximum. The geographical distribution of average annual evaporation from shallow lakes is shown in Figure 4-1. Note that evaporation is higher from small water surface areas than from large water surface areas because of an "oasis effect." The "oasis effect" describes the condition where a small water surface area is surrounded by dry air and consequentially has more evaporation. For a detailed analysis of evaporation from free-water surfaces, see Jones (1992).

Example 4.1

Compute the evaporation for the month of June from a shallow pond if the surface water temperature is 15°C, the average wind speed at 7.6 m height is 1.4 m/s, and the average temperature and relative humidity at 7.6 m height are 22°C and 40 percent, respectively.

Solution. Substituting into Equation 4.1 the saturation vapor pressures for 15 and 22°C are

$$e_s(15) = 0.6108 \exp\left[\frac{17.27 \times 15}{15 + 237.3}\right] = 1.70 \text{ kPa}$$

$$e_s(22) = 0.6108 \exp\left[\frac{17.27 \times 22}{22 + 237.3}\right] = 2.64 \text{ kPa}$$

Now substituting into Equations 4.2 and 4.4 (Meyer equation) where e_a for the air is calculated as $e_s(T_{air}) \times RH/100$

$$E = (112.5 + 25.1 \times 1.4)(1.70 - 2.64 \times 40/100) = 95 \text{ mm/month}$$

Pan Evaporation. Evaporation measurements from free-water surfaces are commonly made using evaporation pans. The Class A pan, accepted as standard by the U.S. Weather Bureau, is 1.21 m in diameter and 250 mm deep. The water level should be kept between 50 and 75 mm below the rim. The pan is supported about 150 mm above the ground so that air may circulate under it, and the materials and color of the pan are specified. The pan is widely used around the world. Description of other styles of pans and correction coefficients for converting evaporation data from a pan of one type to that of another are given by Allen et al. (1998). These pans have higher rates of evaporation than do larger free-water surfaces, a factor of about 0.7 being recommended for converting observed evaporation rates to those for larger surface areas (Meyer, 1942; USGS, 1952). The pan area should be fenced to prevent animals from drinking from the pan. If birds are a problem, a convenient nearby water source can be installed. A screen cover may be placed over the pan, but it will reduce pan evaporation about 10 percent (Allen et al., 1998).

4.2 Evaporation from Land Surfaces

Because of differences in soil texture and in expected soil water movement, it is difficult to generalize on the amounts of evaporation from soil surfaces. For saturated soils, the evaporation may be expected to be similar to that from an open free-water surface. As the water table drops below the soil surface, the evaporation rate will decrease greatly. Prolonged evaporation from the soil surface is generally small at water contents below field capacity, as soil water movement is very slow when the soil surface is relatively dry. Mulches reduce evaporation by restricting air movement, maintaining a high air vapor pressure near the soil surface, and shielding the soil from solar energy. Freezing of a bare soil surface can cause ice to accumulate at the soil surface through condensation of vapor transport from deeper soils, which can greatly increase the evaporation after thawing.

Evapotranspiration

For convenience, evaporation and transpiration are combined into evapotranspiration, ET, also referred to as consumptive use. The various methods for determining evapotranspiration include: (1) tank and lysimeter experiments; (2) field experimental plots where quantity of water applied is controlled to avoid deep percolation losses and surface runoff is measured; (3) soil water studies, with large numbers of samples taken at various depths in the root zone; (4) analysis of climatological data; (5) integration methods where the water used by plants and evaporation from the water and soil surfaces are combined for the entire area involved; and (6) inflow–outflow method for large areas where yearly inflow into the area, annual precipitation, yearly outflow from the area, and the change in ground water level are evaluated.

There are many practical applications for evapotranspiration estimates, but a principal use is to predict soil water deficits for irrigation. Analyzing weather records and estimating evapotranspiration rates, drought frequencies, and excess water periods can show potential needs for irrigation and drainage. Similar studies to determine available tillage and harvesting days can aid in selecting optimum sizes of agricultural equipment. The average daily evapotranspiration during the year obtained from lysimeters at Coshocton, Ohio, is shown in Figure 4-2. Excess soil water at the beginning of the season may delay planting or cause plant diseases. A water deficit at midseason may reduce growth and yield. Reduced ET and excess soil water at the end of the summer may delay maturation of corn and harvesting and tillage operations. Several approaches have been used to develop methods for estimating evapotranspiration.

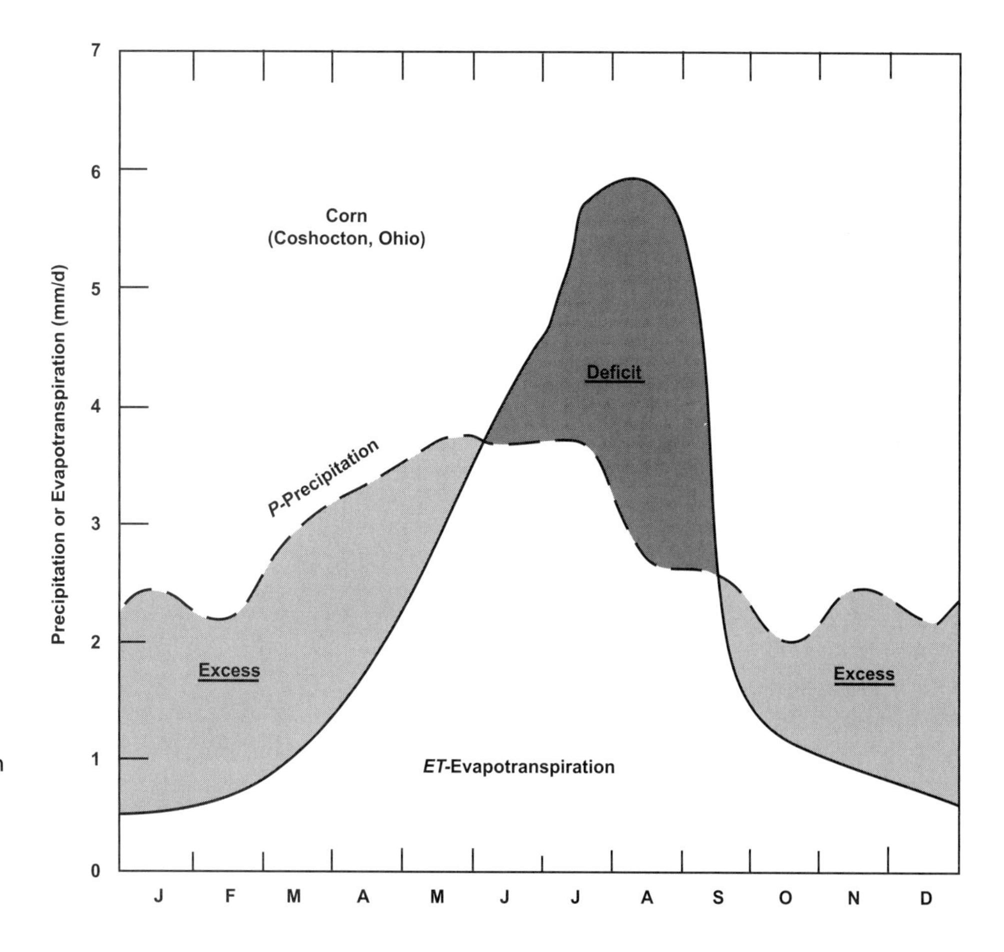

Figure 4–2

Average precipitation and evapotranspiration from corn at Coshocton, Ohio, showing excess and deficit water periods.

4.3 Transpiration Ratio

The effectiveness of the plant's use of water in producing dry matter is often given in terms of its transpiration ratio. This is the ratio of the mass of water transpired to the mass of dry matter in the plant and it varies with the same factors as transpiration. Transpiration ratios for several common crops are 304 for sorghum, 350 for corn, 557 for wheat, 568 for cotton, 575 for potatoes, 682 for rice, and 844 for alfalfa (Howell et al., 1990). This ratio is important, especially where irrigation water is limited.

4.4 Evapotranspiration Definitions

Potential Evapotranspiration. ET_p as defined by Jensen et al. (1990) is "the rate at which water, if available, would be removed from wet soil and plant surfaces expressed as the rate of latent heat transfer per unit area λET_p or as a depth of water per unit time." Potential evapotranspiration is difficult to sustain and measure because of the need to maintain a saturated surface so reference crop *ET* is used as a standard climatic index of evapotranspiration.

Reference Crop Evapotranspiration. ET_{ref} as defined by Jensen et al. (1990) is "the rate at which water, if readily available, would be removed from the soil and plant surfaces expressed as the rate of latent heat transfer per unit area λET_{ref} or expressed as a depth of water evaporated and transpired from a reference crop. The leaf surfaces of the reference crop are typically not wet." Full-cover alfalfa and clipped, cool-season grass are used as reference crops. Both must fully cover the soil surface and be fully transpiring (i.e., not short of water). Alfalfa is a good reference crop because it is aerodynamically more similar to other agricultural crops than grass, it has a deep root system, which makes it less likely to be short of water, and its high *ET* rates are similar to agricultural crops. Alfalfa also has a low leaf resistance to water vapor diffusion. Grass is becoming the standard reference crop under automated weather stations because it is easier to maintain at a nearly constant height. These automated weather stations are capable of measuring the climatic data for the most sophisticated methods for predicting reference crop *ET*. With the establishment of the "standardized" ET_r for alfalfa, it is not necessary to grow alfalfa. Weather measurements taken over grass can be used to calculate the alfalfa reference ET_r (Jensen et al., 1990; Allen et al., 1998). Many states have installed networks of automatic weather stations that measure the data needed to calculate reference *ET* and/or crop *ET*. These *ET* values are disseminated through various media for use by water managers.

Evapotranspiration Estimation Methods

4.5 Evaporation Pan Method

Evaporation pan data E_{pan} can be used to calculate reference *ET* or crop ET_c with the appropriate coefficient. Converting to reference *ET*, as in the following equation, allows using crop coefficients for many crops with one calibration.

$$ET_o = K_{pan} E_{pan} \tag{4.6}$$

where ET_o = grass based reference ET (L/T),
K_{pan} = factor for converting pan evaporation to ET_o,
E_{pan} = measured pan evaporation (L/T).

The best source for K_{pan} is a local or regional calibration. Table 4-1 may be used for a Class A pan if local values for K_{pan} are not available. Values of K_{pan} vary with relative humidity, wind speed, and windward side distance (fetch) of a green crop Case A or dry fallow Case B (Figure 4-3). For desert or semidesert conditions with no agricultural development and bare soils, K_{pan} may need to be reduced up to 20 percent. If the pan is located in tall crops, K_{pan} may need to be increased up to 30 percent (Allen et al., 1998).

Example 4.2

Estimate the grass reference crop ET_0 if pan evaporation for July 10 and 11 was 16 mm. The wind was 3 m/s, relative humidity was 50 percent, and the pan has a 10-m grass fetch.

TABLE 4–1 Pan Coefficients K_{pan} for Class A Pan for Different Pan Siting and Environment Conditions and Different Levels of Mean Relative Humidity and Wind Speed

	Case A: Pan Placed in Short Green Cropped Area				*Case B: Pan Placed in Dry, Fallow Area*			
		RH Mean Percent				*RH Mean Percent*		
		Low < 40	*Medium 40–70*	*High > 70*		*Low < 40*	*Medium 40–70*	*High > 70*
Wind speed (m/s)	*Distance[a] (m)*				*Distance[b] (m)*			
Light	1	0.55	0.65	0.75	1	0.7	0.8	0.85
<2	10	0.65	0.75	0.85	10	0.6	0.7	0.8
	100	0.7	0.8	0.85	100	0.55	0.65	0.75
	1000	0.75	0.85	0.85	1000	0.5	0.6	0.7
Moderate	1	0.5	0.6	0.65	1	0.65	0.75	0.8
2–5	10	0.6	0.7	0.75	10	0.55	0.65	0.7
	100	0.65	0.75	0.8	100	0.5	0.6	0.65
	1000	0.7	0.8	0.8	1000	0.45	0.55	0.6
Strong	1	0.45	0.5	0.6	1	0.6	0.65	0.7
5–8	10	0.55	0.6	0.65	10	0.5	0.55	0.65
	100	0.6	0.65	0.7	100	0.45	0.5	0.6
	1000	0.65	0.7	0.75	1000	0.4	0.45	0.55
Very strong	1	0.4	0.45	0.5	1	0.5	0.6	0.65
>8	10	0.45	0.55	0.6	10	0.45	0.5	0.55
	100	0.5	0.6	0.65	100	0.4	0.45	0.5
	1000	0.55	0.6	0.65	1000	0.35	0.4	0.45

[a]Length of the green crop area upwind from the pan.

[b]Length of dry, fallow area upwind from the pan.

Source: Doorenbos and Pruitt (1977).

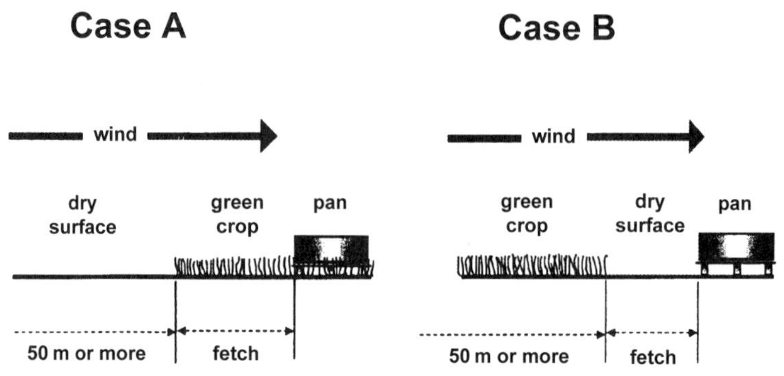

Figure 4–3
Illustration showing two cases of evaporation pan site and environment conditions. (Allen et al., 1998)

Solution. From Table 4-1 for Case A read K_{pan} is 0.7, then substitute into Equation 4.6.

$$ET_o = K_{pan} \times E_{pan} = 16 \times 0.7 = 11.2 \text{ mm or } 5.6 \text{ mm/day}$$

4.6 Penman-Monteith Combination Method

Penman (1948, 1956) first derived a combination equation by combining components for the energy required to sustain evaporation and a mechanism for removal of the vapor. The Penman combination equation combined with aerodynamic and surface resistance terms is called the Penman-Monteith equation (Jensen et al., 1990). The ASCE Standardized Penman-Monteith equation for daily time steps is (ASCE, 2005)

$$ET_{ref} = \frac{0.408\Delta(R_n - G) + \gamma \dfrac{C_n}{T + 273}(e_s - e_a)u_2}{\Delta + \gamma(1 + C_d u_2)} \qquad \textbf{4.7}$$

where ET_{ref} = reference ET for a well-watered crop (mm/day),
Δ = slope of the saturation vapor pressure curve (kPa/°C),
R_n = net radiation at the crop surface (MJ m^{-2} day^{-1}),
G = heat flux density to the soil (MJ m^{-2} day^{-1}) (G is usually small for daily time steps compared to R_n and is neglected),
γ = psychrometric constant (kPa/°C),
T = mean daily temperature at 1.5- to 2.5-m height (°C),
u_2 = mean daily wind speed at 2 m above the soil surface (m/s),
e_s = mean saturation vapor pressure at 1.5- to 2.5-m height (kPa),
e_a = mean actual vapor pressure at 1.5- to 2.5-m height (kPa),
C_n = numerator constant that changes with the reference crop,
C_d = denominator constant that changes with the reference crop.

Values for C_n and C_d are given in Table 4-2.

The following equations and explanations (Allen et al., 1998; ASCE, 2005) assume the Penman-Monteith equation will be applied on a daily basis. The

TABLE 4–2 Values for C_n and C_d in Equation 4.7 with Daily Time Steps

	C_n	C_d
Short reference crop (grass)	900	0.34
Tall reference crop (alfalfa)	1600	0.38

Source: Itenfisu et al. (2003).

procedures and constants are different for other time intervals. The latent heat of vaporization λ varies only slightly and is taken as the value for $T = 20°C$ or 2.45 MJ/kg, which is the reciprocal of 0.408.

The slope of the saturation vapor pressure and temperature curve at a given temperature is computed from the following equation

$$\Delta = \frac{2504 \exp\left(\frac{17.27T}{T + 237.3}\right)}{(T + 237.3)^2} \tag{4.8}$$

where Δ is in kPa/°C and T is the daily mean air temperature in °C obtained by averaging the daily maximum and minimum temperatures.

Net radiation can be obtained from local correlations with solar radiation (Jensen et al., 1990) or calculated from

$$R_n = R_{ns} - R_{nl} = (1 - \alpha)R_s - R_{nl} \tag{4.9}$$

where R_{ns} = net solar or short-wave radiation (MJ m^{-2} day^{-1}),
R_{nl} = net long-wave radiation leaving the earth's surface (MJ m^{-2} day^{-1}),
α = radiation reflection coefficient or albedo = 0.23,
R_s = measured or calculated solar or short-wave radiation received at the earth's surface (MJ m^{-2} day^{-1}).

Solar radiation is generally measured by a weather station. Net long-wave radiation is determined from

$$R_{nl} = \sigma\left[\frac{(T_{max} + 273)^4 + (T_{min} + 273)^4}{2}\right] \times \left[0.34 - 0.14(e_a)^{0.5}\right]\left(1.35\frac{R_s}{R_{so}} - 0.35\right) \tag{4.10}$$

where σ = Stefan-Boltzman constant = 4.903 10^{-9} (MJ K^{-4} day^{-1}),
T_{max} = maximum temperature during the 24-hour period (°C),
T_{min} = minimum temperature during the 24-hour period (°C),
R_{so} = calculated clear-sky radiation (MJ m^{-2} day^{-1}).

The ratio R_s/R_{so} in Equation 4.10 cannot exceed 1.0. Clear-sky radiation can be calculated by

$$R_{so} = (0.75 + 2 \times 10^{-5}z)R_a \tag{4.11}$$

where z in the elevation above sea level in meters and R_a is the extraterrestrial radiation in MJ m^{-2} day^{-1} and given by

$$R_a = \frac{24}{\pi} G_{sc} d_r [\omega_s \sin(\varphi)\sin(\delta) + \cos(\varphi)\cos(\delta)\sin(\omega_s)] \quad \textbf{4.12}$$

where G_{sc} = solar constant = 4.92 (MJ m^{-2} h^{-1}),
d_r = inverse square of relative distance Earth to Sun,
ω_s = sunset hour angle (radians),
φ = latitude (radians),
δ = solar declination (radians).

The inverse square of the relative distance earth to sun d_r is given by

$$d_r = 1 + 0.033\cos\left(2\pi\frac{J}{365}\right) \quad \textbf{4.13}$$

where J is the day of the year given by

$$J = D_M - 32 + Int\left(\frac{275M}{9}\right) + 2Int\left(\frac{3}{M+1}\right) + Int\left(\frac{M}{100} - \frac{Mod(Y,4)}{4} + 0.975\right) \quad \textbf{4.14}$$

where D_M = day of the month,
M = month of the year,
Y = year (4 digits),
Int = function that finds the integer number of the argument [3/(M+1)] by rounding downward,
Mod = function that finds the remainder of the quotient of the argument (Y, 4) or in this case $Y/4$.

Degrees latitude are changed to radians by

$$\text{Radians} = \frac{\pi}{180} \times \text{degrees latitude} \quad \textbf{4.15}$$

The solar declination is

$$\delta = 0.409\sin\left(\frac{2\pi}{365}J - 1.39\right) \quad \textbf{4.16}$$

The sunset angle is given by

$$\omega_s = \arccos[-\tan(\varphi)\tan(\delta)] \quad \textbf{4.17}$$

The psychrometric constant using λ = 2.45 MJ/kg is

$$\gamma = 0.000665P \quad (4.18)$$

where P is the mean atmospheric pressure in kPa at the weather station with elevation z in m above mean sea level. The pressure P in kPa is given by

$$P = 101.3\left(\frac{293 - 0.0065z}{293}\right)^{5.26} \quad (4.19)$$

If the wind speed is measured at a height other than 2 m above the soil surface, the wind speed u_2 at 2 m over a grassed surface can be obtained from

$$u_2 = u_z \frac{4.87}{\ln(67.8z_w - 5.42)} \quad (4.20)$$

where z_w is the height (m) of the wind measurement above the soil surface and u_z is the measured wind speed (m/s) at height z. If the wind speed is measured at a height other than 2 m over a surface having vegetation taller than grass, such as alfalfa, or other vegetation about 0.5 m height, and to be consistent with the standardized ET equation (Equation 4.7), the following equation is used to translate the wind speed to 2 m above the soil surface

$$u_2 = u_z \frac{3.44}{\ln(16.26z_w - 5.42)} \quad (4.21)$$

The saturation vapor pressure is related to the air temperature by Equation 4.1. The mean saturation vapor pressure is the average of the saturation vapor pressures for maximum and minimum air temperatures or

$$e_s = \frac{e_s(T_{max}) + e_s(T_{min})}{2} \quad (4.22)$$

The actual vapor pressure can be calculated from Equation 4.1 using the daily mean dewpoint temperature T_{dew}

$$e_a = e_s(T_{dew}) \quad (4.23)$$

If the dewpoint temperature is not available, e_a can be calculated from the relative humidity using

$$e_a = \frac{e_s(T_{min})\dfrac{RH_{max}}{100} + e_s(T_{max})\dfrac{RH_{min}}{100}}{2} \quad (4.24)$$

where RH_{max} and RH_{min} are maximum and minimum readings of relative humidity during the 24-hour period in percentage.

Values for the constants C_n and C_d are given in Table 4-2 for daily calculations. Values for C_n vary with the aerodynamic roughness of the reference crop. Values for C_d vary with the "bulk" surface resistance and aerodynamic roughness of the surface. Both values were derived by simplifying terms and rounding the result (Allen et al., 1998; ASCE, 2005).

Example 4.3

Compute the grass reference crop ET_o for June 20, 2002, near Bakersfield, California, 35°N, using the Standardized Penman-Monteith equation. Maximum temperature = 38°C, minimum temperature = 22°C, maximum relative humidity = 60 percent, minimum relative humidity = 25 percent, wind speed at 2-m = 1.5 m/s, measured solar radiation = 26 MJ m^{-2} day^{-1}, elevation = 50 m, and assume G = 0.0.

Solution. (1) Begin the calculations by determining the climatic constants. Calculate the mean temperature $T = (38+22)/2 = 30$, then calculate Δ from Equation 4.8

$$\Delta = \frac{2504 \exp\left(\dfrac{17.27 \times 30}{30 + 237.3}\right)}{(30 + 237.3)^2} = 0.243\,\text{kPa}/\,^\circ\text{C}$$

(2) Calculate the saturation vapor pressure at maximum temperature using Equation 4.1

$$e_s(38) = 0.6108 \exp\left(\frac{17.27 \times 38}{38 + 37.3}\right) = 6.625\,\text{kPa}$$

(3) Calculate the saturation vapor pressure at minimum temperature using Equation 4.1

$$e_s(22) = 0.6108 \exp\left(\frac{17.27 \times 22}{22 + 237.3}\right) = 2.644\,\text{kPa}$$

(4) Calculate the mean saturation vapor pressure from Equation 4.22

$$e_s = \frac{6.625 + 2.644}{2} = 4.634\,\text{kPa}$$

(5) Calculate the actual vapor pressure from Equation 4.24

$$e_a = \frac{2.644\dfrac{60}{100} + 6.625\dfrac{25}{100}}{2} = 1.62\,\text{kPa}$$

(6) Calculate the mean atmospheric pressure at the station from Equation 4.19

$$P = 101.3\left(\frac{293 - 0.0065 \times 50}{293}\right)^{5.26} = 100.7\,\text{kPa}$$

(7) Calculate the psychrometric constant from Equation 4.18

$$\gamma = 0.000665 \times 100.7 = 0.067\,\text{kPa}/^\circ\text{C}$$

(8) Determine the day of the year with Equation 4.14

$$J = 20 - 32 + Int\left(\frac{275 \times 6}{9}\right) + 2Int\left(\frac{3}{6+1}\right)$$

$$+ Int\left(\frac{6}{100} - \frac{Mod(2002,4)}{4} + 0.975\right) = 171$$

(9) Determine the latitude in radians from Equation 4.15

$$\varphi = \frac{\pi}{180} \times 35 = 0.611\,\text{radians}$$

(10) Calculate the solar declination from Equation 4.16

$$\delta = 0.409\sin\left(\frac{2\pi}{365}171 - 1.39\right) = 0.409$$

(11) Calculate the sunset angle from Equation 4.17

$$\omega_s = \arccos[-\tan(0.611)\tan(0.409)] = 1.879$$

(12) Calculate the inverse square of the relative distance earth to sun from Equation 4.13

$$d_r = 1 + 0.003\cos\left(\frac{2\pi}{365}171\right) = 0.968$$

(13) Calculate the extraterrestrial radiation from Equation 4.12

$$R_a = \frac{24}{\pi}4.92 \times 0.968[1.879\sin(0.611)\sin(0.409)$$

$$+ \cos(0.611)\cos(0.409)\sin(1.879)] = 41.65\,\text{MJm}^{-2}\,\text{day}^{-1}$$

(14) Calculate the clear-sky radiation from Equation 4.11

$$R_{so} = (0.75 + 2 \times 10^{-5} \times 50)41.65 = 31.28\,\text{MJm}^{-2}\,\text{day}^{-1}$$

(15) Calculate the net long-wave radiation from Equation 4.10

$$R_{nl} = 4.903 \times 10^{-9}\left[\frac{(38 + 273)^4 + (22 + 273)^4}{2}\right][0.34 - 0.14(1.62)^{0.5}]$$

$$\times \left[1.35\frac{26}{31.27} - 0.35\right] = 5.185\,\mathrm{MJ\,m^{-2}day^{-1}}$$

(16) Calculate the net radiation from Equation 4.9

$$R_n = (1 - 0.23)26 - 5.185 = 14.83\,\mathrm{MJ\,m^{-2}day^{-1}}$$

(17) Find the values for C_n and C_d from Table 4-2 for a grass reference

$$C_n - 900 \quad C_d - 0.34$$

(18) Substitute the preceding values into Equation 4.7 to obtain ET_o

$$ET_o = \frac{0.408 \times 0.243(14.83 - 0) + 0.067\dfrac{900}{30 + 273}(4.634 - 1.62)1.5}{0.243 + 0.067(1 + 0.34 \times 1.5)}$$

$$= 6.89\,\mathrm{mm/day}$$

Thus the calculated ET for a grass reference ET_o is 6.89 mm/day, or round to 6.9 mm/day.

4.7 Temperature-Based Methods

Early researchers (Thornthwaite, 1948; Blaney & Criddle, 1950) in water management developed temperature-based methods for estimating consumptive use or ET. These were simple to use and worked reasonably well where they were developed and calibrated. The Blaney-Criddle method was the most popular in the semiarid western parts of the United States and was based on a consumptive use coefficient, mean air temperature, and percentage of annual daylight hours occurring during the period of calculation. This method estimated monthly or seasonal water use. More accurate estimates for evapotranspiration were needed, and the Blaney-Criddle method was revised by the SCS (1970) and by FAO-24 (Doorenbos & Pruitt, 1977). The SCS Blaney-Criddle added a temperature coefficient to the consumptive use ^coefficient but has been superceded by newer methods. FAO-24 made a more fundamental revision to the Blaney-Criddle method by including relative humidity, ratio of actual to possible sunshine hours, and wind speed. This increased the data requirement significantly, and the FAO-24 revision has been dropped in favor of combination methods. It should be noted that when these methods have been locally calibrated, they are reliable for periods of one month or longer.

4.8 Radiation Methods

Empirical radiation methods for estimating potential ET were developed that included an energy term by adding a solar radiation variable. The Jensen-Haise alfalfa-reference radiation method (Jensen & Haise, 1963; Jensen, 1966) was the most widely accepted of these methods. The main variables were mean air temperature, solar radiation, and two constants. The constants were based on elevation and the saturation vapor pressures for the mean maximum and mean minimum temperatures for the warmest month of the year.

The Hargreaves grass-related radiation method (Hargreaves & Samani, 1982, 1985) is based on solar radiation and mean air temperature. Hargreaves & Samani (1982, 1985) recommended estimating solar radiation from extraterrestrial radiation and the difference between mean maximum and mean minimum monthly temperatures (Equation 4.28). In this case, the Hargreaves method becomes essentially a temperature-based method (Jensen et al., 1990). When calibrated to local conditions, these radiation methods have proven reliable in predicting ET.

4.9 Estimating Missing Climatic Data

The ASCE Standardized Penman-Monteith equation requires air temperature, vapor pressure or relative humidity, radiation, and wind speed data. It is normally assumed that these data will be from the area where the ET estimate is required. The quality of the weather data will affect the quality of the reference ET values. "If some of the required data are missing or do not accurately represent an irrigated site/region or are erroneous, then it may be possible that data may be estimated in order to apply the equation" (ASCE, 2005). If reasonably reliable estimates of missing or erroneous data are determined, ET estimates from the ASCE standardized equation are expected to be more reliable than estimates from more empirical methods (ASCE, 2005; Allen et al., 1998).

When estimated rather than measured data are used to estimate ET, the data should be flagged and the estimated parameters should be noted. The following describes procedures for estimating missing or questionable data.

Vapor Pressure. If humidity and dewpoint data are missing or questionable, the actual vapor pressure can be estimated for the site by assuming the dewpoint temperature is near the daily minimum temperature or

$$T_{dew} = T_{min} - K_o \qquad \textbf{4.25}$$

where K_o is approximately 2 to 4°C in arid and semiarid climates and approximately 0°C in humid and subhumid climates (ASCE, 2005). Additional discussion of this assumption is given by Allen et al. (1998). The value of K_o can be estimated or obtained by analyzing the data from a nearby weather station.

Solar Radiation. If solar radiation is not measured, it can be estimated from the hours of sunshine and extraterrestrial radiation by the Angstrom formula (ASCE, 2005)

$$R_s = \left(a_s + b_s \frac{n}{N}\right) R_a \qquad \textbf{4.26}$$

where n = actual duration of sunshine (h),

N = maximum possible duration of sunshine (h),

a_s = fraction of extraterrestrial radiation reaching the earth's surface on overcast days ($n = 0$),

b_s = the additional fraction of extraterrestrial radiation reaching the earth's surface on a clear day,

$a_s + b_s$ = fraction of extraterrestrial radiation reaching the earth's surface on a clear day ($n = N$).

Radiation in Equation 4.26 is expressed in MJ m^{-2} day^{-1}. Values of a_s and b_s vary with atmospheric conditions (dust, humidity) and solar declination (latitude and month). If no actual radiation or no calibration data are available, the values $a_s = 0.25$ and $b_s = 0.50$ are recommended (Allen et al., 1998). The potential daylight hours N are given by

$$N = \frac{24}{\pi}\omega_s \qquad \textbf{4.27}$$

Data from nearby weather stations can be utilized if the climate and physiography are nearly identical. Estimates of ET using estimated radiation data are better when calculated over multiple day periods.

Solar radiation can be estimated from the difference between maximum and minimum temperatures because temperatures are influenced by cloud cover. Hargreaves and Samani (1982) developed an empirical equation for the relationship

$$R_s = k_{R_s}(T_{max} - T_{min})^{0.5}R_a \qquad \textbf{4.28}$$

where k_{R_s} (°C$^{-0.5}$) is an adjustment coefficient and varies for coastal or interior areas. For areas located on or near the coast of a large land mass and where the air masses are influenced by a nearby body of water, $k_{R_s} \approx 0.19$. For interior areas where land mass dominates and air masses are not influenced by a nearby body of water, $k_{R_s} \approx 0.16$ (ASCE, 2005).

Wind Speed. When wind speed data are not available, they may be extrapolated from a nearby agricultural weather station if the airflow conditions are relatively homogeneous. Wind speeds vary throughout the day; however, when averaged over a day or longer the differences between two sites are smaller. If incomplete data are available, a calibration between two sites will improve the accuracy of the estimated speeds. If no data are available, wind speed can be selected from Table 4-3 or a global value of 2 m/s used as a temporary estimate.

Airport wind speeds are typically measured at 10-m heights in the United States. In semiarid and arid areas airport anemometers are generally surrounded by low vegetation. These wind speeds can be adjusted to a 2-m height but will typically exceed the velocity over an irrigated area because of large differences in vegetative roughness and the damping effect caused by the heat sink as water evaporates (ASCE, 2005).

TABLE 4–3 Suggested Mean Monthly Wind Speeds for Various General Classes

Class Description	*Mean Monthly Wind Speed at 2 m*
Light wind	< 1.0 m/s
Light to moderate wind	1—3 m/s
Moderate to strong wind	3—5 m/s
Strong wind	> 5.0 m/s

Source: Allen et al. (1998).

Crop and Landscape Coefficients

4.10 Crop Coefficients

Examples of the daily evapotranspiration from three crops are given in Figure 15-1. Agricultural crops differ in evaporation and transpiration from reference crops, particularly grass, because of differences in ground cover, canopy characteristics, and aerodynamic resistance. These differences are integrated into a single crop coefficient K_c that includes effects of both crop transpiration and soil evaporation. The evapotranspiration from a specific crop ET_c can be estimated from the reference ET and appropriate single crop coefficient K_c by

$$ET_c = K_c \times ET_{ref} \qquad \textbf{4.29}$$

It is important to note that the crop coefficients for alfalfa and grass reference crops are different and cannot be interchanged. Also the crop coefficients may be somewhat different for the same reference crop depending on the location and method used for determining reference ET.

Crop coefficients vary with crop type, climate, and soil evaporation. For a specific crop the coefficient varies with the stage of growth of the plant as shown in Figure 4-4. For annual crops, crop coefficients are lowest at planting, increase as the plants grow, and reach a maximum when the canopy covers the soil surface. As the plants ripen late in the season, the coefficients decrease. In addition to time, crop coefficients also can be expressed as functions of degree-days.

For simplicity, the crop coefficient curve is divided into four straight-line segments for four growth stages (initial, crop development, mid-season, and late season) as shown in Figures 4-4 and 4-5. These segments are defined by three coefficients — $K_{c\ ini}$, $K_{c\ mid}$, and $K_{c\ end}$ — and the number of days in each stage (Figure 4-5). Examples of approximate crop coefficients for a grass reference crop and a few crops are given in Table 4-4. Values of $K_{c\ ini}$ in Table 4-4 are for average soil wetting conditions. The corresponding lengths of each stage are given in Table 4-5. Local crop coefficients are preferred, but if they are not available see Allen et al. (1998), Jensen et al. (1990), Doorenbos & Pruitt (1977), or Pruitt et al. (1987).

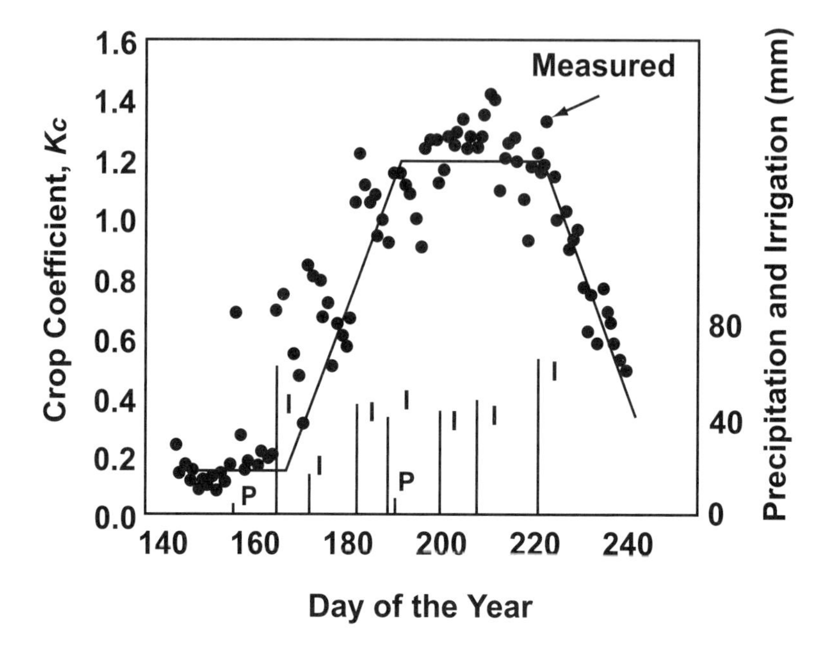

Figure 4–4
Crop coefficients for dry beans at Kimberly, Idaho, as measured by lysimeter, and as represented by four straight-line segments. Precipitation and irrigation have the symbols *P* and *I*, respectively. (Modified from Allen et al., 1998)

Example 4.4 Estimate ET_c for cotton on June 20, 2002, near Bakersfield, California. Assume the planting date was April 1.

Solution. From Example 4.3, ET_0 for Bakersfield on June 20 is 6.9 mm/day. June 20 is 81 days after planting. From Table 4-5 determine that this is mid-season, and from Table 4-4 $K_{c\ mid}$ is 1.2. Substituting into Equation 4.29 yields

$$ET_c = 1.2 \times 6.9 \text{ mm/day} = 8.3 \text{ mm/day}$$

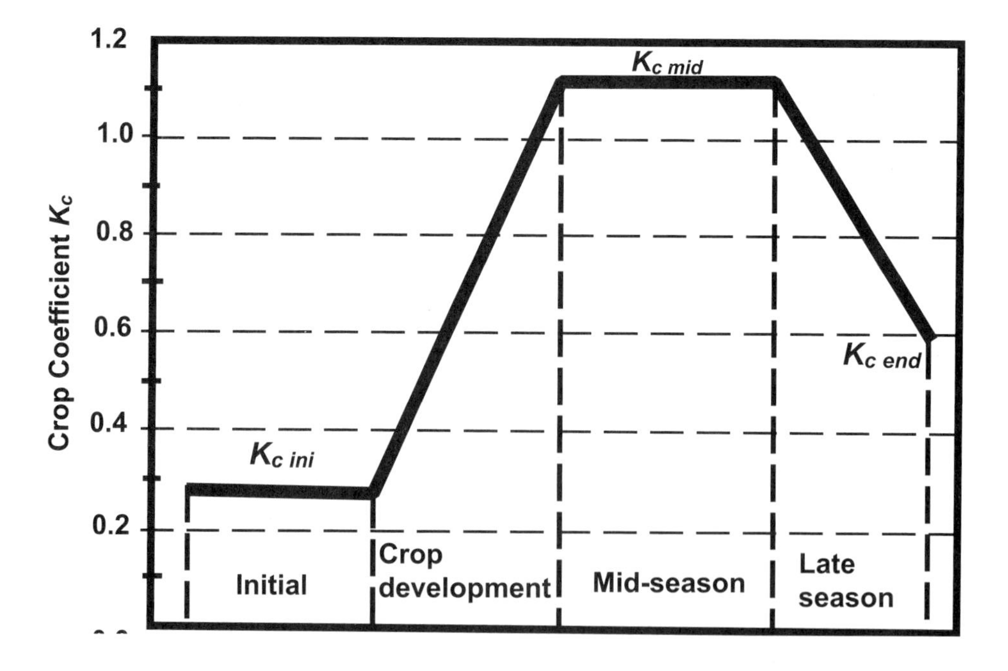

Figure 4–5
Illustration of crop coefficients represented by four line segments for the four major growth stages. (Allen et al., 1998)

TABLE 4–4 Approximate Single Crop Coefficients for a Grass Reference Crop and Mean Maximum Plant Heights for Well-Managed, Nonstressed Crops for Subhumid Regions

Crop	$K_{c\ ini}$	$K_{c\ mid}$	$K_{c\ end}$	*Maximum Crop Height (m)*
Carrots	0.7	1.05	0.95	0.3
Lettuce	0.7	1.00	0.7	0.3
Tomato	0.6	1.15	0.8	0.6
Cantaloupe	0.5	0.85	0.6	0.3
Potato	0.5	1.05	0.95	0.4
Sugar beet	0.35	1.2	0.7	0.5
Soybeans	0.4	1.15	0.5	1.0
Cotton	0.35	1.2	0.7	1.5
Small grain	0.3	1.15	0.4	1
Maize	0.3	1.2	0.5	2
Alfalfa	0.4	0.95	0.9	0.7
Grapes	0.3	0.85	0.45	2
Deciduous orchard	0.5	1.0	0.7	4
Citrus, no ground cover, 50% canopy	0.65	0.6	0.65	3
Turf grass—cool season[a]	0.90	0.95	0.95	[a]
—warm season[b]	0.80	0.85	0.85	0.10

[a]Cool season varieties include dense stands of bluegrass, ryegrass, and fescue. The 0.95 values for cool season grass represent 0.06 to 0.08 m mowing height.

[b]Warm season varieties include bermuda and St. Augustine grass.

Source: Selected values from Allen et al. (1998).

TABLE 4–5 Typical Lengths (Days) of the Four Growing Stages for Selected Crops

Crop	*Initial* (L_{ini})	*Developmental* (L_{dev})	*Mid* (L_{mid})	*Late* (L_{late})
Carrots	30	40	60	25
Lettuce	30	40	30	10
Tomato	30	40	60	30
Cantaloupe	20	50	30	20
Potato	30	30	50	30
Sugar beet	40	50	90	40
Soybeans	20	30	60	25
Cotton	30	50	55	45
Small grain	25	35	60	30
Maize	20	40	50	30
Alfalfa	5	15	10	10
Grapes	20	50	80	60
Deciduous orchard	20	70	120	60
Citrus	60	90	120	95

Source: Selected values from Allen et al. (1998).

Because $K_{c\ ini}$ represents periods of nearly bare soil and varies widely according to the frequency of wetting of the soil surface, values of $K_{c\ ini}$ from Table 4-4 are approximate and recommended for preliminary planning studies (Allen et al., 1998). Improved values can be obtained by adjusting $K_{c\ ini}$ depending on the frequency of rainfall or irrigation and the depth of water infiltrated per event for infiltration depths of 10 mm or less as shown in Figure 4-6. Figure 4-7 shows similar curves for infiltration depths of 40 mm or greater for (1) coarse textured soils and (2) medium and fine textured soils. These curves indicate higher values of $K_{c\ ini}$ for more frequent wetting and larger infiltration depths because soil evaporation is the main component of $K_{c\ ini}$. For average infiltration depths per wetting event between 10 and 40 mm, the value of $K_{c\ ini}$ can be estimated from

$$K_{c\ ini} = K_{c\ ini\ (Fig.\ 4\text{-}6)} + \frac{I - 10}{40 - 10}\left[K_{c\ ini\ (Fig.\ 4\text{-}7)} - K_{c\ ini\ (Fig.\ 4\text{-}6)}\right] \quad \textbf{4.30}$$

where $K_{c\ ini\ (Fig.\ 4\text{-}6)}$ = value of $K_{c\ ini}$ from Figure 4-6,
$K_{c\ ini\ (Fig.\ 4\text{-}7)}$ = value of $K_{c\ ini}$ from Figure 4-7,
I = average infiltration depth (mm).

Example 4.5

Determine $K_{c\ ini}$ if the average infiltration depth per wetting event is 25 mm and the irrigation is applied every 7 days to a medium-textured soil. Estimated ET_o is 5 mm/day.

Solution. From Figure 4-6 read $K_{c\ ini}$ = 0.28, and from Figure 4-7*b* read $K_{c\ ini}$ = 0.67. Substitute into Equation 4.30

$$K_{c\ ini} = 0.28 + \frac{25 - 10}{40 - 10}(0.67 - 0.28) = 0.48$$

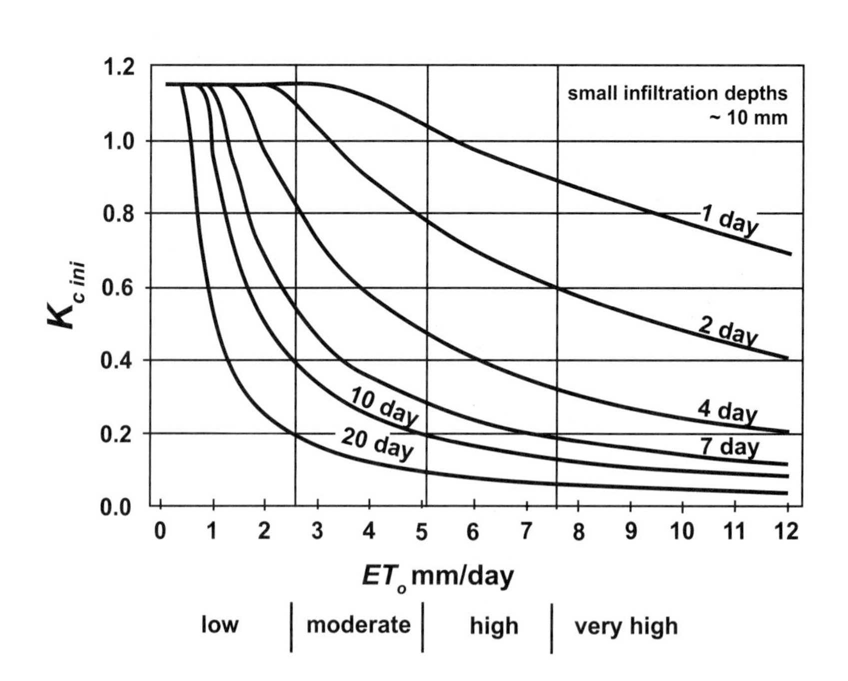

Figure 4–6

Average $K_{c\ ini}$ as related to ET_o and the interval between irrigations and/or significant rain of 3 to 10 mm during the initial growth stage for all soil types. (Allen et al., 1998)

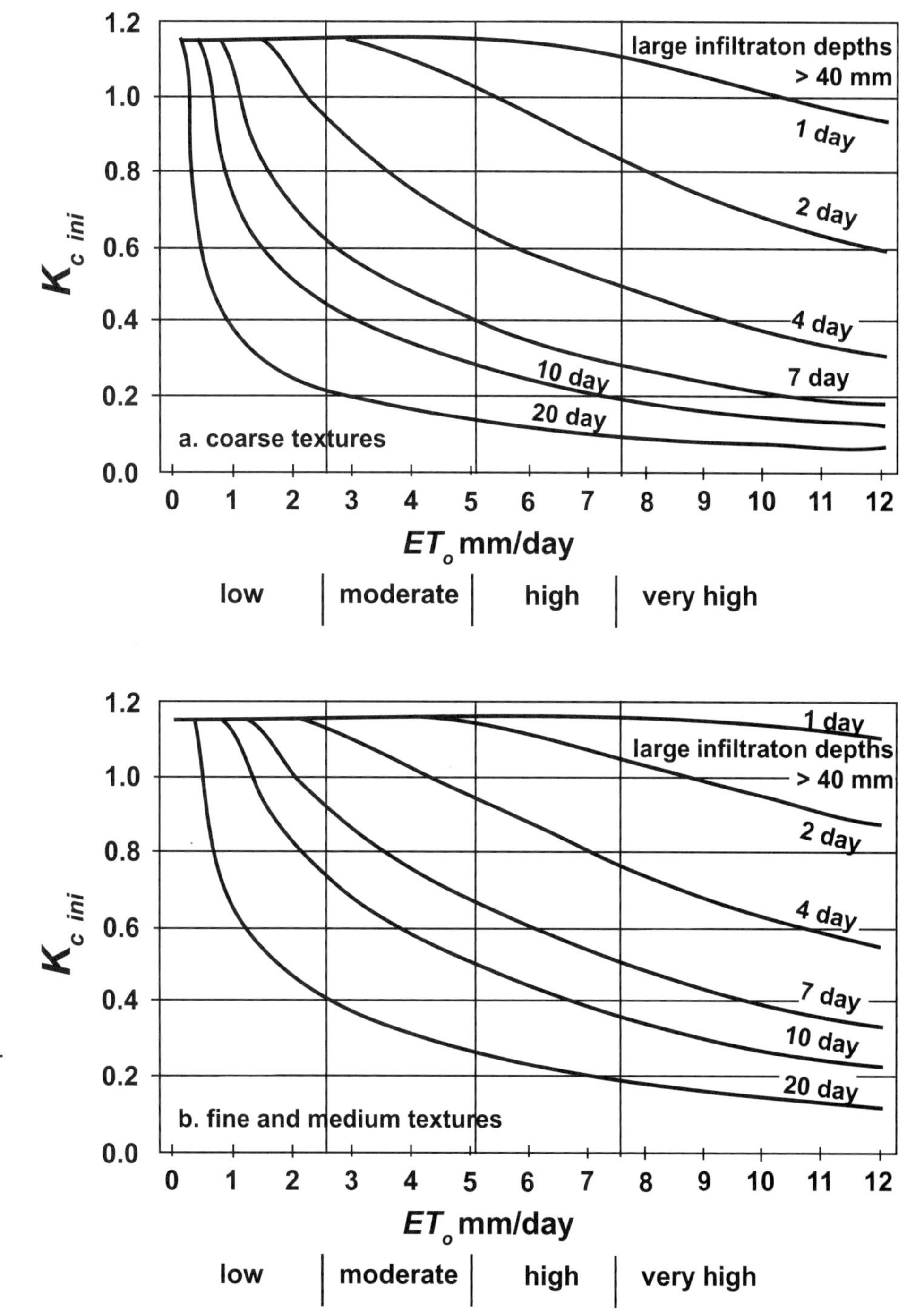

Figure 4–7

Average $K_{c\ ini}$ as related to ET_o and the interval between irrigations greater than or equal to 40 mm per wetting event during the initial growth stage for (a) coarse textured soils and (b) fine and medium textured soils. (Allen et al., 1998)

Dual crop coefficients have K_c as the sum of a basal crop coefficient and a soil evaporation coefficient to account for increased evaporation from wet soil surfaces after irrigation or rainfall, or decreased evaporation and transpiration because soil water is limiting. This procedure is more complicated and recommended when improved estimates of K_c and ET are needed (Allen et al., 1998).

4.11 Landscape Plant Coefficients

The water requirements of landscape plants are handled differently from crop plants because maximum growth is not usually desired. The basic need is to supply sufficient water to maintain appearance, health, and reasonable growth, thus the water requirements are frequently lower than for agricultural crops (Costello & Jones, 2000). Other differences occur because landscape plantings are often composed of more than one species that are irrigated as a unit or zone. The vegetative density may vary from single plants to groups of plants to complete cover. In addition, large trees in some landscape designs will have more leaf area and use more water than a grouping of small plants in the same surface area.

The water use of landscape plants can be estimated by (Costello & Jones, 2000)

$$ET_L = K_L \times ET_o \qquad \textbf{4.31}$$

where ET_L in the estimated water requirement of a landscape planting and K_L is the landscape planting coefficient. The landscape coefficient is estimated from (Costello & Jones, 2000)

$$K_L = k_s \times k_d \times k_{mc} \qquad \textbf{4.32}$$

where K_L = landscape coefficient,
k_s = plant species factor,
k_d = plant density factor,
k_{mc} = microclimate factor.

Estimated values for the plant species, plant density, and microclimate factors are shown in Table 4-6 for California conditions (Costello & Jones, 2000). Until local estimates are available, landscape factors for other geographic locations may be estimated based on evaluation of the values and conditions assumed in developing the values in Table 4-6.

Plants species factors are divided into four categories according to the relative water use of a species and vary from about 0.1 to 0.9. Costello & Jones (2000) assigned categories to over 1800 species based on measurements and observations of plant water needs.

Plant density factors are divided into three categories and vary from 0.5 to 1.3. Immature and sparsely planted landscapes are placed in the Low category because they use less water than mature or full cover plantings. Plantings of one type with full cover are in the Moderate category. Mixed plantings with trees, shrubs, and ground covers use the most water and are placed in the High category.

TABLE 4–6 Estimated Values of Landscape Plant Coefficient Factors

	Very Low	*Low*	*Moderate*	*High*
Species factor, k_s	<0.1	0.1 to 0.3	0.4 to 0.6	0.7 to 0.9
Density factor, k_d		0.5 to 0.9	1.0	1.1 to 1.3
Microclimate factor, k_{mc}		0.5 to 0.9	1.0	1.1 to 1.4

Source: Costello and Jones (2000).

Plant microclimate factors are divided into three categories and vary from 0.5 to 1.4. The Moderate category is similar to open field conditions with no unusual wind or heat sources for the location. For example, most well-vegetated parks, unless exposed to high winds, would fall in the average microclimate category. The High category includes plantings with nearby paved areas, building walls, reflective surfaces, or locations with high winds. Plantings in shade, north-facing slopes, or under building overhangs fall in the Low category.

Example 4.6

Estimate the water required for one citrus tree in a bare area in California. The tree is mature with a canopy diameter of 10 m, irrigated every third day, and ET_o is 7 mm/day.

Solution. From Table 4-6 estimate the following values

k_s = 0.6 for moderate water use, k_d = 0.7 for large canopy and low density, and k_{mc} = 1.3 for the bare soil heat source. Calculate K_L from Equation 4.32

$$K_L = 0.6 \times 0.8 \times 1.4 = 0.62$$

Calculate ET_L from Equation 4.31

$$ET_o = 0.62 \times 7 = 4.3 \text{ mm/day}$$

The estimated volume of water to be delivered per irrigation is

$$\text{Volume} = \frac{4.3 \text{ mm/day}}{1000 \text{ mm/m}} \times 3\,\text{days} \times \pi(5\,\text{m})^2 \times 1000 \text{ L/m}^3 = 1013 \text{ L/irrigation}$$

Thus the irrigation system should be set to deliver about 1000 L of water every three days.

Internet Resources

General reference sites for equipment and information
- http://www.clemson.edu Key search term: irrigation
- http://www.irrigation-mart.com
- http://www.irrigation.org
- http://www.wcc.nrcs.usda.gov/nrcsirrig/

Solar radiation data sources
- http://rredc.nrel.gov/solar/
- http://wrdc-mgo.nrel.gov/

Examples of evapotranspiration data sources
- http://www.usbr.gov/pn/agrimet/ETtotals.html
- http://www.cimis.water.ca.gov
- http://cropwatch.unl.edu/ Key search term: evapotranspiration
- http://ag.arizona.edu/AZMET
- http://water.dnr.state.sc.us Key search term: evaporation

Landscape water use, California DWR publications
- http://www.publicaffairs.water.ca.gov/information/pubs.cfm

References

Allen, R. G., L. S. Pereira, D. Raes, & M. Smith. (1998). *Crop Evapotranspiration—Guidelines for Computing Crop Water Requirements.* FAO Irrigation and Drainage Paper No. 56. Rome: Food and Agriculture Organization.

ASCE Standardization of Reference Evapotranspiration Task Committee (ASCE). (2005). *The ASCE Standardized Reference Evapotranspiration Equation.* Report of Task Committee.

Blaney, H. F., & W. D. Criddle. (1950). *Determining Water Requirements in Irrigated Areas from Climatological and Irrigation Data.* (Litho.). Washington, DC: USDA SCDS-TP-96.

Costello, L. R., and K. S. Jones. (2000). *A Guide to Estimating the Irrigation Water Requirements of Landscape Plantings in California.* Sacramento: California Department of Water Resources.

Doorenbos, J., & W. O. Pruitt. (1977). *Guidelines for Predicting Crop Water Requirements.* FAO Irrigation and Drainage Paper No. 24, 2nd ed. Rome: Food and Agriculture Organization.

Hargreaves, G. H., & Z. A. Samani. (1982). Estimating potential evapotranspiration. Tech. Note. *Journal of Irrigation and Drainage Engineering,* ASCE, *108*(3), 225–230.

Hargreaves, G. H., & Z. A. Samani. (1985). Reference crop evapotranspiration. *Applied Engineering in Agriculture,* ASAE, *1*(2), 92–96. St. Joseph, MI.

Howell, T. A., R. H. Cuenca, & K. H. Solomon. (1990). Crop yield response. In Hoffman, G. J., T. A. Howell, & K. H. Solomon, eds., *Management of Farm Irrigation Systems.* ASAE Monograph. St. Joseph, MI.

Itenfisu, D., R. L. Elliott, R. G. Allen, & I. A. Walter. (2003). Comparison of reference evapotranspiration calculations as part of the ASCE standardization effort. *Journal of Irrigation and Drainage Engineering, 129*(6), 440–448.

Jensen, M. E. (1966, December). Empirical methods of estimating or predicting evapotranspiration using radiation. In ASAE *Conference Proceedings, Evapotranspiration and Its Role in Water Resources Management,* ASAE, 64, 49–53. St. Joseph, MI.

Jensen, M. E., R. D. Burman, & R. G. Allen, eds. (1990). *Evapotranspiration and Irrigation Water Requirements.* New York: ASCE.

Jensen, M. E., & H. R. Haise. (1963). Estimating evapotranspiration from solar radiation. *Proceeding of Irrigation and Drainage Division,* ASCE, *89*(IR4), 15–41.

Jones, F. E. (1992). *Evaporation of Water: With Emphasis on Applications and Measurements.* Chelsea, MI: Lewis Publishers.

Meyer, A. F. (1942). *Evaporation from Lakes and Reservoirs.* St. Paul: Minnesota Resources Commission.

Penman, H. L. (1948). Natural evapotranspiration from open water, bare soil, and grass. *Proceedings of Royal Society of London, 193,* 120–145.

———. (1956). Estimating evapotranspiration. *Transactions of the American Geophysical Union, 37,* 43–46.

Pruitt, W. O., E. Fereres, K. Kaita, & R. L. Snyder. (1987). *Reference Evapotranspiration (ET_0) for California.* Agriculture Experiment Station Bulletin, *1922,* University of California.

Rohwer, C. (1931). *Evaporation from free water surfaces.* USDA Technical Bulletin 271. Washington, DC: U.S. Government Printing Office.

Thornthwaite, C. W. (1948). An approach toward a rational classification of climate. *Geographical Review, 38,* 55–94.

Thornthwaite, C. W., & B. Holzman. (1942). *Measurement of Evaporation from Land and Water Surfaces.* USDA Technical Bulletin 817. Washington, DC: U.S. Government Printing Office.

U.S. Department of Agriculture (USDA). (1981). *Soil and Water Resources Conservation Act, 1980. Appraisal Part I: Soil, Water, and Related Resources in the United States: Status, Condition and Trends.* Washington, DC: USDA.

U.S. Geological Survey (USGS). (1952). *Water-Loss Investigations: Lake Hefner Studies, Technical Report*. Geological Survey Professional Paper 269. Washington, DC: U.S. Government Printing Office.

U.S. Soil Conservation Service (SCS). (1970). *Irrigation Water Requirements*. Tech Release No. 21 (rev.). Washington, DC: USDA – SCS.

Problems

4.1 Compute the daily evaporation from a free-water surface if the wind speed is 4 m/s at 0.15 m height, average water and air temperatures are both 25°C, and the average relative humidity of the air is 50 percent. Atmospheric pressure is 100 kPa. How would your computed value compare with the evaporation from a dry-soil surface? From a Class A Weather Bureau pan?

4.2 For a Class A pan, estimate ET_o for June 1. The measured evaporation from the pan was 11 mm. The wind was 2.5 m/s, and the average minimum relative humidity was 75 percent. The pan is surrounded by 100 m of bare soil.

4.3 Using the Standardized Penman-Monteith equation, estimate ET_o for July 10, 2002, at 40°N, the data were measured over grass and the mean minimum and maximum temperatures are 16°C and 29°C, respectively; maximum and minimum relative humidity are 70 and 40 percent, respectively; measured solar radiation is 27 MJ m^{-2} day^{-1}; 1.7 m/s wind speed at 3 m height; 300 m elevation; and negligible heat flux to the soil.

4.4 From the data in Problem 4.3, estimate the alfalfa-based reference ET_r.

4.5 Assume the only climatic data available from Problem 4.3 are the maximum and minimum temperatures. Estimate ET_o using the equations from Section 4.9 on missing data.

4.6 From Tables 4.4 and 4.5, estimate K_c for soybeans for May 15, June 4, July 20, and September 2. Determine $K_{c\ ini}$ using Equation 4.30 if the typical infiltration depth is 20 mm, a 10-day frequency, the soil is a silt loam, and the planting date is May 1. ET_o is 3 mm/day, 5 mm/day, 6 mm/day, and 4 mm/day for May 15, June 4, July 20, and September 2, respectively.

4.7 Estimate the water requirement for a cluster of small shrubs surrounded by grass in the Central Valley of California in July. ET_o is 10 mm/day, the climate is hot and dry, with light winds, and the plant species factor is 0.5.

CHAPTER

5 Infiltration and Runoff

Infiltration and runoff are two important processes in the hydrologic cycle (Figure 1-1). Infiltration begins when precipitation reaches the land surface. Runoff begins when the precipitation rate exceeds the infiltration rate, and retention and surface storage are filled. The relationships among rainfall, infiltration, and runoff are illustrated in Figure 5-1.

Infiltration is the main source of water for vegetative growth and crop production, provides input to ground water recharge, and transports water-soluble compounds, such as fertilizers, manures, herbicides, and other materials, from the land surface into the soil. Some infiltrated water eventually recharges the ground water. A large fraction of infiltrated water returns to the atmosphere by evapotranspiration. A small fraction of infiltrated water may reappear as surface water and either runoff or infiltrate again.

Surface runoff discharges into channels, streams, rivers, lakes, or other surface water reservoirs. Aquatic life and a large portion of the human population depend on surface water. The quality and quantity of surface water largely depend on runoff quality and quantity. High runoff rates and volumes can cause soil erosion and flooding, damage or destroy structures, and destroy human and animal lives.

Infiltration

The term infiltration is the process of water entry into the soil. The rate at which water infiltrates into the soil is the infiltration rate, which has the dimensions of volume per unit of time per unit of area, which reduces to depth per unit of time. Infiltration rate should not be confused with hydraulic conductivity, which is the ratio of soil water flow rate (flux) to the hydraulic gradient. After water enters the soil, it moves within the soil by a process known as percolation.

Infiltration may be limited by restrictions that often occur at the soil surface or at lower layers of the profile. The major factors influencing the rate of infiltration are the physical characteristics of the soil and the cover on the soil surface, but other

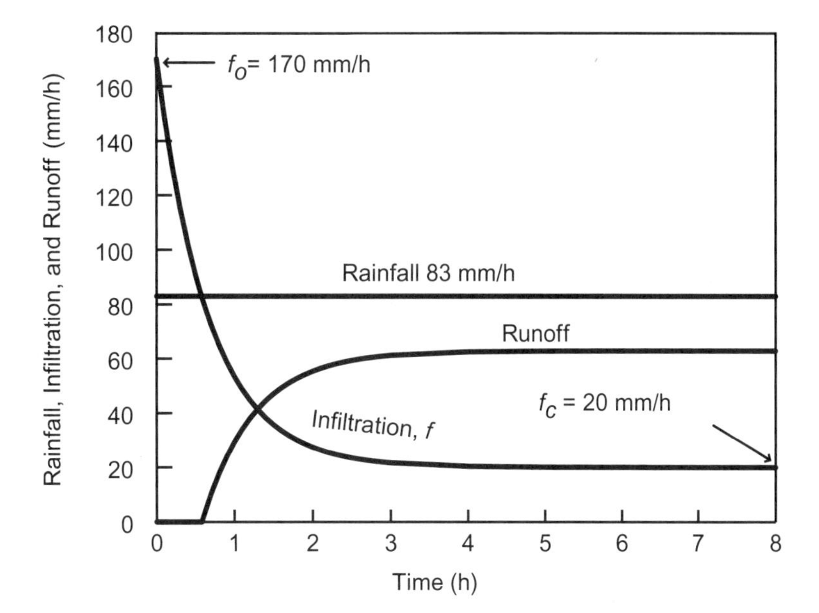

Figure 5–1
Illustration of infiltration and runoff curves for a constant rainfall rate.

factors such as soil water content, temperature, and rainfall intensity are also important.

5.1 Saturated Hydraulic Conductivity—Darcy Equation

The saturated hydraulic conductivity of the soil may limit the infiltration rate of a soil. The one-dimensional flow of water through a saturated soil can be computed from the Darcy equation

$$q = AK\frac{dh}{ds} = AK\frac{\Delta h}{L} \tag{5.1}$$

where q = flux or flow rate (L^3/T),
A = cross-sectional area of the soil through which water is flowing (L^2),
K = saturated hydraulic conductivity of the flow medium (L/T),
dh/ds = hydraulic gradient (L/L) in the direction of flow s,
Δh = total change in head or potential causing flow for a distance L (L).

Hydraulic conductivity is determined from field or laboratory measurements.

The flow path may be downward (as during infiltration), horizontal, or upward. The equation is valid so long as the velocity of flow and the size of soil particles are such that the Reynolds number is less than 1. Hydraulic conductivity K is a function of the effective diameter of the soil pores and of the density and dynamic viscosity of the fluid. It is the average velocity of bulk flow in response to a unit gradient.

Application of the Darcy equation is more difficult for two- and three-dimensional flow systems that have complex boundary conditions. Where water movement is through two soil layers, such as a topsoil layer and a subsoil layer, the composite vertical hydraulic conductivity K can be computed from

$$K = \frac{L}{\frac{L_1}{K_1} + \frac{L_2}{K_2}} \tag{5.2}$$

where L = the total length of flow through all layers (L), subscripts 1 and 2 represent the soil layers 1 and 2, respectively. Another L/K term should be added for each additional layer. The Darcy equation is analogous to Ohm's law for electrical current and to Fourier's law for heat conduction.

Example 5.1

Calculate the composite hydraulic conductivity of a 0.9-m soil profile having three soil layers. The top 0.3-m thick soil layer has a vertical saturated hydraulic conductivity of 15 mm/h, and that of the middle layer (0.2-m thickness) and bottom layer (0.4-m thickness) are 2.0 and 8.0 mm/h, respectively.

Solution. For three soil layers with the thicknesses in mm, use Equation 5.2 in the form

$$K = \frac{L}{\frac{L_1}{K_1} + \frac{L_2}{K_2} + \frac{L_3}{K_3}} = \frac{900}{\frac{300}{15} + \frac{200}{2.0} + \frac{400}{8.0}} = 5.3 \text{ mm/h}$$

Note that the soil layer with the lowest conductivity has a major influence on the composite conductivity.

5.2 Soil Factors

The soil is a pervious medium that contains a large number of micropores. The ease with which water moves through the soil largely depends on the size and permanence of the micropores. The size of the micropores and the infiltration into the soil depend on (1) soil texture, (2) the degree of aggregation between the individual particles, and (3) the arrangement of the particles and aggregates. In general, larger pore sizes and greater continuity of the pores result in higher infiltration rates.

Large openings called soil macropores may be present in the soil. The macropores are mostly interaggregate cavities; however, plant roots, wormholes, soil shrinking during drying, other natural phenomena, and soil tillage can also create these cavities. When macropores are present, soils tend to have higher infiltration rates, but, water moving in macropores may pass through portions of the plant root zone and add little to water storage there.

The maintenance of a porous soil structure, particularly at the soil surface, is critical for infiltration. The infiltration rate is greatly reduced when the uppermost thin layer of the soil surface is relatively sealed or compacted (crusted). This can result from severe breakdown of soil structure caused by heavy equipment traffic, puddling, and other field operations. Surface sealing can also result from the beating action of raindrops and the sorting action of water flowing over the surface. The fine particles can fill the spaces between the large ones to form a relatively impervious seal.

The surface-sealing effect can be largely eliminated when the soil surface is protected by mulch, crop residue, or by some other permeable protection. The effectiveness of such protection is illustrated in Figure 5-2, which shows the measured infiltration rates for covered and uncovered Clay loam, Silt loam, and Sandy

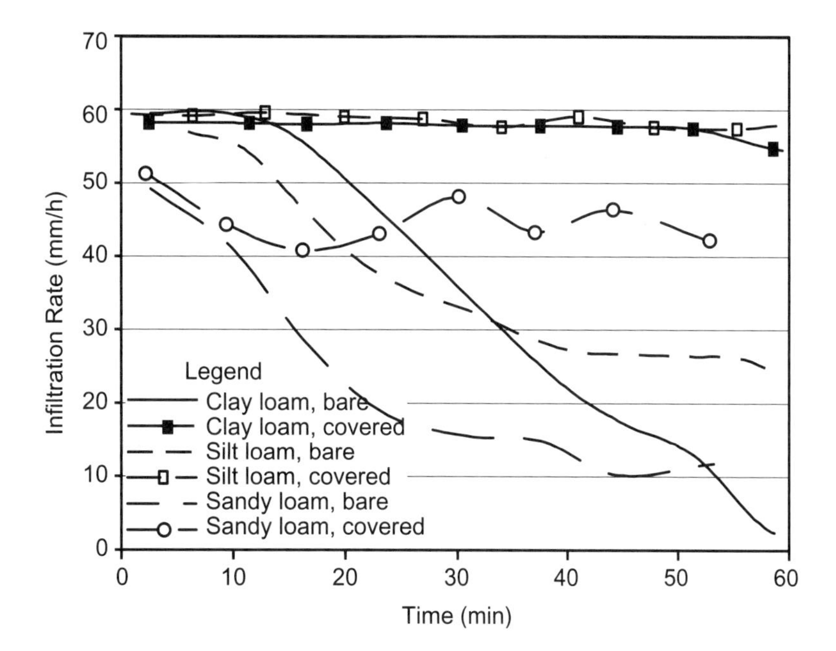

Figure 5–2

Effect of protective cover on infiltration rates. (Data provided by W.J. Elliot)

loam soils. The soils protected by a cover maintained higher infiltration rates than the unprotected soils.

5.3 Vegetation

Vegetation can greatly reduce surface sealing. In general, vegetative and surface conditions have more influence on infiltration rates than soil texture and structure. The protective cover may be grass, other close-growing vegetation, plant residue, and mulch. If the protection of a vegetative cover is lost, surface sealing may occur with drops in infiltration rates similar to those in Figure 5-2. Figure 5-3 illustrates typical infiltration depth curves for a given soil with different surface and vegetative conditions. Infiltration is higher for grass or mulched areas where the soil surface is protected than it is for bare soil conditions. Other soils may have higher or lower depths of infiltration.

5.4 Soil Additives

Chemical additives including fertilizers and manures can change the physical characteristics of the soil, including the infiltration capacity. For example, polyacrylamide can ionically bond soils together to increase the aggregate size, which increases infiltration. In general, chemical additives are one of two types. The first type consists of materials that enhance the stability of the soil aggregates and improve the soil structure. Improved structure can considerably increase infiltration rates. The second type of additive is a wetting agent. It does not change the soil structure but increases the wettability angle of the soil particles, which promotes faster water entry into the soil pores. It may be necessary to reapply wetting agents periodically, because they leach out with continued water application.

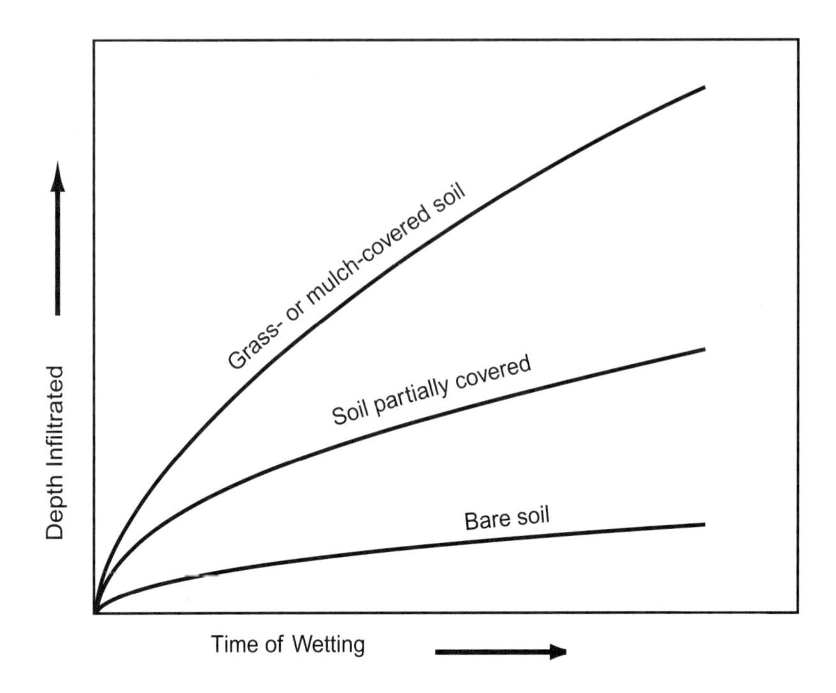

Figure 5–3
Illustration of infiltration depth curves for a soil with various surface covers.

Additives can also decrease infiltration rates. One group of chemical additives causes soil colloidal particles to swell. Swelling clays are sometimes added to soils. These swell and seal soil pores to reduce infiltration rates. Adding sodium salts will disperse the soil aggregates and reduce infiltration. Partial or complete sealants, such as petroleum or plastic films, are applied to soil surfaces to decrease or prevent infiltration. Decreased infiltration is desired to reduce seepage losses from reservoirs, waste storage structures, irrigation canals, or to increase runoff for surface water supplies.

5.5 Other Factors

Other factors affecting infiltration include land slope, antecedent soil water, air entrapment, surface roughness, and temperature (a special case is frozen soil). The effect of slope on infiltration rate is small, but more important on slopes less than 2 percent with high roughness. High soil water contents generally reduce or limit the infiltration rate. The soil matric potential decreases at high soil water content, reducing infiltration rates. The reduction also occurs because water causes some of the colloids in the soil to swell, reducing both the pore size and the rate of water movement. Entrapped air may remain in some of the soil pores and reduce both the infiltration rate and hydraulic conductivity. Field infiltration tests are customarily conducted to observe infiltration rates, once when the soil is dry and once when it is wet. The rate of change in infiltration values and the minimum infiltration values obtained from these tests may be used in a design process. In a completely saturated soil underlain by an impervious layer, infiltration is negligible.

The effect of water temperature on infiltration is not significant, perhaps because the soil changes the temperature of the entering water and the size of the pore spaces may change with temperature changes. Although freezing of the soil surface greatly reduces its infiltration capacity, freezing does not necessarily render the soil impervious. Near freezing the water viscosity increases and infiltration may be reduced.

5.6 Infiltration Curves

Infiltration data are commonly expressed as infiltration rate (Figure 5-1) or infiltration depth (Figure 5-3). In Figure 5-1 the initial infiltration rate exceeds the rate of water application; however, as the soil pores are filled with water and as surface sealing occurs, the rate of water intake gradually decreases, asymptotically approaching a constant value that is known as final infiltration rate or steady-state infiltrability (Hillel, 1998). Figure 5-1 shows that when the initial infiltration rate exceeds the application rate, the actual infiltration rate is equal to the application rate. When the infiltration rate fell below the rainfall rate, water began to accumulate on the surface, creating the potential for runoff. Several empirical and theory-based equations have been developed and used for predicting the curves for infiltration depth or rate as a function of time (Hillel, 1998).

5.7 The Kostiakov Equation

Kostiakov (1932) proposed the equation

$$I = kt^a \tag{5.3}$$

where I is the cumulative depth of infiltration at time t and k and a are constants. The constants are obtained from measured infiltration data and have no particular physical meaning. This equation is only valid for short infiltration times because it does not account for the final infiltration rate. Additional terms have been added to alleviate this shortcoming (Chapter 16).

5.8 The Horton Equation

The Horton (1940) equation is a common model for infiltration rate, which is expressed as

$$f = f_c + (f_o - f_c)e^{-kt} \tag{5.4}$$

where f = infiltration rate (L/T) at time t (T),
f_c = steady-state infiltration rate at large times (L/T),
f_o = initial infiltration rate at time $t = 0$ (L/T),
k = constant for a given soil and initial condition (1/T).

The values of f_o, f_c, and k can be determined experimentally for any soil. The steady-state infiltration rate f_c can be approximated by the saturated hydraulic conductivity for a given soil. An example of Horton's equation is included in Figure 5-1 with f_c = 20 mm/h, f_o = 170 mm/h, and k = 1.5/h.

Example 5.2

Use Horton's equation to estimate the infiltration rate of a soil after 20 min of infiltration. The initial infiltration rate of the soil is 75 mm/h, the steady-state infiltration rate is 10 mm/h, and the constant k is 12.5 per hour.

Solution. In Horton's equation, time $t = 20$ min = 0.33 h, $f_o = 75$ mm/h, $f_c = 10$ mm/h, and $k = 12.5$ per hour. Substituting in Equation 5.4,

$$f = 10 + (75 - 10)e^{-(12.5)0.33} = 11.0 \text{ mm/h}$$

5.9 The Philip Equation

Philip (1957) derived the following equation for infiltration into a uniform soil from a theoretical analysis of vertical, one-dimensional flow

$$I = St^{1/2} + At \tag{5.5}$$

where I is the depth (L) of infiltration in time t (T), S is a sorptivity term (L/T$^{1/2}$), and A is a conductivity term (L/T).

5.10 The Green-Ampt Equation

Another widely used empirical equation for predicting infiltration was developed by Green & Ampt (1911). The equation was developed by applying the Darcy equation (Equation 5.1) to the wetted soil zone and assuming vertical flow, uniform water content, and uniform soil hydraulic conductivity (near saturation). The form of the Green-Ampt equation for cumulative infiltration depth I is given as

$$K_e t = I - N_s \ln\left(1 + \frac{I}{N_s}\right) \tag{5.6}$$

where K_e = effective hydraulic conductivity (L/T),
T = time (T),
I = cumulative infiltration depth (L),
N_s = effective matric potential (L).

The wetting front is modeled as an abrupt interface between saturated and unsaturated zones, and is characterized by the effective matric potential N_s, which can be expressed as

$$N_s = (\eta_e - \theta)\psi \tag{5.7}$$

where θ is the soil water content (L^3/L^3), η_e is the effective porosity (L^3/L^3), and ψ is the matric potential (L).

The parameters in the Green and Ampt model (effective porosity, capillary potential, and hydraulic conductivity) can be estimated from readily measured soil properties, using equations developed by Rawls and co-workers (Rawls & Brakensiek, 1989; Rawls et al., 1982). The drainage model DRAINMOD and the erosion prediction model WEPP both incorporate the Green-Ampt infiltration equation.

Because of the difficulty in evaluating infiltration, the SCS (1972) divided all soils into four hydrologic groups—A, B, C, and D—on the basis of infiltration rates. Table 5-1 gives the Hydrologic Soil Groups, and descriptions, and their corresponding infiltration rates. The procedure for applying infiltration data to obtain runoff is discussed in the next section.

TABLE 5–1 Hydrologic Soil Group Descriptions

Soil Group	*Description*	*Basic Infiltration Rate f_c (mm/h)*
A	*Lowest Runoff Potential.* Includes deep sands with very little silt and clay, also deep rapidly permeable loess.	8–12
B	*Moderately Low Runoff Potential.* Mostly sandy soils less deep than A, and loess less deep or less aggregated than A, but the group as a whole has above-average infiltration after thorough wetting.	4–8
C	*Moderately High Runoff Potential.* Comprises shallow soils and soils containing considerable clay and colloids, though less than those of group D. The group has below-average infiltration after presaturation.	1–4
D	*Highest Runoff Potential.* Includes mostly clays of high swelling percentage, but the group also includes some shallow soils with nearly impermeable subhorizons near the surface.	0–1

Source: SCS (1972).

Runoff

Much of the water in ponds, lakes, and reservoirs comes from runoff. Quantity and quality of surface water, soil erosion, and stability of channels, streambanks, and other structures are affected by runoff volume and rate. Runoff constitutes the hydraulic "loading" that conservation structures or channels must withstand. The design of soil conservation structures, reservoirs, spillways, and channels must be based on runoff rate and/or volume. Therefore, understanding the runoff processes and their estimation techniques is important.

5.11 Definition

Runoff is that portion of the precipitation that flows overland toward stream channels, lakes, or oceans after the demands of interception, evapotranspiration, infiltration, surface storage, and surface detention are satisfied. In areas with significant land slopes, runoff may include near-surface flow that moves laterally beneath the soil surface and exfiltrates at some point downhill, often called interflow, to become surface runoff. This is particularly important in forested watersheds where old root channels provide conduits for flow. Appropriate knowledge of peak runoff rates, runoff volumes, and their spatial and temporal distributions is required for design and analysis.

5.12 The Runoff Process

Runoff can occur only when the rate of precipitation exceeds the soil infiltration rate after the demands for interception and surface storages are fulfilled. Interception by dense covers of forest or shrubs may be as much as 25 percent of the annual precipitation. A good stand of mature corn may have a net interception storage capacity of 0.5 mm per storm. Trees such as willows may intercept nearly 13 mm from a long, gentle storm. Interception also has a detention storage effect, delaying the progress of precipitation that reaches the soil surface only after running down the plant or dropping from the leaves. Runoff may not start even if the precipitation amount exceeds interception.

After the interception amount is met and the infiltration rate is exceeded, water begins to fill the depressions on the soil surface, which is called depression storage. After the depressions are filled, a thin static layer of water builds up on the soil surface beyond which the water layer starts moving overland. This thin layer of water is called surface detention. After surface detention storage is satisfied, overland flow or runoff begins. The depth of the water layer continues to build up on the surface until runoff rate is in equilibrium with the rate of precipitation less infiltration and interception. After precipitation ceases, the water in surface storage eventually infiltrates or evaporates.

Factors Affecting Runoff

Most of the factors that affect infiltration rate also affect runoff. The factors affecting runoff may be broadly categorized into (1) precipitation characteristics, and (2) watershed characteristics. In addition, climatic variables (temperature, relative humidity, wind speed, and direction) have some effect on runoff.

5.13 Precipitation Characteristics

Rainfall amount, duration, intensity, and distribution pattern (Chapter 3) influence the rate and volume of runoff. Total runoff for a storm event is clearly related to the rainfall duration and intensity. Infiltration is high in the initial stages of a storm and decreases with time. Thus, a storm of short duration may produce no runoff, whereas a storm of the same intensity, but of longer duration, can produce runoff.

Rainfall intensity influences both the rate and the volume of runoff. An intense storm exceeds the infiltration rate by a greater margin than does a gentle rain; thus, the volume of runoff is greater for the intense storm even though the precipitation totals for the two rainfall events are the same. The intense storm may actually decrease the infiltration rate because of its destructive action on the soil structure at the surface.

Rate and volume of runoff from a watershed are influenced by the spatial distribution of rainfall amount and intensity over the watershed. Generally the maximum runoff rate will occur when the entire watershed contributes to runoff from a uniform rainfall; however, an intense storm on a portion of the watershed could result in a greater runoff rate than a moderate storm over the entire watershed.

Frozen soil conditions significantly impact surface runoff. Winter flooding occurs frequently on frozen lands east of the Cascade Mountains in the Pacific Northwest and in the northern parts of the Intermountain West. Rainfall intensities in these areas during winter are relatively low and runoff events usually occur from November through March (McCool et al., 2000) while the soil is frozen. During this period, rainfall on snow accompanied by high wind speeds and warm, moist Pacific air masses provide high dewpoint temperatures and accelerate the snowmelt rate. Since the infiltration rate of frozen soil is almost zero, almost all of the rain and snowmelt contributes to surface runoff.

5.14 Watershed Characteristics

Watershed factors affecting runoff are (1) size, shape, and orientation, (2) topography, (3) soil type, and (4) land use and land management conditions. Both runoff volumes and rates increase as watershed size increases; however, both rate and volume per unit of watershed area decrease as the runoff area increases. Watershed size may determine the season at which high runoff may be expected to occur. On watersheds in the Ohio River basin, 99 percent of the floods from drainage areas of 260 ha occur in May through September; 95 percent of the floods on drainage areas of 26 million ha occur in October through April.

Long, narrow watersheds are likely to have lower runoff rates than do more compact watersheds of the same area. The runoff from the former does not concentrate as quickly as it does from the compact areas, and long watersheds are less likely to be covered uniformly by intense storms. When the long axis of a watershed is parallel to the storm path, storms moving upstream cause a lower peak runoff rate than storms moving downstream. For a storm moving upstream, runoff from the lower end of the watershed may be diminished before the runoff contribution from the upstream area arrives at the outlet; however, a storm moving downstream causes a higher runoff rate because runoff from the lower portion flows concurrently with runoff arriving from the upstream area.

Topographic features, such as slopes of upland areas and channels, channel morphology, and the extent and number of depressed areas, affect rates and volumes of runoff. Watersheds having extensive flat areas or depressed areas without surface outlets have lower runoff than areas with well-defined surface drainage. The geological formation and soil types determine, to a large degree, the infiltration rate, and thus affect runoff.

Land cover and land use practices influence infiltration. Vegetation retards overland flow and increases surface detention and infiltration rates and volumes. Various tillage and land management conditions that affect infiltration directly affect runoff rates and volumes. Structures such as dams, levees, bridges, and culverts all influence runoff rates and volumes.

Estimating Runoff Volume

Total volume of runoff is important in the design of wetlands, retention ponds, reservoirs, and flood control dams. It is necessary to predict the total volume of runoff that may come from a watershed during a design flood so that the structures can be built to control the runoff. The total volume is also needed for estimating the total maximum daily load (TMDL) of pollutants to surface water sources and to develop criteria for protecting water quality.

5.15 Soil Conservation Service (SCS) Method for Runoff Volume Estimation

The Soil Conservation Service (SCS) method (also known as the **curve number method**) for predicting runoff volume was primarily developed from many years of storm flow records for agricultural watersheds in many parts of the United States. With proper modifications and assumptions, the method has also been used to estimate runoff from urban areas (SCS, 1986). The runoff volume is usually expressed in the units of depth similar to the precipitation units. An average runoff depth from the entire watershed area is usually considered. If the watershed area is known, the runoff in the units of volume can be obtained by multiplying the watershed area with the runoff depth. The basic runoff equation is

$$Q = \frac{(I - 0.2S)^2}{I + 0.8S} \tag{5.8}$$

where Q = direct surface runoff depth (mm),
I = storm rainfall depth (mm) (Chapter 3),
S = maximum potential difference between rainfall and runoff in (mm).

The variable S includes both surface storage and infiltration potential of a watershed. Runoff decreases as S or infiltration increases. The initial abstraction I_a consists of interception losses, surface storage, and infiltration prior to runoff; it is assumed to be 0.2 S. The SCS (1972) developed a relationship between S and a variable called the curve number (CN) as

$$S = \frac{25\,400}{CN} - 254 \tag{5.9}$$

The curve number CN varies from 0 to 100. A higher CN value means a smaller initial abstraction (for example, wet surface condition) and higher runoff. Conversely, a smaller CN value would give a higher initial abstraction (for example, dry surface condition) and smaller runoff. Thus if $CN = 100$, then $S = 0$ and $Q = I$.

Curve numbers depend on soil type, land use, ground cover, and soil water conditions. Typical CN values are presented in Tables 5-2, 5-3, and 5-4 for average soil

TABLE 5–2 Runoff Curve Numbers (CN) for Urban Areas with Average Runoff Conditions and $I_a = 0.2S$

Cover Description		***Curve Numbers for Hydrologic Soil Group***			
Cover Type and Hydrologic Condition	***Average Percentage Impervious Area***[a]	A	B	C	D
Fully developed urban areas (vegetation established)					
Open space (lawns, parks, golf courses, cemeteries, etc.)[b]					
Poor condition (grass cover < 50%)		68	79	86	89
Fair condition (grass cover 50% to 75%)		49	69	79	84
Good condition (grass cover > 75%)		39	61	74	80
Impervious areas:					
Paved parking lots, roofs, driveways, etc. (excluding right-of-way)		98	98	98	98

(Continued)

Cover Description		*Curve Numbers for Hydrologic Soil Group*			
Cover Type and Hydrologic Condition	***Average Percentage Impervious Area***[a]	A	B	C	D
Streets and roads:					
Paved; curbs and storm sewers (excluding right-of-way)		98	98	98	98
Paved; open ditches (including right-of-way)		83	89	92	93
Gravel (including right-of-way)		76	85	89	91
Dirt (including right-of-way)		72	82	87	89
Western desert urban areas:					
Natural desert landscaping (pervious areas only)[c]		63	77	85	88
Artificial desert landscaping (impervious weed barrier, desert shrub with 1- to 2-inch sand or gravel mulch and basin borders)		96	96	96	96
Urban districts:					
Commercial and business	85	89	92	94	95
Industrial	72	81	88	91	93
Residential districts by average lot size:					
0.05 ha or less (town houses)	65	77	85	90	92
0.1 ha	38	61	75	83	87
0.13 ha	30	57	72	81	86
0.2 ha	25	54	70	80	85
0.4 ha	20	51	68	79	84
0.8 ha	12	46	65	77	82
Developing urban areas					
Newly graded areas (pervious areas only, no vegetation)[d]		77	86	91	94

Source: Adapted from SCS (1986).

[a]The average percentage impervious area shown was used to develop the composite CNs. Other assumptions are as follows: Impervious areas are directly connected to the drainage system, impervious areas have a *CN* of 98, and pervious areas are considered equivalent to open space in good hydrologic condition.

[b]CNs shown are equivalent to those of pasture. Composite CNs may be computed for other combinations of open space cover type.

[c]Composite CNs for natural desert landscaping should be computed using procedures from SCS (1986) based on the impervious area percentage (*CN* = 98) and the pervious area CN. The pervious area CNs are assumed equivalent to desert shrub in poor hydrologic condition.

[d]Composite CNs to use for the design of temporary measures during grading and construction should be computed using procedures from SCS (1986) based on the degree of development (impervious area percentage) and the CNs for the newly graded pervious areas.

water conditions. These values apply to antecedent rainfall condition *II*, which is an average value for annual floods. Correction factors for other antecedent rainfall conditions are listed in Table 5-5. Antecedent rainfall condition *I* is for low runoff potential with soil having a low antecedent water content, generally suitable for cultivation. Antecedent rainfall condition *III* is for wet conditions prior to a storm

TABLE 5–3 Runoff Curve Numbers *CN* for Agricultural Lands with Average Runoff Conditions and $I_a = 0.2S$

Land Use or Crop	*Treatment or Practice*[a]	*Hydrologic Condition*[b]	*Hydrologic Soil Group* A	B	C	D
Fallow	Bare soil		77	86	91	94
	Crop residue (CR)	Poor	76	85	90	93
		Good	74	83	88	90
Row crops	Straight row (SR)	Poor	72	81	88	91
		Good	67	78	85	89
	SR + CR	Poor	71	80	87	90
		Good	64	75	82	85
	Contoured (C)	Poor	70	79	84	88
		Good	65	75	82	86
	C + CR	Poor	69	75	82	86
		Good	64	74	81	85
	C +Terraced (T)	Poor	66	74	80	82
		Good	62	71	78	81
	C + T + CR	Poor	65	73	79	81
		Good	61	70	77	80
Small grain	Straight row	Poor	65	76	84	88
		Good	63	75	83	87
	SR + CR	Poor	64	75	83	86
		Good	60	72	80	84
	Contoured	Poor	63	74	82	85
		Good	61	73	81	84
	C + T	Poor	61	72	79	82
		Good	59	70	78	81
	C + T + CR	Poor	60	71	78	81
		Good	58	69	77	80
Close-seeded legume or rotation meadow	Straight row	Poor	66	77	85	89
		Good	58	72	81	85
	Contoured	Poor	64	75	83	85
		Good	55	69	78	83
	C + T	Poor	63	73	80	83
		Good	51	67	76	80

Source: SCS (1972, 1986).

[a]Crop residue cover applies only if residue is on at least 5% of the surface throughout the year.

[b]Hydrologic condition is based on combination factors that affect infiltration and runoff, including (1) density and canopy of vegetative areas, (2) amount of year-round cover, (3) amount of grass or close-seeded legumes, (4) percentage of residue cover on the land surface (good ≥ 20%), and (5) degree of surface roughness.

event. As indicated in Table 5-5, no upper limit for antecedent rainfall is intended. The limits for the dormant season apply when the soils are not frozen and when no snow is on the ground.

Since the duration of a storm affects the amount of rainfall, runoff volume must be evaluated for each design application. The time of concentration (see Section 5.16)

TABLE 5–4 Runoff Curve Numbers (CN) for other Agricultural, Arid, and Semiarid Lands with Average Runoff Conditions and $I_a = 0.2S$

Cover Description		*Curve Numbers for Hydrologic Soil Groups*			
Cover Type	*Hydrologic Condition*	*A*	*B*	*C*	*D*
Other Agricultural Lands					
Pasture or range—continuous forage for grazing[a]	Poor	68	79	86	89
	Fair	49	69	79	84
	Good	39	61	74	80
Meadow (permanent) protected from grazing	Good	30	58	71	78
Brush—brush-weed-grass mixture with brush the major element[a]	Poor	48	67	77	83
	Fair	35	56	70	77
	Good	30	48	65	73
Woods-grass combination (orchard or tree farm)[b]	Poor	57	73	82	86
	Fair	43	65	76	82
	Good	32	58	72	79
Woods or forest land[c]	Poor	45	66	77	83
	Fair	36	60	73	79
	Good	30	55	70	77
Farmsteads		59	74	82	86
Arid and Semiarid Rangelands[d]					
Herbaceous—mixture of grass, weeds, and low-growing brush (brush the minor element)	Poor	[e]	80	87	93
	Fair	[e]	71	81	89
	Good	[e]	62	74	85
Oak-aspen-mountain brush	Poor	[e]	66	74	79
	Fair	[e]	48	57	63
	Good	[e]	30	41	48
Pinyon, juniper, or both; grass understory	Poor	[e]	75	85	89
	Fair	[e]	58	73	80
	Good	[e]	41	61	71
Sagebrush with grass understory	Poor	[e]	67	80	85
	Fair	[e]	51	63	70
	Good	[e]	35	47	55
Desert shrub	Poor	63	77	85	88
	Fair	55	72	81	86
	Good	49	68	79	84

Source: SCS, (1986).

[a]Poor, < 50% ground cover; Fair, 50 to 75% ground cover; Good, > 75% ground cover.

[b]Computed for 50% woods and 50% grass. Use CNs for woods and pasture for other percentages.

[c] *Poor*: Forest litter, small trees, and brush are destroyed by heavy grazing or regular burning.
Fair: Woods are grazed but not burned, and some forest litter covers the soil.
Good: Woods are protected from grazing, and litter and brush adequately cover the soil.

[d]Poor, < 30% ground cover (litter, grass, and brush overstory); Fair, 30 to 70% ground cover; Good, > 70% ground cover.

[e]Curve numbers not developed.

TABLE 5–5 Antecedent Rainfall Conditions and Curve Numbers (for $I_a = 0.2S$)

Curve Number for Condition II	*Factor to Convert Curve Number from Condition II to*	
	Condition I	*Condition III*
10	0.40	2.22
20	0.45	1.85
30	0.50	1.67
40	0.55	1.50
50	0.62	1.40
60	0.67	1.30
70	0.73	1.21
80	0.79	1.14
90	0.87	1.07
100	1.00	1.00

Condition	*General Description*	*5-Day Antecedent Rainfall (mm)*	
		Dormant Season	*Growing Season*
I	Optimum soil condition from about lower plastic limit to wilting point	<13	<36
II	Average value for annual floods	13–28	36–53
III	Heavy rainfall or light rainfall and low temperatures within 5 days prior to the given storm	>28	>53

Source: SCS (1972).

is not a good criterion for the determination of storm volume because a short-duration high-intensity storm may produce the largest peak flow for a given watershed, but not necessarily the maximum runoff volume. The SCS has established 6 hours as the minimum storm duration for runoff control structures, but this time is modified for conditions where a greater runoff may result.

Example 5.3

Estimate the volume of runoff during the growing season for a 50-yr return period that may be expected from a 40-ha watershed by O'Hare airport at Chicago, Illinois. Assume that antecedent rainfall during the last 4 of the 5 days prior to the storm was 40 mm and the critical duration of the storm is 6 h. The watershed has the following characteristics:

Subarea (ha)	*Topography (% slope)*	*Hydrologic Soil Group*	*Land Use, Treatment, and Hydrologic Condition*
24	0–5	C	Row crop, contoured, good
16	5–10	B	Woodland, good

Solution. From Chapter 3, the 6-h rainfall for a 50-year return period at O'Hare airport is 122.3 mm. Since the percentage reduction for converting point rainfall to areal rainfall is less than 1 percent (see Chapter 3) in this case, no correction need be made. Because the 5-day rainfall prior to the event was 40 mm during the growing season, antecedent rainfall condition *II* applies (Table 5-5). From Tables 5-3 and 5-4 for antecedent rainfall condition *II*, read the appropriate curve numbers and calculate the weighted value as follows:

Subarea (ha)	*Hydrologic Soil Group*	*Land Use, Treatment, and Condition*	*Curve No. (CN)*	*CN × A*
24	C	Row crop, contoured, good	82	1968
16	B	Woodland, good	55	880
Total 40 ha				Total 2848

Weighted $CN = 2848/40 = 71.2$ or use 71. Substituting in Equation 5.9,

$$S = \frac{25\,400}{71} - 254 = 103.7 \text{ mm}$$

$$Q = \frac{(122.3 - 0.2 \times 103.7)^2}{122.3 + (0.8 \times 103.7)} = 50.3 \text{ mm or } \frac{50.3 \times 40 \times 10^4}{1000} = 20\,120 \text{ m}^3$$

Therefore, the estimated runoff from the 40-ha watershed is 50.3 mm (in depth) or 20 120 m^3 (in volume). The 50.3 mm runoff can also be expressed as 0.0503 m × 40 ha = 2.01 ha-m (in volume).

Example 5.4

Assume the information from Example 5.3 except that the watershed has the following characteristics and curve numbers (Table 5-2).

Subarea (ha)	*Hydrologic Soil Group*	*Land Use, Treatment, and Condition*	*Curve No. (CN)*	*CN × A*
24	C	Industrial area	91	2184
16	B	Town houses	85	1360
Total 40 ha				Total 3544

Solution. Weighted $CN = 3384/40 = 88.6$ or use 89. Substituting into Equations 5-9 and 5-8 yields

$$S = \frac{25\,400}{89} - 254 = 31.4 \text{ mm}$$

$$Q = \frac{(122.3 - 0.2 \times 31.4)^2}{122.3 + (0.8 \times 31.4)} = 91.3 \text{ mm or } \frac{91.3 \times 40 \times 10^4}{1000} = 36\,520 \text{ m}^3$$

The runoff is about 80 percent higher than for the row crop and wooded watershed in Example 5.3.

Example 5.5

If 60 mm of rainfall occurs the day after the 50-yr storm in Example 5.3, what is the expected runoff?

Solution. From Table 5-5, antecedent rainfall condition *III* applies because the 5-day prior rainfall was 40 + 122.3 = 162.3 mm, which exceeds the value of 53 mm. From Table 5-5, interpolate a correction factor of 1.20 for $CN = 71$. The new curve number is = 71 × 1.20 = 85.4 or use 85. From Equation 5.9 for $CN = 85$, calculate $S = 44.8$ mm.

$$S = \frac{25\,400}{85} - 254 = 44.8$$

Substituting into Equation 5.8, for Q

$$Q = \frac{(60 - 0.2 \times 44.8)^2}{60 + 0.8 \times 44.8} = 27.2 \text{ mm}$$

Although the rainfall of 60 mm was about half the 122.3 mm in Example 5.3, the runoff was about 54 percent (27.2/50.3) of the previous amount, illustrating the importance of antecedent rainfall.

Design Runoff Rates

The design runoff rate is the maximum runoff rate that will occur from a storm of a specific duration and recurrence period. Structures and channels are designed for a specific return period and must withstand the runoff rates produced by events of that magnitude. Vegetated and temporary structures such as vegetative waterways, earthen channels, and filter strips are usually designed for the maximum runoff rate that may be expected once in 10 years. Expensive, permanent structures such as dams and reservoirs are designed for runoff expected to occur once in 50 or 100 years. Selection of the design return period, also called recurrence interval, depends on the economic balance between the cost of periodic repair or replacement of the facility and the cost of providing additional capacity to reduce the frequency of repair or replacement. In most instances, the potential damage from failure of the structure dictates the selection of the return period of the design storm.

A relationship between the return period (T) and the acceptable probability (P_r) of the design capacity being exceeded during the design life of a structure can be used to calculate the required design period (Haan et al., 1994)

$$P_r = 1 - (1 - 1/T)^n \qquad \textbf{5.10}$$

where n is the design life of the structure in years. According to Equation 5.10, for a structure designed for a 10-yr return period, the probability that the design capacity will be exceeded at least once during the 10-yr life of the structure is $P_r = 1-(1 - 1/10)^{10} = 0.65$ or 65 percent. If this risk is reduced to $P_r = 10$ percent, the 10-yr structure should be designed on the basis of a $T = 95$-yr return period, which may be very expensive. The criteria on which the risk and return period are selected should be based on the consequences of the design capacity being exceeded.

There are several methods for estimating a design runoff rate. These methods make simplifying assumptions regarding the influence of some factors and necessarily neglect other factors. Methods presented here are applicable to watersheds less than a few hundred hectares.

5.16 SCS-TR55 Method for Estimating Peak Runoff Rate

The SCS-TR55 method has been widely used to estimate peak runoff rates from small rural and urban watersheds (SCS, 1986). This method of estimating peak runoff rate is applicable to watersheds that are smaller than 900 ha and with average slopes greater than 0.5 percent with one main channel or two tributaries with nearly the same time of concentration. The peak runoff rate equation was developed from the analysis of hydrographs by the SCS-TR20 computer program (SCS, 1983), and is given by

$$q = q_u A Q F_p \tag{5.11}$$

where q = peak runoff rate (m^3/s),
q_u = unit peak flow rate (m^3/s per ha per mm of runoff),
A = watershed area (ha),
Q = runoff depth from a 24-h storm of the desired return period (Equation 5.8) (mm),
F_p = pond and swamp adjustment factor from Table 5-6.

Before Equation 5.11 is utilized for estimating peak runoff rate, the *time of concentration* of a watershed must be calculated. The time of concentration of a watershed is the time required for water to flow from the most hydraulically remote (in time of flow) point of the watershed to the outlet once the soil has become saturated and minor depressions are filled. It is assumed that, when the duration of a storm equals the time of concentration, all parts of the watershed are contributing simultaneously to the discharge at the outlet. The time of concentration may be obtained from the equation (SCS, 1990)

$$T_c = L^{0.8}\left[\frac{\left(\dfrac{1000}{CN} - 9\right)^{0.7}}{4407(S_g)^{0.5}}\right] \tag{5.12}$$

where T_c = time of concentration (h),
L = longest flow length (from the most remote point to the outlet) (m),
CN = runoff curve number (Section 5.15),
S_g = average watershed gradient (m/m).

After calculating the time of concentration from Equation 5.12, unit peak flow rate q_u is obtained from Figure 5-4 using T_c and the ratio of initial abstraction (I_a) to 24-h rainfall (P) with a return period equal to the return period of the peak flow. This ratio (I_a/P) represents the fraction of rainfall that occurs before runoff begins. The initial abstraction is usually taken as I_a = 0.2 S, and S is calculated using Equation 5.9. The curves in Figure 5-4 apply only for the Type II rainfall distribution, which is applicable for the unshaded areas of the United States shown in Figure 5-5. Curves for other types of rainfall are given by SCS (1986, 1990) and McCuen (1989). Application of the method is shown in Example 5.6.

TABLE 5–6 Adjustment Factor F_p for Pond and Swamp Areas That Are Spread Throughout the Watershed

Percentage of Pond and Swamp Areas	F_p
0	1.0
0.2	0.97
1.0	0.87
3.0	0.75
5.0	0.72

Source: SCS, (1986).

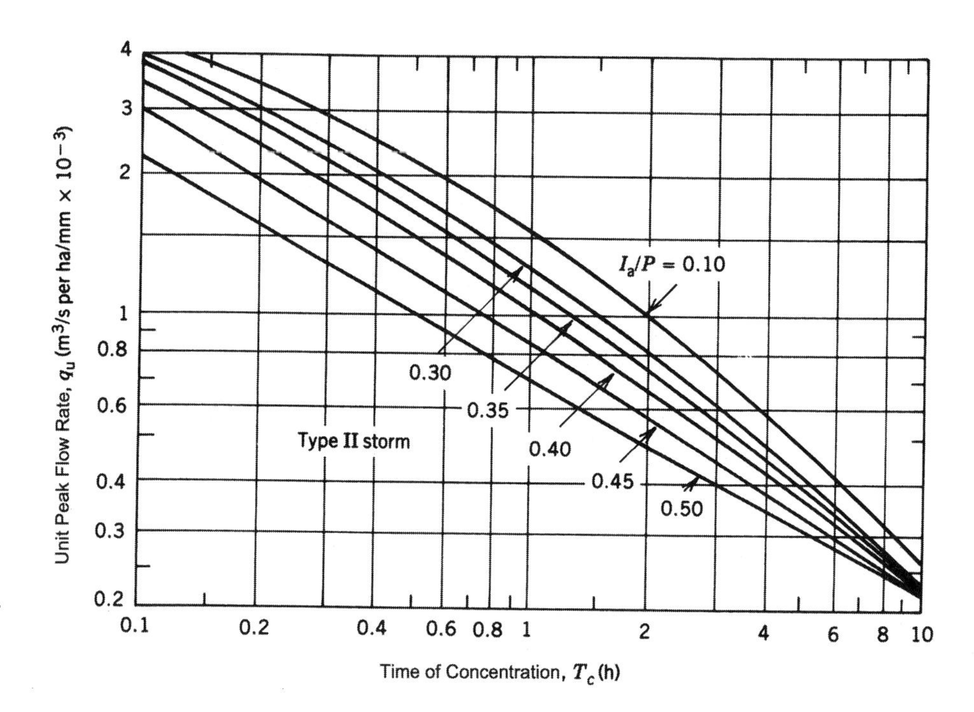

Figure 5–4
Unit peak runoff rates for SCS type II rainfall distribution. (Revised from SCS, 1990)

Example 5.6

Determine the peak runoff rate from a 100-ha watershed from a 120-mm, 24-h storm that produced 10-mm depth of runoff. Assume a flow length of 1500 m, antecedent rainfall conditions II, weighted average curve number of 75, an average watershed gradient of 0.02 m/m, and 0.2 percent pond and swamp areas.

Solution. Substitute in Equation 5.9.

$$S = (25\,400/75) - 254 = 85$$

From Equation 5.12,

$$T_c = 1500^{0.8}\left[\frac{\left(\frac{1000}{75} - 9\right)^{0.7}}{4407(0.02)^{0.5}}\right] = 1.56\text{h}$$

Taking $I_a = 0.2S$, $= 0.2 \times 85 = 17$ mm

$$I_a/P = 17/120 = 0.14$$

Read from Figure 5-4

$$q_u = 1.2 \times 10^{-3} = 0.0012 \text{ m}^3\text{/s per ha per mm}$$

Find $F_p = 0.97$ for 0.2 percent pond and swamp areas in the watershed from Table 5-6. Substitute these values into Equation 5.11.

$$q = 0.0012 \times 100 \times 10 \times 0.97 = 1.16 \text{ m}^3\text{/s}$$

5.17 Rational Method for Estimating Peak Runoff Rate

The rational method is one of the simplest methods for estimating design peak runoff rate from watersheds of less than 800 ha. This method assumes that the frequencies of rainfall and runoff are similar, which was confirmed by Larson & Reich (1973). The method greatly simplifies a complicated process; however, the method is considered sufficiently accurate for runoff estimation in the design of relatively inexpensive structures where the consequences of failure are limited. Details of the rational method are provided in Haan et al. (1994).

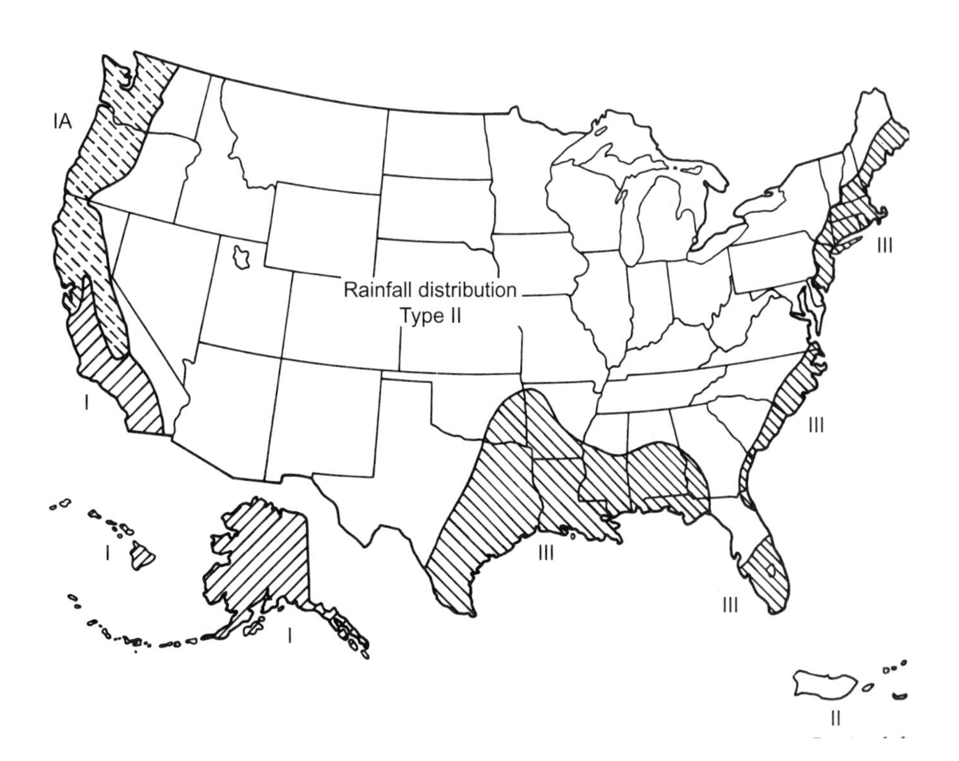

Figure 5–5

Approximate geographic boundaries for SCS rainfall distributions. (Revised from SCS, 1990)

5.18 Flood Frequency Analysis Method

Another method of runoff rate estimation, called flood frequency analysis, depends on the historical records from the drainage area under study. These records constitute a statistical array that defines the probable frequency of recurrence of floods of given magnitudes. Extrapolation of the frequency curves can be used to predict flood peaks for a range of return periods. The procedure for this method is the same as that for rainfall described in Chapter 3.

5.19 Computer Model Prediction of Runoff

Numerous computer models have been developed to predict storm runoff, many for special applications. The hydrologic and erosion models CREAMS (Knisel, 1980), WEPP (Nearing et al., 1989), and AnnAGNPS (Bingner et al., 2001); the drainage model DRAINMOD (Skaggs, 1982); the soil and water assessment model SWAT (Arnold et al., 1993); the storm water management model SWMM (Metcalf et al., 1971); and the water quality model RZWQM (Ahuja et al., 1999) include runoff predictions within the models. Many other models have been developed with runoff prediction capabilities for such applications as forests, water quality, frozen and thawing soils, wetlands, surface mines, wind erosion prediction, and plant growth applications (Goodrich & Woolhiser, 1991). Several hydrologic models have been integrated with GIS to account for spatial variability of soils, cover, land use, and land management conditions.

When selecting a runoff model, the user should generally select the model that best suits the purpose. Generally, more sophisticated models require larger input files, and obtaining the necessary input data can be time consuming and difficult. Models allow the user the opportunity to compare the effects of different land use and land management practices on runoff, and other watershed responses and to select best management practices (BMP) for a given situation.

Runoff Hydrographs

A hydrograph represents flow rate with respect to time. A stream flow hydrograph may have four components: (1) base flow, (2) direct runoff, (3) interflow, and (4) channel precipitation. Base flow is the local ground water flow component that discharges into the stream. Direct runoff is the surface runoff component that reaches the stream and is often the most significant component of the hydrograph after a precipitation event. Interflow moves laterally in the soil at shallow depths and discharges into the streams. In flat watersheds with pipe drainage systems, interflow may move through the pipe and discharge into the stream. Channel precipitation is the flow contribution from precipitation directly on the stream. In hydrologic analysis, channel precipitation and interflow are most often considered together with the direct runoff. Since the base flow component is almost constant, runoff dictates the shape of a hydrograph. From stream flow records, the base flow component can be subtracted to obtain ordinates of the surface runoff hydrograph. Hydrograph analysis is a convenient method for determining peak flow rates.

Figure 5-6 shows the measured precipitation rate and runoff hydrograph from a 123-ha agricultural watershed having woods, pasture, cropland, and hay near Coshocton, Ohio. The element of the hydrograph representing the increasing flow rate is called the rising limb; the element representing flow rate from the peak to the end of the flow period is called the falling or recession limb. In Figure 5-6, both the precipitation rate and the runoff rate show two peaks with runoff slightly delayed. The area under the hydrograph represents the volume of runoff. For the storm, 35.6 mm of rainfall occurred and the runoff was 19.1 mm. Water samples were collected and analyzed. The results for one sample taken during a high-flow period are given in Table 5-7.

Small agricultural watersheds rarely have adequate stream flow records for hydrograph development. Methods are available for developing hydrographs when the actual flow record is not available. These methods are based on some common characteristics. One of the characteristics of hydrographs for a given watershed is that the duration of flow is nearly constant for individual storms regardless of the peak flow.

Several theoretical hydrographs have been proposed based on different statistical frequency distributions. Dodge (1959) developed a unit hydrograph from the Poisson probability function, Gray (1973) employed a two-parameter gamma distribution, and Reich (1962) investigated a three-parameter Pearson type III function. Linsley et al. (1982) described several methods of developing unit hydrographs and a hydrograph for overland flow.

5.20 Triangular Hydrograph

The simplest form of runoff hydrograph is the triangular hydrograph as shown by the dashed lines in Figure 5-7. Except for the tail end, it is a good approximation of an actual hydrograph. As shown in Figure 5-7, the time-to-peak T_p and the peak flow rate for the triangular hydrograph are the same as those of the dimensionless hydrograph. When constructing a triangular hydrograph, the altitude of the triangle should be the peak flow rate and the base of the triangle taken as 2.67 T_p. These are

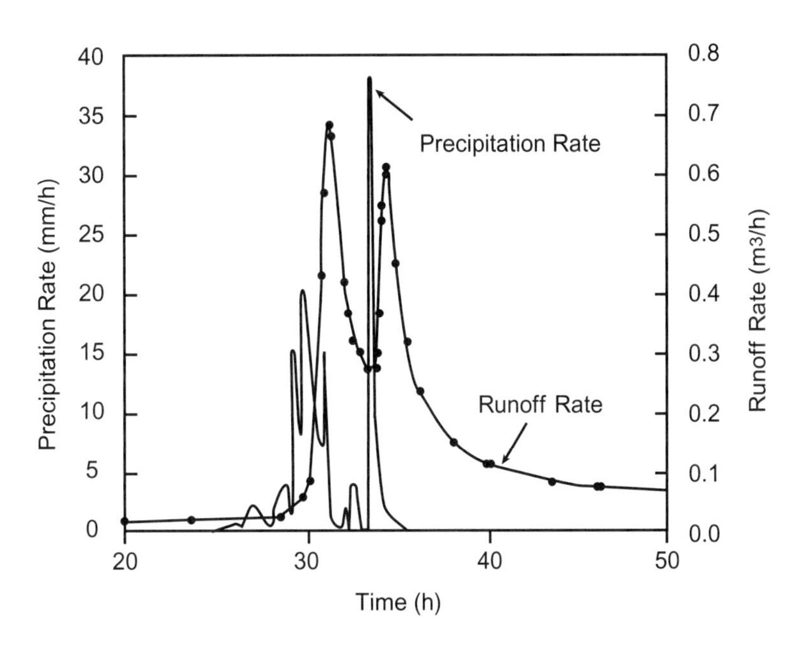

Figure 5–6
Example of a measured runoff hydrograph and precipitation from Watershed 196 of the North Appalachian Experimental Watershed, Coshocton, Ohio. (Adapted from data supplied by USDA, ARS, NAEW, 2004)

TABLE 5–7 Results of Analysis of Runoff from Watershed 196 of the North Appalachian Experimental Watershed, Coshocton, Ohio.

Cl (mg/L)	Br (mg/L)	NO3-N (mg/L)	PO4-P (mg/L)	SO4 (mg/L)	Na (mg/L)	NH4-N (mg/L)
8.2	0.0	1.0	0.0	22.2	8.2	0.0
K (mg/L)	**Mg (mg/L)**	**Ca (mg/L)**	**TOC (mg/L)**	**pH**	**Sediment (mg/L)**	
3.4	8.2	16.6	10.4	7.8	336	

Sources: Data provided by USDA, ARS, NAEW, (2004).

applied to (1) subareas of a watershed or (2) time increments of a rainstorm. In either case, the flow rates at a given time are added from several triangular hydrographs, thus producing a curvilinear hydrograph.

5.21 Dimensionless Hydrograph

Commons (1942) and later others developed dimensionless hydrographs as shown by the smooth curve in Figure 5-7. The shape approximates the flow from an intense storm from a small watershed. It has an idealized shape, and is sometimes called a synthetic hydrograph, that can be used to develop approximate design hydrographs for any small watershed for which flow records are not available. The dimensionless hydrograph in Figure 5-7 divides the peak flow rate into 100 flow units and divides the duration of flow into 100 units of time with a total area of 2620 volume units under the hydrograph. To develop a design hydrograph for a watershed from the dimensionless hydrograph, the peak flow rate (q) and the total runoff volume (Q) for the desired return period storm must be known. The values of q and Q can be determined for any desired storm for a watershed by methods described in Sections 5.15 and 5.16.

The design hydrograph is developed from the dimensionless hydrograph by using appropriate conversion factors. The factor u is the ratio of the total runoff volume from the desired storm to the area under the dimensionless hydrograph. Since the area under the dimensionless hydrograph is 2620 volume units and the design storm has a total runoff volume Q, each volume unit under the dimensionless hydrograph has a value of

$$u = Q/2620 \tag{5.13}$$

in the design hydrograph. The factor w is the ratio of peak runoff rate (q) for the design storm to the peak flow of 100 on the dimensionless hydrograph. Thus, each unit of flow on the dimensionless hydrograph has a value of

$$w = q/100 \tag{5.14}$$

in the hydrograph of the design storm. The factor k is the value that each unit of time on the dimensionless hydrograph represents in the design hydrograph. On the design hydrograph, 1/100 of the peak flow times 1/100 of the duration of runoff must equal 1/2620 of the total runoff volume. Since w is equal to 1/100 of the design peak flow,

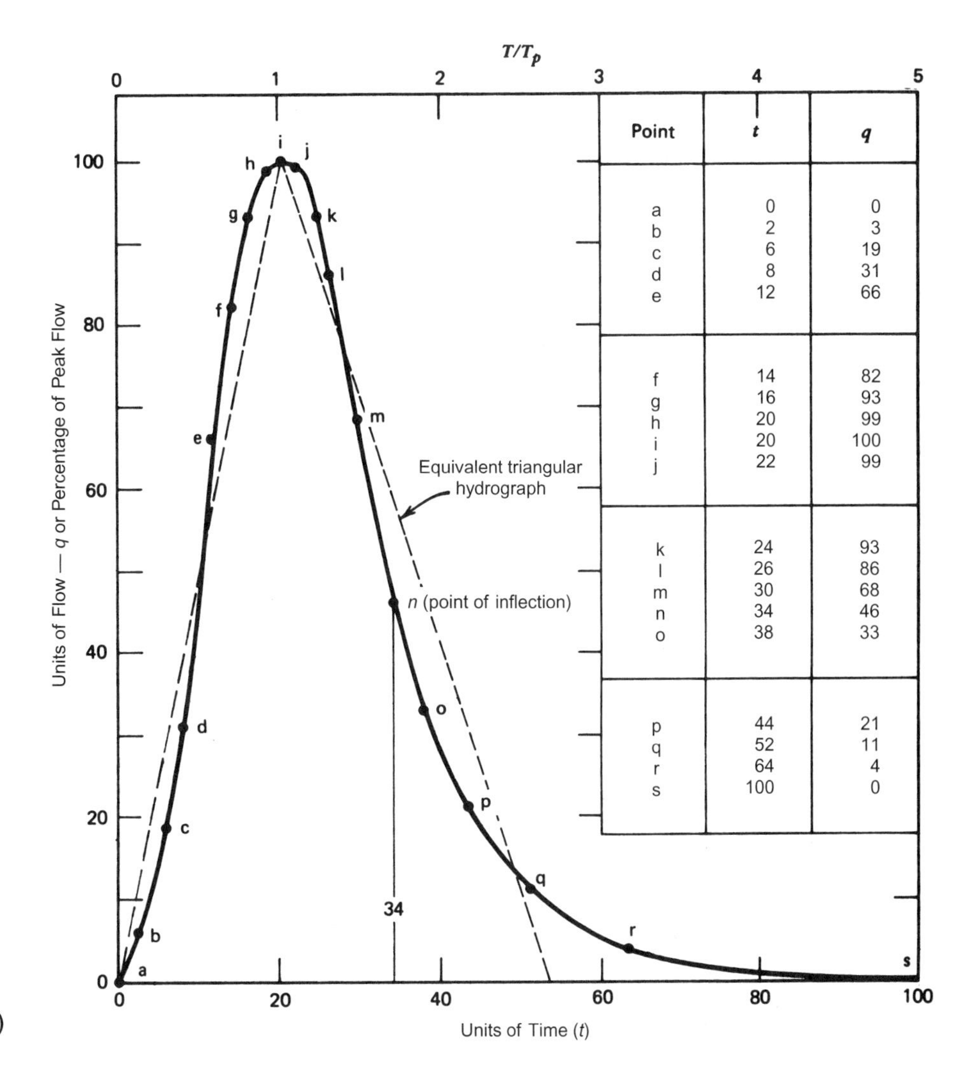

Point	t	q
a	0	0
b	2	3
c	6	19
d	8	31
e	12	66
f	14	82
g	16	93
h	20	99
i	20	100
j	22	99
k	24	93
l	26	86
m	30	68
n	34	46
o	38	33
p	44	21
q	52	11
r	64	4
s	100	0

Figure 5–7
Dimensionless and triangular flood hydrograhs. (Adapted from SCS, 1972)

k must be equal to 1/100 of the design hydrograph duration, and u is 1/2620 of the design runoff volume. Therefore,

$$wk = u \text{ or } k = u/w \tag{5.15}$$

When runoff rate q is measured in m^3/s, runoff volume Q is measured in ha-m, and time is measured in minutes, u will be in ha-m per unit, w will be in m^3/s per unit, and

$$k = \frac{u(\text{ha} - \text{m}) \times 10\,000(\text{m}^2/ha)}{w(\text{m}^3/s) \times 60(\text{s/min})} = 167\frac{u}{w} \text{minutes per unit} \tag{5.16}$$

The coordinates of the design hydrograph are obtained by multiplying the ordinates and abscissas of the dimensionless hydrograph by w and k, respectively. Example 5.7 illustrates the development of a design hydrograph from the dimensionless hydrograph.

Example 5.7 Using the dimensionless hydrograph, develop a design runoff hydrograph for a return period of 50 years for the watershed in Example 5.3. Assume a peak runoff rate of 7.0 m^3/s.

Solution. From Example 5.3, the runoff volume is 2.01 ha-m. From Equations 5-13, 5-14, and 5-16,

$$u = 2.01/2620 = 0.000767 \text{ ha-m/unit}$$

$$w = 7.0/100 = 0.07 \text{ m}^3\text{/s per unit}$$

$$k = 167 \times 0.000767/0.07 = 1.83 \text{ min/unit}$$

Ordinates and abscissas of the design runoff hydrograph are obtained by multiplying the values of q and t from Figure 5-7 by w and k, respectively. The calculated coordinates are as follows:

Point	$k \times t$ (min)	$w \times q$ (m^3/s)	Point	$k \times t$ (min)	$w \times q$ (m^3/s)
a	0	0	j	39.6	6.93
b	3.6	0.21	k	43.2	6.51
c	10.8	1.33	l	46.8	6.02
d	14.4	2.17	m	54.0	4.76
e	21.6	4.62	n	61.2	3.22
f	25.2	5.74	o	68.4	2.31
g	28.8	6.51	p	79.2	1.47
h	32.4	6.93	q	93.6	0.77
i	36.0	7.0	r	115.2	0.28
			s	180	0

If these points are plotted, the hydrograph will have the shape shown in Figure 5-7 and point i is at T_p with a peak flow of 7.0 m^3/s.

5.22 Unit Hydrograph

The unit hydrograph developed by Sherman (1932), as described by Haan et al. (1994), is "...a hydrograph of runoff resulting from a unit of excess rainfall occurring at a uniform rate, uniformly distributed over a watershed in a specified duration of time." In addition to the restrictions in the definition, the unit hydrograph is assumed to reflect all watershed characteristics so that the runoff rate is proportional to the runoff volume for a rainfall excess of a specific duration.

Based on the above definition, every watershed has a unit hydrograph for a specific duration of rainfall excess. Durations of 20 minutes, 1 h, 2 h, 6 h, or 24 h are typical examples. Theoretically a unit hydrograph could be constructed for an infinite number of durations for every watershed. For practical reasons, the unit hydrograph is applied to rainfall excesses of durations up to 25 percent different from the duration of the unit hydrograph (Haan et al., 1994).

Like the dimensionless hydrograph, a unit hydrograph can also be used to construct a design runoff hydrograph for an ungauged watershed if the rainfall excess depth for any specific storm duration is known. The difference between these two methods is that the shape of the dimensionless hydrograph is fixed regardless of watershed location, whereas the shape of a unit hydrograph of any duration can be specific to a particular watershed. Therefore the design hydrograph constructed from a unit hydrograph of a watershed can be more accurate for that watershed than that constructed from the idealized dimensionless hydrograph. In this case, however, the unit hydrograph for a specific rainfall excess duration for the specific watershed must be available to construct a design hydrograph for that watershed.

Plots of unit hydrographs show flow rate versus time. The flow rate is typically shown as volume per time. Because a unit hydrograph has a depth associated with it, the flow rate is actually volume per unit time per unit depth of excess rainfall. To obtain the ordinates of the design runoff hydrograph from a unit hydrograph, the corresponding ordinate of the unit hydrograph should be multiplied by the rainfall excess depth. The following example illustrates a simple procedure for constructing a unit hydrograph and the application of the unit hydrograph for developing a design runoff hydrograph.

Example 5.8

Actual runoff rates resulting from a rainfall excess of 2-hour duration for a particular watershed are given as follows:

Time, h	0	1	2	3	4	5	6	7
Flow rate, m^3/s	0	2	5	3	2	1	0.5	0

The total runoff volume for this event was 35 mm. Construct a 2-h unit hydrograph. Apply the unit hydrograph to develop a design runoff hydrograph for this watershed that will result from two successive 2-h rainfall runoff events (rainfall excess duration is 4-h) of 15- and 25-mm, respectively.

Solution. First, the ordinates of the unit hydrographs are calculated by dividing the actual runoff rates by the total runoff depth of 35-mm to obtain a 2-h, 1-mm unit hydrograph. The values are shown in column (2). For example, dividing the flow rate of 2 m^3/s at 1-h time by 35 mm, the unit hydrograph ordinate = 2/35 = 0.057 m^{3}s^{-1}mm^{-1}. Details of the calculations are tabulated below:

Time (h)	Unit Hydrograph Ordinates (m^{3}s^{-1}mm^{-1})	Rain Event	Runoff (mm)	Hydrograph Ordinates for Rain 1 (m^3/s)	Hydrograph Ordinates for Rain 2 (m^3/s)	Ordinate for the Design Hydrograph (m^3/s)
(1)	(2)	(3)	(4)	(5)	(6)	(7)
0	0	1	15	0	—	0
1	0.057			0.855	—	0.855
2	0.143	2	25	2.145	0	2.145
3	0.086			1.290	1.425	2.715
4	0.057			0.855	3.575	4.430
5	0.029			0.435	2.150	2.585
6	0.014			0.210	1.425	1.635
7	0			0	0.725	0.725
8				—	0.350	0.350
9				—	0	0

Column (1) contains the time steps and column (2) has the corresponding ordinates for a 2-h unit hydrograph. These 2-h unit hydrograph ordinates were utilized to construct a design runoff hydrograph for excess rainfall duration of 4 hours. The first excess rainfall of 15-mm lasted for 2 hours, and the hydrograph ordinates for this event were obtained by multiplying the unit hydrograph ordinates from column (2) by 15-mm. These values are shown in column (5). A second rainfall excess of 25-mm occurred for the next 2 hours, and the hydrograph ordinates for this rain were obtained by multiplying the unit hydrograph ordinates in column (2) by 25-mm. These values are shown in column (6) and start at the third hour and end at the ninth hour after the start of runoff. The values of the flow in columns (5) and (6) were added to obtain the direct runoff ordinates in column (7). Column (7) values can be plotted against time (column 1) to obtain the design runoff hydrograph. This plot will show that the peak runoff rate is about 4.5 m^3/s and occurs early in the fourth hour.

Events of different durations cannot be added to obtain unit hydrographs. The unit hydrographs of multiple events of the same duration can be added to calculate new unit hydrograph. For example, three 1-h unit hydrographs can be added, but a 1-h and a 2-h cannot be added for a 3-h unit hydrograph. After the three 1-h unit hydrographs are added, the sum must be divided by 3 to obtain the 3-h unit hydrograph.

Internet Resources

NRCS source for TR-55
www.wcc.nrcs.usda.gov/hydro/hydro-tools-tr55.html

NRCS source for TR-20
www.wcc.nrcs.usda.gov/hydro/hydro-tools-tr20.html

References

Ahuja, L. R., K. W. Rojas, J. D. Hanson, M. J. Shaffer, & L. Ma. (1999). *Root Zone Water Quality Model.* Highlands Ranch, CO: Water Resources Publications, LLC.

Arnold, J. G., P. M. Allen, & G. A. Bernhardt. (1993, February). Comprehensive surface-groundwater flow model. *Journal of Hydrology,* 142(1–4), 47–69.

Bingner, R. L., F. D. Theurer, R. G. Cronshey, & D. W. Darden. (2001). *AGNPS 2001*. Web site. Online at http://www.sedlab.olemiss.edu/

Commons, G. G. (1942). Flood hydrographs. *Civil Engineering, 12*, 571–572.

Dodge, J. C. I. (1959). A general theory of the unit hydrograph. *Journal of Geophysical Research, 64*, 241–256.

Goodrich, D. C., & D. A. Woolhiser. (1991, April). Catchment hydrology. *Review of Geophysics Supplement.*

Gray, D. M., ed. (1973). *Handbook on the Principles of Hydrology*. Port Washington, NY: Water Information Center.

Green, W. H., & G. A. Ampt. (1911). Studies in soil physics I, the flow of air and water through soils. *Journal of Agricultural Science, 4*, 1–24.

Haan, C. T., B. J. Barfield, & J. C. Hayes. (1994). *Design Hydrology and Sedimentology for Small Catchment.* New York: Academic Press.

Hillel, D. (1998). *Environmental Soil Physics*. New York: Academic Press.

Horton, R. E. (1940). An approach toward a physical interpretation of infiltration capacity. *Soil Science Society of American Proceedings, 5,* 399–417.

Knisel, W. G., ed. (1980). *CREAMS: A Field-Scale Model for Chemicals, Runoff, and Erosion from Agricultural Management Systems*. Cons. Res. Rep. No. 26. Washington, DC: USDA—Science and Education Administration.

Kostiakov, A. N. (1932). On the dynamics of the coefficient of water percolation in soils and the necessity of studying it from a dynamic view for the purposes of amelioration. *Transactions of the 6th Congress of International Society of Soil Science,* Russian part A;17–21. Reference from W. A. Jury, W. R. Gardner, & W. H. Gardner. (1991). *Soil Physics,* 5th ed. New York: Wiley.

Larson, C. L., & B. M. Reich. (1973). Relationship of observed rainfall and runoff recurrence intervals. In E. F. Schulz et al., eds., *Flood and Droughts. Proceedings of 2nd International Symposium in Hydrology,* September 1972. Ft. Collins, CO: Water Research Publications.

Linsley, R. K., M. A. Kohler, & J. L. H. Paulhus. (1982). *Hydrology for Engineers,* 3rd ed. New York: McGraw-Hill.

McCool, D. K., C. D. Pankuk, K. E. Saxton, & P. K. Kalita. (2000). Winter runoff and erosion on northwestern USA cropland. *International Journal of Sediment Research, 15*(2), 149–161.

McCuen, R. H. (1989). *Hydrologic Analysis and Design*. Englewood Cliffs, NJ: Prentice Hall.

Metcalf and Eddy, Inc., & University of Florida, and Water Resources Engineers, Inc. (1971). *Storm Water Management Model*. Vol. I, *Final Report,* pp. 5-28 11024DOC07/71 (NTIS PB-203289). Washington, DC: U.S. Environmental Protection Agency.

Nearing, M. A., G. R. Foster, L. J. Lane, & S. C. Finkner. (1989). A process-based soil erosion model for USDA-water erosion prediction project technology. *ASAE Transactions, 32,* 1587–1593.

Philip, J. R. (1957). The theory of infiltration. 4. Sorptivity and algebraic infiltration equations. *Soil Science, 84,* 257–178.

Rawls, W. J., & D. L. Brakensiek. (1989). Estimation of Soil Water Retention and Hydraulic Properties. In H. J. Morel-Seytoux, ed., *Unsaturated Flow in Hydrologic Modeling, Theory, and Practice.* NATO Series C, Mathematical and Physical Science, Vol. 275, pp. 275–300. London: Kluwer Academic.

Rawls, W. J., D. L. Brakensiek, & K. E. Saxton. (1982). Estimation of soil water properties. *ASAE Transactions, 25,* 1316–1320, 1328.

Reich, B. M. (1962, July). *Design Hydrographs for Very Small Watersheds from Rainfall*. Civil Engineering Section. Ft. Collins, CO: Colorado State University.

Sherman, L. K. (1932). Stream flow from rainfall by unit graph method. *Engineering News-Record, 108,* 501–505.

Skaggs, R. W. (1982). Field evaluation of a water management simulation model. *ASAE Transactions, 25,* 666–674.

U.S. Soil Conservation Service (SCS). (1972). Hydrology. In *National Engineering Handbook,* Sect. 4. Washington, DC: U.S. Government Printing Office.

——— .(1983). *Computer Program for Project Formulation—Hydrology*. Tech. Release 20. Washington, DC: U.S. Department of Agriculture.

——— .(1986). *Hydrology for Small Watersheds*. Tech. Release 55. Springfield, VA: National Technical Information Service.

——— .(1990). *Engineering Field Manual,* Chap. 2. (Litho.). Washington, DC: USDA-SCS.

Problems

5.1 Assuming that the Horton infiltration (Equation 5.4) is valid, determine the constant infiltration rate if f_0 = 50 mm/h, f at 10 min is 13 mm/h, and k = 12.9 h^{-1}. What is the infiltration rate at 20 min? Determine the infiltrated depth after 2 hours of wetting.

5.2 Estimate the runoff volume in mm of depth from 80 mm of rainfall on a 90-ha watershed during the growing season. Assume Antecedent Rainfall Condition II, 50 ha of row crop contoured with terraces and poor condition, soil group C, and 40 ha of alfalfa, straight and good condition, soil group B.

5.3 Estimate the runoff volume in mm, m^3, and ha-m for the watershed and conditions in Problem 5.2, except that a 60-mm storm occurred 3 days before the 80-mm rainfall.

5.4 Estimate the runoff volume in mm for the watershed and conditions in Problem 5.2, except that no rainfall occurred for 6 days before the 80-mm rainfall.

5.5 Estimate the runoff volume in mm for a 6-h, 50-yr return period storm at your location. Assume the watershed conditions in Problem 5.2.

5.6 Estimate the runoff volume in mm of depth from 80-mm rainfall on a 90-ha watershed during the growing season. Assume Antecedent Rainfall Condition II, 50 ha of houses on 0.1-ha lots, soil group B, and 40 ha of businesses, soil group B.

5.7 Estimate the peak runoff rate from a 150-ha watershed from 100-mm, 24-h storm that produced 8-mm depth of runoff. Assume Antecedent Rainfall Condition II, CN = 70, flow length = 2000 m, average gradient = 0.01 m/m, and 1 percent pond and swamp.

5.8 Estimate the design peak runoff rate for a 6-h, 25-yr storm. The watershed consists of 60 ha, one-third in good permanent meadow with no grazing and the remainder in row crops on the contour in good condition. The hydrologic soil group for the area under meadow is B and that for the row crop is C. The watershed is near St. Louis, Missouri. The maximum length of flow of water is 1000 m, and the fall along this path is 5.0 m.

5.9 If the watershed in Problem 5.8 is in western Tennessee at 36°N 89°W, estimate peak runoff rate. If the return period was decreased to 2 years, what would be the rate of runoff?

5.10 Estimate the runoff volume and peak runoff rate from a 6-h, 25-y storm on a 200-ha watershed near Benson, Arizona. The watershed has desert shrub in fair condition, and the soil is hydrologic group C. Assume an average slope of 0.02 m/m, a 15-mm rainfall occurred 3 days prior to this event, and the maximum length of flow is 3000 m.

5.11 Assuming the dimensionless hydrograph is applicable, determine the duration of flow if the peak runoff is 4.0 m^3/s and the volume of runoff is 50 mm for a 50-year storm from an 85-ha watershed. What are the coordinates for point n on the design hydrograph?

5.12 By frequency analysis methods discussed in Chapter 3, determine the estimated maximum annual discharge for return periods of 5 and 100 years from the following 18 years of maximum annual floods (1979–1996) from a gauged watershed; 50, 92, 108, 1, 0.8, 1.5, 20, 36, 0.5, 0.3, 56, 38, 2, 8, 14, 5, 0.4, and 3 mm. Is the length of record adequate for these estimates to be reliable 90 percent of the time?

5.13 The flow rates from a 1-h rainfall excess of 40-mm in an agricultural watershed were 0, 3, 8, 15, 7, 5, 3, 1, and 0 m^3/s at 0, 1, 2, 3, 4, 5, 6, 7, and 8 hr, respectively. By the unit hydrograph method, construct the design runoff hydrograph for the watershed for a total excess rainfall of 90 mm that fell in the following order: 30 mm in the first hour, 50 mm in the second hour, and 10 mm in the third hour.

5.14 Construct a 2-h unit hydrograph from the data in Problem 5.13.

CHAPTER

6 Open Channel Flow

Open channel flow occurs when a free water surface in a channel is at atmospheric pressure. Common examples of open channel flow are rivers, streams, drainage ditches, and irrigation canals. Open channel flow may also occur in pipes if the pipe is not flowing full and the water surface is at atmospheric pressure.

Open channel design is common in many applications of soil and water conservation, including drainage and irrigation ditches, grassed waterways, reservoir spillways, and large culverts. In all these applications, the designer must consider channel shape, slope, hydraulic roughness or resistance to flow, and in many cases, channel resistance to erosion.

Channels may be earth or concrete lined, vegetated, lined with impervious material such as rubber or fabric, or lined with erosion-resistant material such as large rocks or high-strength geotextile materials. Channels may be left in a natural condition, shaped to achieve a desired capacity, or designed to minimize bed erosion. In some cases, channels may be confined by vertical sides made from materials that are resistant to erosion or sloughing. A properly designed, earth-lined open channel should provide (1) velocity of flow such that neither serious scouring nor sedimentation will result, (2) sufficient capacity to carry the design flow, (3) hydraulic grade at the proper depth for good water management, (4) sideslopes that are stable, and (5) minimum initial and maintenance costs. Additional requirements must be met for carrying irrigation water, such as low seepage loss. Details on open channel flow and design can be found in Chow (1959), French (1985), and Henderson (1966).

States of Flow

The states of flow vary with viscous, inertial, and gravitational forces. The Reynolds and Froude numbers define the states of flow.

6.1 Reynolds Number

The state of flow in open channels may be laminar or turbulent. In laminar flow, viscous forces predominate over inertial forces. In turbulent flow, inertial forces predominate over viscous forces. The state of flow is determined by the Reynolds number, which is the ratio of the inertial forces and viscous forces, and for open channels is given by

$$R_e = \frac{\text{inertial forces}}{\text{viscous forces}} = \frac{\rho v R}{\mu} \tag{6.1}$$

where R_e = Reynolds number,
ρ = fluid density (M/L^3),
v = velocity (L/T)
R = hydraulic radius (L),
μ - dynamic viscosity ($M\,L^{-1}\,T^{-1}$).

The hydraulic radius is the cross-sectional area of the flow divided by the channel wetted perimeter P (Figure 6-1). The wetted perimeter is the length of the channel cross section in contact with water.

$$R = \frac{\text{cross-sectional area of flow}}{\text{wetted perimeter of the channel}} = \frac{A}{P} \tag{6.2}$$

Note that the Reynolds number uses the hydraulic radius for open channel flow, whereas the diameter is used for pipe flow. Reynolds number less than 500 are laminar flow and numbers over 1000 are turbulent flow in channels. A transition range exists between 500 to 1000 where the state of flow is uncertain; however, flows in the transition region are not common in engineering applications.

6.2 Froude Number

The state of flow in channels may be subcritical, critical, or supercritical depending on the ratio of inertial to gravitational forces. The determination is made by the Froude number F_r, which is

$$F_r = \frac{\text{inertial forces}}{\text{gravitational forces}} = \frac{v}{(gy)^{1/2}} \tag{6.3}$$

where g = the gravitational acceleration (L/T^2),
y = the depth of flow (L).

The Froude number describes the following states of flow

$F_r < 1$ the flow is subcritical (quiet),
$F_r = 1$ the flow is critical,
$F_r > 1$ the flow is supercritical (rapid).

Most flow in open channels is subcritical. The most common exceptions are spillways and flow over weirs and through measuring flumes where the flow is critical.

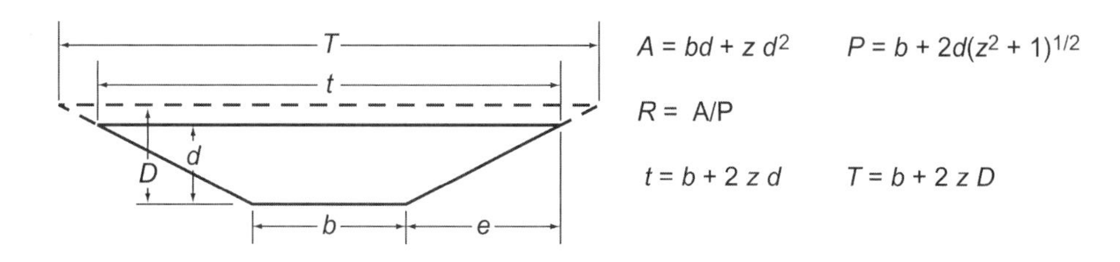

(*a*) Trapezoidal cross section, $z = e/d$

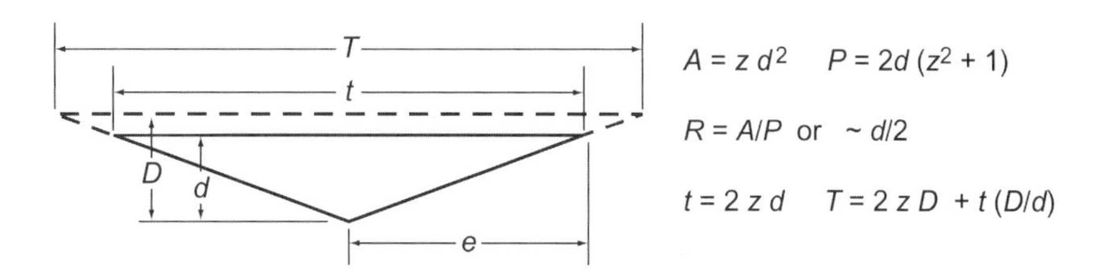

(*b*) Triangular cross section, $z = e/d$

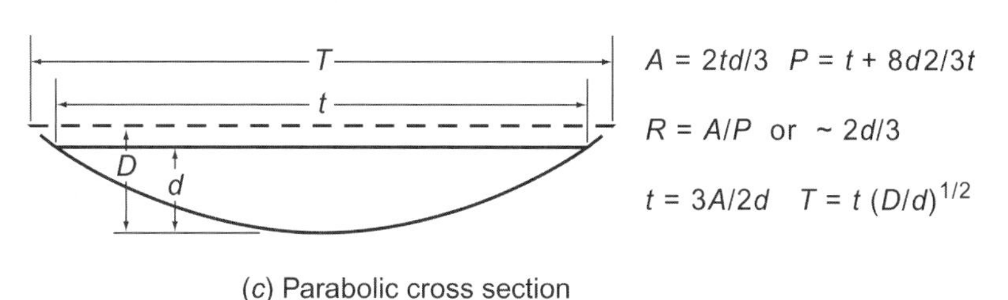

(*c*) Parabolic cross section

Figure 6–1
Channel cross section, wetted perimeter, hydraulic radius, and top width formulas for trapezoidal, triangular, and parabolic shapes.

Equations of Flow

6.3 Continuity Equation

The continuity equation is based on the conservation of mass and in its simplest form for an incompressible fluid is given by

$$Q = Av \tag{6.4}$$

where Q is the flow rate (L^3/T), A is the area of flow perpendicular to the average velocity (L^2), and v is the average velocity (L/T).

6.4 The Bernoulli Equation

The Bernoulli equation is based on the conservation of energy and for open channels is described by

$$y_1 + z_1 + \frac{v_1^2}{2g} = y_2 + z_2 + \frac{v_2^2}{2g} + (h_f)_{1-2} \tag{6.5}$$

where y = depth of flow above the channel bottom (L),
z = distance of channel bottom above a datum (L),
v = average flow velocity (L/T),
h_f = head loss due to friction (L),
$1, 2$ = subscripts denoting two channel locations.

Figure 6-2 illustrates these variables for uniform flow in an open channel.

6.5 Specific Energy

A given discharge of water in an open conduit may flow at two depths (called alternate depths) having the same total energy head. For convenience in design of control structures, specific energy is often used rather than total energy (Henderson, 1966). Specific energy (or specific head) is the energy (or head) of flow with respect to the channel bottom. It disregards elevation potential, since elevation changes are negligible in short reaches of mildly sloped conduits. The specific energy of flow in a channel having a rectangular cross section may be expressed as

$$H_e = y + \frac{v^2}{2g} = y + \frac{q^2}{2a^2g} = y + \frac{q^2}{2b^2y^2g} \tag{6.6}$$

where H_e = specific energy (L),
y = depth of flow (L),
v = average velocity of flow (L/T),
g = gravitational acceleration (L/T^2),
q = discharge (L^3/T),
a = cross-sectional area of flow (L^2),
b = width of flow (L).

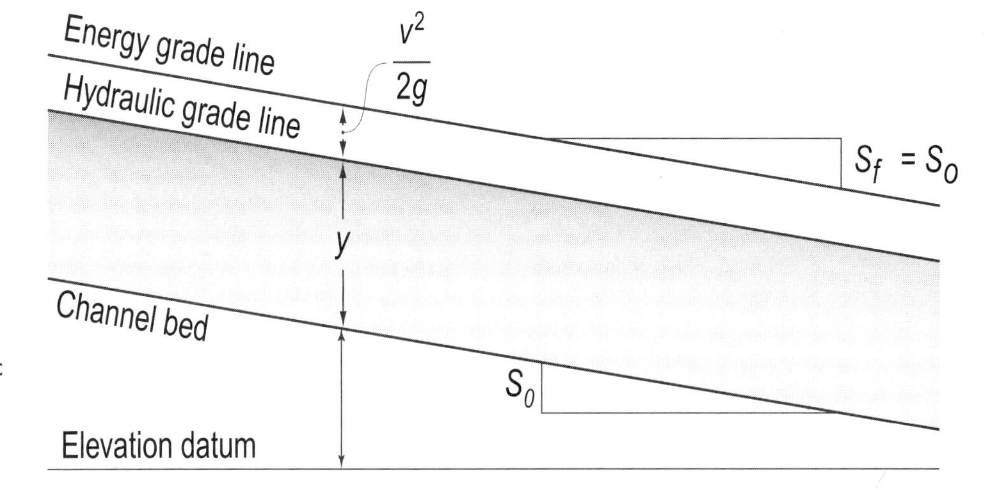

Figure 6–2
Steady uniform flow in an open channel with bottom slope, hydraulic grade line, and energy grade line with the same slopes.

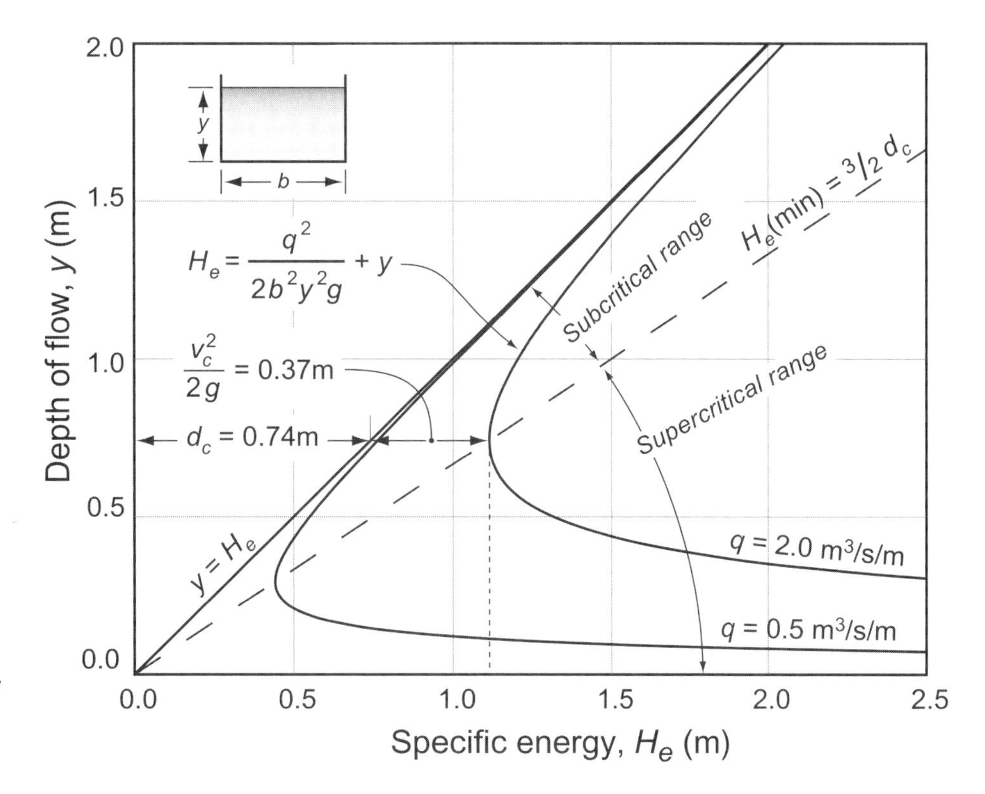

Figure 6–3
Depth of flow and specific energy for two flow rates in a rectangular channel.

Figure 6-3 shows the relationship of depth of flow to specific energy in a rectangular channel for two discharges. For any discharge, specific energy has a minimum value that corresponds to a unique depth of flow, which is called the critical depth. Most control structures include a section at which flow passes through the critical depth. Being a unique value, it is useful for determination of discharge capacities and is therefore frequently used in the design equations for certain components of control structures.

Example 6.1

Determine the specific energy of a channel with $y = 0.5$ m, $b = 1.0$, and $q = 0.7$ m^3/s.

Solution. Use Equation 6.6 to calculate H_e.

$$H_e = y + \frac{q^2}{2b^2y^2g} = 0.5 + \frac{0.7^2}{2 \times 1.0^2 \times 0.5^2 \times 9.81} = 0.6 \text{ m}$$

6.6 Critical Depth

To determine the critical depth, differentiate Equation 6.6 with respect to y and set $dH_e/dy = 0$. The value for y where H_e is a minimum is the critical depth, d_c.

$$\frac{dH_e}{dy} = 1 - \frac{q^2}{b^2y^3g} = 0$$

$$d_c = \left(\frac{q^2}{b^2g}\right)^{1/3} \quad \textbf{6.7}$$

Equation 6.6 may also be used to determine the depth at which maximum discharge will occur for a given specific energy. Solve for q in Equation 6.6.

$$q = [2y^2b^2g(H_e - y)]^{1/2} \quad \textbf{6.8}$$

Assume H_e is constant and set $dq/dy = 0$ to get

$$\frac{dq}{dy} = \frac{2b^2g(2yH_e - 3y^2)}{2[2y^2b^2g(H_e - y)]^{1/2}} = 0 \quad \textbf{6.9}$$

If $y = d_c$, then solving Equation 6.9 gives

$$d_c = \frac{2}{3}H_e \quad \textbf{6.10}$$

For a given specific energy, the maximum discharge occurs at the critical depth, that is, where $H_e = (3/2)y$. Equation 6.8 is plotted in Figure 6-4 for $H_e = 1.0$ m. The critical depth is 0.67 m, which corresponds to the maximum discharge of 1.83 m^3/s per meter of channel width. The plot shows how the discharge drops off for any depth greater or less than the critical depth.

6.7 Hydraulic Jump as an Energy Dissipater

A change in flow depth from a depth less than the critical depth (supercritical) to a depth greater than the critical depth (subcritical) is called a hydraulic jump. Where the change in depth is great and the transition occurs very quickly, the jump is called a direct jump. Direct jumps are very turbulent and dissipate an appreciable fraction of the energy of the flow (Chow, 1959). The profile and depth–energy relationships of a hydraulic jump are shown in Figure 6-5. Note that the specific energy of flow after the jump is less than the specific energy of the flow entering the jump. Many control structures are designed so that a hydraulic jump forms within the structure, dissipating some of the energy of flow and reducing the flow velocity to a nonerosive value before exiting the downstream portion of the structure. The incoming depth of flow in Figure 6-5 must be in the supercritical range for a hydraulic jump to occur.

Types of Flow

The flow in open channels can be categorized according to the type of flow. The simplest case is steady, uniform flow (Figure 6-2). Uniform flow occurs in channels

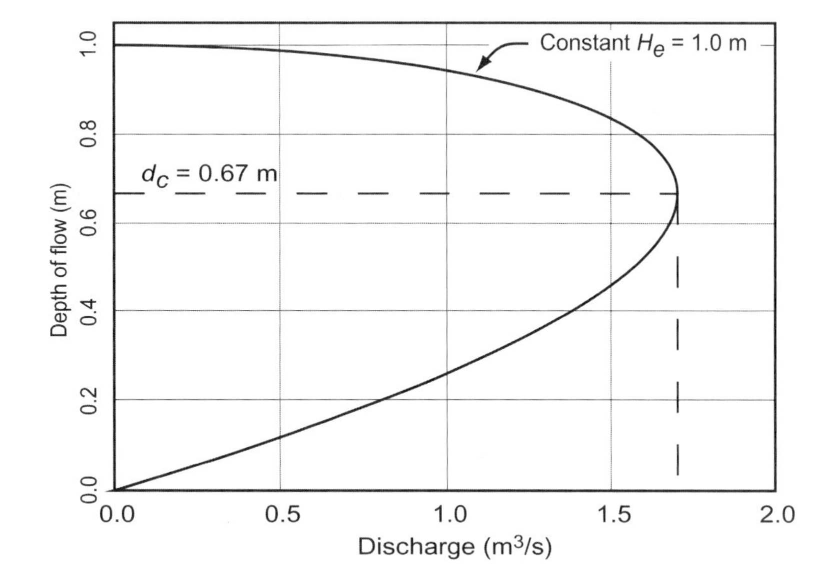

Figure 6–4

Flow rate and depth of flow in a rectangular channel with a constant H_e of 1.0 m.

with uniform cross section, slope, flow depth, and flow rate. The flow is varied if any of the channel characteristics change along the channel length. Gradually varied flow occurs when the change in characteristics with length is small (Figure 6-6). Rapidly varied flow occurs when the change with length is rapid, such as at a sharp change in slope, over a weir, or a free overfall. With unsteady flow the depth and/or flow rate vary with time along the channel. Unsteady flow is nearly always nonuniform and much more difficult to analyze. Fortunately, the majority of varied flows are steady and uniform in a piece-wise manner. This chapter emphasizes steady, uniform flow.

Uniform Flow

In uniform flow the slope of the energy grade line, the water surface slope and the bottom slope are equal. Uniform flow occurs in long channels of uniform shape, constant flow rate, constant slope, and no obstructions (Figure 6-2). The depth of flow in such channels is known as the normal depth. If the normal depth is greater than the critical depth, the flow is subcritical ($F_r < 1$). If the normal flow depth is less than the critical depth, the flow is supercritical ($F_r > 1$).

Special equations have been developed for uniform flow conditions. These equations simplify the design and analysis of channels with uniform flow.

6.8 The Manning Equation

The Manning equation is widely used because of its simplicity and accuracy and is expressed as

$$v = \frac{C_u R^{2/3} S_f^{1/2}}{n} \qquad \textbf{6.11}$$

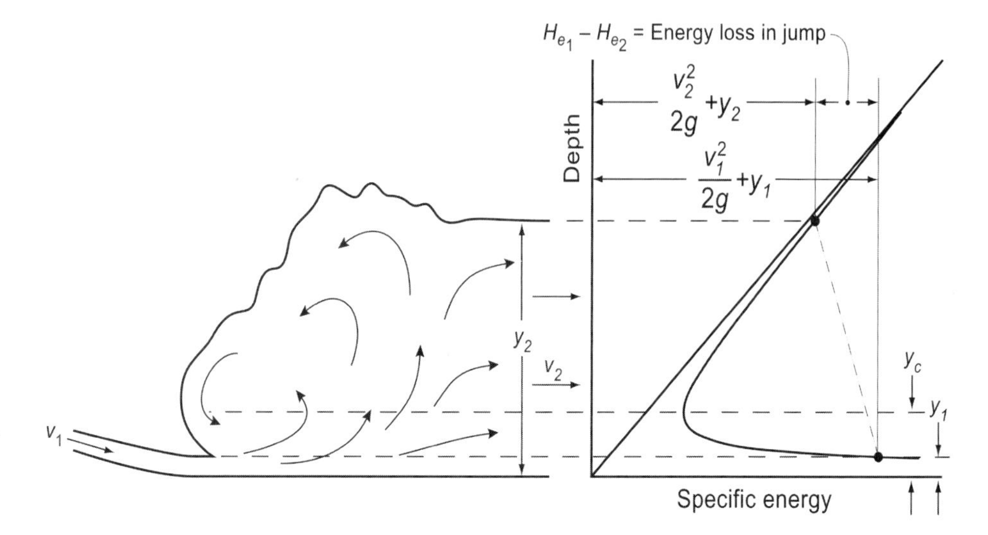

Figure 6–5

Flow depths and energy relationships in a hydraulic jump.

where v = average flow velocity (L/T),

C_u = 1.0 $m^{1/2}$/s for units in meters and seconds,

R = hydraulic radius (flow area divided by wetted perimeter) (L) (see Figure 6-1 for equations),

S_f = friction slope (L/L),

n = Manning roughness coefficient, including soil and vegetative effects ($m^{1/6}$) (see Appendix B for recommended values).

The Manning equation is also combined with the continuity equation (Equation 6.6) as follows

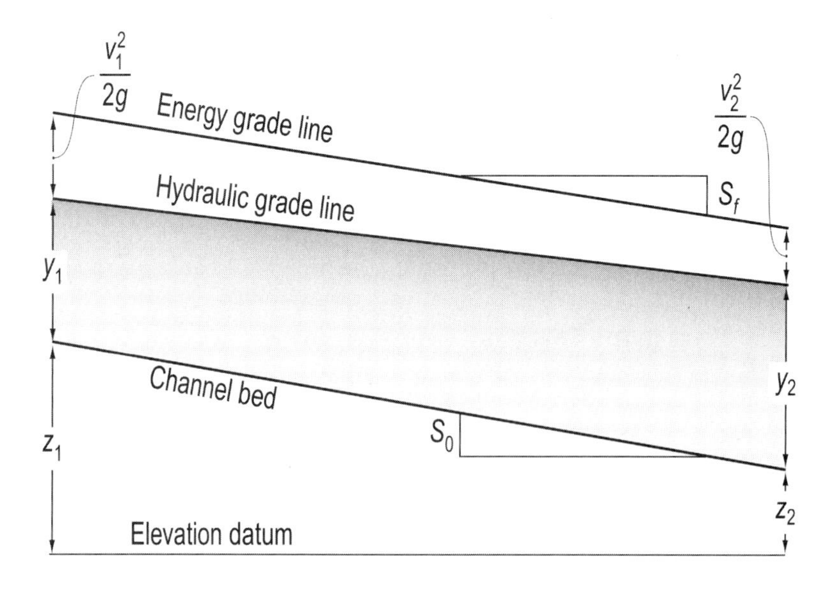

Figure 6–6

Gradually varied flow in an open channel with bottom slope, hydraulic grade line, and energy grade line with different slopes.

$$Q = \frac{C_u A^{5/3} S_f^{1/2}}{nP^{2/3}} \tag{6.12}$$

where Q is the flow rate (L/T^3).

Example 6.2 What is the normal flow depth in a rectangular concrete-lined channel, with $b = 2$ m wide, slope = 0.001 m/m, flow rate is 1 m^3/s? What is the Froude number? Is the flow subcritical, critical, or supercritical?

Solution. Assume a depth, calculate Q, and compare to the required discharge. Substitute the equations for A and P into Equation 6.12 and solve for y by iteration. From Appendix B, read n for concrete = 0.015 and take $C_u = 1$ for SI units.

$$Q = \frac{1}{n}\frac{(b\,y)^{5/3}}{(b + 2y)^{2/3}} S_f^{1/2} = \frac{1}{0.015}\frac{(2y)^{5/3}}{(2 + 2y)^{2/3}} 0.001^{1/2}$$

Assume $y = 0.3$ m

$$Q = \frac{1}{0.015}\frac{(2 \times 0.3)^{5/3}}{(2 + 2 \times 0.3)^{2/3}} 0.001^{1/2} = 0.48 \text{ m/s}$$

This is below the actual flow of 1.0 m^3/s. Increase y and iterate to find that $y = 0.496$ m gives $Q = 1.0$ m^3/s. The velocity is $1.0/(2 \times 0.496) = 1.01$ m/s. The Froude number is

$$F_r = \frac{v}{(g\,y)^{1/2}} = \frac{1.01}{(9.81 \times 0.496)^{1/2}} = 0.46$$

The normal depth is 0.496 m and the flow is subcritical because $F_r < 1$.

Channel Design

The major objective of open channel design is to provide a channel that has adequate capacity and is stable. Generally, one of the first steps in any channel design is a field survey to determine the limitations the grade places on the channel design. Wherever practical, the channel should be designed for a high hydraulic efficiency.

6.9 Channel Grade

The engineer frequently has little choice in the selection of grades for channels, because the grade is determined largely by the outlet elevation, elevation and distance to the beginning of the channel, and the channel depth. An allowance may be made for accumulation of sediment, depending on channel velocities and soil conditions. Structures may be required to limit grade, or specialized designs may be necessary to elevate the channel to maintain a uniform grade in uneven terrain.

In earth channels the stability of the soil places limitations on channel grade and sideslopes. The topography and desired water levels may limit the design grade and the velocity of flow.

6.10 Channel Roughness

The Manning roughness coefficient n for open channels varies with the height, density, and type of any vegetation; physical roughness of the bottom and sides of the channel; and variation in channel size and shape of the cross section, alignment, and hydraulic radius. Primarily because of differences in vegetation, the roughness coefficient may vary from season to season. In general, conditions that increase turbulence increase the roughness coefficient. Values of the Manning roughness n can be obtained from Appendix B or hydraulics books.

Experience and knowledge of local conditions are helpful in the selection of the roughness coefficient, but it is one of the most difficult values to define in channel design. In general, a value of 0.035 is satisfactory for medium-sized earth ditches with bottom widths of 1.5 to 3 m, and 0.04 is suitable for smaller bottom widths.

6.11 Channel Cross Sections

Natural channels tend to be parabolic in cross section. Constructed channels may be rectangular, trapezoidal, triangular, or parabolic. Under the normal action of channel flow, deposition, and bank erosion, unlined trapezoidal and triangular channels tend to become parabolic. The geometric characteristics of the trapezoidal, triangular, and parabolic channels are given in Figure 6-1. The figure defines the cross sections and presents the formulas to calculate area, wetted perimeter, hydraulic radius, and top width for each of the three.

Earth channels and lined canals are normally designed with trapezoidal cross sections (Figure 6-1). Waterways and streams are frequently designed with parabolic cross sections. The size of the cross section will vary with the velocity and quantity of water to be removed. Designers are often only able to vary the sideslopes and bottom width of a channel to meet the capacity requirements.

Channel dimensions are generally determined by an iterative analysis with a programmable calculator, spreadsheet, computer equation solver, or the NRCS (NRCS, 2002) computer programs to determine channel dimensions. The depth of the constructed channel should provide a freeboard allowance ($D - d$ in Figure 6-1). Freeboard increases the depth to provide a safety factor to prevent overtopping and to allow a reasonable depth for sediment accumulation. Freeboard is typically 20 percent of the total depth, D. On large earthen irrigation canals the freeboard generally varies from 30 to 35 percent, and for lined canals it ranges from 15 to 20 percent. The freeboard for lined canals refers only to the top of the lining. The total freeboard is the same for lined and unlined channels. The freeboard should be increased on the outside edge of curves.

Sideslopes. Channel sideslopes are determined principally by soil texture and stability. The most critical condition for sloughing occurs after a rapid drop in the flow level that leaves the banks saturated, causing a positive pore water pressure that makes the banks highly unstable. For the same sideslopes, the deeper the ditch the more likely it is to slough. Sideslopes should be designed to suit soil conditions and not the limitations of construction equipment. Suggested sideslopes are shown in Table 6-1. Whenever possible, these slopes should be verified by experience and local practices. Narrow ditches should have slightly flatter sideslopes than wider ditches because of greater reduction in capacity in the narrow ditches if sloughing occurs.

Bottom Width. The bottom width for the most hydraulic-efficient cross section and minimum volume of excavation can be computed from the formula

$$b = \frac{2d}{z + \sqrt{z^2 + 1}} \qquad \textbf{6.13}$$

where b = bottom width (L),
d = design depth (L),
z = sideslope ratio (L/L) (horizontal/vertical expressed as z:1).

For any sideslope it can be shown mathematically that, for a bottom width computed from Equation 6.13, the hydraulic radius is equal to one half the depth. The minimum bottom width of larger channels should be 1.2 m. It is not always possible to design for the most efficient cross section because of construction equipment limitations, site restrictions, allowable velocities, or increased maintenance.

Example 6.3

Determine the most hydraulically efficient cross section for a channel with Q = 4.4 m³/s, S_o = 0.2 percent, z = 2:1, and n = 0.02.

Solution. Perform an iterative solution where b is a function of d using Equations 6-12 and 6-13, and Figure 6-1. Assume an initial flow depth of 1.5 m with the variables in Figure 6-1.

Estimate	*Equation 6.13*	*Figure 6-1*	*Figure 6-1*	*Equation 6.12*	
d	**b**	**A**	**R**	**Q**	***Comment***
1.5	0.71	5.56	0.75	10.27	Q too high
1.1	0.52	2.99	0.55	4.49	Q satisfactory

The most efficient cross section has a bottom width less than 1.2 m, which may be difficult to construct with some machinery.

6.12 Channel Velocity and Tractive Force

In earth or vegetated channels, a channel shape is generally selected that will minimize the risk of channel erosion. The desired shape may be based on maximum velocities or on the shear strength of the channel material.

TABLE 6–1 Maximum Recommended Sideslopes for Open Channels

Material	*Sideslope (Horizontal/Vertical)*
Solid rock, cut section	0.25:1
Loose rock or cemented gravel, cut section	0.75:1
Heavy clay, cut section	1:1
Heavy clay, fill section	2:1
Heavy clay in CH classification	4:1
Sand or silt with clay binder, cut or fill section	1.5:1
Loam	2:1
Peak, muck, and sand	1:1
Silts and sands with high water table	3.5:1

Source: NRCS (2001).

When designing for maximum velocity, no definite optimum velocity can be prescribed, but minimum and maximum limits can be approximated. Velocities should be low enough to prevent scour but high enough to prevent sedimentation. Usually an average velocity of 0.5 to 1.0 m/s for shallow channels is sufficient to prevent sedimentation. In channels that flow intermittently, vegetation may retard the flow to such an extent that adequate velocities at low discharges are difficult to maintain, if not impossible. For this reason maximum or even higher velocities may be desired, since scouring for short periods at high flows may aid in maintaining the required cross section.

A typical velocity distribution in an open channel is shown in Figure 6-7. The maximum velocity occurs near the center of the channel and slightly below the surface. The average velocity is generally about 70 percent of the maximum velocity near the surface. The maximum velocity and the velocity contours move toward the outer bank on channel bends.

Limiting Velocity Method. Maximum recommended velocities for earth-lined channels are presented in Table 6-2. These velocities may be exceeded for some soils where the flow contains sediment because deposition may produce a channel bed resistant to erosion. Where an abrasive material, such as coarse sand, is carried in the water, these velocities should be reduced by 0.15 m/s. For depths over 0.9 m, velocities can be increased by 0.15 m/s. If the channel is winding or curved, the limiting velocity should be reduced about 25 percent. Limiting velocities for grassed waterway designs are a specialized case and are discussed in Chapter 8.

Tractive Force Method. Tractive force is based on the hydraulic shear stress from the flowing water on the periphery of the channel. The method can be applied to earthen channels, or used to evaluate various channel liners (MoDOT, 2003). The critical tractive force is the maximum hydraulic shear that the channel lining can withstand before eroding. In a uniform channel of constant slope and constant flow, the water flow is steady state and uniform (without acceleration) because the force tending to prevent particle motion is equal to the force causing motion. For small channel slopes, hydraulic shear stress can be calculated from

$$\tau = \rho g R S_f \quad (6.14)$$

TABLE 6–2 Critical Shear and Velocity for Selected Channel Lining Materials

Category	*Critical Shear (Pa)*	*Maximum Velocity (m/s)*
Soils		
Fine sand (colloidal)	1.2	0.46
Sandy loam	1.7	0.53
Alluvial silt (noncolloidal)	2.3	0.61
Alluvial silt (colloidal)	12.5	1.14
Silty loam (noncolloidal)	2.3	0.61
Firm loam	3.6	0.76
Fine gravels	3.6	0.76
Stiff clay	12.5	1.16
Graded loam to cobbles	18.2	1.14
Graded silts to cobbles	20.6	1.22
Shales and hardpan	32.1	1.83
Gravel/Cobble		
25-mm	15.8	1.14
50-mm	32.1	1.37
150-mm	95.8	1.75
300-mm	191.5	2.67
Vegetation		
Long native grasses	70	1.52
Short native and bunchgrasses	40	1.07
Degradable Linings		
Jute net	22	0.53
Straw with net	80	0.61
Coconut fiber with net	110	1.07
Fiberglass roving	96	1.45
Soil Bioengineering		
Live fascine	104	2.1
Willow stakes	125	2.0
Hard Surfacing		
Gabions	480	5.0
Concrete	600	5.5

Source: Fischenich (2001).

where τ = hydraulic shear stress (Pa),
ρ = density of water (1000 kg/m^3),
g = gravitational acceleration (9.81 m/s^2),
R = hydraulic radius (Equation 6.3) (m),
S_f = hydraulic gradient, generally assumed to be the channel slope (m/m).

When designing channels with the tractive force method, the channel is designed to ensure that the tractive force does not exceed a critical shear value for the bed of the channel. Critical shear depends on lining material properties. For cohesive soils, it has been correlated with the plasticity index, dispersion ratio, particle size, clay content, and void ratio. In noncohesive coarse soils, it approaches a value approximately equal to the diameter of the bed material in millimeters. For example, a gravel channel with bed material about 10 mm dia. has an estimated critical tractive force of about 10 Pa. Typical critical shear values for natural and artificial channel linings are presented in Table 6-2. Local design practices and critical shears for channel lining materials should be followed if they are available.

Example 6.4

Using the same conditions as described in Example 6.3, use the Tractive Force method to design an open channel with an 8-mm-dia. gravel lining to carry 4.4 m^3/s with a slope of 0.2 percent. Assume a trapezoidal cross section with sideslopes of 2:1 and $n = 0.020$.

Solution. For 8-mm-dia. gravel, the critical shear value can be estimated as 8 Pa. Solving Equation 6.14 for the hydraulic radius where the critical shear is 8 Pa gives

$$R = \frac{8.0}{1000 \times 9.8 \times 0.002} = 0.41 \text{ m}$$

Set up a spreadsheet or equation solver for an iterative solution. Starting with the solution to Example 6.2, decrease the depth and increase the width until a solution is obtained that provides the desired flow rate for the least width, to minimize excavation.

d	*b*	*A*	*R*	*v*	*Q*	*Comment*
1.1	0.52	2.99	0.55	1.50	4.49	*R* too high in Example 6.2
0.46	7.0	3.64	0.40	1.22	4.44	*R* and *Q* satisfactory

The depth of 0.46 m and bottom width of 7.0 m meet the design requirements. The velocity of 1.22 m/s is also within the range of small gravel as recommended in Table 6-2.

Gradually Varied Flow

Thus far, a straight, uniform-sloped channel has been assumed. For simple design problems the depth of flow and the velocity can be assumed constant. For this uniform flow condition the energy grade line, hydraulic grade line, and the channel bottom are all parallel (Figure 6-2). For gradually varied flow the channel characteristics change slowly with distance and the bottom slope, hydraulic grade line, and energy grade line are not parallel (Figure 6-8).

For gradually varied flow, special design considerations are needed. For example, if an obstruction, such as a culvert or a weir, is placed in the channel with subcritical flow, a concave water surface called a "backwater curve" develops upstream of the structure. A change from a gentle to a flatter grade could produce the same

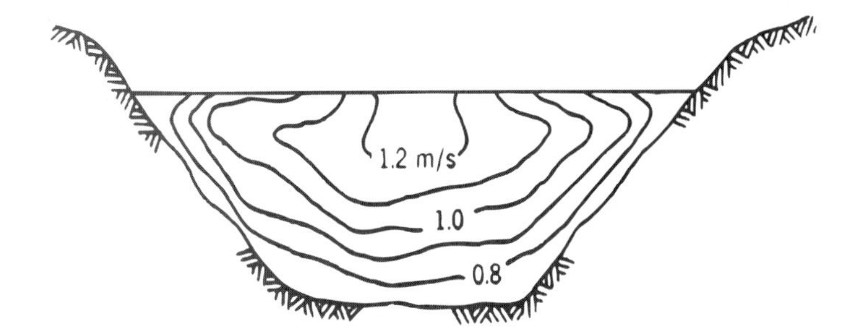

Figure 6–7
Velocity distribution in an open channel.

effect. If the channel slope changed from a flat to a steep grade, a convex water surface would develop. These examples illustrate two common types of gradually varied flow conditions.

The equation for gradually varied flow in a rectangular channel can be obtained from the following form of the Bernoulli equation (Henderson, 1966)

$$H = z + y + \frac{v^2}{2g} \tag{6.15}$$

where H is the total energy. Differentiating Equation 6.15 with respect to distance x and writing in finite difference form yields

$$\frac{\Delta H}{\Delta x} = \frac{\Delta z}{\Delta x} + \frac{\Delta}{\Delta x}\left(y + \frac{v^2}{2g}\right) = \frac{\Delta z}{\Delta x} + \frac{1}{\Delta x}\left(y_2 + \frac{v_2^2}{2g} - y_1 - \frac{v_1^2}{2g}\right) \tag{6.16}$$

Substituting $\frac{\Delta H}{\Delta x} = -S_f$ and $\frac{\Delta z}{\Delta x} = -S_o$ into Equation 6.16 and solving for Δx gives the equation for gradually varied flow

$$\Delta x = \frac{\left(y_2 + \frac{v_2^2}{2g}\right) - \left(y_1 + \frac{v_1^2}{2g}\right)}{S_o - S_f} = \frac{H_{e2} - H_{e1}}{S_o - S_f} \tag{6.17}$$

where S_o is the bottom slope of the channel, S_f is the friction slope calculated from the Manning equation, assuming uniform flow over the channel segments, and the specific energy is determined from Equation 6.6.

6.13 Backwater Curve

Equation 6.17 can be numerically solved to determine the shape of the water surface for gradually varied flow in channels. The most common example is a backwater curve formed when a flow restriction (Figure 6-8) is placed in channel with a mild slope $F_r < 1$. The solution process for subcritical flow begins at the obstacle and proceeds upstream.

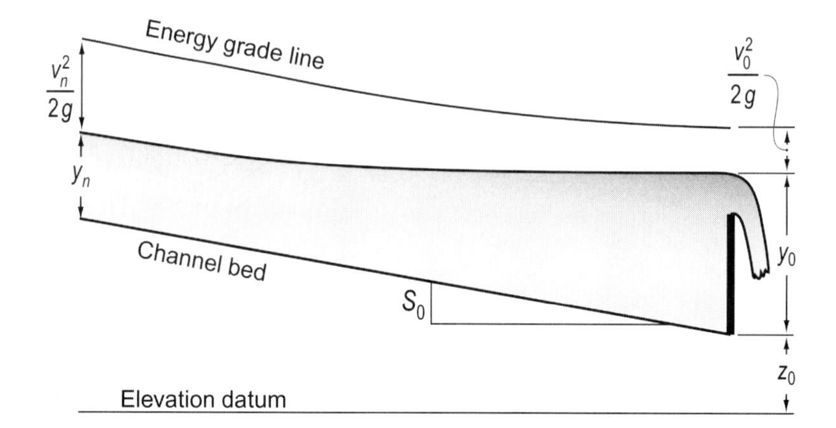

Figure 6–8
Example of the backwater curve caused by a weir in a rectangular channel.

Example 6.5

Assume a weir is to be placed in the channel in Example 6.2. If the weir will raise the flow depth 0.1 m, how far upstream will the flow depth be 0.5 m?

Solution. (1) Calculate area, wetted perimeter, and velocity of flow at $y = 0.596$.

$$A = by = 2.0 \times 0.596 = 1.192 \text{ m}^2$$

$$P = b + 2y = 2.0 + 2 \times 0.596 = 3.192 \text{ m}$$

$$v = Q/A = 1.0/1.192 = 0.8389 \text{ m/s}$$

(2) Calculate S_f from Equation 6.12.

$$S_f = \left(\frac{QnP^{2/3}}{A^{5/3}}\right)^2 = \left(\frac{1.0 \times 0.015 \times 3.192^{2/3}}{1.192^{5/3}}\right)^2 = 0.000589$$

(3) Determine H_e from Equation 6.6.

$$H_e = y + \frac{v^2}{2g} = 0.596 + \frac{0.8389^2}{2 \times 9.81} = 0.6319 \text{ m}$$

(4) Reduce y by 0.01 m to 0.586 m and repeat steps 1, 2, and 3 to obtain

$$A = 1.171 \text{ m}^2,\ P = 3.172 \text{ m},\ v = 0.8532 \text{ m/s},\ S_f = 0.000618,\ H_e = 0.6231 \text{ m}$$

(5) Δx is now determined from Equation 6.17.

$$\Delta x = \frac{0.6231 - 0.6319}{0.001 - (0.000589 + 0.000618)/2} = -22.2 \text{ m}$$

Note that S_f is taken as the mean over the interval.

(6) Results from a spreadsheet for these calculations are

y (m)	*A* (m²)	*P (m)*	*v (m/s)*	S_f	H_e (m)	Δx (m)	*x (m)*
0.596	1.192	3.192	0.8389	0.000589	0.6319		0
0.586	1.172	3.172	0.8532	0.000618	0.6231	−22.2	−22
0.576	1.152	3.152	0.8681	0.000649	0.6144	−23.7	−46
0.566	1.132	3.132	0.8834	0.000682	0.6058	−25.8	−72
0.556	1.112	3.112	0.8993	0.000718	0.5972	−28.5	−100
0.546	1.092	3.092	0.9158	0.000756	0.5887	−32.2	−133
0.536	1.072	3.072	0.9328	0.000798	0.5804	−37.6	−170
0.526	1.052	3.052	0.9506	0.000842	0.5721	−46.1	−216
0.516	1.032	3.032	0.9690	0.000890	0.5639	−61.2	−277
0.506	1.012	3.012	0.9881	0.000941	0.5558	−95.8	−373
0.5	1	3	1.0000	0.000974	0.5510	−112.8	−486

The depth reaches 0.5 m at a distance of 486 m upstream from the weir. This demonstrates that in channels with small slopes, a partial obstruction will affect the flow depth for large distances upstream.

Water Measurement

Good water management requires that flow rates and volumes be quantified. The most common measurements are flow velocity and stage. These are used in conjunction with knowledge of the shape of a channel or impoundment to calculate the flow rate or storage volume. Determination of a volume of flow requires a series of flow rate measurements that are integrated over the time of interest.

6.14 Units of Measurement

For volume flow rates, units in common use are liters per second or cubic meters per second. Units for volume are liters, cubic meters, or particularly for irrigation purposes, hectare-millimeters or hectare-meters. Flow velocity is typically in meters per second.

6.15 Float Method

A crude estimate of the velocity of water in an open channel can be made by determining the velocity of an object floating with the current. Select and mark a reasonably straight and uniform channel section about 100 meters long. Measure the time required for an object floating on the surface to travel the selected distance and calculate its velocity. Repeat this several times and calculate an average surface

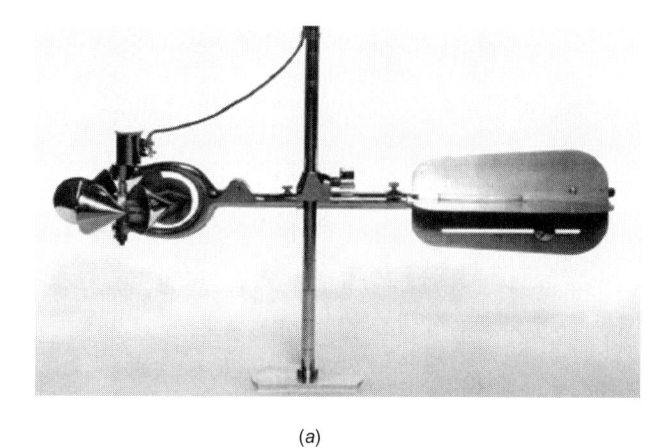

(a)

(b)

Figure 6–9
Current meters: (a) The Price type AA has a vertical axis with cups and (b) Dumas meter has an impeller on a horizontal axis. Hoff and Ott meters are similar. (USBR, 2001)

velocity. The mean velocity of the stream can be estimated at 0.8 to 0.9 times the average surface velocity.

6.16 Impeller Meters

Many instruments employ an impeller that rotates at speeds proportional to the velocity of the water. The relationship between rotational speed and velocity of water is usually not linear, so a calibration curve must be employed. Most modern impeller meters integrate this calibration into the electronic readout signal. Some impeller meters are designed for use in open channels (often called *current meters*), whereas others are designed for installation in pipes. Impeller meters function well in clear or sediment-laden water, but are not suitable for applications with large suspended solids or fibrous materials that could jam or tangle in the mechanism. Figure 6-9 shows examples of current meters.

Current meters measure the flow velocity at a specific point in the stream. It is not possible, however, to predetermine a single point where a measurement would represent the average velocity of the stream. To obtain a good estimate of the total

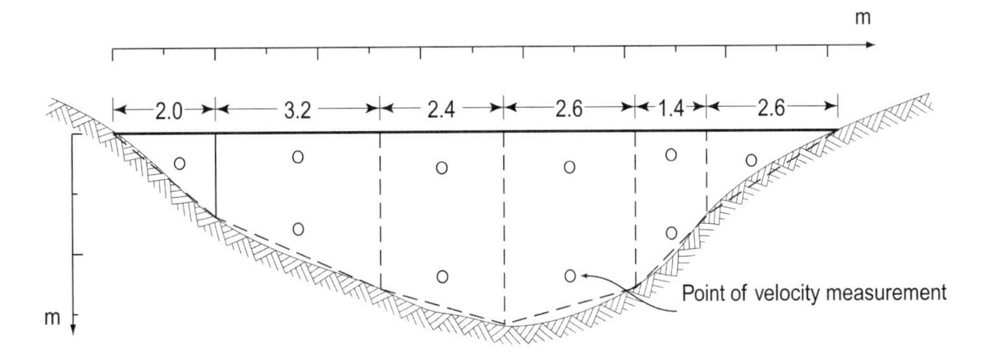

Figure 6–10
Cross section of a stream with subdivisions for current meter measurements.

stream flow, the cross section can be divided into a number of subsections in which measurements are taken to represent the velocity in each subsection (Figure 6-10). The average of readings taken at 0.2 and 0.8 times the depth provides a good estimate of the average velocity in a deep subsection. Since impeller-type current meters should not be used within 2–3 impeller diameters of the stream boundary (USBR, 2001), where the stream is too shallow to allow the reading at 0.8 times the depth, the velocity at 0.6 times the depth from the water surface may be taken as the average velocity. Summing the products of average velocity and cross-sectional area for each subsection gives the total stream flow. This method is demonstrated below.

Example 6.5

Determine the discharge for the stream shown in Figure 6-10. The measured velocities are given in the following table.

Solution. Begin by dividing the cross section into sections that can be approximated by triangles or trapezoids, as shown by the dashed lines in the figure. For each section, find its width and average depth. For a shallow section, calculate the measurement distance from the water surface as 0.6 times the average depth for that section. For deeper sections, calculate the measurement depths as 0.2 and 0.8 times the average depth of each section. Measure the velocities at those points. Then find the average velocity for each of the deeper sections as the average of the two measurements. Finally, multiply the area of each section by its average velocity and add them to obtain total discharge. The measurements and calculations for the stream are given in tabular form below.

Section	*Width (m)*	*Avg. Depth (m)*	*Meas. Depths (m)*[a]	*Measured Velocity (m/s)*	*Average Velocity (m/s)*	*Area (m²)*	*Discharge (m³/s)*
1	2.0	0.9	0.54	0.09	0.09	1.80	0.162
			0.40	0.29			
2	3.2	2.0			0.27	6.40	1.728
			1.60	0.25			
			0.60	0.33			
3	2.4	3.0			0.30	7.20	2.160
			2.40	0.27			

Continue

Section	Width (m)	Avg. Depth (m)	Meas. Depths (m)[a]	Measured Velocity (m/s)	Average Velocity (m/s)	Area (m^2)	Discharge (m^3/s)
			0.62	0.32			
4	2.6	3.1			0.30	8.06	2.418
			2.48	0.28			
			0.42	0.28			
5	1.4	2.1			0.27	2.94	0.794
			1.68	0.26			
6	2.6	0.7	0.42	0.11	0.11	1.82	0.200
Total							**7.46**

[a]Distance from the water surface where velocities were measured.

The number of sections to use for a particular stream is a matter of judgment. Where cross sections have more complex shapes, use more sections.

A variation of the impeller meter is the paddle-wheel meter. A small paddle wheel with a shield that protects half of it from exposure to the flow is positioned inside a pipe. The speed of rotation is proportional to the flow velocity.

6.17 Doppler Meters

Doppler meters emit acoustic signals into a stream and measure the apparent frequencies of the reflections. Although the concept is simple, the actual workings are not, and will not be detailed here. Doppler meters require something in the water to reflect the signal, such as suspended solids or bubbles. Some are designed for use in open water as current meters. Others are intended for use in well-defined sections such as culverts. The meter lies on the bottom of the channel, has an integrated pressure sensor for flow depth measurement, and can be programmed to calculate discharges directly. Such a configuration is not suitable for a stream with a high sediment load that could bury the meter.

6.18 Magnetic Meters

Magnetic meters are based on the principle of Faraday's law, that the voltage induced in a conductor moving at right angles to a magnetic field is proportional to the velocity of the conductor. Since natural waters contain some dissolved salts, they are conductive and work with this method. The moving water (conductor) causes a distortion in an induced magnetic field that is sensed by the meter and converted to a velocity readout.

Magnetic meters are used with pipe flow. Various models are available that either clamp onto the outside of the pipe or are fabricated as short in-line sections.

6.19 Pitot Meters

Pitot tubes are used in a variety of meters for flow velocity. Pitot tubes are probes with openings positioned to sense the velocity head of the flow. Modern pitot meters couple pressure transducers with microprocessors to sense the differential pressures and calculate flow velocities and/or discharges.

6.20 Slope Area

The principles of open channel flow can be used to estimate velocity and discharge. The reach of channel should be uniform in cross section and, if possible, up to 300 meters long. Given data for the shape of the cross section, the gradient, and the channel roughness, the Manning equation can be applied. This method gives relatively crude estimates with natural channels because of the difficulty of selecting an appropriate value for the roughness coefficient.

6.21 Orifice Meters

Orifice meters are usually circular openings in a plate placed perpendicular to the flow stream to measure the flow rate. The orifice flow equation is based on the conservation of energy with an orifice discharge coefficient accounting for the head loss through the orifice and the velocity of the upstream flow. The equation for flow through a sharp-edged orifice is

$$Q = CA(2gh)^{1/2} \tag{6.18}$$

where Q = flow rate (L^3/T),
C = discharge coefficient,
A = area of the orifice (L^2),
g = acceleration due to gravity (L/T^2),
h = head on the orifice (L).

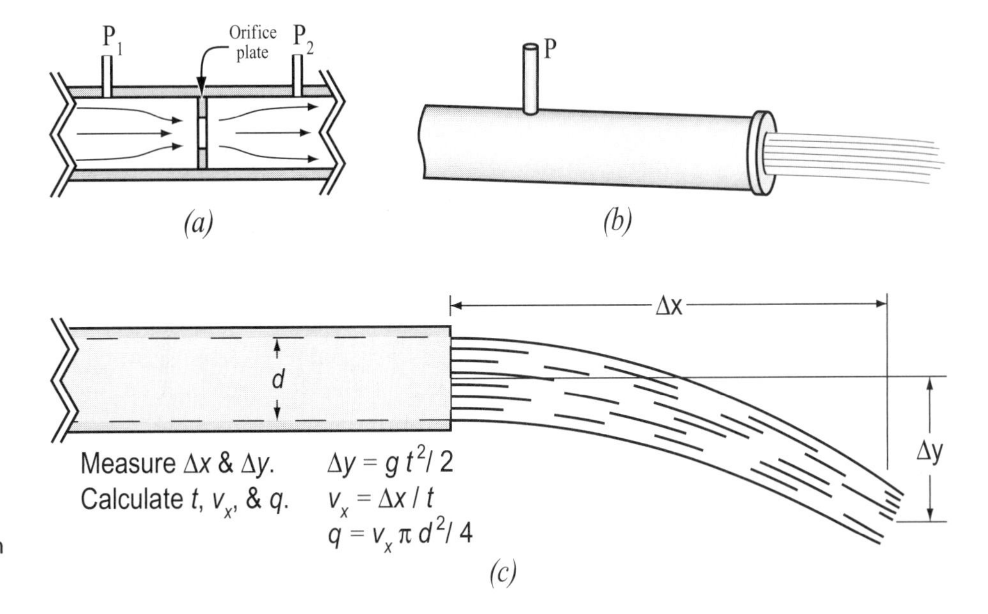

Figure 6–11
(a) Orifice plate,
(b) end-cap orifice, and
(c) coordinate method for flow measurement in pipes.

The value of C is 0.61 to 0.63 if the orifice area is small relative to the upstream flow area. The head is measured from the center of the orifice to the upstream water surface, or the head difference if the orifice outlet is submerged. As the orifice size becomes larger relative to the upstream flow area, the value of C increases.

If an orifice plate is installed in a pipe as shown in Figure 6-11a, the differential head measurements (upstream one pipe diameter and downstream one half the pipe diameter) are used to calculate the flow rate. If an orifice is configured as an end-cap on a horizontal pipe (Figure 6-11b), a head measurement taken about 0.6 m upstream can be used to calculate the discharge. The orifice diameter should be 0.5–0.8 times the pipe diameter. The orifice coefficient must be determined for each configuration and can be obtained from hydraulic handbooks.

6.22 Coordinate Method

The coordinate method can be used where water discharges freely from the end of a horizontal pipe. Measurements of the distance the stream has fallen at a given distance from the end of the pipe are used to calculate the exit velocity (Figure 6-11c). Discharge is estimated as the exit velocity times the cross-sectional area of the pipe.

6.23 Level Sensors

Many applications require measurement of water levels or depths. Methods range from something as simple as a staff gauge to ultrasonic-level sensors. There are more methods available than can be discussed here. The best choice of sensor will depend on the application, data requirements, power available at the site, temperature, consequences of failure, etc. A few methods for sensing water levels are mentioned below.

Float. A number of devices use a float to ride the water surface, providing a mechanical link to some type of position sensor. Floats linked to pens that traced lines on moving paper charts were the standard for much of the 1900s. Floats have also been used to carry small magnets that activate reed switches or otherwise influence accessory circuits.

Capacitance. The dielectric constant of water is significantly different from that of air. As a water level rises and falls around a probe, it changes the capacitance of the probe, which can be detected and converted to a level readout.

Pressure transducer. Pressure transducers can be used in a variety of ways. The simplest is to submerge the transducer with its sensing element at a known position. A pressure signal can be easily converted to a depth of water above the sensor. Indirect linkage can be provided in a bubbler arrangement. A gas-filled tube is connected to the pressure transducer and positioned with the open end at a known elevation. As long as the water does not intrude into the tube, the pressure at the transducer will be the same as the pressure at the end of the tube. That value can be converted to a depth readout, as above.

A wide range of pressure transducers is available. Unit costs have declined as production methods improved to the point that many are both inexpensive and robust. Pressure transducers are probably the most widely used method for level sensing, especially for research or feedback/control applications.

Ultrasonic. In some applications (e.g., a tank or a manhole), an ultrasonic unit can be positioned some distance above the water surface. An acoustic signal is emitted and the time of return of the reflected signal is used to calculate the distance to the water surface. Ultrasonic sensors require a fairly clean and quiet surface. Foam or floating debris can absorb or scatter the signal.

6.24 Weirs and Flumes

For open channel flow measurements, the best accuracy requires structures having known hydraulic behavior. Weirs and flumes cause water to pass through a critical depth, which is unique for a given structure and flow rate.

Weirs. Weirs are the simplest, least expensive, and most common devices used for flow measurements in open channels (Figure 6-12). A weir is some type of barrier placed in the channel that restricts the flow, causing it to flow through a well-defined section. The most common are sharp-crested weirs, which are very thin in the direction of flow so that the flow will spring free of the weir crest. Weirs are further classified according to the shape of the opening or notch and whether the weir is contracted, i.e., narrower than the channel. Rectangular-notch weirs are most common. For accurate measurement of small flows, V-notch weirs are best. Trapezoidal notches provide a combination of the characteristics of the rectangular and V-notch designs. A number of more exotic shapes have been developed to provide specific stage-discharge relationships. For design details, rating curves, etc., see a hydraulics reference such as Brater et al. (1996).

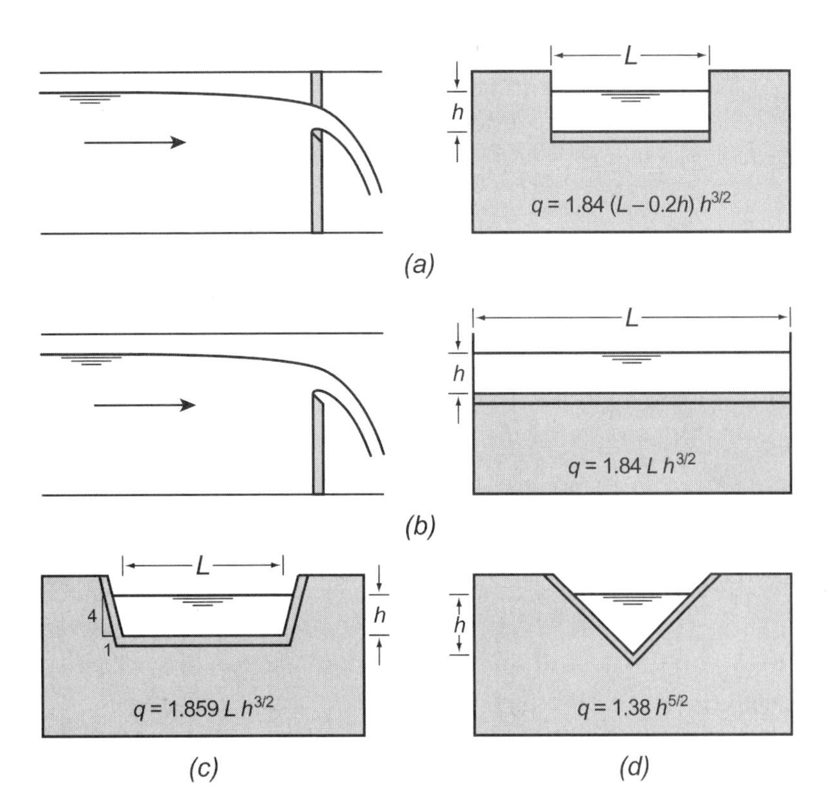

Figure 6–12
Common sharp-crested weirs and their flow equations for h and L in m and q in m^3/s: (a) contracted rectangular, (b) suppressed rectangular, (c) Cipolletti, and (d) triangular or V-notch.

Weirs can be designed for a wide range of flow rates. Equations for discharge for several common weirs are shown in Figure 6-12 for lengths in m and flow rate in m^3/s. For rectangular and trapezoidal weirs, discharge is proportional to $h^{3/2}$. For V-notch weirs, the exponent on h is 2.5. The weir coefficient C depends on the specific weir configuration. (To obtain SI coefficients from English values, multiply by 0.552.)

Head measurements should be taken upstream of the weir, at a distance at least 3–4 times the maximum height of flow over the weir. A limitation on the application of weirs is that they require significant upstream–downstream differences in water levels to work properly. Common problems with weirs are the tendencies for sediment deposition in front of the weir and for debris to snag on the weir crest. A detailed presentation on weirs can be found in USBR (2001).

Flumes. Flumes are stable, specially shaped channel sections that cause flow to pass through well-defined configurations. Compared to weirs, they pass debris much more easily and do not require as much head change. Parshall (1950) developed a common measuring flume (Figure 6-13). Designs are available for throat widths from 25 mm to 3.66 m, with flow rates from 0 to 7.67 m^3/s. Design specifications and discharge relationships may be found in Parshall (1950) or hydraulics handbooks. Various flow calculators are available on the Internet.

Simpler flumes that are easier to construct have been developed. These include the cutthroat flume described by Skogerboe (1973) and the long-throated flume designed by Replogle (Clemmens et al., 2001) shown in Figure 6-14. The Bureau of Reclamation recommends the use of long-throated flumes for their many advantages (USBR, 2001) and distributes the design software WinFlume via the Internet

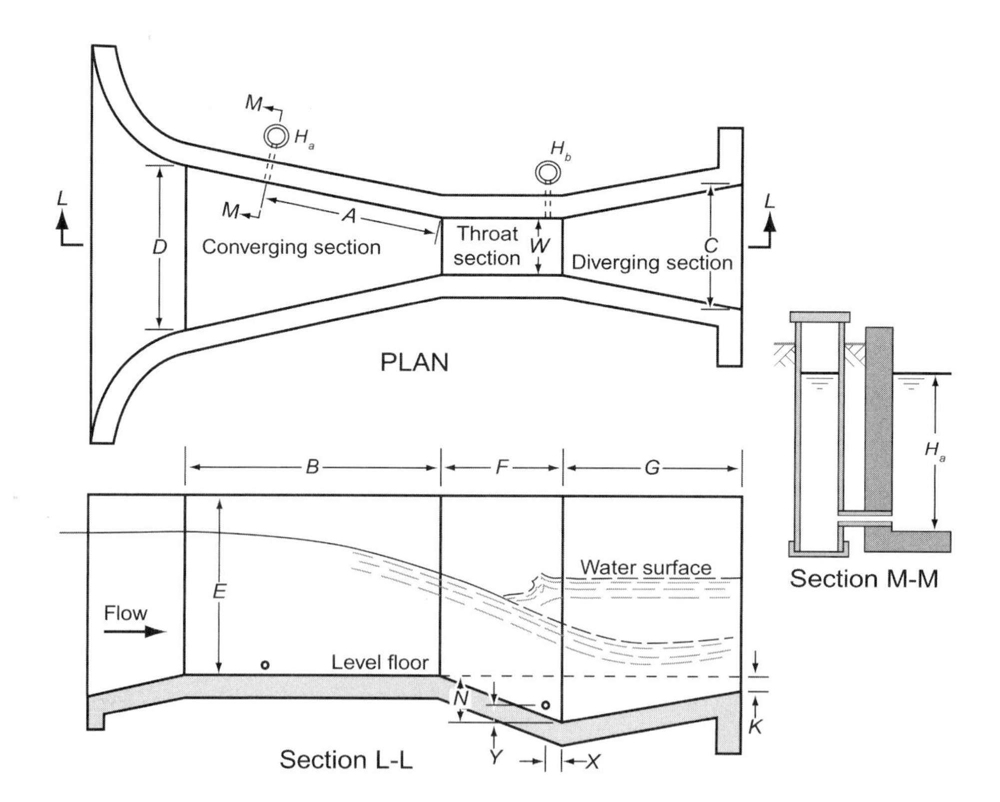

Figure 6–13
Parshall measuring flume. (Redrawn from Parshall, 1950)

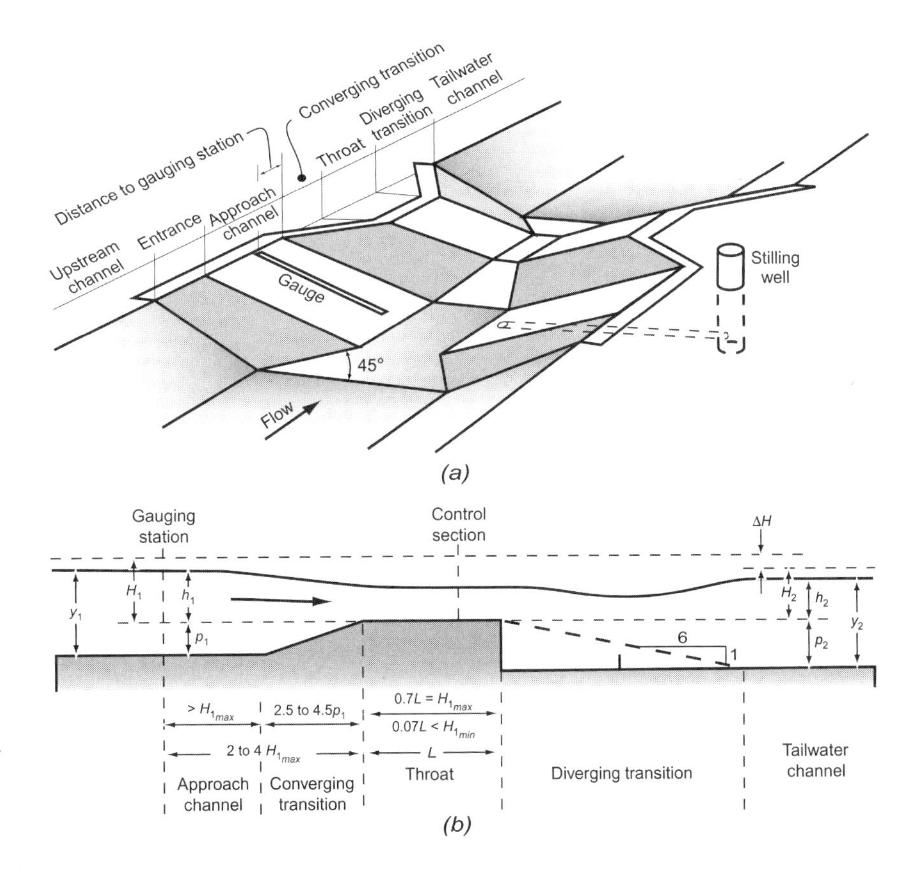

Figure 6–14

Long-throated flume in a trapezoidal channel: (a) perspective view, and (b) cross section. (Redrawn from Wahl, 2001)

(Wahl, 2001). For applications such as runoff measurement from small plots, the H-flume developed by the USDA Soil Conservation Service in the 1930s provides good accuracy at low flows as well as having high capacity. H-flumes (Figure 6-15) have flat bottoms and require very little head loss. They pass debris well.

There are three H-flume variants: HS for flows up to 0.023 m^3/s, H for flows up to 0.88 m^3/s, and HL for flows up to 3.3 m^3/s. As with weirs, the water level must be measured a short distance upstream from a broad-crested or H-flume. The discharge can then be determined from a rating table. Computer programs are also available for discharge calculation. Clemmens et al. (2001) provide a thorough treatment of the use of flumes and weirs.

Figure 6–15

Runoff from a no-till plot measured with an H-flume. (Photo by O. R. Jones, USDA-ARS)

Internet Resources

U. S. Army Waterways Experiment Station library
http://libweb.wes.army.mil

U. S. Bureau of Reclamation library
http://www.usbr.gov/main/library/

Source of pictures and Manning *n* values for channels
http://manningsn.sdsu.edu

References

Brater, E. F., H. W. King, J. E. Lindell, & C. Y. Wei. (1996). *Handbook of Hydraulics*, 7th ed. New York: McGraw-Hill.

Clemmens, A. J., T. L. Wahl, M. G. Bos, & J. A. Replogle. (2001). *Water Measurement with Flumes and Weirs*. Wageningen, The Netherlands: International Institute for Land Reclamation and Improvement.

Chow, V. T. (1959). *Open-Channel Hydraulics*. New York: McGraw-Hill.

Fischenich, C. (2001). *Stability thresholds for stream restoration materials*. U.S. Army Corps of Engineers. Online at http://www.wes.army.mil. Accessed January 2004.

French, R. H. (1985). *Open-Channel Hydraulics*. New York: McGraw-Hill.

Henderson, F. M. (1966). *Open Channel Flow*. New York: Macmillan.

Lane, E. W. (1955). Design of stable channels. *Transactions of the ASCE, 120*, 1234–1260.

Missouri Department of Transport (MoDOT). (2003). Chapter 9, *Hydraulics and drainage*, Section 9-04. Online at http://www.modot.state.mo.us. Accessed January 2004.

Natural Resource Conservation Service (NRCS). (2001). Part 650: Engineering Field Handbook. *National Engineering Handbook*, Chapter 14. Online at http://www.info.usda.gov/CED/ftp/CED/EFH-Ch14.pdf. Accessed January 2004.

———. (2002). Ohio Engineering Software. Online at http://www.zert.info. Accessed December 2003.

Parshall, R. L. (1950). *Measuring Water in Irrigation Channels with Parshall Flumes and Small Weirs*. USDA Agriculture Cirular 843. Washington, DC: USDA.

Skogerboe, G. V. (1973). *Selection and Installation of Cutthroat Flumes for Measuring Irrigation and Drainage Water*. Colorado State University Engineering Experiment Station Technical Bulletin 120.

U.S. Bureau of Reclamation (USBR). (1987). *Design of Small Dams*, 3rd ed. Washington, DC: U.S. Government Printing Office.

———. (2001). *Water Measurement Manual*, 3rd ed., rev. Denver, CO: U.S. Government Printing Office.

Wahl, T. L. (2001). *WinFlume User's Manual*. U.S. Department of the Interior, Bureau of Reclamation. Online at http://www.usbr.gov/wrrl/winflume. Accessed January 2004.

Problems

6.1 What is the flow rate in a concrete-lined trapezoidal channel if the depth of flow is 0.5 m, slope is 0.002 m/m, bottom width is 1 m, and sideslope is 1:1? Is the flow subcritical?

6.2 What is the normal depth of flow in the channel in Problem 6.1 if the flow rate is 0.5 m^3/s?

6.3 Determine the velocity of flow in a parabolic, a triangular, and a trapezoidal waterway, all having a cross-sectional area of 2 m^2, a depth of flow of 0.3 m, a channel slope of

4 percent, and a roughness coefficient of 0.04. Assume 4:1 side slopes for the trapezoidal cross section.

6.4 Compute the most efficient bottom width for a trapezoidal channel ($z = 2$) in a silty loam soil with a flow depth of 0.75 m. What are the velocity and the channel capacity if the hydraulic gradient is 0.1 percent and $n = 0.04$? Is the velocity below the critical velocity for this channel lining?

6.5 Determine the specific energy for the channel in Problem 6.4. Is the flow subcritical?

6.6 If the critical shear on the bottom of a channel in cohesive soil is 10 Pa, what is the maximum average velocity of flow at maximum slope for a trapezoidal channel where the depth of flow is 1.2 m, bottom width 1.5 m, $n = 0.04$, and sideslopes 1:1?

6.7 A rectangular channel must carry 5 m^3/s of clear water. The slope of the channel is 1 percent. What width of channel is necessary to limit the shear to prevent channel erosion if the bottom is (a) noncolloidal silt and (b) fine gravel?

6.8 Specify a nonerodible channel lining for the channel in Example 6.3, using the tractive force method.

6.9 Determine the backwater curve for a concrete-lined rectangular channel 3 m wide with a flow rate of 1.5 m^3/s and bottom slope is 0.002 if an obstruction is added that raises the water depth by 0.1 m.

6.10 Flow depth and velocity measurements, as in Example 6.5 were repeated for several stages of flow as follows: 2.5 m, 4 m^3/s; 3.5 m, 10 m^3/s; 4.4 m, 25 m^3/s. Plot a stage-discharge curve for this channel. Explain the shape of the curve.

6.11 What is the flow in a channel if a contracted rectangular weir with a 1-m crest length is placed in the channel and has a head of 0.4 m?

6.12 Find the recommended dimensions for the height of the weir crest above the channel bottom and the width of the side contractions for the weir in Problem 6.11.

CHAPTER 7

Soil Erosion by Water

Erosion is one of the most important agricultural and natural resource management problems in the world. Water erosion is the detachment and transport of soil from the land by water, including runoff from melted snow and ice. It reduces soil productivity and is a primary source of sediment that pollutes streams and fills reservoirs. In the 1970s, soil erosion estimates in the United States were as high as 4 billion Mg of soil annually. This amount declined to about 2 billion Mg in 1997, mainly because of the increased use of conservation or Best Management Practices (BMPs), (Iivari & Kertis, 2001), and is unlikely to change greatly from that level.

Since the early 1970s, greater emphasis has been given to erosion as a contributor to nonpoint source pollution. In this chapter, "nonpoint source" refers to erosion from the land surface rather than from channels and gullies. Eroded sediment can carry nutrients, particularly phosphates, to waterways, and contribute to eutrophication of lakes and streams. Adsorbed pesticides and microorganisms are also carried with eroded sediments, adversely affecting surface water quality.

The two major types of erosion are geological erosion and erosion from human or animal activities. Geological erosion includes soil-forming as well as soil-eroding processes that maintain the soil in a favorable balance, suitable for the growth of most plants. Geological erosion is a process in the formation of soils and their distribution on the surface of the earth. This long-time eroding process caused most of the present topographic features, such as canyons, stream channels, and valleys. Disturbances due to human or animal influences can reduce vegetative cover and compact soil. Human disturbances include agricultural, mining, forestry, and construction activities. Following disturbances, runoff and soil erosion rates increase well above geological levels.

Erosion Processes

7.1 Factors Affecting Erosion by Water

The major factors affecting soil erosion are climate, soil properties, vegetation characteristics, and topography. These factors are not totally independent, as geology affects topography, which can influence climate. Climate and soils influence the vegetation, which may influence climate and snow accumulation and melt rates. Human disturbances, such as tillage and construction, and natural disturbances, such as severe weather or fire, dramatically increase erosion. Of these, the vegetation and some disturbances, and, to a lesser extent, the soil and topography can be managed to reduce erosion.

Climate. Climatic attributes affecting erosion are precipitation, temperature, wind, humidity, and solar radiation. Temperature and wind are most evident through their effects on evaporation and transpiration. However, wind also changes raindrop velocities and the angle of impact. Humidity and solar radiation are somewhat less directly involved in that they are associated with temperature and rate of soil water depletion.

The relationships among precipitation characteristics, runoff, and soil loss are complex. Rainfall amount, intensity, and energy all impact erosion rates. Studies on individual erosion processes have found that erosion due to raindrop splash and shallow overland flow varies with the rainfall intensity to a power varying from 1.56 to 2.09 (Watson & Laflen, 1986), with the power of 2 generally being accepted. Such erosion may also be affected by raindrop energy, drop diameter, and runoff rates, particularly on permeable soils. Concentrated flow erosion is a function of runoff rate, which depends on both rainfall intensity and soil infiltration rates. Gullying and channel erosion processes are also dominated by runoff rates.

Climate also affects erosion from snowmelt processes. Generally, snowmelt erosion rates are low compared to rainfall erosion, but rainfall events on melting snow, particularly if the soil is frozen or saturated beneath the snow, can result in high erosion rates in some areas. In climates where the majority of the precipitation is from snow, the amount of snow accumulation and the melt rate will dominate the hydrologic and erosion processes.

Soil. Physical properties of soil affect the infiltration capacity and the extent to which particles can be detached and transported. In general, soil detachability increases as the size of the soil particles or aggregates increase, and soil transportability increases with a decrease in the particle or aggregate size. That is, clay particles are more difficult to detach than sand, but clay is more easily transported. The properties that influence erosion include soil structure, texture, organic matter, water content, clay mineralogy, and density, as well as chemical and biological characteristics of the soil. In nonagricultural soils, the type and amount of vegetation or disturbance can also influence erodibility. No single soil characteristic or index has been identified as a satisfactory means of predicting erodibility.

Vegetation. The major effects of vegetation in reducing erosion are (1) interception of rainfall by absorbing the energy of the raindrops on plant canopy and surface residue, reducing surface sealing and runoff, (2) retardation of erosion by resisting erosive forces, (3) physical restraint of soil movement, (4) improvement of aggregation and porosity of the soil by roots and plant residue, (5) increased biological activity in the soil, and (6) transpiration, which decreases soil water, resulting in increased storage capacity and less runoff. These influences vary with vegetation characteristics, soil, and climate.

When crop residues are left in the field after harvest, or when natural vegetation senesces and drops leaves and needles, the potential for erosion is decreased. Residue and tillage management practices can have a dramatic effect on soil erosion, potentially greater than soil properties. Any practice that maintains a residue cover on the soil surface decreases erosion potential, whereas any practice that removes, burns, or buries vegetative residue increases the potential for soil erosion. Conventional tillage practices generally bury all residue. Reduced or minimum tillage practices, with fewer operations or less aggressive machines, leave more residue on the soil surface, decreasing runoff and erosion rates. The residue itself will decay with time, ranging from under a year for residues such as soybeans, several years for herbaceous residue, to several decades for wood and woody debris.

Topography. Topographic features that influence erosion are slope length and steepness, shape (including concave, uniform, or convex) and size and shape of the watershed. On steep slopes, soil is more easily detached and transported downslope. On longer slopes, an increased accumulation of overland flow tends to increase concentrated flow erosion (Section 7.5). Concave slopes, with flatter slopes at the foot of the hill, deliver less sediment than convex slopes.

Disturbances. Disturbances can be natural, such as prolonged periods of wet weather or particularly severe storms, or human induced, such as construction or tillage. In many climates, erosion occurs as the result of only a few critical runoff events each decade, which overshadow year-to-year small but chronic erosion. A field may appear stable for a number of years, only to have a single erosion event cause severe gullying once during a landowner's lifetime.

The undisturbed rangeland and forests that covered North America before development had extremely low erosion rates. Fires were the greatest natural disturbances in these ecosystems. A severe fire leading to significant erosion may have occurred only once per century, or even less frequently. Such a fire may not necessarily be followed by a wet year, so the impact of the fire could be minimal. In contrast, agricultural practices may disturb the soil every year, so that it is always susceptible to the occasional severe erosion events whenever they occur.

7.2 Raindrop Erosion

Raindrop erosion is soil detachment and transport resulting from the impact of water drops directly on soil particles or on thin water surfaces. Tremendous quantities of soil are splashed into the air, most particles more than once. The impact of raindrops may not splash soil on streams greater than two- to three-drop diameters in depth, but raindrops increase turbulence, providing a greater sediment-carrying capacity. The relationship between erosion, rainfall momentum, and energy is determined by raindrop mass, size distribution, shape, velocity, and direction.

Factors affecting the direction and distance of soil splash are slope, wind, surface condition, and impediments to splash such as vegetative cover and mulches. On sloping land, the splash moves farther downhill than uphill, not only because the soil particles travel farther, but also because the angle of impact causes the splash reaction to be in a downhill direction. Components of wind velocity up or down the slope also affect soil movement by splash. Surface roughness and impediments to splash tend to counteract the effects of slope and winds. Contour furrows and ridges break up the slope and cause more of the soil to be splashed uphill. If raindrops fall on crop residue or growing plants, the energy is dissipated and thus soil splash is reduced. Raindrop impact on bare soil not only causes splash but also breaks down soil aggregates and causes deterioration of soil structure. This leads to increased surface sealing, runoff, and erosion.

7.3 Sheet Erosion

Sheet erosion is considered to be a uniform removal of soil in thin layers from sloping land, resulting from sheet or overland flow. This type of erosion rarely occurs because minute channels (rills) form almost simultaneously with the first detachment and movement of soil particles. The constant meander and change of position of these rills may obscure their presence from normal observation, hence establishing the false concept of sheet erosion.

The beating action of raindrops combined with surface flow causes initial rilling. Raindrops detach the soil particles, and the detached sediment can reduce the infiltration rate by sealing the soil pores. The eroding and transporting ability of overland flow depends on the rainfall intensity, infiltration rate, slope steepness, soil properties, and vegetative cover.

7.4 Interrill Erosion

Splash and sheet erosion are sometimes combined and called interrill erosion. Research has shown interrill erosion to be a function of soil properties, rainfall intensity, runoff rate, and slope. The relationships among these parameters can be expressed as (Flanagan & Nearing, 1995)

$$D_i = K_i\, i\, q\, S_f\, C_v \qquad \textbf{7.1}$$

where D_i = interrill erosion rate (kg m^{-2} s^{-1}),
K_i = interrill erodibility of soil (kg s m^{-4}) (Table 7-1),
i = rainfall intensity (m/s),
q = runoff rate (m/s),
S_f = interrill slope factor = $1.05 - 0.85e^{-4\sin\theta}$,
C_v = cover adjustment factor, including effects of soil surface cover, canopy cover, root characteristics and/or sealing and crusting; values range from near zero to 1.0,
θ = interrill slope angle.

TABLE 7–1 Typical RUSLE *K*-Factors and WEPP Erodibility Values for a Number of Cropland, Rangeland, and Forest Soils

State	*Soil*	*Texture*	*USLE K*[a]	K_i[b]	K_r[a]	τ_c[a]
Croplands						
CA	Academy	l	0.057	4.15E+06	0.0057	1.6
CA	Los Baños	c	0.026	3.64E+06	0.0011	2.9
CA	Whitney	sa l	0.032	1.83E+06	0.0233	4.7
GA	Bonifay	sa	0.008	5.84E+06	0.0179	1.0
GA	Cecil	sa l	0.033	3.94E+06	0.0084	2.2
GA	Hiwassee	sa l	0.022	3.55E+06	0.0103	2.3
GA	Tifton	l sa	0.018	5.28E+05	0.0113	3.5
IA	Clarion	l	0.032	2.37E+06	0.0046	0.4
IA	Monona	si l	0.065	4.76E+06	0.0076	2.8

State	*Soil*	*Texture*	*USLE K*[a]	K_i[b]	K_r[a]	τ_c[a]
ID	Portneuf	si l	0.079	5.44E+06	0.0307	3.1
IN	Lewisburg	cl	0.042	4.82E+06	0.0059	3.4
IN	Miami	si l	0.059	4.78E+06	0.0095	3.3
MD	Frederick	si l	0.054	5.14E+06	0.0084	6.6
MD	Opequon	cl	0.026	3.86E+06	0.0035	6.3
MD	Manor	l	0.025	4.09E+06	0.0054	3.6
ME	Caribou	g l	0.034	4.78E+06	0.0045	4.3
MN	Barnes	l	0.033	4.92E+06	0.0063	4.0
MN	Sverdrup	sa l	0.012	3.44E+06	0.0033	1.4
MO	Mexico	si l	0.050	4.66E+06	0.0036	0.7
MS	Grenada	si l	0.058	4.94E+06	0.0073	4.5
MT	Zahl	l	0.040	5.13E+06	0.0123	3.5
NC	Gaston	c l	0.021	4.17E+06	0.0049	4.4
ND	Barnes	l	0.021	5.17E+06	0.0033	2.3
ND	Williams	l	0.028	4.99E+06	0.0045	3.4
NE	Sharpsburg	si c	0.036	3.84E+06	0.0053	3.2
NE	Hersh	sa l	0.037	9.18E+06	0.0112	1.7
NY	Collamar	si l	0.072	5.23E+06	0.0241	6.4
OH	Miamian	l	0.037	4.36E+06	0.0096	5.4
OK	Woodward	si l	0.066	10.2E+06	0.0250	1.3
SD	Pierre	c	0.029	3.33E+06	0.0117	4.8
TX	Amarillo	l sa	0.026	6.78E+06	0.0453	1.7
TX	Heiden	c	0.025	3.23E+06	0.0089	2.9
WA	Nansene	si l	0.079	5.34E+06	0.0307	3.1
WA	Palouse	si l	0.053	4.95E+06	0.0066	0.7
Rangelands[a]						
AZ	Forest	sa l	0.037	9.00E+05	0.00035	1.4
CA	Apollo	l	0.037	5.77E+05	0.00004	0.03
CO	Degater	si c l	0.038	2.60E+06	0.00162	4.4
MT	Vida	l	0.005	7.35E+05	0.00032	0.8
NM	Querencia	sa l	0.072	6.87E+05	0.00017	0.6
NV	Durorthid	sa l	0.066	3.34E+05	0.00046	0.1
OK	Grant	sa l	0.047	5.23E+05	0.00015	1.2
TX	Purves	co c	0.008	4.00E+05	0.0001	2
Forest[c]						
Native road		sa l	0.01	2.00E+06	0.0004	2.0
Native road		si l	0.02	2.00E+06	0.0003	2.0
Gravelled road		g si l	0.014	2.00E+06	0.0003	2.0
Undisturbed forest		sa l	0.002	0.40E+06	0.0005	1
Low severity fire		sa l	0.003	1.00E+06	0.0006	1
High severity fire		sa l	0.015	2.00E+06	0.0007	1

[a]Laflen et al. (1991).
[b]From 1995 WEPP release soils database.
[c]Values available within WEPP:Road and Disturbed WEPP interfaces at http://forest.moscowfsl.wsu.edu. Search for Forest Service WEPP.

7.5 Rill Erosion

Rill erosion is the detachment and transport of soil by a concentrated flow of water. Rills are eroded channels that are small enough to be removed by normal tillage operations. Rill erosion is the predominant form of surface erosion.

Rill erosion is a function of the flow rate or hydraulic shear τ of the water flowing in the rill, the soil's rill erodibility K_r, and critical shear τ_c, the shear below which soil detachment is negligible (Flanagan & Livingston, 1995). The relationship between rill erosion and the hydraulic shear of water in the rill is

$$D_r = K_r(\tau - \tau_c)\left(1 - \frac{Q_s}{T_c}\right) \tag{7.2}$$

where D_r = rill detachment rate (kg m^{-2} s^{-1}),
K_r = rill erodibility, due to shear (s/m) (Table 7-1),
τ = hydraulic shear of flowing water (Pa) (Equation 7.3),
τ_c = critical shear below which no rill erosion occurs (Pa) (Table 7-1),
Q_s = rate of sediment flux in the rill (kg m^{-1} s^{-1}),
T_c = sediment transport capacity of rill (kg m^{-1} s^{-1}).

The hydraulic shear τ is defined as

$$\tau = \gamma R S \tag{7.3}$$

where γ = specific weight of water (N m^{-3}), about 9810 N m^{-3},
R = hydraulic radius of the rill (m),
S = hydraulic gradient of rill flow (m/m).

The sediment transport capacity can be estimated from the relationship (Flanagan & Nearing, 1995)

$$T_c = B\tau^{1.5} \tag{7.4}$$

where T_c = transport capacity per unit width (kg m^{-1} sec^{-1}),
B = transport coefficient based on soil and water properties generally between 0.01 and 0.1 (dimensionless) (Elliot et al., 1989).

The interrill and rill erosion processes are used in several process-based erosion prediction computer models, including the Water Erosion Prediction Project (WEPP) model (Flanagan & Livingston, 1995).

Example 7.1

Determine the rill erodibility (K_r) from the following observations: Observed runoff rate = 1.0 L s^{-1}; sediment concentration in runoff = 0.12 kg L^{-1}; rill length = 20 m; width = 0.15 m; gradient (S) = 0.07; hydraulic radius (R) = 0.01 m; soil transport coefficient (B) = 0.1 kg Pa$^{-1.5}$; soil critical shear (τ_c) = 2 Pa.

Solution.

Sediment delivery = runoff × concentration = 1.0 × 0.12 = 0.12 kg/s

$$Q_s = \text{sediment delivery}/(\text{rill width}) = 0.12/0.15 = 0.8 \text{ kg m}^{-1}\text{ s}^{-1}$$

$$D_r = \text{sediment flux}/(\text{rill length}) = 0.8/20 = 0.04 \text{ kg m}^{-2}\text{ s}^{-1}$$

$$\tau = 9810 \times 0.01 \times 0.07 = 6.87 \text{ Pa}$$

$$T_c = B\,\tau^{1.5} = 0.1\,(6.87)^{1.5} = 1.80 \text{ kg m}^{-1}\text{ s}^{-1}$$

$$K_r = D_r/[(\tau - \tau_c)(1 - Q_s/T_c)] = \frac{0.04}{\left[(6.87 - 2)\left(1 - \frac{0.8}{1.8}\right)\right]} = 0.015 \text{ s m}^{-1}$$

7.6 Gully Erosion

Gully erosion produces channels larger than rills. These ephemeral channels carry water during and immediately after rains. Gullies are distinguished from rills in that gullies cannot be obliterated by tillage. The amount of sediment from gully erosion is usually less than from upland areas, but the nuisance from having fields or developed areas divided by large gullies is often a greater problem.

The rate of gully erosion depends primarily on the runoff-producing characteristics of the watershed, the drainage area, soil characteristics, the slope in the channel, and the alignment, size, and shape of the gully (Bradford et al., 1973). A shallow depth to the water table and the presence of an impeding layer are also common in gully-prone areas. A gully develops by processes that may take place either simultaneously or during different periods of its growth. These processes are (1) waterfall erosion or headcutting at the gully head, (2) erosion caused by water flowing through the gully or by raindrop splash on exposed gully sides, (3) alternate freezing and thawing of the exposed soil banks, and (4) slides or mass movement (sloughing) of soil into the gully. Runoff then removes loose soil from the gully floor.

7.7 Stream Channel Erosion

Stream channel erosion includes soil removal from streambanks and soil scour of the channel bed. Bank erosion can also lead to stream meandering and rechannelization, resulting in major erosion and deposition within the floodplain. Stream channel erosion and gully erosion are distinguished primarily in that stream channel erosion applies to the lower end of headwater tributaries, and to streams that have nearly continuous flow and relatively flat gradients, whereas gully erosion generally occurs in intermittent or ephemeral streams or channels near the upper ends of headwater tributaries.

Streambanks erode either by runoff flowing over the side of the streambank, or by scouring and undercutting below the water surface. Streambank erosion, less serious than scour erosion, is often increased by the removal of vegetation, overgrazing, tilling too near the banks, or straightening the channel. The velocity and direction of flow, depth and width of the channel, and soil texture influence scour erosion. Channel disturbances and the presence of obstructions such as sandbars increase meandering, the major cause of erosion along the bank.

Alluvial plains can be an area of deposition for eroded upland sediments. Erosion in the nineteenth and twentieth centuries resulted in major deposition of sediments in many streams. With current conservation practices, relatively clean water leaving upland areas is now transporting these deposited sediments farther downstream (Trimble, 1999). This ability to store sediment confounds human attempts to reduce sediment yield at a watershed scale since anticipated reductions in sediment yield from watersheds with restored upland areas may not occur for many years as the stream flow slowly moves "legacy" sediment through the stream network.

7.8 Sediment Transport

Numerous methods for predicting sediment transport capacity of channels have been based on channel hydraulic shear (Equation 7.3), flow rate, velocity, and sediment properties. Equation 7.4 is commonly used to estimate transport capacity of individual rills as well as larger channels.

Soil Losses

Soil losses vary considerably with the erosion factors. Figure 7-1 presents some typical soil erosion rates as influenced by crops and regions of the United States. Note the large differences due to conservation practices and the large differences between regions. The importance of soil losses is indicated by the effect of erosion phase on crop yield, as shown in Figure 7-2. Schertz et al. (1989) reported that much of the reduced yield was due to a decrease in the amount of water available

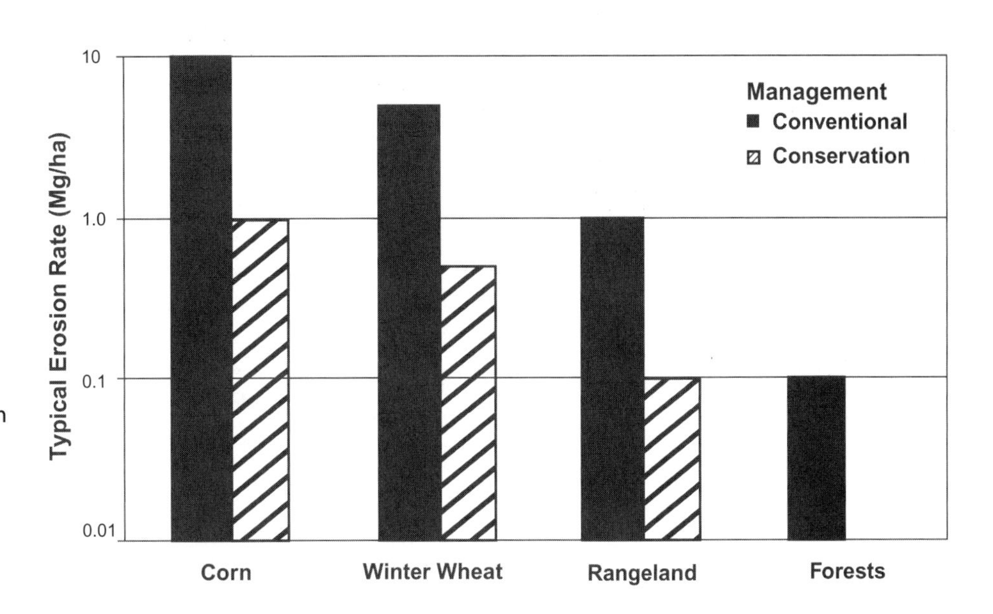

Figure 7–1
Some typical U.S. erosion rates for corn in the Midwest, winter wheat in the Great Plains, western rangeland, and northwestern forests.

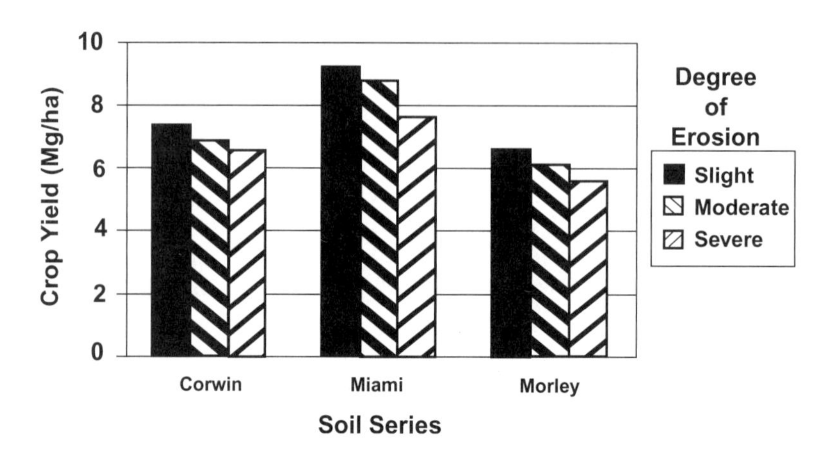

Figure 7–2
Effect of degree of erosion and soil series (Typic Argiudoll, Typic Hapludalf, and Typic Hapludalf, respectively) on crop yield. (Adapted from Schertz et al., 1989)

to the plant on eroded soils. On some soils, these crop yield decreases can be largely overcome by higher fertilization levels. On other soils, particularly more shallow soils on sloping terrain, erosion may completely destroy productivity if appropriate conservation practices are not initiated (USDA, 1989).

Off-site impacts of eroded sediment may be more important than loss of productivity, particularly on deeper, fertile soils. Eroded sediments can alter channel shapes and alignment, increasing the frequency and severity of flooding. Sediment can also fill reservoirs, shortening their useful lives. Sediment in irrigation ditches increases maintenance costs. Eroded sediments carry undesirable pollutants such as phosphates, which alter downstream nutrient balances, or pesticides, which can have a wide range of adverse impacts on plants, animals, and humans throughout the food chain.

Soil losses, or relative erosion rates for different management systems, are estimated to assist farmers and government agencies to evaluate existing farming systems or to compare alternative management strategies to reduce soil loss. In the period from 1945 until 1965, a method of estimating losses based on statistical analyses of field plot data from small plots located in many states was developed, which resulted in the Universal Soil Loss Equation (USLE). A revised version of the USLE (RUSLE) was later developed as a computer application. When predicting erosion, RUSLE allows more detailed consideration of management practices, rangeland, seasonal variation in soil properties and topography than the USLE (Renard et al., 1997).

Since the 1960s, scientists have been developing process-based computer models that estimate soil loss by considering the processes of infiltration, runoff, detachment, transport, and deposition of sediment. One of these is the Water Erosion Prediction Project (WEPP) (Flanagan & Livingston, 1995).

7.9 Predicting Upland Erosion

Most recent erosion prediction methods, such as RUSLE and the WEPP model, are computer based, and for some applications, can be run over the Internet. To demonstrate erosion prediction principles, this chapter will present the USLE technology in SI units, with RUSLE enhancements where possible, and will also discuss the extended prediction capabilities of RUSLE (Renard et al., 1997), RUSLE2 (Foster et al., 2000), and WEPP (Flanagan & Livingston, 1995). Readers are encouraged to obtain the most recent computer software and compare the predictions from that software with the examples presented here. The USLE addresses most of the important erosion factors and mitigation practices (Wischmeier & Smith, 1978). It is useful for comparing management systems and predicting upland erosion. It is unable to predict sediment delivery without additional assumptions and calculations, some of which are incorporated into the latest RUSLE and RUSLE2 programs. The average annual soil loss is estimated from the equation

$$A = R\,K\,L\,S\,C\,P \tag{7.5}$$

where A = average annual soil loss (Mg/ha),

R = rainfall and runoff erosivity index for geographic location (Figure 7-3),

K = soil erodibility factor (Equation 7.6, Table 7-1),

L = slope length factor (Equation 7.7),

S = slope steepness factor (Equation 7.8),

C = cover management factor (Table 7-2),

P = conservation practice factor (estimate with RUSLE).

The R-factor is derived from long-term recording rain gauge data. The average annual rainfall and runoff erosivity index R is shown in Figure 7-3 for the continental United States. More detailed maps are available in the RUSLE documentation

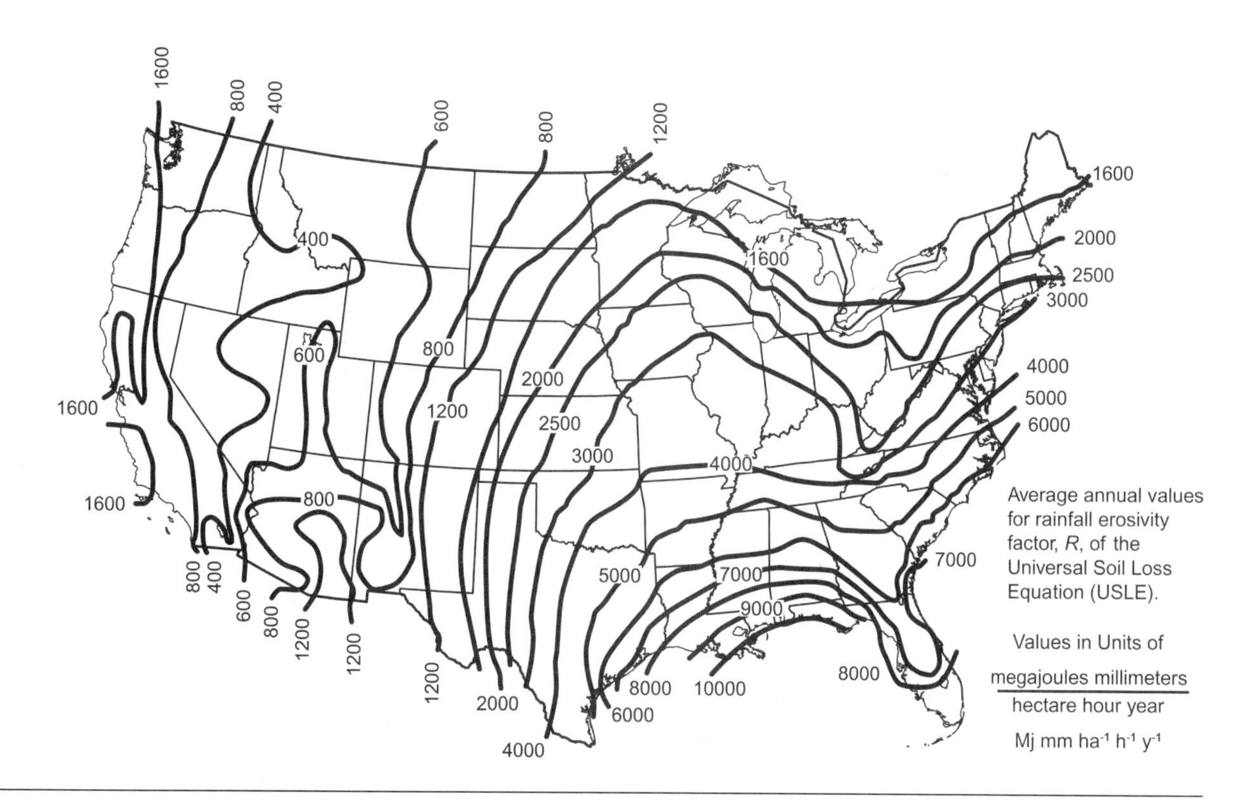

Figure 7–3 Rainfall and runoff erosivity R-factor by geographic location. (Adapted from Foster et al., 1981)

(Renard et al., 1997). The RUSLE and WEPP climate databases both contain hundreds of specific local climates, which include the annual distributions of precipitation. Estimates for R have been made for many other locations around the world (Lal, 1994).

The soil erodibility K-factor encompasses a wide range of soil erodibility properties, including the soil runoff potential, the ease of soil detachment, and soil transport properties. In RUSLE, K varies to account for seasonal variation in soil erodibility. Values for K were initially measured on a series of benchmark soils from direct soil loss measurements on fallow plots in many locations (Figure 7-4). Table 7-1 presents some typical K-factors. The SI K-factor can be calculated from the regression equation

$$K = 2.8 \times 10^{-7} M^{1.14} (12 - a) + 4.3 \times 10^{-3} (b - 2) + 3.3 \times 10^{-3} (c - 3) \quad \textbf{7.6}$$

where M = particle size parameter,
= (% silt + % very fine sand) × (100 – % clay),
a = organic matter content (percentage),
b = soil structure code: very fine granular, 1; fine granular, 2; medium or coarse granular, 3; blocky, platy, or massive, 4,
c = profile permeability class: rapid, 1; moderate to rapid, 2; moderate, 3; slow to moderate, 4; slow, 5; very slow, 6.

The K-factors in numerous soil databases have been calculated from Equation 7.6, or its equivalent in English units. Most state NRCS offices have tables of K-factors for all the dominant soils in their states. If using R-factors or K-factors from other sources, such as Wischmeier & Smith (1978) or RUSLE (Renard et al., 1997), the units will be in the English system. Compatible units of K and R factors must be used, both English or both SI. The WEPP model requires or estimates five soil erodibility variables: hydraulic conductivity, interrill erodibility, rill erodibility, critical shear, and a transport coefficient (Equations 7.1, 7.2, and 7.4). Databases distributed with WEPP contain estimates of the values for hundreds of U.S. soils.

Figure 7–4

Typical USLE fallow soil erosion plot, near Pullman, Washington. (Courtesy of D. McCool, USDA-ARS)

Guidelines for estimating the variables are given in the WEPP documentation (Flanagan & Livingston, 1995).

The topographic factors, *L* and *S*, adjust the predicted erosion rates to give greater erosion rates on longer and/or steeper slopes, when compared to the USLE "standard" slope steepness of 9 percent and length of 22 m. These factors address the increasing rill erosion rates as more runoff accumulates with longer slopes, and the hydraulic shear in runoff increases on steeper slopes. The *L* factor can be calculated from the equation

$$L = \left(\frac{l}{22}\right)^b \tag{7.7a}$$

where L = slope length factor,
l = slope length (m),
b = dimensionless exponent.

For conditions where rill and interrill erosion are about equal on a 9 percent, 22-m-long slope, then b is

$$b = \frac{\sin\theta}{\sin\theta + 0.269(\sin\theta)^{0.8} + 0.05} \tag{7.7b}$$

where θ = field slope angle = $\tan^{-1}(S)$,
S = slope steepness (m/m).

For most conditions where rill erosion is greater than interrill erosion (such as soils with a large silt or fine sand content), b should be increased up to 75 percent. Where rill erosion is less than interrill erosion (on short slopes or low-gradient soils with ridge tillage, or soils generating little runoff), b should be decreased as much as 50 percent. RUSLE2 makes this calculation internally.

The *S*-factor depends on the length and steepness category of the slope. For slopes less than 4 m long,

$$S = 3.0\,(\sin\theta)^{0.8} + 0.56 \tag{7.8a}$$

For slopes greater than 4 m long and steepness less than 9 percent,

$$S = 10.8\sin\theta + 0.03 \tag{7.8b}$$

For slopes greater than 4 m long and steepness greater than or equal to 9 percent,

$$S = 16.8\sin\theta - 0.50 \tag{7.8c}$$

The slope length is measured from the point where soil erosion begins (usually near the top of the ridge) to the outlet channel or a point downslope where deposition begins for the USLE. RUSLE also considers nonuniform concave or convex slopes. The WEPP model and RUSLE2 consider the entire hillslope, including areas of deposition.

The cover management *C*-factor includes the effects of cover, crop sequence, productivity level, length of growing season, tillage practices, residue management, and the expected time distribution of erosive events. Table 7-2 presents some typical *C*-factors for a number of cropping systems, as a ratio of soil loss from the selected system to soil loss from a fallow plot. These values vary with geographic location. Most NRCS offices have *C*-factors for common cropping systems in their respective states. RUSLE is programmed to calculate *C*-factors from user-specified crop yields and dates of tillage, planting, and harvest. The WEPP model also uses specified dates and yields. WEPP addresses vegetation by modeling the daily growth of vegetation, leaf drop, and residue accumulation and decomposition.

The conservation practice *P*-factor should be assigned a value of 1.0 if farming operations are up and down the slope. If conservation practices such as contour farming, strip cropping, and terracing are considered, then RUSLE2 or WEPP should be used to evaluate the practice. The *P*-factor is generally around 0.5 for contouring

TABLE 7–2 Typical *C*-Factors for a Number of Land Cover Conditions

Vegetation	***Tillage Practice***			
Corn Belt	***Autumn Conventional***	***Spring Conventional***	***Spring Conservation***	***No-Till***
Continuous corn or soybeans	0.40	0.36	0.27	0.10
Corn and soybean rotation	0.40	0.35	0.24	0.10
Corn–corn–oats–meadow	0.14	0.12	0.11	0.07
Corn–oats–meadow–meadow	0.06	0.05	0.03	0.03
Permanent pasture (good)				0.01
Permanent pasture (poor)				0.04
Wheat Belt		***Conventional***	***Conservation***	***No-Till***
Winter wheat—fallow		0.20	0.17	0.04
Spring barley		0.06	0.02	
Wheat–barley–fallow rotation		0.21	0.13	0.02
Other Crops		***Conventional***	***Conservation***	
Winter wheat and pea rotation		0.11	0.02	
Cotton		0.40	0.30	
Peanuts—use soybeans				
Sorghum—use corn				
Nonagricultural		***Good***	***Poor***	
Rangeland		0.01	0.15	
Forest		0.001	0.002	
Forest after fire		0.01	0.15	
Unpaved road		0.35	0.45	
Construction site		0.50	1.00	

Note: These values are for example only. Contact local agencies for local *C*-factors.

only, and 0.25 for contour strip cropping. One of the effects of terraces is that the slope length is shortened to the terrace spacing. This effect can be considered in the USLE with the *L*-factor. Chapter 8 discusses erosion and sediment delivery processes associated with terraces. The *P*-factor is likely to be the least reliable of all the USLE factors, as the impacts of conservation practices vary with climate, soils, vegetation, and topography of the field, contoured rows, and terrace channels. RUSLE2 incorporates more complex topographic and climatic considerations and calculates the *P*-factor internally. The WEPP model addresses contouring through a management input, and informs the user when contour ridges overtop. If contour ridges have overtopped, the user needs to increase the ridge height, model the site without contouring, or alter the slope of the contour channels to obtain an erosion estimate. Terraces are modeled as small watersheds in WEPP, allowing detailed analysis of upland erosion and terrace channel processes for a specific site.

Example 7.2

Determine the soil loss for the following conditions: location, Memphis, Tennessee; soil, Grenada silt loam similar to that in Mississippi; $l = 50$ m; $S = 5$ percent; $C = 0.14$ (approximate for corn–corn–oats–meadow rotation with good management); and the field has no conservation practices ($P = 1$).

Solution.

(1) From Figure 7-3, read $R = 5000$ as the erosion index.

(2) From Table 7-1, read $K = 0.058$.

(3) Calculate slope angle in degrees:

$$\theta = \tan^{-1}(0.05) = 2.86°$$

(4) From Equation 7.7b, assuming a moderate relationship between rill and interrill erosion, calculate b:

$$b = \frac{\sin(2.86)}{\sin(2.86) + 0.269\,[\sin(2.86)]^{0.8} + 0.05} = 0.40$$

(5) From Equation 7.7a, calculate the L-factor:

$$L = \left(\frac{50}{22}\right)^{0.4} = 1.4$$

(6) From Equation 7.8c, calculate the S-factor:

$$S = 16.8 \sin(2.86) - 0.5 = 0.34$$

(7) Substituting in Equation 7.5,

$$A = 5000 \times 0.058 \times 1.4 \times 0.34 \times 0.14 \times 1 = 19 \text{ Mg ha}^{-1}\text{ yr}^{-1}$$

Running the online WEPP Model at http://octagon.nserl.purdue.edu for a corn–soybeans–wheat–alfalfa rotation, and the Memphis climate, Grenada soil on a uniform slope will predict a similar erosion rate of 25 Mg ha^{-1} for conservation tillage.

There are several reasons for the differences in the predicted erosion rates in Example 7.2 between the WEPP website and RUSLE. The WEPP website did not have the same rotation as was assumed for the RUSLE C-factor. The RUSLE C-factor was from Table 7-2 for a state further north (e.g., Ohio or Indiana), and it may be higher in Tennessee. The WEPP prediction was for a uniform hillslope. If a reduced slope at the top and the bottom of the hill were assumed, then the WEPP predictions would be lower. Generally, any predicted erosion rate is only accurate to about plus or minus 50 percent, which means that the ranges of these two predicted values overlap. These differences in predictions demonstrate some of the limitations when trying to predict absolute erosion rates.

By evaluating the factors in Equation 7.5, the predicted soil loss can be determined for a given set of conditions. If the soil loss is higher than the minimum required to maintain productivity, changing the cover management practices, or the conservation practices may reduce it.

Example 7.3

If the soil loss for the conditions in Example 7.2 is to be limited to 10 Mg ha^{-1}, what practices could be adopted to accomplish this level?

Solution. Since C, P, and L are the only variables that can be changed, the following combinations are possible by substituting the appropriate factors in Equation 7.5.

Conservation Practice	*L*	*C*[a]	*P*[b]	*A* Mg ha^{-1}	*Remarks*
(1) None	1.4	0.14	1	19	Soil loss too high
(2) Contoured	1.4	0.14	0.55	10	Satisfactory
(3) Contoured and spring tillage	1.4	0.12	0.55	9	Satisfactory
(4) Conservation tillage, no contours	1.4	0.11	1	15	Soil loss too high
(5) Terrace at 24 m spacing	1.0	0.14	0.33	4.5	Satisfactory
(6) Terrace with conservation tillage	1.0	0.11	0.33	3.5	Satisfactory

[a] From Table 7-2.
[b] From RUSLE program.

Note that any practice that includes contouring will be adequate to reduce soil erosion to acceptable levels. The low C-factor due to the incorporation of oats and meadow in the rotation offers considerable conservation benefits in this case.

Because of the complexity of using the WEPP model and limitations associated with the USLE, some simplified models have been developed. These include a simplified Internet interface to the WEPP model for agricultural and rangeland conditions (USDA ARS, 2003), forest conditions (Elliot, 2004), and another model for rangeland erosion (Lane, 2002). Readers should check current availability of erosion-prediction Internet sites for other conditions. These sites have made numerous assumptions in their databases for addressing specific applications to gain ease of operation, but modeling flexibility is lost.

Example 7.4

An "insloping" forest road near Mt. Shasta, California, has a gradient of 6 percent, a width of 4 m, and a road length between culverts of 200 m. The soil is best described as a sandy loam and the ditch is covered with vegetation. The road is native-surfaced (it has no gravel). It is located 45 m from the nearest stream, and the slope steepness between the road and the stream is 35 percent. What effect will adding gravel have on the sediment leaving the road, and how much is delivered to the stream each year?

Solution. For this specialized application requiring sediment delivery, the Internet application of the WEPP model for forest roads (WEPP:Road) is accessed at http://forest.moscowfsl.wsu.edu/fswepp/. The specified conditions are entered, using the default values for the road fill slope, into the WEPP:Road interface, giving the following results for a 30-year run:

Surface	*Sediment Leaving Road (kg)*	*Sediment Leaving Buffer (kg)*
Native	1586	1343
Gravel	983	780

Thus the addition of gravel reduces sediment leaving the road by 38 percent, and reduces sediment entering the stream by 42 percent.

For erosion analysis, both physical and economic factors, as well as social aspects, need to be considered in establishing soil loss tolerances, sometimes called T values. These values vary with topsoil depth and subsoil properties from 2 to 10 Mg ha^{-1}. They are considered the maximum rates of soil erosion that will permit a high level of crop productivity to be sustained economically and indefinitely. Criteria for control of sediment pollution may dictate lower tolerance values, particularly in forested watersheds, or watersheds that have streams with beneficial uses such as water supply, swimming, fishing, or providing habitat for endangered species.

7.10 Sediment Delivery

Sediment delivery downstream in a watershed may be estimated from the USLE and a sediment delivery ratio. The soil loss equation estimates gross sheet and rill erosion but does not account for sediment deposited en route to the place of measurement, nor for gully or channel erosion downstream. The sediment delivery ratio is defined as the ratio of sediment delivered at a location in the stream system to the gross erosion from the drainage area above that point. This ratio varies widely with size of area, steepness, density of drainage network, and many other factors. The sediment delivery ratio varies roughly as the inverse of the drainage area to the power of 0.2. For watersheds of 2.6, 130, 1300, and 26 000 ha, ratios of 0.65, 0.33, 0.22, and 0.10, respectively, were suggested as average values by Roehl (1962).

The WEPP watershed version predicts upland erosion and channel sediment delivery for watersheds up to about 200 ha. The SWAT model was developed to model larger watersheds up to hundreds of square kilometers (Arnold et al., 1998). It uses modified USLE technology to estimate erosion rates which considers runoff as well as precipitation in predicting sediment delivery from hillslopes. Curve Number methods are used to estimate storm flows, with options for more sophisticated runoff prediction. Stream gauging records are used to calibrate stream flows. GIS linkages have been developed for both the WEPP watershed version (Renschler,

2004) and the SWAT model (Arnold et al., 1998). As the size of a watershed increases, channel processes become more important, and upland processes less important in the delivery of sediment.

Example 7.5

The predicted annual erosion rate for a typical hillslope in a sensitive watershed is 3.2 Mg ha^{-1}. The watershed is about 1000 ha in size. Estimate the amount of sediment delivered from this watershed.

Solution. Using Roehl's assumptions, the delivery ratio is about 0.22. Using this value to calculate the sediment delivery gives

$$\text{Sediment delivery} = 3.2 \times 0.22 = 0.7 \text{ Mg ha}^{-1}$$

$$0.7 \text{ Mg ha}^{-1} \times 1000 \text{ ha} = 700 \text{ Mg per year}$$

Erosion Control Practices

In this chapter, the primary emphasis is on describing and estimating erosion by water from cultivated farm land and other nonpoint sources. Vegetated waterways and terraces are important control measures, and they will be discussed in Chapter 8, since specialized design procedures are required.

7.11 Contouring

Contouring is the practice of performing field operations, such as plowing, planting, cultivating, and harvesting, parallel to elevation contours. It reduces surface runoff by impounding water in small depressions, and decreases the development of rills. Contouring is most widely practiced where the benefits include both erosion reduction and conservation of water.

The relative effectiveness of contouring for controlling erosion can range from preventing all erosion to increasing hillside erosion by concentrating runoff that may initiate gullying. If crops are planted on ridges, runoff is retained in the furrows between the ridges and erosion is reduced.

Contouring on steep slopes or under conditions of high rainfall intensity and soil erodibility will increase the risk of gullying because row breaks may release the stored water. Breakovers cause cumulative damage as the volume of water increases with each succeeding row. To reduce the damage caused by breakovers, machines are available to form small dams or dikes about every 2 m in furrows. This prevents ponded water from following small gradients along a furrow to a breakover point.

The effectiveness of contouring is also impaired by changes in the infiltration capacity of the soil due to surface sealing. Depression storage is reduced after tillage operations cease and consolidation takes place, reducing ridge height.

Because of nonuniform slopes in most fields, all crop rows cannot be on the true contour. To establish row directions, a guide line (true contour) is laid out at one or more elevations in the field. On small fields of uniform slope, one guide line may be sufficient. Another guide line should be established if the slope along the row direction exceeds 1 to 2 percent in any row laid out parallel to the guide line.

A small slope along the row is desirable to prevent runoff from a large storm breaking over the ridges. Where practical, field boundaries for contour farming should be relocated on the contour or moved. This eliminates odd-shaped fields with short, variable-length rows (point rows).

Contouring is more likely to fail in climates with high-intensity spring and summer storms, or on sites with steeper slopes (over about 4 percent). It is most beneficial in climates with lower precipitation, or when used in conjunction with other practices such as strip cropping, terracing, or ridge tillage.

7.12 Strip Cropping

Strip cropping is the practice of growing alternate strips of different crops in the same field (Figure 7-5). For controlling water erosion, the strips are on the contour, but in dry regions, strips are placed normal to the prevailing wind direction for wind erosion control (Chapter 20). The greatest concentration of strip cropping is in drier, wheat-growing areas where wind erosion is prevalent, with alternating strips of small grain and fallow. One of the major reasons strip cropping reduces water erosion is that managers generally plant alternative strips to forage rather than row crops. Any practice increasing the number of years when a field is not in a row crop will reduce soil erosion, regardless of tillage direction. Table 7-2 shows that changing from continuous–corn to a corn–meadow rotation significantly decreases erosion potential before considering tillage direction.

Three types of strip cropping are shown in Figure 7-6. In contour strip cropping, layout and tillage are held closely to the contour and the crops follow a definite rotational sequence. With field strip cropping, strips of uniform width are placed across the general slope. When used with adequate grassed waterways (Chapter 8), the strips may be placed where the topography is too irregular to make contour strip cropping practical. Field strip cropping may also be used for wind erosion control. Buffer strip cropping has strips of a grass or legume crop between contour strips of crops in the regular rotations. Buffer strips may be even or irregular in width, or placed on critical slope areas of the field. Their main purpose is to give protection from erosion or allow for areas of deposition. Buffer strips can also serve as wildlife refuges or provide agroforestry products.

The methods for laying out contour strip cropping are (1) both edges of the strips on the contour, (2) one or more strips of uniform width laid out from a key or base contour line, and (3) alternating uniform-width and variable-width buffer strips. Contour strip width should be convenient for multiple-row equipment operation.

Figure 7–5

Contour strip-cropping in southeastern Wisconsin. (Photograph by W. Elliot)

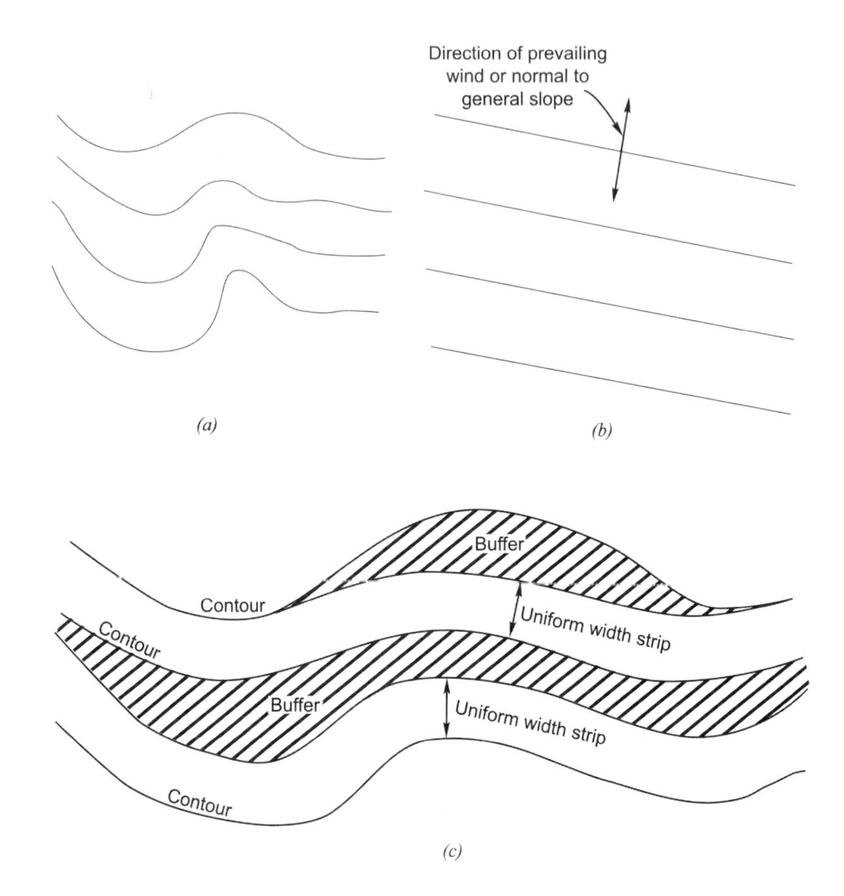

Figure 7–6

Three types of strip-cropping: (a) contour, (b) field, and (c) buffer.

In some areas, permanent contoured vegetative filters are effective as conservation tools (Figure 7-7). Stiff-stemmed tropical grasses or leguminous shrubs have been effective in the tropics and in warmer areas of the United States. A study with stiff-stemmed warm-season grasses in western Iowa showed that such filters could reduce erosion by about 50 percent (Kramer & Alberts, 2000). The forage from these

Figure 7–7

Grass contour strips in western Uganda. (Photograph by W. Elliot)

filter strips is sometimes harvested for livestock so that no land is lost to production. A variation of this principle is to have the vegetated filters include productive plants such as bananas or coffee or other agroforestry products in the tropics. Such systems are not common in the United States because high costs of labor for installation and maintenance make small-plot management uneconomical.

The WEPP model can simulate up to 10 strips of crops along a hillslope. It can predict the distribution of erosion and deposition along the hillslope. This makes it possible to compare rotations, strip widths, and tillage practices for the climate, soil, and topography of interest to determine an optimum strip-cropping system. Figure 7-8 shows an example distribution of soil loss or deposition during one season of the practice, with erosion in the 30-m-wide corn strips and deposition in the 30-m-wide alfalfa strips. The first 30 m of the hillslope is in corn, but does not have sufficient overland flow to initiate rilling, hence the erosion rate is a nearly constant from interrill erosion, a function of rainfall intensity only. The second strip in Figure 7-8 is alfalfa, and is mainly an area of deposition, with a small amount of erosion evident starting part way through the strip. The third strip is corn, and it experiences a very high level of erosion from the relatively clean water leaving the alfalfa strip, with the erosion rate gradually decreasing because detachment of the rill flow is decreased as sediment in transport is increased (Equation 7.2). The final strip of alfalfa is an area of very large deposition due to increased infiltration, decreasing sediment transport capacity, and decreased detachment due to greater amounts of surface cover by vegetation.

7.13 Tillage Practices

Tillage management is an important conservation tool. Tillage should provide an adequate soil and water environment for the plant. Its role as a means of weed control has diminished with increased use of herbicides and improved timing of operations. The effect of tillage on erosion depends on such factors as surface

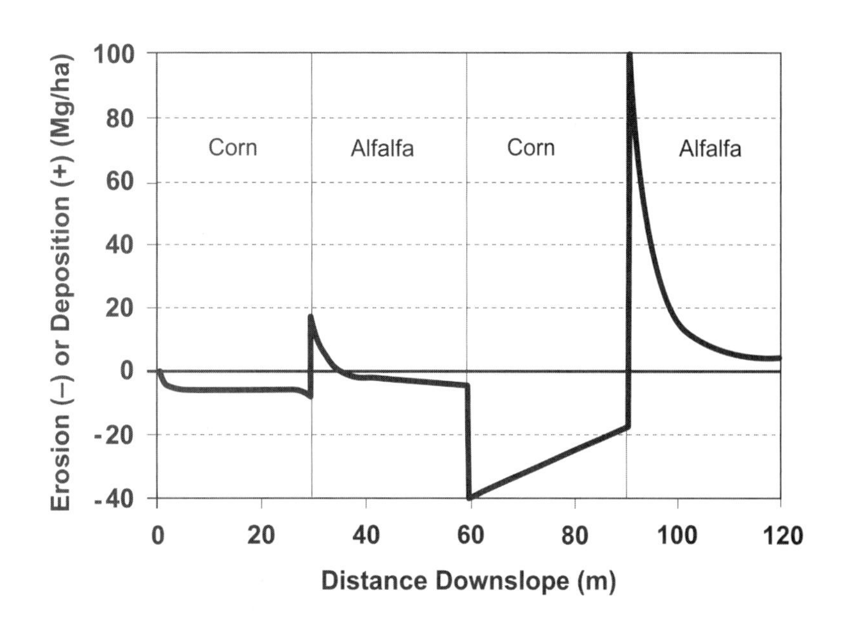

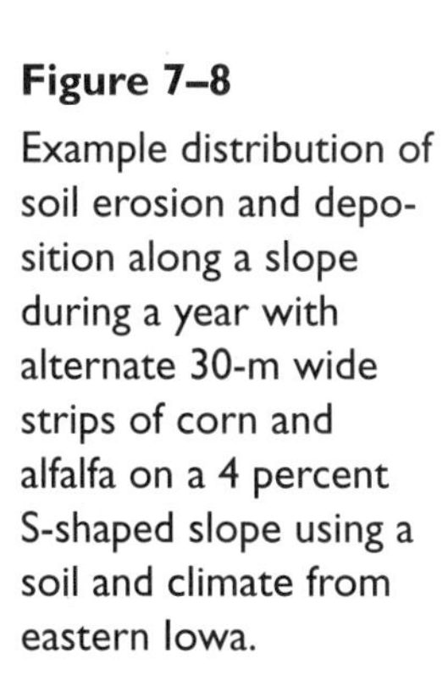

Figure 7–8

Example distribution of soil erosion and deposition along a slope during a year with alternate 30-m wide strips of corn and alfalfa on a 4 percent S-shaped slope using a soil and climate from eastern Iowa.

residue, aggregation, surface sealing, infiltration, and resistance to wind and water movement. Excessive tillage destroys structure, increasing the susceptibility of the soil to erosion.

One of the major benefits of minimum tillage is increased residue on the surface. Such residue is extremely effective in reducing erosion. Typically, conservation tillage reduces runoff by 60 percent and erosion by 90 percent compared to conventional tillage (Table 7-2).

7.14 Erosion on Nonagricultural Sites

The USLE, RUSLE, and WEPP models have all been applied to construction, forestry, and rangeland applications. On construction sites, and forest or range sites that have been disturbed by fire, an average annual erosion value has less meaning because the site is only susceptible to erosion until some cover is reestablished. This cover may be a grass crop, straw or similar mulch, a synthetic blanket, or some combination of treatments. Wischmeier and Smith (1978) suggest considering single storm-risk values for such sites. RUSLE does not directly address such recovering conditions. The WEPP model can address changing conditions, but care is needed in interpreting the results because the storms that cause severe erosion may not be predicted in the first year of the stochastic climate files that drive WEPP. The Forest Service has developed an Internet interface to predict the probability of a given amount of annual erosion occurring the first year after a disturbance for forests that have been burned or harvested (Elliot, 2004). This interface can also be used to estimate soil erosion risks on construction sites.

Example 7.6

What is the likelihood that soil erosion will exceed 50 Mg/ha in northern Georgia following a wildfire? The hillsides are generally 250 m long with an average steepness of about 20 percent, decreasing to 10 percent at the bottom. The soils are sandy loam. Following the fire, the surface cover averages about 65 percent.

Solution. The Forest Service Internet site has an interface to predict the likelihood of a given amount of erosion occurring in the year following a severe forest disturbance. The above values are entered into the Disturbed:WEPP interface at http://forest.moscowfsl.wsu.edu/fswepp/. The fire is assumed to be "low severity" because there is still about 65 percent surface cover. The Birmingham, Alabama, climate is selected as the nearest, and the model is run for 100 possible years of climate to get the following results:

Return Period (years)	*Precipitation (mm)*	*Runoff (mm)*	*Erosion (Mg ha^{-1})*
100	2236	218	92
50	2014	158	89
20	1751	123	64
10	1660	93	55
5	1534	66	39

From this table, it appears that there is about a 1-in-10 chance that erosion will exceed 50 Mg ha^{-1} from this site.

7.15 Erosion Control Techniques

In addition to the agricultural practices to reduce erosion discussed in this chapter and Chapter 8, there are commercial products available to reduce erosion. They are generally too expensive for agricultural applications but are common on construction sites in sensitive watersheds. The effectiveness of these products in reducing erosion can be predicted by determining the amount of surface cover generated, or the amount of sediment retained behind the structure.

One product is an erosion blanket. These blankets are generally made from a plastic or jute mesh that degrades within a year or two, and straw. Many mats also contain seeds and may contain fertilizers. Some manufacturers will allow customers to specify the desired seed mixtures.

Another set of products restricts overland flow carrying silts. These include silt fences, plastic mesh tubes of straw about 200 mm in diameter called straw logs or wattles, and tubes of geotextile fabric filled with sand. Straw bales are also commonly used to detain runoff, particularly in channels. These products tend to provide short-term solutions on severely disturbed sites, such as construction sites. Correct installation of these products is essential. In some cases, ongoing maintenance is necessary to ensure their effectiveness. More permanent structures to collect sediment or reduce channel erosion are discussed in Chapter 9.

Products to enhance vegetation establishment are also common in industrial applications. Machines to blow straw, or spray water mixed with seed, fertilizer, and other chemicals to decrease erosion or increase the chances of successful regeneration, are often used to protect bare slopes following construction activities.

7.16 Soil and Water Conservation Districts and Groups

The purpose of the soil and water conservation district in the United States is to provide a local organization to promote the conservation of soil, water, and related resources, and better land use. Districts may request technical assistance from such agencies as the Natural Resources Conservation Service, State Cooperative Extension Service, and county officials for carrying out erosion control and other land-use management activities. They may have the power (1) to carry out surveys and research relating to soil erosion programs; (2) to conduct demonstration projects; (3) to carry out preventive and control measures on the land; (4) to furnish technical and financial aid; (5) to sell or rent machinery, equipment, fertilizers, and so on; (6) to develop conservation plans for farms; (7) to take over government erosion-control projects; (8) to lease, purchase, or acquire property; and (9) to sue or to be sued in the name of the district. In many districts, the principal activities have been to assist in the development of conservation farm plans and to provide technical assistance in adopting conservation practices.

To address water quality issues on a watershed or basin scale, advisory groups were set up in the 1990s. These groups are made up of various stakeholders within a given watershed or basin, including farmers, timber owners, industrial representatives, and local and state government agency personnel. Such groups collectively determine watershed- or basin-wide management objectives and practices to improve water quality within their areas. Frequently, such objectives are in terms of defining total maximum daily loads (TMDLs) of pollutants such as sediments or pesticides. Once the TMDLs are defined, the advisory group identifies sources of pollutants and develops options to reduce total watershed or basin loads.

Internet Resources

RUSLE and RUSLE2 model
http://www.ars.usda.gov/Research/docs.htm?docid=5971

WEPP model, documentation, and Internet cropland and rangeland interface
http://topsoil.nserl.purdue.edu. Key search term: software

GeoWEPP GIS interface for WEPP Model
http://www.geog.buffalo.edu. Key search term: geowepp.

SWAT basin sedimentation model and Internet interfaces
http://www.brc.tamus.edu. Key search term: soil & water assessment tool.

Forest soil erosion prediction with the WEPP model
http://forest.moscowfsl.wsu.edu/fswepp/

Rangeland erosion prediction from runoff estimates
http://eisnr.tucson.ars.ag.gov/HillslopeErosionModel/

References

Arnold, J. G., R. Srinivasan, R. S. Muttiah, & J. R. Williams. (1998). Large area hydrologic modeling and assessment. Part I: Model development. *Journal of the American Water Resources Association, 34*(1), 73–89.

Bradford, J. M., D. A. Farrell, & W. E. Larson. (1973). Mathematical evaluation of factors affecting gully stability. *Soil Science Society America Proceeding, 37,* 103–107.

Elliot, W. J., A. M. Liebenow, J. M. Laflen, & K. D. Kohl. (1989). *A Compendium of Soil Erodibility Data from WEPP Cropland Soil Field Erodibility Experiments 1987 & 88.* NSERL Report No. 3. Online at http://topsoil.nserl.purdue.edu. Search: WEPP Compendium. Accessed July 2004. West Lafayette, IN: USDA-ARS, National Soil Erosion Research Laboratory.

Elliot, W. J. (2004). WEPP internet interfaces for forest erosion prediction. *Journal of the America Water Resources Association, 40*(2), 299–309.

Flanagan, D. C., & S. J. Livingston, eds. (1995). *WEPP User Summary.* NSERL Report No. 11. Online at http://topsoil.nserl.purdue.edu. Search: WEPP User Summary. Accessed July 2004. West Lafayette, IN: USDA National Soil Erosion Research Laboratory.

Flanagan, D. C., & M. A. Nearing. (1995). *USDA-Water Erosion Prediction Project Hillslope Profile and Watershed Model Documentation.* NSERL Report No. 10. West Lafayette, IN: USDA National Soil Erosion Research Laboratory.

Foster, G. R., D. C. Yoder, D. K. McCool, G. A. Weesies, T. J. Toy, & L. E. Wagner. (2000). Improvements in science in RUSLE2. Paper No. 002147. Presented at the Annual International Meeting of the ASAE, Milwaukee, WI, July 9–12, 2000. St. Joseph, MI: ASAE.

Foster, G. R., D. K. McCool, K. G. Renard, & W. C. Moldenhauer. (1981). Conversion of the universal soil loss equation to SI metric units. *Journal of Soil and Water Conservation, 36*(6), 355–359.

Iivari, T. A., & C. A. Kertis. (2001). Status and trends of national erosion rates on non-federal lands 1982–1997. (Poster). *Proceedings of the Seventh Federal Interagency Sedimentation Conference, March 25–29, Reno, NV,* 41–44.

Kramer, L. A., & E. E. Alberts. (2000). Grass hedges for erosion control on a small HEL watershed. Paper No. 002082. Presented at the ASAE Annual International Meeting, Milwaukee, WI, July 9–12. St. Joseph, MI: ASAE.

Laflen, J. M., W. J. Elliot, J. R. Simanton, C. S. Holzhey, & K. D. Kohl. (1991). WEPP soil erodibility experiments for rangeland and cropland soils. *Journal of Soil and Water Conservation, 46*(1), 39–48.

Lal, R., ed. (1994). *Soil Erosion Research Methods.* Delray Beach, FL: St. Lucie Press.

Lane, L. J. (2002). Hillslope erosion model. Online at http://eisnr.tucson.ars.ag.gov/HillslopeErosionModel/. Accessed September 2004.

Renard, K. G., G. R. Foster, G. A. Weesies, D. K. McCool, & D. C. Yoder, coordinators. (1997). *Predicting Soil Erosion by Water: A Guide to Conservation Planning with the Revised Universal Soil Loss Equation (RUSLE)*. U.S. Department of Agriculture, Agriculture Handbook No. 703. Washington, DC: U.S. Government Printing Office.

Renschler, C. S. (2004). The Geo-spatial interface for the water erosion prediction project (GeoWEPP). Online at http://www.geog.buffalo.edu Search: geowepp. Accessed June 2004.

Roehl, J. N. (1962). Sediment source areas, delivery ratios and influencing morphological factors. International Association for Scientific Hydrology, Commission of Land Erosion. Publ. No. 59, Wallingford Oxfordshire, U.K.

Schertz, D. L., W. C. Moldenhauer, S. J. Livingston, G. A. Weesies, & E. A. Hintz. (1989). Effect of past soil erosion on crop productivity in Indiana. *Journal of the Soil and Water Conservation Society, 44*(6), 604–608.

Trimble, S. W. (1999). Decreased rates of alluvial sediment storage in the Coon Creek Basin, Wisconsin, 1975–1993. *Science, 285*, 1244–1246.

U.S. Department of Agriculture (USDA). (1989). *The Second RCA Appraisal, Soil, Water, and Related Resources on Nonfederal Land in the United States, Analysis of Condition and Trends.* Washington DC: U. S. Government Printing Office.

U.S. Department of Agriculture, Agricultural Research Service (USDA ARS). (2003). WEPP Web Prototype. Online at http://octagon.nserl.purdue.edu. Accessed June 2004.

Watson, D. A., & J. M. Laflen. (1986). Soil strength, slope, and rainfall intensity effects on interrill erosion. *Transactions of the American Society of Agricultural Engineers, 29*(1), 98–102.

Wischmeier, W. H., & D. D. Smith. (1978). *Predicting Rainfall Erosion Losses—A Guide to Conservation Planning*. USDA Handbook 537. Washington, DC: U.S. Government Printing Office.

Problems

7.1 If the predicted soil loss at Memphis, Tennessee, for a given set of conditions is 12 Mg ha^{-1}, what is the expected soil loss at your present location if all factors are the same except the rainfall factor?

7.2 On a 60-m long slope, what is the relative difference in water erosion between slopes of 2 and 10 percent, assuming other factors are constant (compare USLE equations and WEPP Internet interfaces)?

7.3 If the soil loss for a given set of conditions is 4.5 Mg/ha for a 60-m length of slope, what soil loss could be expected for a 240-m slope length if the slope is 5 percent?

7.4 Given your present location, $K = 0.015$, $l = 91$ m, $S = 10$ percent, $C = 0.2$, and up- and downslope farming is practiced. (a) Determine the soil loss. (b) What conservation practice should be adopted if the soil loss is to be reduced to 5 Mg/ha?

7.5 If the soil loss for up- and downslope farming at your present location is 80 Mg/ha from a field having a slope of 6 percent and a slope length of 120 m, what will be the soil loss if the field is terraced with a horizontal spacing of 24 m? Assume that the cover management conditions remain unchanged and the *P*-factor is 0.5 for contour farming.

7.6 A farmer wishes to install stiff-grassed buffers on an 8-percent slope. What width of cropping strip is necessary to allow 3 rounds (6 passes) with 6-row equipment with a row spacing of 0.75 m? What width of crop sprayer boom is necessary to cover the strip in 2 rounds (4 passes)?

7.7 A rill channel is observed at two locations during a storm, 20 m and 60 m from the top of the hill. The slopes are 5 percent and 4 percent, the rill widths are 150 mm and 220 mm, and the hydraulic radii are estimated to be 0.01 and 0.04 m, respectively. The sediment transport rate is measured and found to be 0.2 kg/m-s and 0.8 kg/m-s. The rill erodibility (K_r) is estimated to be 0.004 s/m, critical shear (τ_c) to be 2.5 Pa, and transport coefficient (B) to be 0.1. Assume that the density of water is 9810 N/m3. Calculate the total sediment transport capacity and erosion rate for each position.

7.8 Use the Forest Service Internet site (http://forest.moscowfsl.wsu.edu/fswepp/) to estimate the road erosion and sediment delivery for a road located near Flagstaff, Arizona, for a graveled road segment that is 250 m long, with a gradient of 5 percent, a 4-m long fillslope with a steepness of 50 percent, and a 35-m long buffer with a steepness of 40 percent. The width of the road is 5 m, and the soil is a sandy loam. Compare the runoff and erosion from a rutted road to a road that is maintained to be outsloped.

7.9 A consultant is evaluating three treatments for an excavated slope behind a new superstore near Charleston, West Virginia. The soil is a clay loam, with a steepness of 50 percent and a length of 40 m. The three treatment options are simple seeding expected to give 30 percent cover the first year, hydroseeding expected to provide 60 percent cover the first year, and a straw mat containing seeds, which will provide 100 percent cover the first year. Use the Forest Service Disturbed WEPP interface (http://forest.moscowfsl.wsu.edu/fswepp/) to estimate the amount of sediment delivery that may be exceeded for each option if the first year is the most erosive in ten years.

7.10 A current rangeland hillslope has a steepness of 20 percent and a length of 150 m. The canopy and ground cover are about 15 percent, and the soil is sandy loam. Use the EISNR interface (http://eisnr.tucson.ars.ag.gov/HillslopeErosionModel/) to determine what amount of cover is necessary to reduce erosion by 30 percent from a storm that generates 12 mm of runoff.

7.11 Use the USLE to estimate the soil erosion rate from an unpaved road on a silt loam soil in "good" condition in northern California if the length of road segment is 60 m and the road grade is 4 percent.

7.12 A research plot in a Montana forest is 1 m wide and 1 m long. The slope is 40 percent. Rainfall is applied at 60 mm per hr. When runoff reaches equilibrium, 0.8 liters of runoff are collected in 20 seconds. (a) What is the infiltration rate? (b) The sediment concentration in the runoff sample is 15 g L^{-1}. If K_i and cover are lumped into a single value (K_iC_v), calculate K_iC_v for Equation 7.1.

CHAPTER 8

Terraces and Vegetated Waterways

Terraces, vegetated or grassed waterways, and a range of open channels may all be part of a watershed drainage system. Terraces and waterways are two designs requiring application of the principles of channel design, whereas all three watershed elements may need special features to minimize soil erosion. Vegetated waterways and terraces are special conservation structures for erosion and runoff control, whereas other open channels may be constructed, or may have been natural, but now require improvement to continue their function in watersheds that have been disturbed by agriculture and development.

Terraces

Terracing is a practice to reduce runoff, soil erosion, and sediment delivery from upland areas by constructing broad channels across the slope of rolling land. The first terraces consisted of large steps or level benches. For several thousand years, bench terraces have been constructed all over the world (Figure 8-1c). During the latter part of the nineteenth century, farmers in the southern United States constructed ditches across the slopes of cultivated fields that functioned as terraces. As technology has advanced, terrace design has been adapted to the hydrologic and erosion control needs of a number of different farming systems. Cross sections have been modified to become more compatible with modern mechanization. During the 1940s and 1950s broadbase terraces (Figure 8-1a) with vegetated outlets were commonly installed. Since the 1970s, parallel terraces with steep, grassed backslopes have become popular (Figure 8-2).

In the United States, most terraces are found in the Midwest and South Central states as well as in the Southeast. Crop rows are usually parallel to the terrace channel, so terracing includes contouring as a conservation practice (Figures 8-2 and 8-3). Since terracing requires additional investment and causes some inconvenience in farming, it should be considered only where other cropping and soil management practices will not provide adequate erosion control or water management.

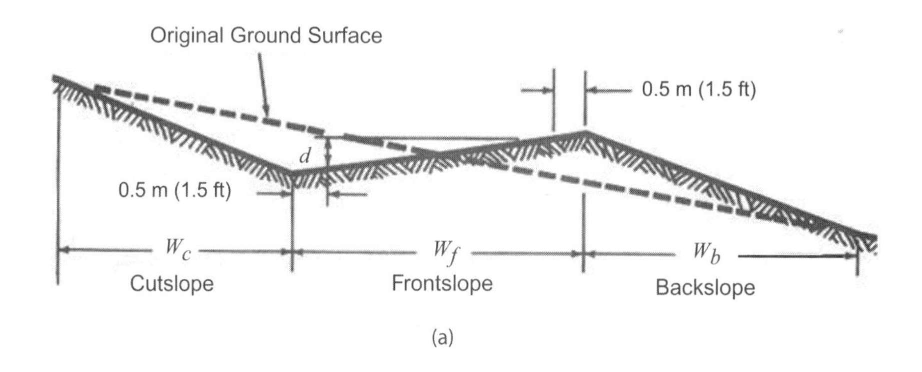

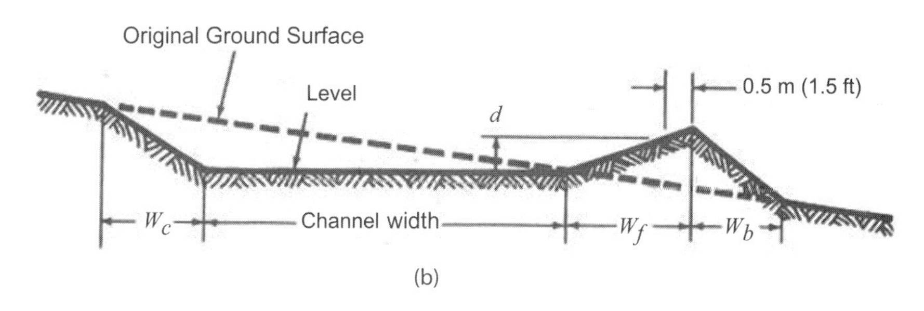

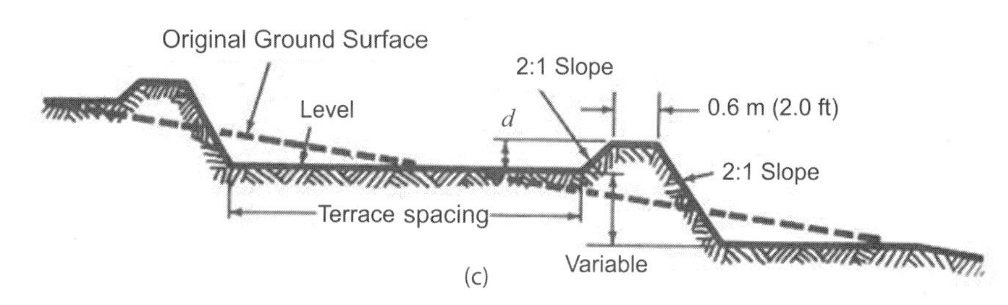

Figure 8–1

Classification of terrace shapes. (a) Three-segment section broadbase; (b) Conservation bench broadbase; (c) Bench. (ASAE, 2003a)

Figure 8–2

Grassed backslope terrace in northeast Iowa. (Photograph by W. Elliot)

Functions of Terraces

Terraces decrease the length of the hillside slope, reducing rill erosion. By diverting or attenuating overland flow, they prevent the formation of gullies, allow sediment

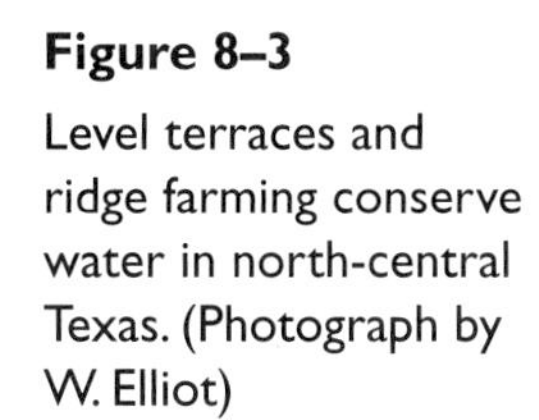

Figure 8–3
Level terraces and ridge farming conserve water in north-central Texas. (Photograph by W. Elliot)

to settle from runoff water, and improve the quality of runoff water leaving the field.

In drier areas, terraces serve to retain runoff and increase the amount of water available for crop production (Figures 8-1b and 8-3). Such retention of water also reduces the risk of wind erosion. Terracing can aid in surface irrigation on steeper land (Chapter 16), particularly in paddy rice production.

Diversions, a form of terrace, can protect bottomlands, buildings, and special areas from unwanted overland flow and erosion from hillsides. Similar structures are also employed in landfills, mining, and forests to reduce soil erosion and offsite sedimentation. In sensitive forest watersheds, similar structures may be constructed after a severe wildfire to reduce upland erosion and downstream sedimentation.

Terrace Classification

8.1 Classification by Alignment

Terrace alignment can either be nonparallel or parallel. Nonparallel terraces follow the contour of the land regardless of alignment. Some minor adjustments are frequently made to eliminate sharp turns and short rows by installing additional outlets, using variable grade, and installing vegetated turning strips. Nonparallel terraces are best suited to applications other than row crop farming, such as small-grain agriculture or pastures. Parallel terraces are preferred for row-crop farming operations. They generally require greater cut and fill volumes during construction than nonparallel systems.

8.2 Classification by Cross Section

There are numerous shapes of terrace cross sections. Bench terraces (Figure 8-1c) are built on steep (20–30 percent) slopes where labor is cheap or land in short supply. They are more efficient at distributing water under both irrigated and dryland production. Bench terraces on very steep slopes may be too narrow or too inaccessible for mechanized farming systems.

The broadbase shapes include the three-segment section (Figures 8-1a and 8-4), the conservation bench (Figures 8-1b, 8-3, and 8-5b), and the grassed backslope terrace (Figures 8-2 and 8-4c). The three-segment section terrace is more com-

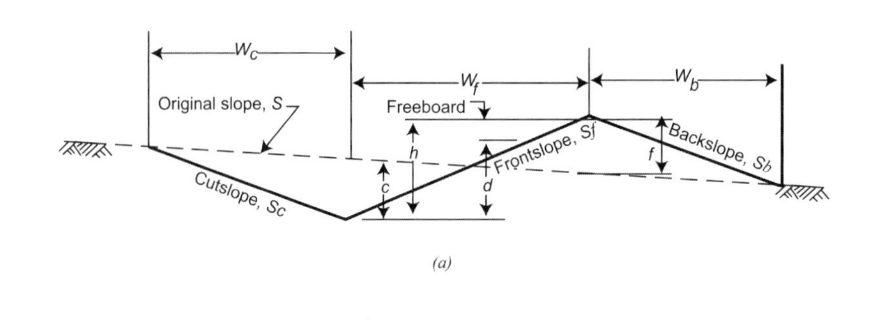

(a)

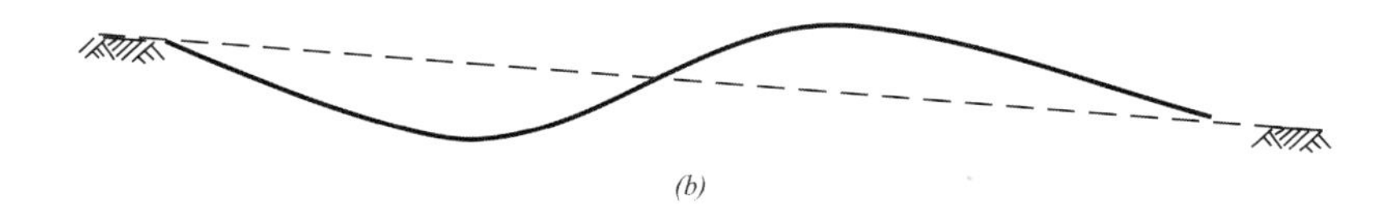
(b)

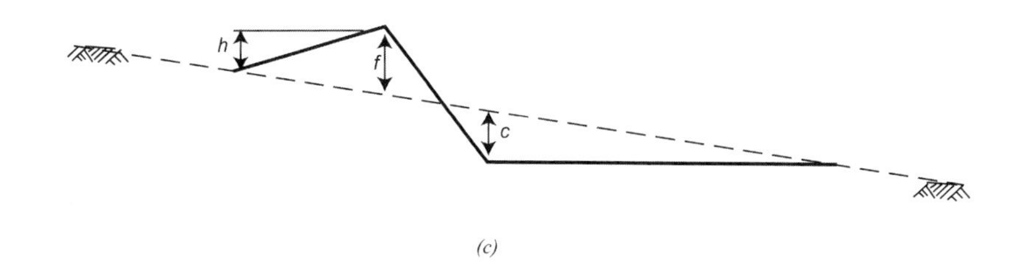

(c)

Figure 8–4

(a) Design of a three-segment broadbase cross section.
(b) Broadbase cross section after 10 years of farming. (c) Grassed backslope cross section.

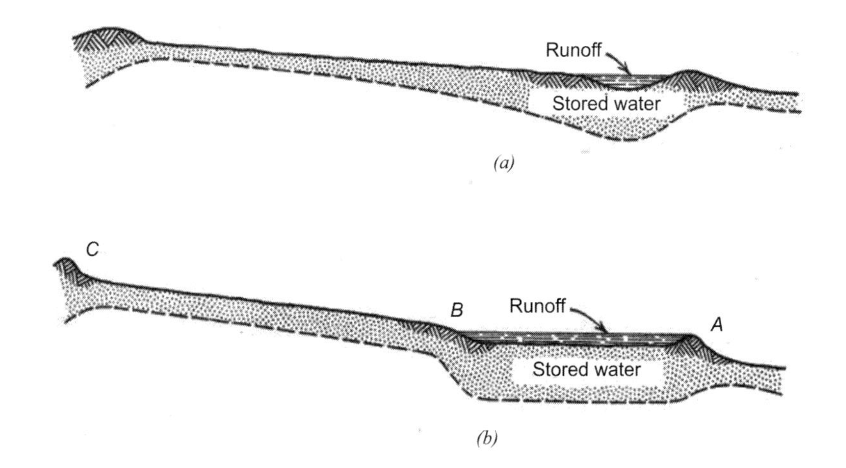

Figure 8–5

Comparison of cross sections of (a) three-segment section level terrace and (b) conservation bench terrace showing stored soil water. Ratio of storage area *AB* to runoff area *BC* may be varied to suit soil, cover, and topographic conditions. (After Hauser et al., 1962)

mon in mechanized farming systems on moderate slopes (6–8 percent). All slopes on the three-segment section broadbase are sufficiently flat for the operation of farm machinery. Lengths of each side slope are designed to match the width of equipment that operates on those slopes. The conservation bench variation incorporates a wide, flat channel uphill of the embankment to provide a maximum area for infiltration of runoff. A comparison of the infiltration abilities between the conservation bench and the three-segment section is shown in Figure 8-5. The grassed backslope terrace is constructed with a 2:1 backslope that is usually seeded to permanent grass because it is too steep to farm. Fill soil may be obtained from the lower side of the terrace, which tends to reduce the land slope between terraces, improving farmability. When field slopes are uniform, a constant terrace cross section is recommended. With nonuniform slopes, some sections will be predominantly fill and some will be predominantly cut (Figure 8-6).

8.3 Classification by Grade

The channels in terraces can be graded toward an outlet, or level. Graded or channel-type terraces control erosion by reducing the hillside slope length of overland flow, and then by conducting the intercepted runoff to a safe outlet at a nonerosive velocity. The reduced flow velocities in the channel minimize channel erosion and promote deposition of eroded sediment. Because of the importance of constructing and maintaining a satisfactory channel, graded terraces should not be built on soils that are too stony or steep, or that have topsoil too shallow to permit adequate construction. Water conservation may be a primary or secondary benefit of a graded terrace system.

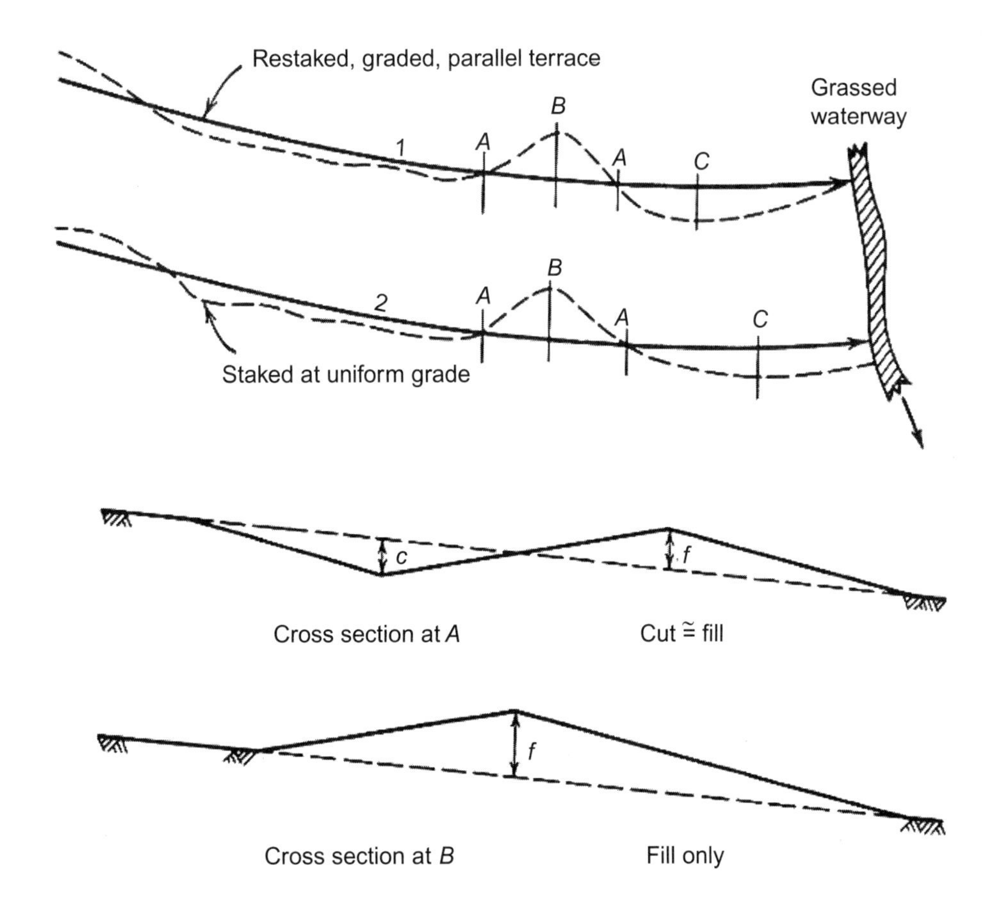

Figure 8–6

Layout of parallel terraces with varying cuts and fills.

Level terraces are constructed to conserve water and control erosion (Figure 8-3). In low- to moderate-rainfall regions they trap and hold rainfall for infiltration into the soil profile. They may be suitable for this same purpose on permeable soils in high-rainfall areas. Frequently, it is necessary to excavate soil from both sides of the embankment to achieve sufficient height to store the design runoff without overtopping or piping through the embankment by the entrapped water. The channel is level and is sometimes closed at both ends. It may have dams installed in it to assure maximum water retention and minimize damage if a section overtops during a severe runoff event. On slopes over 2 percent, water in the channel is spread over a relatively small area, thus limiting the effect on crop yield. The conservation bench cross section (Figures 8-1b and 8-5b) was designed to overcome this deficiency.

8.4 Classification by Outlet

Terrace outlets may be classified as blocked (all water infiltrates in the terrace channel), permanently vegetated (grassed waterway or a vegetated area), or piped (water is removed through subsurface pipe drains). Combinations of outlets may be employed to meet specific conditions.

Planning the Terrace System

The terrace system should be coordinated with the complete water management system for the farm or watershed, giving adequate consideration for proper land use. Terrace systems should be planned by watershed areas, and designs should consider all terraces that may be constructed at a later date. Factors such as fence and road locations must be considered.

8.5 Selection of Outlets

One of the first steps in planning is the selection of outlets or disposal areas. Outlet types include natural drainage ways, constructed channels, sod flumes, permanent pasture or meadow, stable road ditches, waste land, concrete or stabilized channels, pipe drains, and stabilized gullies.

Natural drainage ways, where properly vegetated, provide a desirable and economical outlet. Where these channels are inadequate, constructed waterways along field boundaries or pipe outlets may be considered. The design, construction, and maintenance of vegetated outlets and watercourses as discussed later are applicable for terrace outlets. Terrace outlets on pastureland should be staggered by increasing the length of each terrace a few meters, starting with the lowest terrace (Figure 8-7 terraces 2c and 3c). Sod flumes and concrete channels can be expensive and may require excessive maintenance. Road ditches and active gullies may scour or enlarge if terrace runoff is added. Some highway agencies do not allow outlets to road ditches.

8.6 Terrace Location

After a suitable outlet system is selected, the next step is to determine the location of the terraces. Factors that influence terrace location include (1) land slope; (2) soil conditions, such as degree and extent of erosion; (3) proposed land use; (4) boulders,

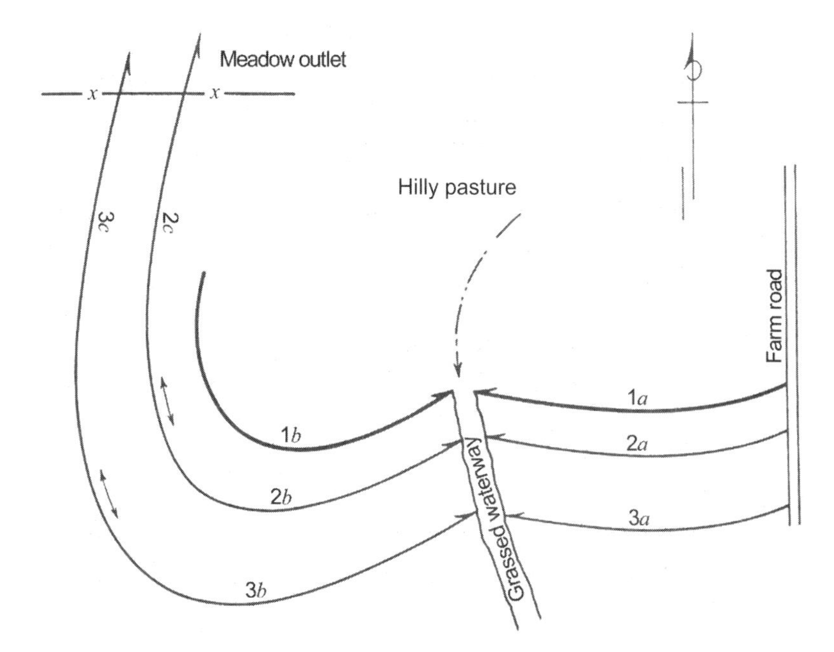

Figure 8–7
Typical layout for broadbase graded terraces.

trees, gullies, and other impediments to construction or cultivation; (5) roads and boundaries; (6) fences; (7) row layout; (8) type of terrace; and (9) outlet. Minimum maintenance, ease of farming, and adequate control of erosion are the criteria for good terrace location. Better alignment of terraces can usually be obtained by placing the terrace ridge just above eroded spots, gullies, and abrupt changes in slope. Satisfactory locations for roads and fences are on the ridge, on the contour, or on the spoil beside the outlet.

The terrace spacing should be calculated (Section 8.7) and the top and bottom terraces located using the multiples of the design or adjusted interval (ASAE, 2003a). For parallel terraces, the second or third terrace is considered the "key" terrace, and its location is determined by surveying (Section 8.13). On long nonuniform slopes, additional key terraces may be required to maintain the desired grade. The exact location of the other terraces can then be located relative to the key terrace(s).

A typical terrace layout is shown in Figure 8-7. The top terrace (1a and 1b) is a diversion that intercepts runoff from the pasture and prevents overland flow crossing the cultivated land below. Since the slope below terrace 1a is uniform, terraces 1a and 3a are laid out parallel to 2a. Because terraces 2bc and 3bc are each longer than 500 m, outlets are provided at each end.

Level terraces are located in much the same manner as graded terraces. In drier climates, level terraces are sometimes constructed so that runoff is allowed to flow from one terrace to the next by opening alternate ends of the terraces.

Terrace Design

The design of a terrace system includes specifying the proper spacing and location of terraces, the design of a channel with adequate capacity, and the development of a stable and sometimes farmable cross section. For the graded terrace, runoff must be removed at nonerosive velocities in both the channel and the outlet. Soil characteristics, cropping and soil management practices, and climatic conditions are the most important considerations in terrace design.

8.7 Terrace Intervals

Spacing is expressed as the vertical distance between the channels of successive terraces. For the top terrace, the spacing is the vertical distance from the top of the hill to the bottom of the channel. This vertical distance is commonly known as the vertical interval or *VI*. The horizontal interval *HI* is found by dividing *VI* by the slope (m/m). The *HI* for parallel terraces is often specified as an even multiple of the row-crop equipment width (such as 2, 4, or 6 widths). The vertical interval is more convenient for terrace layout and construction with surveying equipment.

Example 8.1 If the soil loss was 16 Mg/ha at Memphis, Tennessee, for $K = 0.1$, $1 = 120$ m, $S = 8$ percent, $C = 0.2$, and $P = 0.6$ (Contouring) in the USLE, what is the maximum slope length and corresponding terrace spacing to reduce the soil loss to the terrace channel to 7 Mg/ha?

$$\theta = \tan^{-1}\left(\frac{8}{100}\right) = 4.57\,^{\circ}$$

Solution. From Equations 7-7,

$$b = \frac{\sin(4.57)}{\sin(4.57) + 0.269[\sin(4.57)]^{0.8} + 0.05} = 0.48$$

$$L = \left(\frac{120}{22}\right)^{0.48} = 2.26$$

The maximum L factor to reduce loss to 7 Mg/ha is

$$L = 2.26 \times \frac{7}{16} = 0.99$$

Calculate slope length l to achieve the above L factor value:

$$l = 0.99^{1/0.48} \times 22 = 21.5 \text{ m}$$

Calculate the vertical interval *VI* for the above slope length:

$$VI = \frac{8}{100} \times 21.5 = 1.7 \text{ m}$$

Graded. The graded terrace VI is often expressed as a function of land slope by the empirical formula

$$VI = XS + Y \qquad \textbf{8.1}$$

where VI = vertical interval between corresponding points on consecutive terraces or from the top of the slope to the bottom of the first terrace in m,

X = constant for geographical location (Figure 8-8),

Y = constant for soil erodibility and cover conditions during critical erosion periods, or

= 0.3, 0.6, 0.9, or 1.2, with the low value for highly erodible soils with no surface residue, and the high value for erosion resistant soils with conservation tillage (ASAE, 2003a),

S = average land slope above the terrace, in percent.

Spacings thus computed may be varied as much as 25 percent to allow for soil, climatic, and tillage conditions. Terraces are seldom recommended on slopes over 20 percent, and in many regions slopes from 10 to 12 percent are considered the maximum.

Where soil loss data are available, spacing should be based on slope lengths using the contouring P-factor and the appropriate cover management C-factor for USLE technology, or similar analyses with other erosion prediction methods. This will result in soil losses within the tolerable loss as outlined in Chapter 7.

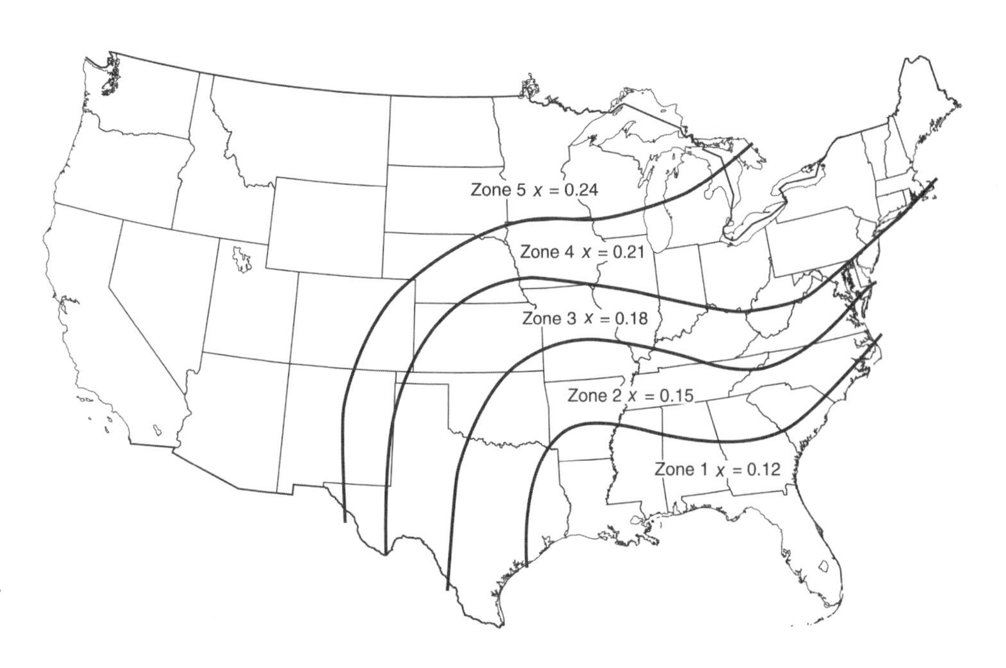

Figure 8–8
Values of X in the terrace spacing equation (Equation 8.1). (ASAE, 2003a)

Example 8.2

Compute the terrace spacing for Example 8.1 from Equation 8.1 assuming that the soil has a low intake rate and that good cover conditions exist.

Solution. From Figure 8-8, read $X = 0.15$ and assume $Y = 0.6$. Substituting in Equation 8.1,

$$VI = 0.15 \times 8 + 0.6 = 1.8 \text{ m}$$

Since the permissible spacing based on soil loss takes into account more of the erosion variables, a vertical interval of 1.7 m as computed in Example 8.1 is preferred.

Level. The horizontal interval for level terraces is a function of channel infiltration and runoff; however, in more humid areas, where erosion control is important, the slope length may limit the spacing. The storage capacity of the terrace should be adequate to prevent overtopping from upslope runoff, and the infiltration rate in the channel should be sufficiently high to prevent serious damage to crops. Horizontal intervals vary widely in different parts of the country, and the NRCS or other agencies should be consulted to determine local practices.

8.8 Terrace Grades

The gradient in the channel must be sufficient to provide adequate drainage while removing the runoff at nonerosive velocities. The minimum slope is desirable from the standpoint of soil loss. Grades may be uniform or variable.

In the uniform-graded terrace, the slope remains constant throughout its entire length. A grade of 0.4 percent is common in many regions; however, grades may range from 0.1 to 0.6 percent, depending on soil and climatic factors. Generally, the steeper grades are recommended for impervious soils and short terraces.

ASAE (2003a) recommends maximum velocities of 0.5 m/s for highly erodible soils and 0.6 m/s for most other soils, when the roughness coefficient in the Manning formula is taken as 0.035 (Appendix B). Recommended minimum and maximum grades are given in Table 8-1.

The variable-graded terrace is more effective because the capacity increases toward the outlet with a corresponding increase in runoff. The grade usually varies from a minimum at the upper portion to a maximum at the outlet end. The resulting reduced velocity in the upper reaches provides for greater absorption of runoff and more deposition of sediment than with a uniformly graded terrace. A variable gradient also increases design flexibility. For instance, varying the grade in the channel can provide either constant velocity or constant capacity. Such designs are sometimes required, particularly in large diversion terraces.

8.9 Terrace Length

Size and shape of the field, outlet possibilities, rate of runoff as affected by rainfall and soil infiltration, and channel capacity are factors that influence terrace length. The number of outlets should be a minimum consistent with good layout and design. Extremely long graded terraces should be avoided. Lengths may be reduced in some terraces by dividing the flow midway in the terrace length and draining the runoff to outlets at both ends of the terrace (Figure 8-7). The length should be such that erosive

TABLE 8–1 Maximum and Minimum Terrace Grades

Terrace Length or Length from Upper End of Long Terraces (meters)	*Maximum Slope[a] (percent)*
30 or less	2.0
31 to 60	1.2
61 to 150	0.5
151 to 365	0.35
366 or more	0.3
	Minimum Slope[b] (percent)
Soils with slow internal drainage	0.2
Soils with good internal drainage	0.0

Source: [a]Beasley (1963). [b]ASAE (2003a).

velocities and large cross sections are not required. Permeable soils can have longer terraces than impermeable. The maximum length for graded terraces generally ranges from about 300 to 500 m, depending on local conditions. The maximum applies only to that portion of the terrace that drains toward one of the outlets.

There is no maximum length for level terraces, particularly where dams are placed in the channel about every 150 m. These dams prevent total loss of water from the entire terrace and reduce gully damage should a break occur. The ends of the level terrace may be left partially or completely open to prevent overtopping in case of excessive runoff.

8.10 Terrace Cross Section

The terrace cross section should provide adequate capacity, may have broad farmable sideslopes, and should be economical to construct with available equipment. For design purposes, the cross section of a broadbase terrace can be considered a triangular channel as shown in Figure 8-4a. The flow depth d is the height h to the top of the ridge minus about 0.08 m for freeboard. After smoothing, the ridge and bottom widths will be about 1 m, which will give a cross section that approximates the shape of a terrace after 10 years of farming (Figure 8-4b).

In designing the cross section, the frontslope width W_f is specified to be equal to the machinery width ordinarily used for farming operations. The depth of flow is determined from the runoff rate for a 10-year return period storm or for the required runoff volume for storage-type terraces. When the side slope widths are equal ($W_c = W_f = W_b = W$), cuts and fills from the geometry are

$$c + f = h + S \times W \tag{8.2}$$

where c = cut (L),
f = fill (L),
h = depth of channel including freeboard (L),

S = original land slope (L/L),

W= width of side slope (L).

For a balanced cross section, cut and fill are equal. Larson (1969) has developed similar equations for grassed backslope terraces (Figure 8-4c).

Example 8.3 For a channel depth h of 0.33 m, and for 8-row (0.75-m row width) equipment 6 m wide on 7 percent land slope, compute the cut and fill heights and the slope ratios for the frontslope and the backslope assuming a balanced cross section.

Solution. From Equation 8.2, letting the cut equal the fill

$$c = f = (h + S \times W)/2$$

$$c = f = (0.33 + 0.07 \times 6)/2 = 0.38 \text{ m}$$

By geometry for the frontslope

$$S_f = 6/0.33 = 18.2 \text{ or } 18{:}1 \text{ slope ratio}$$

Similarly, for the backslope or cutslope

$$S_b = \frac{6}{0.33 + 0.07 \times 6} = 8.0 \text{ or } 8{:}1 \text{ slope ratio}$$

For practical reasons, a terrace is usually constructed with a uniform cross section from the outlet to the upper end, although this construction results in overdesign of the upper portion of the channel. On the conservation bench terrace (Figure 8-5b) the ridge is generally built up to provide a settled height of 0.3 m above the level of the bench. The ends of the bench are blocked to retain 0.15 m of water on each bench before they overtop into a stable drainage system.

8.11 Pipe Outlets

Pipe outlet terraces shown in Figure 8-9 were known in the early 1900s, but the present version was developed in the 1960s. These outlets eliminate the need for grassed waterways. Pipe outlets allow straightening the terrace at natural channels with an earth fill, so it is easier to make terraces parallel. The pipe outlet as shown in Figure 8-9b has an orifice plate to restrict the outflow. This restriction ensures that the subsurface drains are not overloaded, and that sediment in the runoff has time to settle in the terrace channel, improving the quality of the runoff water. To provide for storage in the terrace, the top of the terrace ridge may be constructed at the same elevation along its length even though the bottom of the channel may have a slope to the pipe inlet. These terraces are often constructed with a grassed backslope. The outlet pipe should not be perforated and should be able to withstand static soil loads and dynamic construction and farm machinery loads.

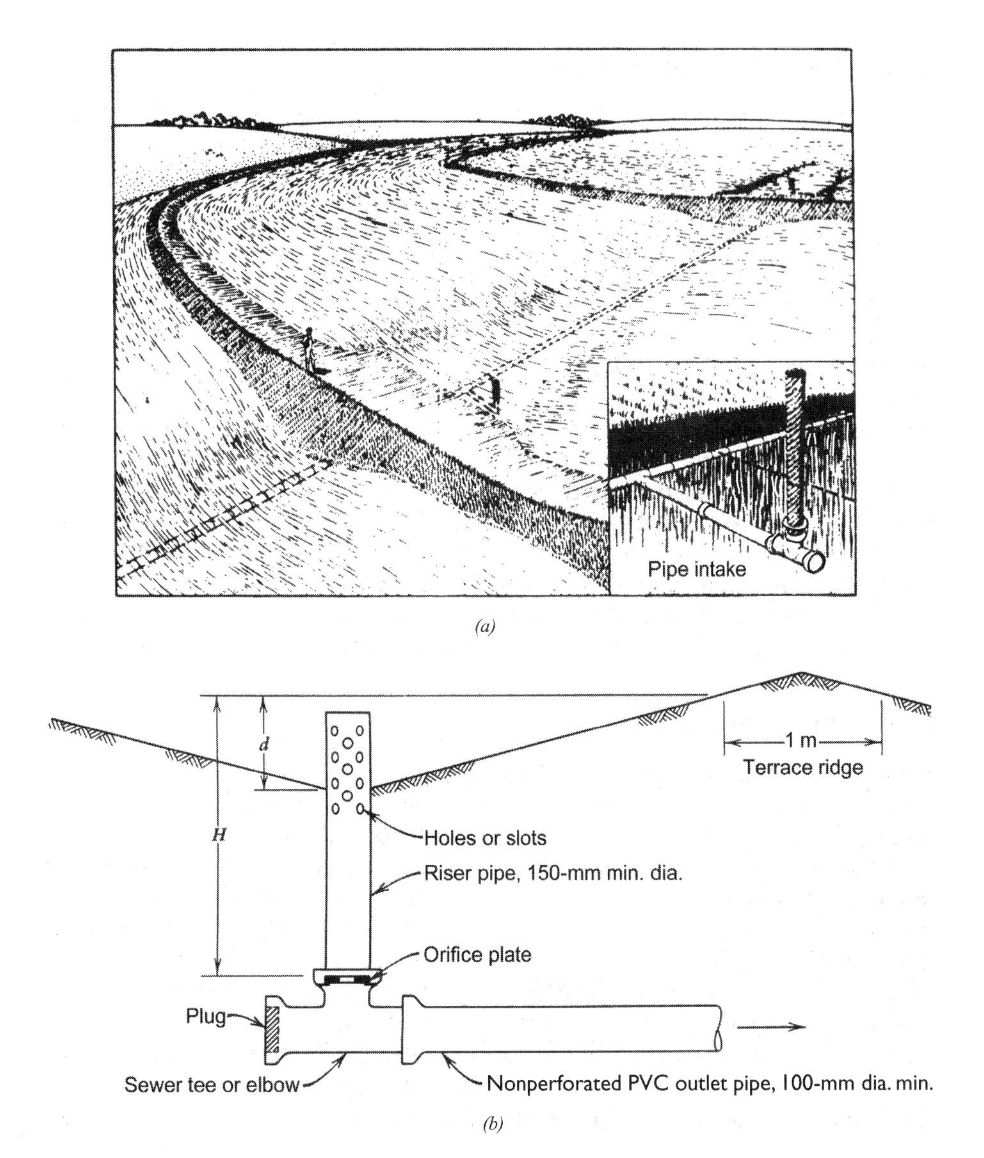

Figure 8–9 (a) Grassed backslope and pipe outlet graded terraces and (b) details of controlled-flow pipe intake. (Redrawn from SCS, 1979)

A variation of the pipe outlet terrace is a noncontinuous sediment and water control basin (ASAE, 2003b). These structures can be less expensive and better suited to greater variations in topography than terrace systems. They are constructed across waterways to prevent or reclaim gullies, reduce the sediment leaving the farm, reduce peak runoff rates, or to conserve water. Slopes may be farmable, or have a steep, grass-covered, backslope. The flow capacity of the outlet may be restricted, or may be designed to accommodate runoff from larger areas, following design guidelines in Chapter 9.

8.12 Terrace Channel Capacity

With graded terraces, the rate of runoff is more important than total runoff, whereas both rate and total runoff influence the design of level, pipe outlet, and conservation bench terraces. Graded terraces are designed as drainage channels or waterways, and level terraces function as storage reservoirs. The terrace channel acts as a temporary storage reservoir subjected to unequal rates of inflow and outflow. The Manning velocity equation (Equations 6-11 and 6-12) is suitable for design. A roughness coefficient of 0.06 is recommended (ASAE, 2003a) to ensure that the channel will carry the design runoff under the most severe channel conditions without overtopping. The maximum design velocity will vary with the erodibility of the soil but should rarely exceed 0.6 m/s for soil devoid of vegetation. The channel depth should permit a freeboard of about 20 percent of the total depth after allowing for settlement of the fill of about 10 percent.

For graded terraces, the design peak runoff rate should be based on a 24-h 10-yr storm (Chapter 3). The runoff volume for level, pipe outlet, and conservation bench terraces should also be based on a 24-h 10-yr storm. The pipe outlet orifice and the subsurface pipe are usually selected so that the 24-hr design storm will be removed in 24 hours for most crops or within 48 hours for flood-tolerant crops like corn (ASAE, 2003b). Subsurface pipe drain capacity is discussed in Chapter 14. To restrict the flow in to the pipe drain, the orifice opening is determined from the orifice equation:

$$q = CA\sqrt{2gh} \tag{8.3}$$

where q = flow (L^3/T),
C = orifice coefficient,
= 0.5,
A = orifice area (L^2),
g = acceleration due to gravity (L/T^2),
h = head of water above orifice (L).

8.13 Layout Procedure

When available, topographic maps or high-resolution digital elevation models (DEMs) are used to develop an initial plan (Wittmus, 1988; ASAE, 2003a). The plan should specify the approximate locations of terraces, drainage structures, and other important features. After the plan is developed, a surveying level and chain or tape are used to set out the terrace location. Field modifications to the plan may be necessary to address topographic features not shown on the contour map.

Use of wide farm equipment makes the elimination of sharp turns and point rows important. In many instances, with a relatively small amount of land forming, the topography of a field can be sufficiently changed to permit parallel terraces. Pipe outlets can simplify designs for parallel terrace systems.

Terrace Construction

8.14 Construction Equipment

A variety of equipment is available for terrace construction, including bulldozers, scrapers, motor graders, and hydraulic excavators (Figure 8-10). Smaller equipment,

Figure 8–10 Terrace construction equipment (a) bulldozer; (b) hydraulic excavator; (c) elevating scraper; and (d) motor grader. (From: http://www.cat.com/)

such as moldboard and disk plows are suitable for slopes of less than about 8 percent, but the rate of construction is much slower than with heavier machines. Soil and crop conditions are likely to be most suitable for construction in the spring and fall.

8.15 Settlement of Terrace Ridges

The amount of settlement in a newly constructed terrace ridge depends largely on soil and water conditions, type of equipment, construction procedure, and amount of vegetation or crop residue. The settlement based on unsettled height will vary from 5 percent or less for a motor grader to 10 to 20 percent for a bulldozer. These values are applicable for soils in good tillable condition with little or no vegetation or residue, and for normal construction procedures. In general, machines that compact the loose fill during construction, like a motor grader, result in less settling than those that carry the soil to the ridge, like a hydraulic excavator.

Terrace Maintenance

Proper maintenance is as important as the original construction of the terrace. However, it need not be expensive, since normal farming operations will usually suffice.

Any breakovers should be repaired as soon as possible. The terrace should be watched more carefully during the first year after construction, and any excessive settlement, failures, or cracking repaired. Channels may occasionally need to be cleared of deposited sediment or ridges rebuilt.

8.16 Tillage Practices

In a terraced field, all farming operations should be carried out as nearly parallel to the terrace as possible (Figures 8-2 and 8-3). The most evident effect of tillage operations after several years is the increase in the base width of the terrace. Reversible plows can be used to increase ridge heights or to redistribute soil that has accumulated in the channel.

Diversions

Diversions or diversion terraces effectively protect bottomland from hillside runoff and divert water away from buildings, cropped fields, and other special-purpose areas. A typical application is shown in Figure 8-11 where a diversion was constructed to allow farming a low area with poor natural drainage. Diverting runoff from above a gully and causing it to flow in a controlled manner to some suitably protected outlet may often be the most effective way to control gullies. Diversions are also common in urban areas for protecting high-value structures and equipment.

Figure 8–11
Typical diversion to protect bottomland crops from upland surface runoff. (SCS, 1979)

8.17 Design of Diversions

A diversion is a channel constructed around the slope and given a slight gradient to cause water to flow to a suitable outlet. The capacity of diversion channels should be based on estimates of peak runoff for the 10-year return period if they are to empty into a vegetated waterway. More severe storms may be considered if the diversion is protecting high-value areas such as building sites. The design procedures for diversions are the same as for vegetated waterways discussed later in this chapter.

Cross section design may vary to suit soil, land slope, and maintenance needs. Sideslopes of 4:1 with bottom widths that permit mowing are frequently used. Since sediment deposition is often a problem in diversions, the designed flow velocity should be kept as high as the channel protection will permit. If the channel has been designed to permit cultivation, the velocity of flow must be based on bare soil conditions, that is, a maximum of about 0.5 m/s.

Vegetated Waterways

Vegetated waterways are surface drains that convey natural concentrations of runoff or carry the discharge from terrace systems, contour furrows, poorly drained low areas (see Chapter 13), diversion channels, or emergency spillways for farm ponds or other structures. Vegetated waterways are installed in channels susceptible to concentrated flow erosion, which in the extreme, may lead to gully formation. Properly designed channels protected by vegetation may completely solve a problem of gully formation. For large runoff volumes or steep channels, it may be necessary to supplement the vegetated watercourse by permanent grade control or gully control structures (see Chapter 9). Waterways may also be used to route storm runoff from structures including building roofs, highways, or other developed areas. Vegetated waterways should not be used for continuous flows, such as discharge from pipe drains, as prolonged wetness in the waterway will result in poor vegetal protection.

The basic approaches for controlling channel erosion include reduction of peak flow rates through the channel with upland protection practices, and provision of a stable channel that can carry the flow that remains. Often this stabilization is accomplished with vegetal protection for the channel together with designing a channel cross section and grade to limit the flow velocities to a rate that the soil with its vegetation cover can withstand.

The design of vegetated waterways is more complex than the design of channels lined with concrete or other stable material because of variation in the roughness coefficient with depth of flow, stage of vegetal growth, hydraulic radius, and velocity. In these waterways, the channel should be designed to carry the runoff at a permissible velocity under conditions of minimum retardance (short vegetation) that may be encountered during the runoff season, and the design then adjusted to ensure adequate capacity under conditions of maximum retardance (tall vegetation). The minimum retardance condition establishes the basic proportions of the channel, for example, the bottom width of a trapezoidal channel. Additional depth is then added to the channel to provide adequate capacity under conditions of maximum retardance. A freeboard of 0.1 to 0.15 m is then added to the design depth. Example 8.4 illustrates the design procedure.

8.18 Vegetated Waterway Design

In the design of a vegetated watercourse, design flow, soil properties, and topography should be determined and then the channel shape designed for these conditions. The capacity of the channel should be based on the estimated runoff from the contributing drainage area. The 10-year return period storm is generally adequate for vegetated waterway design, except in flood spillways for dams or other high-value or high-risk applications. For exceptionally long watercourses it may be desirable to estimate the flow for each of several reaches of the channel to account for changing flow rates. For short channels the estimated flow at the waterway outlet is adequate.

8.19 Shape of Waterway

The cross-sectional shape of the waterway may be parabolic, trapezoidal, or triangular. The parabolic cross section approximates that of natural channels. Under the normal action of channel flow, deposition, and bank erosion, the trapezoidal and triangular sections tend to become parabolic. In some waterways no earthwork is necessary; the natural drainageway is adequate, and only the width and location of the waterway is determined.

A number of factors influence the choice of the shape of cross section. Channels built with a blade-type machine may be trapezoidal if the bottom width of the channel is greater than the minimum width of the blade. Triangular channels may also be readily constructed with blade equipment. Trapezoidal channels having bottom widths less than a mower swath are difficult to mow. Triangular or parabolic channels with sideslopes of 4:1 or flatter may be easily maintained by mowing. Sideslopes of 4:1 or flatter are also desirable for crossing with farm machinery or other off-road vehicles.

Broad-bottom trapezoidal channels require less depth of excavation than do parabolic or triangular shapes. During low-flow periods, sediment may be deposited in trapezoidal channels with wide, flat bottoms. Uneven sediment deposition may result in meandering of higher flows and development of turbulence and eddies that may damage the channel. Triangular channels reduce meandering, but high velocities may damage the bottom.

Parabolic cross sections should usually be selected for natural waterways. A trapezoidal section with a slight V bottom is common where the channel is artificially located as in a terrace outlet along a fence line. The geometric characteristics of three shapes of cross section are given in Figure 6-1.

8.20 Selection of Suitable Vegetation

Soil and climatic conditions are primary factors in the selection of vegetation. Vegetation recommended for various regions of the United States is indicated in Table 8–2. Other factors to be considered are duration, quantity, and velocity of runoff, ease of establishment of vegetation, time required to develop a good protective cover, and risk of sedimentation from excessive vegetation retardance. The landowner may have additional considerations, including utilization of the vegetation as a seed or hay crop, wildlife forage or cover, spreading of vegetation to adjoining areas, and cost and availability of seed. If herbicides are applied near the waterway, it may be necessary to select herbicide-tolerant vegetation.

TABLE 8–2 Vegetation Recommended for Grassed Waterways

Geographical Area of United States	***Vegetation***[a]
Northeastern	Kentucky bluegrass, red top, tall fescue, white clover
Southeastern	Kentucky bluegrass, tall fescue, Bermudagrass, brome, Reed canary
Upper Mississippi	Brome, Reed canary, tall fescue, Kentucky bluegrass
Western Gulf	Bermudagrass, King Ranch bluestem, native grass mixture, tall fescue
Southwestern	Intermediate wheatgrass, western and tall wheatgrass, smooth brome
Northern Great Plains	Smooth brome, western wheatgrass, red top, switchgrass, native bluestem mixture

[a]Recommended vegetation does not necessarily apply to all areas in the region.

8.21 Design Velocity

The ability of vegetation to resist erosion is limited. The permissible velocity in the channel is dependent on the type, condition, and density of vegetation and the erosive characteristics of the soil. Uniformity of cover is very important, as the stability of the most sparsely vegetated areas controls the stability of the channel. Permissible velocities for bunchgrasses or other nonuniform covers are lower than those for sod-forming grasses (Table 8–3). Bunchgrasses produce nonuniform flow with highly localized erosion. Their open roots do not bind the soil firmly against erosion.

8.22 Roughness Coefficient

Slope and hydraulic radius are readily calculated from the geometry of the channel; however, the roughness coefficient is more difficult to evaluate. Figure 8-12 illustrates the complexity of the problem. The roughness coefficient varies with the depth of flow. Shallow flows encounter a maximum resistance because the vegetation is upright in the flow. The slight increase in resistance as the flow increases is due to the greater bulk of vegetation encountered with increasing depth. Intermediate flows bend over and submerge some of the vegetation, and resistance drops off sharply as more and more vegetation is submerged.

Resistance to flow is also influenced by the gradient of the channel. Decreasing resistance results from higher velocities on steeper slopes with an accompanying greater flattening of the vegetation. Type and condition of vegetation have a great influence on the retardance. Newly mown grass offers less resistance than taller growth. Long plants, stems, and leaves tend to whip and vibrate in the flow, introducing and maintaining considerable turbulence.

The product vR, velocity multiplied by hydraulic radius, serves as an index of channel retardance for design purposes. Vegetation has been grouped into five retardance categories designated A through E. Table 8–4 gives a portion of this classification of vegetation by degree of retardance, and Figure 8-13 shows the n–vR curves for five retardance categories. Design considerations generally include the

TABLE 8–3 Permissible Velocities for Vegetated Channels

Cover	Slope Range[a]	Permissible Velocity (m/s)[b] Erosion-Resistant Soils	Easily Eroded Soils
Bermudagrass	0–5	2.4	1.8
	5–10	2.1	1.5
	>10	1.8	1.2
Bahia Buffalo grass Kentucky bluegrass Smooth brome Blue grama Tall fescue	0–5	2.1	1.5
	5–10	1.8	1.2
	>10	1.5	0.9
Grass mixtures	0–5	1.5	1.2
Reed canary grass	5–10	1.2	0.9
Lespedeza sericea Weeping lovegrass Yellow bluestem Redtop Alfalfa Red fescue	0–5[c]	1.0	0.8
Common lespedeza[d] Sudangrass[d]	0–5	1.0	0.8

[a]Do not use on slopes steeper than 10 percent except for vegetated side slopes in combination with a stone, concrete, or highly resistant vegetative center section.

[b]Use velocities exceeding 1.5 m/s only where good covers and proper maintenance can be obtained.

[c]Do not use on slopes steeper than 5 percent except for vegetated side slopes in combination with a stone, concrete, or highly resistant vegetative center section.

[d]Annuals—use on mild slopes or as temporary protection until permanent covers are established.

Source: SCS (1979).

vegetation and flow conditions, especially for long, large channels where refinements in design may result in lower construction costs.

Example 8.4

Design a trapezoidal grassed waterway to carry 5 m^3/s on a 3 percent slope on erosion-resistant soil. The vegetation is a grass-legume mixture, and the channel should have 4:1 sideslopes.

Solution. From Table 8-3, the permissible velocity for grass mixtures on erosion resistant soils with a slope between 0 and 5 percent slopes is 1.5 m/s. Table 8-4 shows the grass mix in retardance Class D in the fall and in Class B during other times.

An initial estimate of hydraulic radius of 0.3 m leads to a vR value of 0.45. From the equation in Figure 8-13, for retardance Class D and a vR of 0.45, calculate a roughness of about 0.036.

Solving the Manning equation (Equation 6.11) for R gives

$$R = \left[\frac{vn}{S^{0.5}}\right]^{3/2}$$

$$R = \left[\frac{1.5 \times 0.036}{0.03^{0.5}}\right]^{3/2} = 0.17 \text{ m}$$

From Equation 6.4, the cross-sectional area must be

$$A = \frac{5}{1.5} = 3.33 \text{ m}^2$$

From the cross-sectional area equation for a trapezoidal channel in Figure 6-1a, the relationship between depth, bottom width, and area is

$$b = \frac{A - zd^2}{d}$$

A spreadsheet or equation solver can now be used to determine the correct depth and width for this channel. For an initial estimate, assume a depth of 0.2 m.

A	z	d	b	R	Comment
Given	Given	Est	Above Eq	Fig. 6–1a	
3.33	4	0.2	15.85	0.19	*R* is too high
3.33	4	0.18	17.78	0.17	*R* is acceptable

The dimensions on the second line will provide a stable channel with v = 1.5 m/s.

The design depth must now be increased when the grass is long with retardance Class B because the velocity is reduced. Another spreadsheet solution is developed to determine the new depth at the higher retardance. An estimate of n, assuming an R of 0.4 and a velocity of 1.5, for Class B is 0.056.

z	d	b	A	R	n	s	v	Q	Comment
4	0.18	17.78	3.33	0.17	0.034	0.03	1.58	5.26	Class D condition
4	0.18	17.78	3.33	0.17	0.056	0.03	0.96	3.19	Class B condition, *Q* too small
4	0.24	17.78	4.50	0.23	0.056	0.03	1.15	5.18	Class B, *Q* is correct

This solution calculates a depth of 0.24 m for the Class B condition. A freeboard of 0.1 m should be included. This increases the depth to 0.34 m. The equation for top width (T) from Figure 6-1a calculated the constructed top width to be

$$T = 17.78 + (2 \times 0.34 \times 4) = 20.5 \text{ m}$$

The NRCS grassed waterway design program (NRCS, 2002) calculated similar values, estimating a higher Manning's n of 0.0839, and thus recommending a larger bottom width of 20.18 m, and a depth of 0.277 m for the long grass condition.

The bottom width is determined by the permissible velocity under the mowed condition of minimum retardance, and the depth is determined by the need to provide capacity under conditions of maximum retardance and to allow for freeboard.

Immediately after construction, the channel may need to carry runoff under conditions of little or no vegetation. It is not practical to design for this extreme condition. In many channels, it may be practical and desirable to divert flow from the channel until vegetation is established. In others, the possibility that high runoff will occur before vegetation is established is accepted as a calculated risk. In some cases, it may be desirable to line the channel with a geotextile fabric or mulch to minimize erosion until the vegetation is established.

8.23 Waterway Drainage

Waterways that are located in wet draws or below seeps, springs, or pipe outlets will be wet for long periods. The wet condition will inhibit the development and

TABLE 8–4 Classification of Vegetal Cover According to Retardance

Retardance	*Cover*	*Condition*	*Height (m)*
A	Reed canary	Excellent stand, tall	0.9
	Yellow bluestem *Ischaemum*		
B	Smooth brome	Good stand, mowed	0.3–0.4
	Bermuda	Good stand, tall	0.3
	Native grass mixture	Good stand, unmowed	0.3
	Tall fescue	Good stand, unmowed	0.5
	Lespedeza sericea	Good stand, not woody, tall	0.5
	Grass-legume mixture	Good stand, uncut	0.5
	Reed canary	Good stand, mowed	0.3–0.4
	Tall fescue with bird's-foot trefoil or ladino	Good stand, uncut	0.5
	Blue grama	Good stand, uncut	0.3
C	Bahia	Good stand, uncut	0.2
	Bermuda	Good stand, mowed	0.15
	Redtop	Good stand, headed	0.4–0.5
	Grass-legume mix—summer	Good stand, uncut	0.2
	Centipede grass	Very dense cover	0.15
	Kentucky bluegrass	Good stand, headed	0.2–0.3
D	Bermuda	Good stand, cut	0.1
	Red fescue	Good stand, headed	0.3–0.5
	Buffalograss	Good stand, uncut	0.1–0.2
	Grass-legume mixture—fall	Good stand, uncut	0.2
	Lespedeza sericea	After cutting	0.1
E	Bermudagrass	Good stand	0.1

Source: SCS (1979).

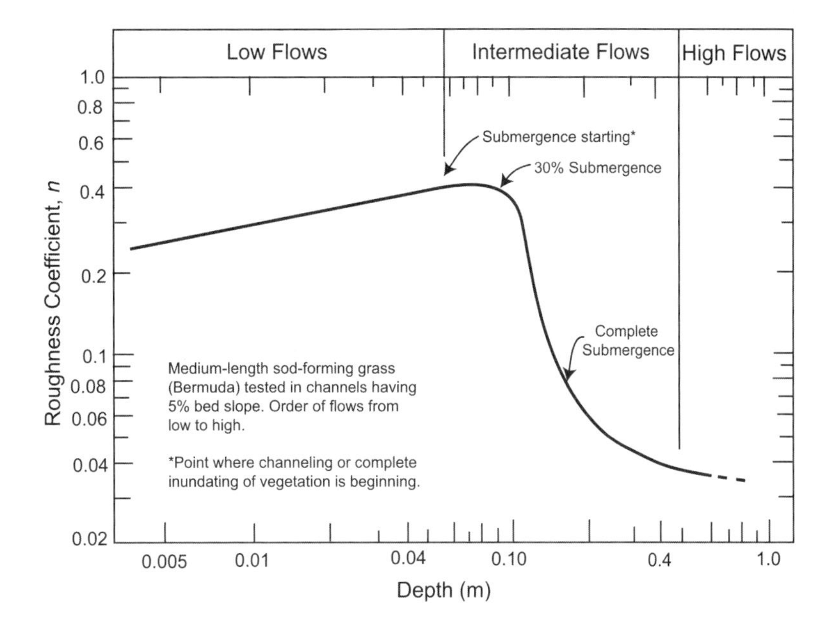

Figure 8–12 Hydraulic behavior of a medium-length sod-forming grass. (Adapted from Ree, 1949)

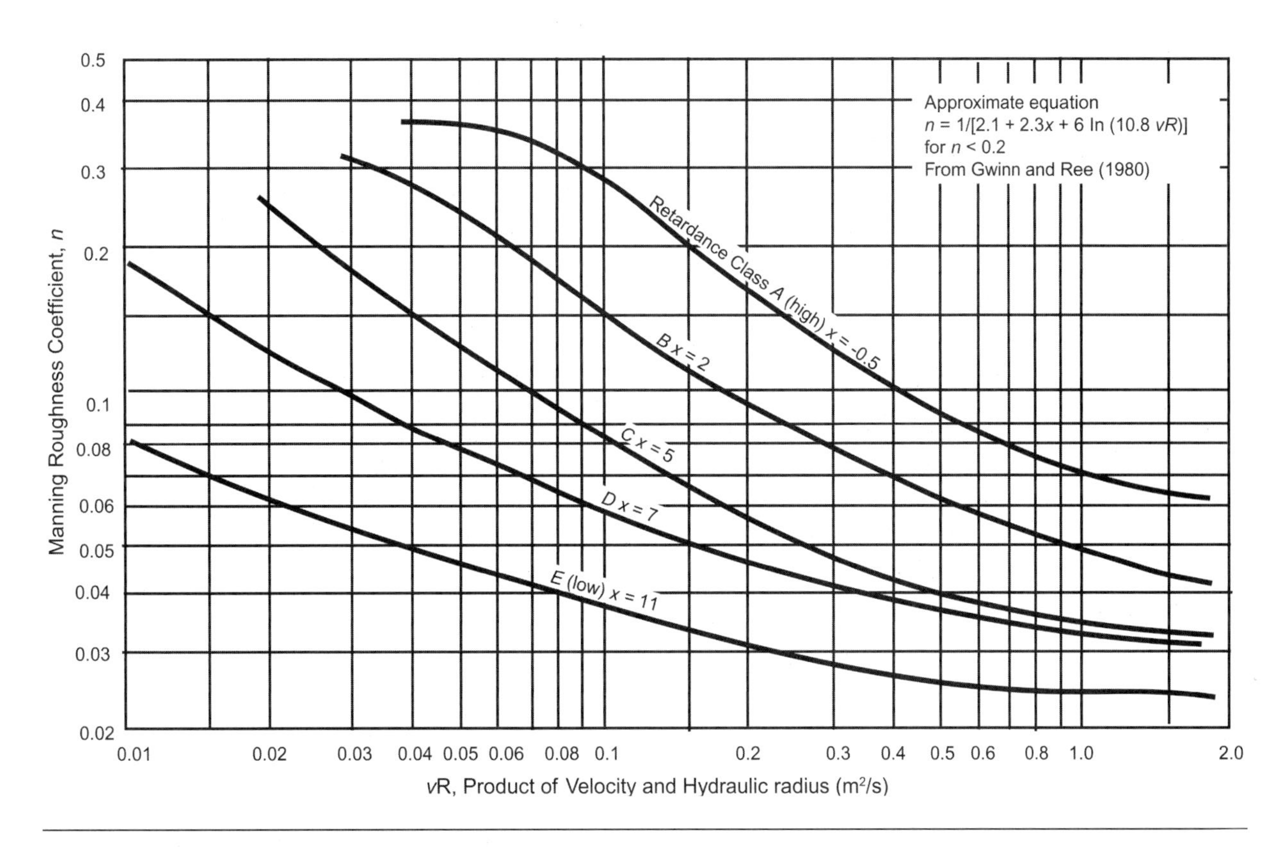

Figure 8–13 Roughness coefficient as a function of *vR* for various retardance classes of vegetation. (Redrawn from SCS, 1979)

maintenance of a good vegetal cover and may cause the soil to be in a weak, erodible condition. Subsurface drainage or diversion of such flow may be necessary for the waterway.

Designs for sites having prolonged flows, a high water table, or seepage problems should include subsurface drains, underground outlets, stone center waterways, or other suitable measures to avoid saturated conditions (NRCS, 2000). Subsurface drains may be installed to intercept seepage along the sides or upper end of the waterway. Drains should be placed on one side of the center of the waterway to reduce the chance of erosion leading to exposure of the drain should the waterway fail.

Construction

Channel construction methods depend on topography and available equipment. Most channels require earth-moving equipment, such as bulldozers or excavators (Figure 8-10). Large projects that require transport distances greater than 50 m may require scrapers for efficient construction. Motor graders, scrapers, and bulldozers can all be equipped with laser-guided grade depth control for channels requiring precision grades. For small waterways or drainage channels, agricultural equipment such as plows or blades may be adequate for construction. For waterways located in a natural waterway or meadow outlet where there is little gullying, only smoothing and normal seedbed preparation may be required.

8.24 Shaping Channels

Correct shaping of channels will improve the hydraulic characteristics, facilitate farming or other operations, and reduce maintenance. If the waterway is to reclaim an established gully, considerable earthwork may be required. Small waterways may be easily shaped with farm equipment or small construction equipment. Large gully reclamation, however, will require a bulldozer, excavator, or other heavy earth-moving equipment (Figure 8-10).

8.25 Establishment of Vegetation

In grassed waterways and some surface drainage ditches, the channel is vegetated. In these cases, soil in the channel should be adequately tilled for the seeding equipment. Direct drills will require less tillage than conventional drills or broadcast seeding. It may be necessary to incorporate adequate chemical fertilizer or manure into the seedbed to provide both plant nutrients and organic matter to increase resistance to erosion.

Channel-seeding mixtures should include some quick-growing annual grasses for temporary cover as well as a mixture of hardy perennials for permanent protection (Table 8-2). Seed should be either broadcast or drilled nonparallel to the direction of flow. Mulching after seeding may help to secure a good stand. If it is likely that the channel may carry high flows before seedlings can become established, special materials can be applied to control infiltration and runoff and facilitate revegetation. Such materials include chemical soil stabilizers, plastic, fiber, or other mesh or net covers, asphalt mulches, and plastic or other surface covers. These materials reduce the rate of drying and crusting and absorb the energy of overland flow, raindrops, and/or wind, thus protecting the soil and increasing the likelihood of good vegetation establishment.

Channel Maintenance

8.26 Causes of Failure

Grassed waterways and other channels may fail from insufficient capacity, excessive velocity, insufficient resistance of surface treatment, or inadequate vegetal cover. The first three failure modes are largely a matter of design. The condition of the vegetation, however, is influenced not only by the initial preparation of the waterway but also by the subsequent management. Use of any channel, especially in wet weather, as a lane, stock trail, or pasture may damage the channel shape or vegetation, leading to failure.

Grassed waterways are particularly susceptible to damage. Terraces that empty into a waterway at too steep a grade may erode into the terrace channel, damaging both terrace and channel. Careless handling of machinery in crossing a channel may injure the surface treatment. When land adjacent to the waterway is tilled, the ends of furrows abutting against the vegetated strip should be staggered to prevent flow concentration down the edges of the watercourse.

8.27 Controlling Vegetation

Nonvegetated channels may need occasional cleaning, generally by a tractor backhoe or hydraulic excavator (Figure 8-10b) to remove unwanted vegetation as well as accumulated sediment. Other channels may need mowing to prevent growth of brush or trees that may limit channel capacity. Grassed waterways should be mowed several times a season to stimulate new growth and control weeds. Any breaks in the sod should be repaired. Rodents that are damaging channels should be controlled.

8.28 Sediment Accumulation

Good conservation practices on the watershed are the most effective means of controlling sedimentation. Accumulated sediments restrict the capacity of channels and smother vegetation in waterways. Extending vegetal cover well up the sideslopes of waterways and into the outlets of terrace channels helps to reduce channel sedimentation. Removal of vegetation from earthen channels and control of vegetation in grassed waterways to prevent a rank, matted growth reduces the accumulation of sediment. Higher design velocities, particularly on low-gradient channels, also decrease sedimentation on erosion-resistant soils. In some cases, culverts or other downstream constrictions can cause unwanted sediment to accumulate in channels. In some cases, it may be necessary to redesign these structures.

Vegetation of Large Gullies

The discussion to this point has been applicable mainly to channels and vegetated waterways and to the stabilization of small gullies. Larger gullies may be controlled by reduction of the surface inflow, by shaping and intensive natural or artificial revegetation, or by the installation of control structures (see Chapter 9). Reclaiming an established gully may require considerable earthwork. In some cases, the gully may be filled and the channel cross section properly shaped; in others, the gully sides and bottom may be reshaped to minimize further erosion or side sloughing, and revegeated.

8.29 Control of Inflow

Large gullies generally have a large contributing watersheds with a high runoff potential. To facilitate the vegetation and control of the gully, the normal conservation practices that will protect small gullies must be replaced with more effective methods. Diversion terraces, constructed above the gully head and carefully laid out on a grade that will resist channel erosion, can be used to intercept runoff from the watershed area above the gully and then convey it to a safely stabilized outlet area (Section 8.17).

8.30 Sloping Gully Banks

Bank sloping should be done only to the extent required for establishment of vegetation or for facilitating tillage operations. Where trees and shrubs are to be established, rough sloping of the banks to about 1:1 should be sufficient. On unstable soils, a stability analysis may be necessary to determine a stable slope for the gully sides. Where gullies are to be reclaimed as grassed waterways, sloping of banks to 4:1 or flatter may be desired.

8.31 Revegetation

If the runoff that has caused the gully is diverted, and livestock are fenced from the gullied area, plants will frequently reestablish naturally. A gradual succession of plant species eventually will protect the gullied area with grasses, vines, shrubs, or trees native to the area. In some gullied areas, the development of vegetation may be stimulated by fertilizing and by spreading mulch to conserve water and protect young volunteer plants.

In some cases, it may be desirable to artificially revegetate the gully. Grasses and legumes may be planted if the vegetation is to be used for a hay or pasture crop. Where gullies are reclaimed as grassed waterways, sod-forming vegetation should be selected to allow crossing with farm machines. In some areas, trees and shrubs are easier to establish in gullies than are grasses, particularly if the gully is not regraded to allow mechanized seeding. Specific shrubs may be desirable for establishing the gullied area as a wildlife refuge, and local practices should be followed in selection and planting.

Internet Resources

Waterway design program
http://www.cis.ksu.edu Search site for channel design sites-ide.

Ohio Engineering Software, including grassed waterway and open channel design tools
http://www.oh.nrcs.usda.gov/technical/engineering/engineering_software.html

References

American Society of Agricultural Engineers (ASAE). (2003a). *Design, Layout, Construction and Maintenance of Terrace Systems*. ASAE Standard: S268.4, 811–816. St. Joseph, MI: Author.

American Society of Agricultural Engineers (ASAE). (2003b). *Water and Sediment Control Basins*. ASAE Standard: S442, 913–915. St. Joseph, MI: Author.

Beasley, R. P. (1963). *A New Method of Terracing*. Missouri Agricultural Experiment Station Bulletin 699.

Hauser, V. L., & M. B. Cox. (1962). A comparison of level and graded terraces in the southern high plains. *Transactions of the American Society of Agricultural Engineers, 5*, 75–77.

Larson, C. L. (1969). Geometry of broad-based and grassed-backslope terrace cross sections. *American Society of Agricultural Engineers Transactions, 12*(4), 509–511.

Natural Resource Conservation Service (NRCS). (2000). *Natural Resources Conservation Service Conservation Practice Standard Grassed Waterway*. Online at http://www.nrcs.usda.gov/technical/Standards/nhcp.html. Accessed August 2004.

———. (2002). *Ohio Engineering Software*. Online at http://www.oh.nrcs.usda.gov/technical/engineering/engineering_software.html. Accessed January 2005.

Ree, W. 0. (1949). Hydraulic characteristics of vegetation for vegetated waterways. *Agricultural Engineering, 30*, 184–187, 189.

U.S. Soil Conservation Service (SCS). (1979). *Engineering Field Manual for Conservation Practices*. Washington, DC: Author.

Wittmuss, H. (1988). A study of time required for planning, staking, and designing parallel terrace systems. *Transactions of the American Society of Agricultural Engineers*(30), 1076-1081.

Problems

8.1 A diversion terrace is needed to protect a gully for a flow of 0.85 m^3/s. Assume a poor stand of grass, and use an allowable velocity of 0.9 m/s. Determine the total depth, including freeboard and the required slope using a 2.4-m bottom width. Allow 0.1 m of freeboard with long grass (Class D). Use a trapezoidal cross section with 3:1 sideslopes.

8.2 Using a Web browser, find three sources for geotextiles suppliers whose products can be used for channel protection. Determine the important differences among the three products that you have found.

8.3 On one graph, plot two curves with the slope in percent (0 to 10) on the *x* axis, and the vertical interval and the horizontal interval for graded terraces on the *y* axis. Follow recommendations specified for your area for a resistant soil with good cover.

8.4 Compute the cut volume and the fill volume for a 200-m long terrace on a 4 percent hillslope with a depth of flow of 0.3 m. Assume the three slope widths are 4.3 m, the freeboard is 0.8 m, and the cross section is balanced (cut volume = fill volume).

8.5 Determine the vertical interval from Equation 8.1 for graded terraces on resistant soil and with good cover on a 6 percent hillslope at your present location. What is the peak runoff from the second terrace if the terrace length is 350 m?

8.6 After a fire, a level terrace system is to be excavated in a forest to stop overland flow in western Montana. The system is to store the runoff from a 5-year, 24-hour storm. If the terrace horizontal spacing is 40 m, hillslopes are 40 percent, and terrace sideslopes are

2:1, what is the depth and height of the terrace, assuming a balanced cross section (cut volume = fill volume)?

8.7 If a subsurface outlet terrace is to store 50 mm of runoff from a 10-year return period, 24-h duration storm, determine the orifice opening size to remove this volume in 48 hours. Assume an average pressure head of 1.5 m at the orifice, an orifice discharge coefficient of 0.6, and a runoff area above the terrace of 1 ha.

8.8 For an implement width of 5 m, a depth of flow of 0.25 m, and a 4 percent slope, what are the slope ratios of the cutslope, frontslope, and backslope of a three-section broad-base terrace?

8.9 What are the slope ratios for Problem 8.8 if the land slope is 10 percent?

8.10 A farmer in your area desires to drain 250 ha of alluvial sandy loam land. A topographic survey indicates that the maximum slope is 0.2 percent. Assuming cultivated land, design the ditch cross section, including the spoil bank.

8.11 Design an open ditch to carry the runoff from a 3600-ha watershed in your area. The slope of the land along the route of the ditch is 0.3 percent, the average slope of the watershed is 0.5 percent, and the soil is heavy clay. Assume that the ditch should provide sufficient depth for pipe drains (about 0.4 m) and the land is cultivated.

8.12 Design a parabolic-shaped grassed waterway to carry 1.5 m^3/s. The soil is easily eroded; the channel has a slope of 4 percent; and a good stand of Bermudagrass, cut to 60 mm, is to be maintained in the waterway.

8.13 Design a trapezoidal-shaped waterway with 5:1 sideslopes to carry 0.6 m^3/s where the soil is resistant to erosion and the channel has a slope of 5 percent. Bromegrass in the channel may be either mowed or long when maximum flow is expected.

8.14 Design a parabolic-shaped waterway to carry 1.8 m^3/s from a terraced field where the soil is resistant to erosion and the channel slope is 6 percent. Tall fescue is normally mowed, but the channel should have adequate capacity when the grass is long.

8.15 Design a trapezoidal-shaped emergency spillway for a farm pond to carry 1.8 m^3/s with retardance Class C condition. The maximum velocity in the spillway is 1.5 m/s. Around the end of the dam the channel slope is to be 2 percent, but it is increased to 7 percent from this section to the stream channel below. The depth of flow should not exceed 0.3 m. Determine the bottom width, design depth, and recommend design slopes for both the 2 and 7 percent sections.

CHAPTER 9 Water and Sediment Control Structures

In recent years, considerable emphasis has been placed on "natural" engineering of water conveyances and structures. Such methods are useful and beneficial in many situations (see Chapter 10); however, there are situations where "soft" structures are ineffective or uneconomical. For example, where energy must be dissipated quickly, hard structures usually provide the most practical solution. This chapter presents several examples of hard structures used for conveyance, flow control, and energy dissipation.

Earthen embankments are used in various storage and flood control structures. The ready availability and low cost of earthen materials make them an ideal choice for construction of dams and dikes for ponds, reservoirs, waste-holding structures, and flood control levees. Another important application is in the design of sedimentation basins. Many jurisdictions now require sediment control practices on disturbed lands, such as construction sites, open mines, and agricultural lands, which tend to be highly erodible. This chapter presents fundamentals of design of earthen structures and introduces their use as sedimentation basins.

Water Control Structures

9.1 Temporary and Permanent Structures

Temporary Structures. Temporary structures should be recommended only where inexpensive labor and materials are available. Increasing mechanization and higher labor costs have resulted in a decline in the practicality of temporary channel stabilization structures. Practices that make use of temporary materials such as logs and root wads can be effective if combined with channel modifications that will result in a stable stream (Chapter 10). Without such modifications, the problems are likely to recur, progressively degrading the land.

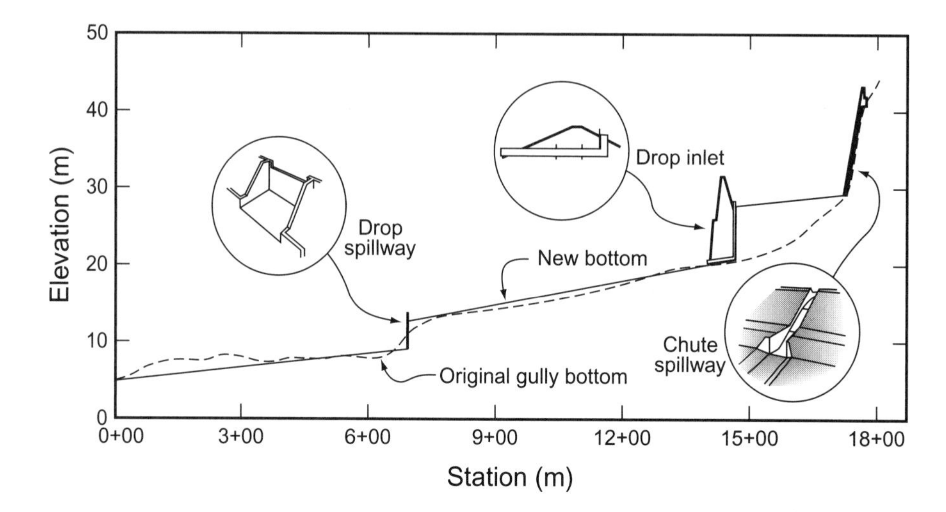

Figure 9–1

Profile of a gully stabilized by three types of permanent structures.

Permanent Structures. Permanent structures of hard materials may be required to dissipate the energy of the water; for example, where a vegetated waterway discharges into a drainage ditch, at the head of a large gully, or in a channel reach where the grade is too steep to be stable. Where flow velocities must exceed the maximum values for nonerosive conditions (Chapter 8), an erosion-resistant lining may be required. Figure 9-1 shows the profile of a gully that has been reclaimed by methods involving the use of several types of permanent structures. Standard designs are available from SCS (1984), USBR (1987), and Brater & King (1996).

9.2 Functional Requirements of Control Structures

The design of control structures must address two primary requirements: (1) adequate capacity to pass the design discharge, and (2) dissipation of the energy of the water within the structure in a manner that protects both the structure and the downstream channel from damage or erosion. The main causes of failure of permanent control structures are insufficient hydraulic capacity and insufficient energy dissipation capacity. All permanent structures require maintenance, though it may be infrequent. Where maintenance is neglected, small problems can grow and eventually lead to total failure.

9.3 Design Features of Control Structures

The basic components of a hydraulic structure are the inlet, the conduit, and the outlet. Structures are classified and named in accordance with the form of these three components. Figure 9-2 shows various types of inlets, conduits, and outlets in common use. In addition to these hydraulic features, the structure must include suitable wing walls, side walls, head wall extensions, and toe walls to prevent seepage under or around the structure and to prevent damage from local erosion. These structural components are identified in Figure 9-3 for one common type of structure. It is important that a firm foundation be secured for permanent structures. Wet foundations should be avoided or provided with adequate artificial

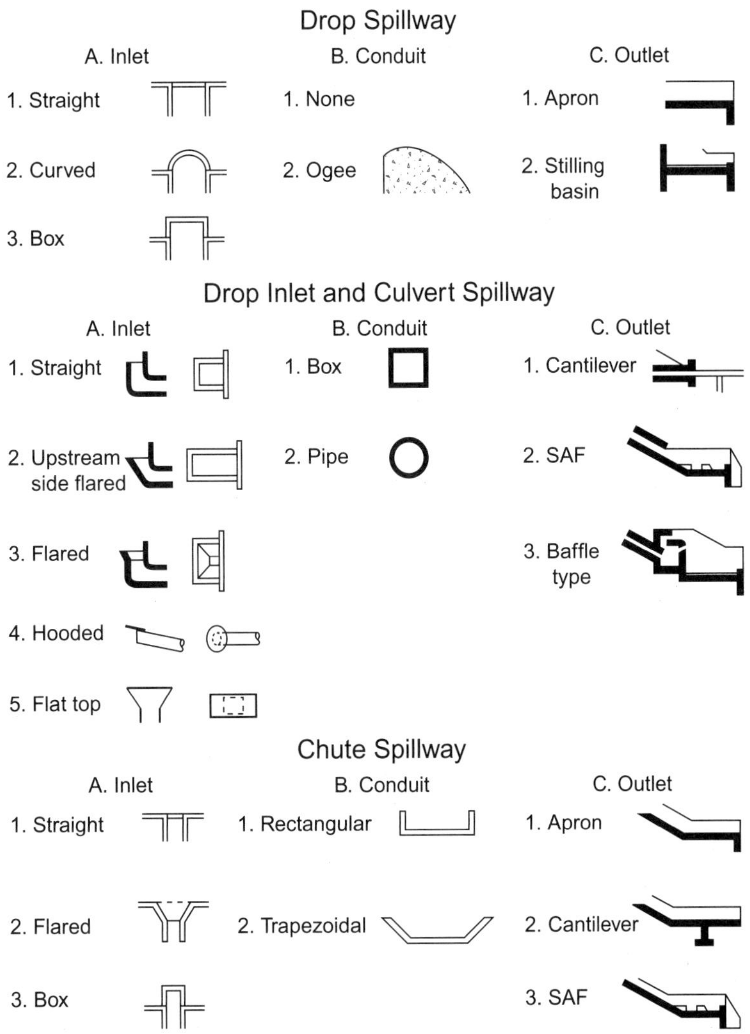

Drop Spillway

A. Inlet
1. Straight
2. Curved
3. Box

B. Conduit
1. None
2. Ogee

C. Outlet
1. Apron
2. Stilling basin

Drop Inlet and Culvert Spillway

A. Inlet
1. Straight
2. Upstream side flared
3. Flared
4. Hooded
5. Flat top

B. Conduit
1. Box
2. Pipe

C. Outlet
1. Cantilever
2. SAF
3. Baffle type

Chute Spillway

A. Inlet
1. Straight
2. Flared
3. Box

B. Conduit
1. Rectangular
2. Trapezoidal

C. Outlet
1. Apron
2. Cantilever
3. SAF

Figure 9–2
Classification of components of hydraulic structures.
(Adapted from SCS, 1984)

drainage. Topsoil and organic material should be removed from the site to allow a good bond between the structure and the foundation material.

Many energy dissipation structures make use of a hydraulic jump, which is a transition from a relatively shallow and rapid flow to a relatively deep and slow flow. Flow in the transition zone is highly turbulent and dissipates some of the energy of the water. See Chapter 6 for discussions of states and energies of flow.

Drop Spillways

A typical drop spillway is shown in Figure 9-3. Drop spillways may have a straight, arched, or box-type inlet. The energy dissipater may be a straight apron or some type of stilling basin.

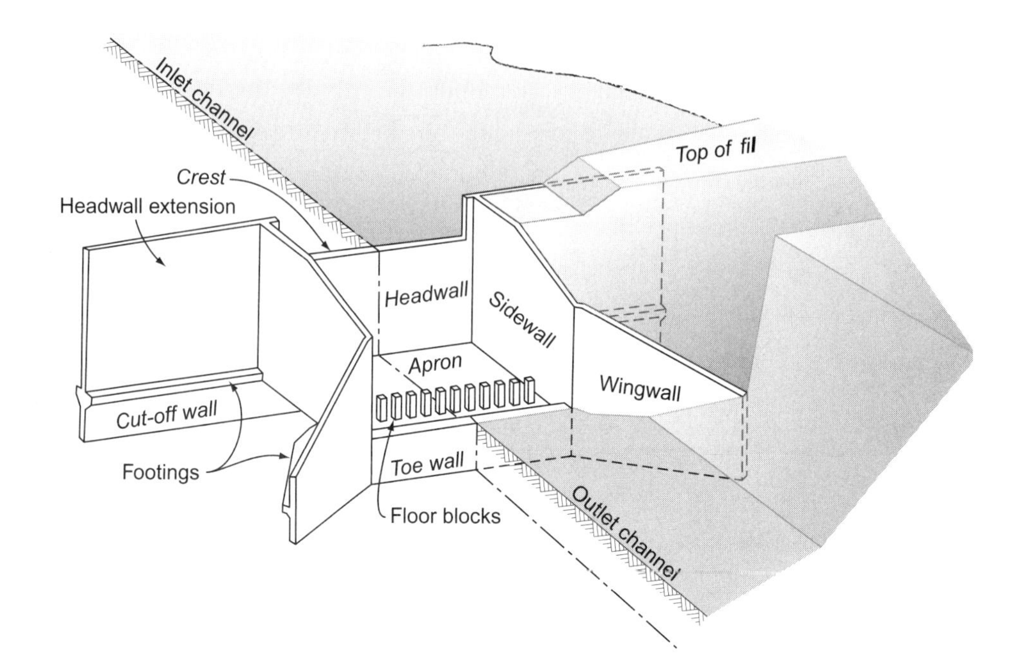

Figure 9–3
Straight drop spillway showing structural components. (Redrawn from SCS, 1984)

9.4 Function and Limitations of Drop Spillways

Drop spillways are installed in channels to establish permanent control elevations below which an eroding stream cannot lower the channel floor. The structures control the stream grade from the spillway crest through the entire ponded reach upstream. Drop structures placed at intervals along the channel can stabilize it by changing its profile from a continuous steep gradient to a series of more mildly sloping reaches. Where relatively large volumes of water must flow through a narrow structure at low head, the box-type inlet is preferred. The curved inlet serves a similar purpose and also gives the advantage of arch strength where masonry construction is used. Drop spillways are usually limited to drops of 3 m; flumes or drop-inlet pipe spillways are used for greater drops.

9.5 Design Features of Drop Spillways

Capacity. The free flow (i.e., with no submergence) capacity for drop spillways is given by the weir formula (Chapter 6)

$$Q = CLh^{3/2} \tag{9.1}$$

where Q = discharge (L^3T^{-1}),
C = weir coefficient ($L^{1/2}T^{-1}$),
L = weir length (L),
h = depth of flow over crest (L).

The length L is the sum of the lengths of the three inflow sides of a box inlet, the circumference of an arch inlet, or the crest length of a straight inlet. Some references provide weir coefficients only for English units (foot, second). Values of C can be

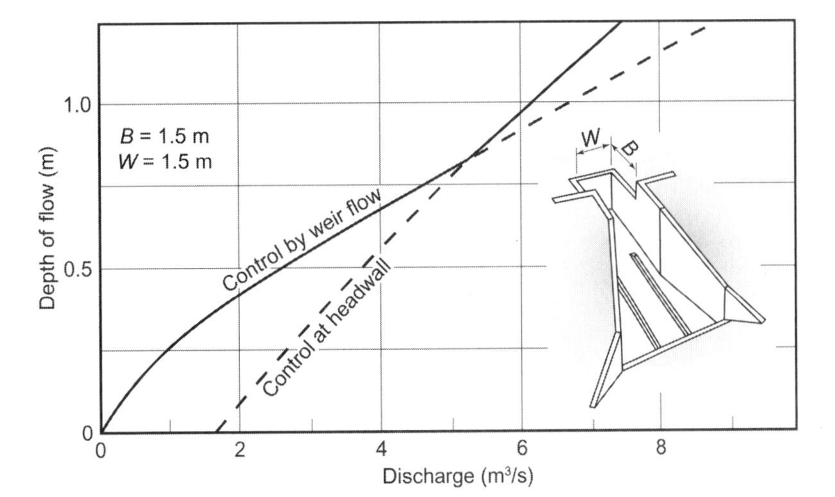

Figure 9–4

Discharge-head relationships of a box-inlet drop spillway. (Redrawn from Blaisdell & Donnelly, 1951)

converted from English units to SI (meter, second) by multiplying by 0.552. The value of C varies considerably with the entrance conditions. Blaisdell & Donnelly (1966) provide correction charts to modify C for a wide range of conditions of entrance and crest geometry for box inlets. Where the ratio of head to box width is 0.2 or greater, the ratio of the width of the approach channel to the total length L is greater than 1.5, and no dikes or other obstacles are within $3h$ of the crest, a value of $C = 1.8$ (SI) may be used with an accuracy of ±20 percent. Discharge characteristics of a typical box-inlet drop spillway are given in Figure 9-4. Using $C = 1.8$ will also give satisfactory results for the straight inlet or the control section of a flood spillway. The inlet should have a freeboard of 0.15 m above h, the height of the water surface.

Whenever the tailwater is nearly up to or above the crest of the inlet section, submergence decreases the capacity of the structure. When such conditions occur, other design equations apply (beyond the scope of this text).

Apron Protection. The kinetic energy gained by the water as it falls from the crest must be dissipated and/or converted to potential energy before the flow exits the structure. For straight-inlet drop structures the dissipation and conversion of energy are accomplished in either a straight apron or a Morris & Johnson (1942) stilling basin. Dimensions for the Morris & Johnson stilling basin are given in Figure 9-5. For larger structures, the Morris & Johnson outlet is preferred, as it results in a shorter apron and the transverse sill induces a hydraulic jump at the toe of the structure. The longitudinal sills serve to straighten the flow and prevent transverse components of velocity from eroding the side slopes of the downstream channel. The flow pattern through a Morris & Johnson stilling basin is shown in dimensionless form in Figure 9-6. A stilling basin for the box-inlet drop spillway is shown in Figure 9-7.

Chutes

Chutes are designed to carry flow down steep slopes through a concrete-lined channel rather than by dropping the water in a free overfall.

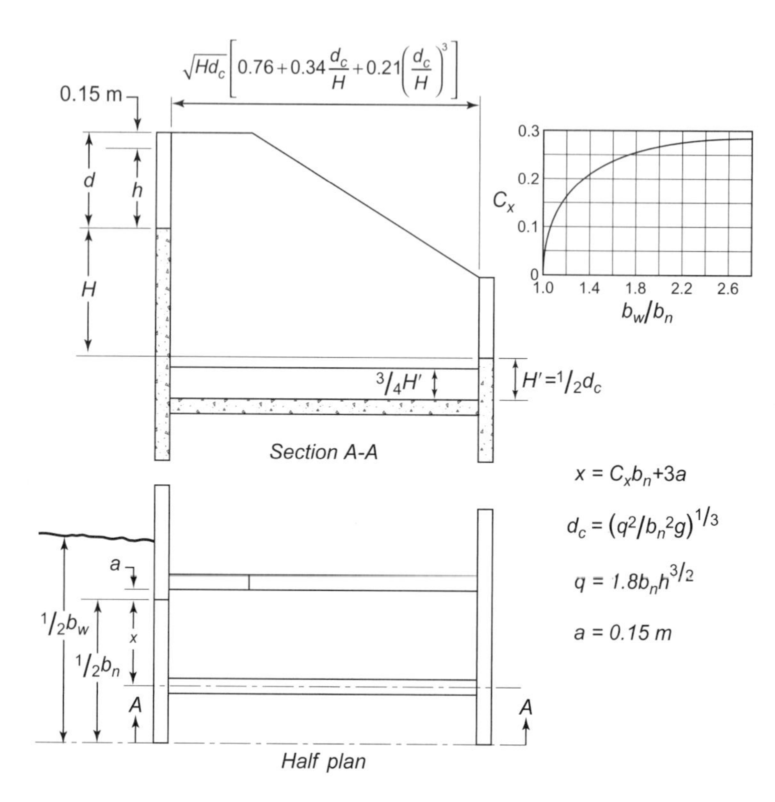

Figure 9–5
Design dimensions for a drop spillway with straight inlet and Morris & Johnson outlet.
(Redrawn from Morris & Johnson, 1942)

9.6 Function and Limitations of Chutes

Chutes may be used for the control of elevation changes up to 6 m. They usually require less concrete than drop-inlet structures of the same capacity and elevation change; however, there is considerable danger of burrowing animals undermining the structure, and in poorly drained locations seepage may threaten foundations. Where there is no opportunity to provide temporary storage above the structure, the inherent high capacity of the chute makes it preferable to the drop-inlet pipe spillway. The capacity of a chute is not decreased by sedimentation at the outlet.

9.7 Design Features of Chutes

Capacity. Chute capacity normally is controlled by the inlet section. Inlets may be similar to those for straight-inlet or box-inlet drop spillways, for which the capacity formulas already discussed will apply.

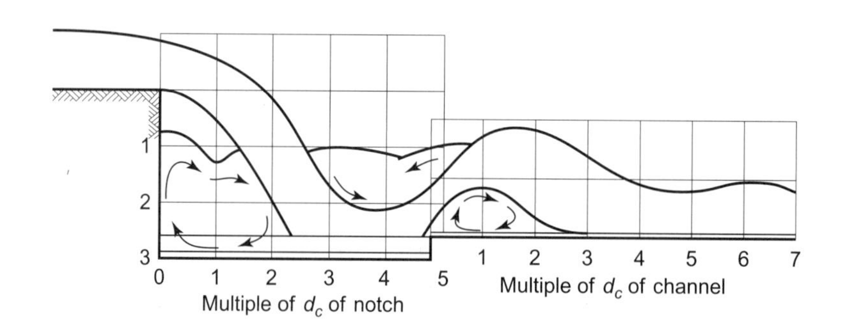

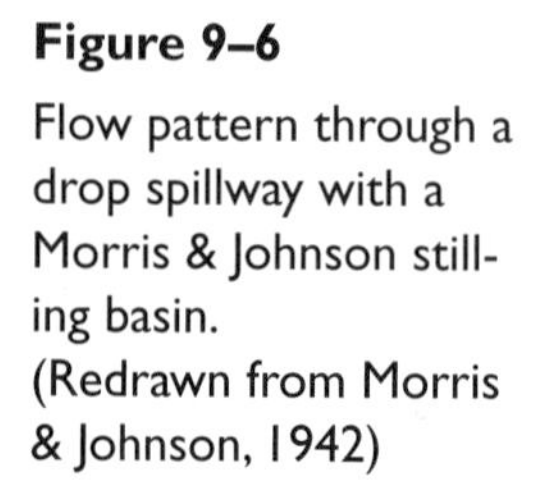
Figure 9–6
Flow pattern through a drop spillway with a Morris & Johnson stilling basin.
(Redrawn from Morris & Johnson, 1942)

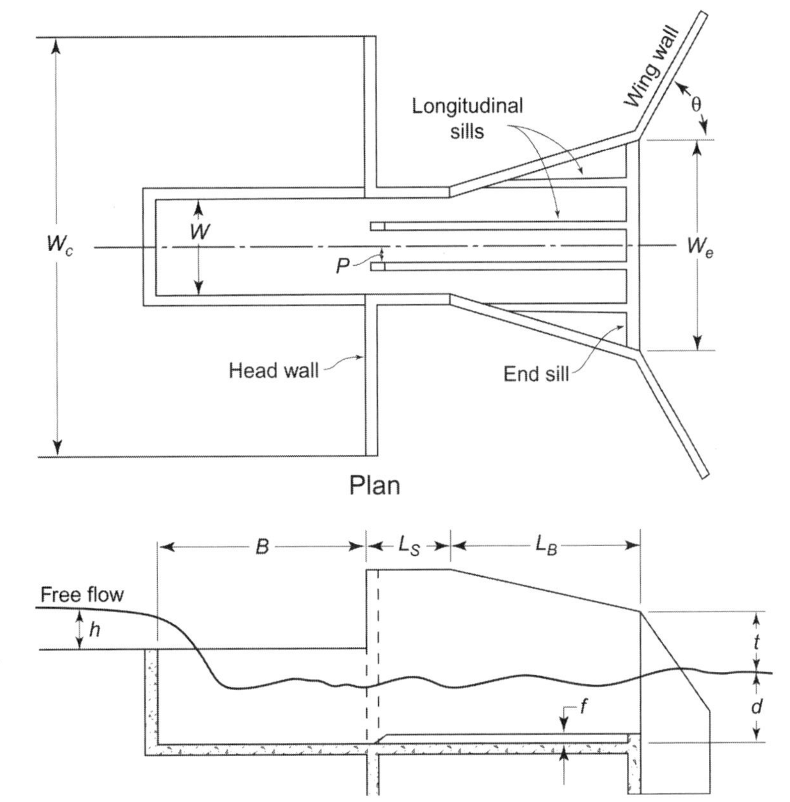

Figure 9–7
Box-inlet drop spillway. See reference for design equations. (Redrawn from Blaisdell & Donnelly, 1966)

Outlet Protection. The cantilever-type outlet should be used where the channel grade below the structure is unstable. In other situations, either the straight-apron or St. Anthony Falls (SAF) outlet is suitable. The straight apron is applicable to small structures. Figure 9-8 shows dimensions of the SAF type of stilling basin.

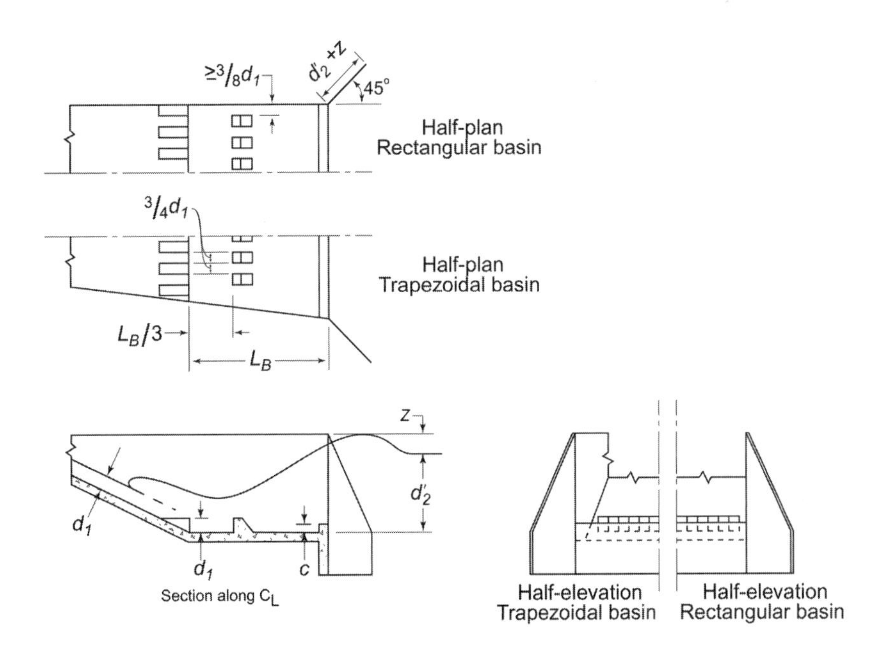

Figure 9–8
The Saint Anthony Falls (SAF) stilling basin. See reference for design equations.
(Redrawn from Blaisdell, 1959)

Formless Flume

9.8 Function and Limitations of Formless Flumes

This structure has the advantage of low-cost construction. It may replace drop spillways where the fall does not exceed 2 m and the width of notch required does not exceed 7 m. The flume is constructed by shaping the soil to conform to the shape of the flume and applying a 0.13-m layer of concrete reinforced with wire mesh. Since no forms are needed, the construction is simple and inexpensive. The formless flume should not be used where water is impounded upstream (because of the danger of undermining the structure by seepage) or where freezing occurs at great depth.

9.9 Design Features of Formless Flumes

Figure 9-9 shows the design features and dimensions of the formless flume. The capacity is given by Equation 9.1 using $C - 2.2$ (SI). This weir coefficient accounts for the increased cross-sectional area because the sides of the weir slope outward rather than vertically and the entrance is rounded. The depth of the notch, D, is h plus a freeboard of 0.15 m.

Pipe Spillways

Pipe spillways may take the form of a simple conduit under a fill (Figure 9-10a) or they may have a riser on the inlet end and some type of structure for outlet protection (Figure 9-10b). The pipe in Figure 9-10c, called an inverted siphon, is often

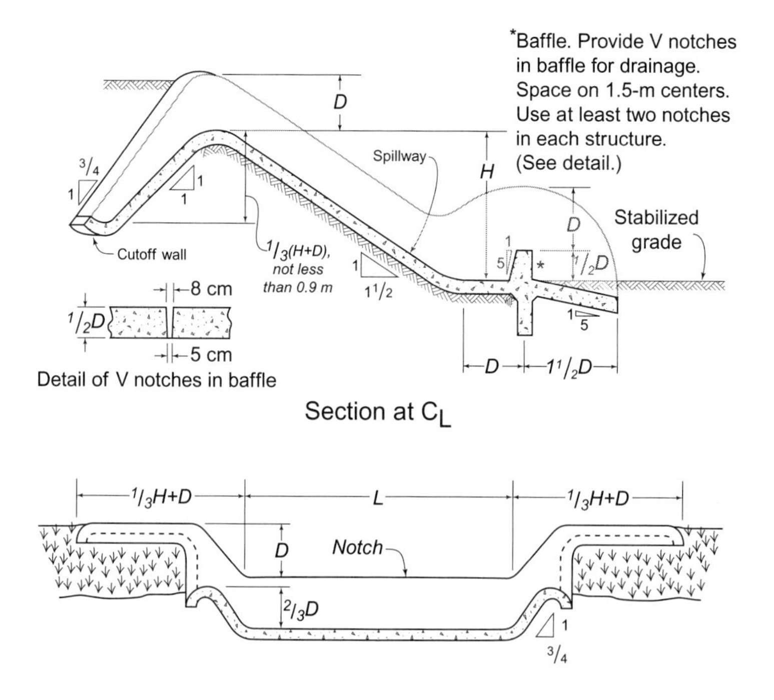

Figure 9–9
A formless flume. (Based on design by Wooley et al., 1941)

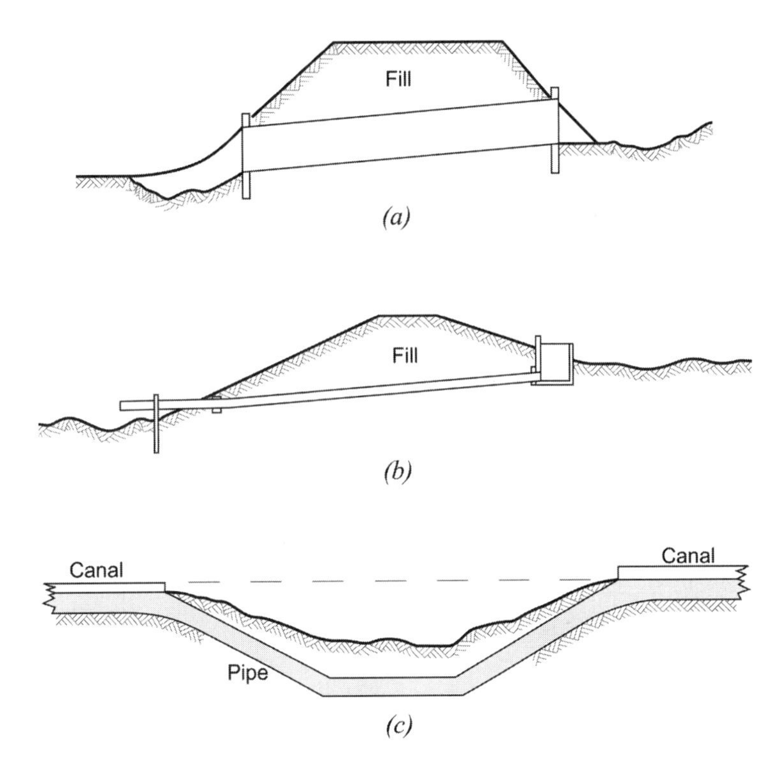

Figure 9–10

Types of pipe spillways: (a) simple culvert; (b) drop-inlet pipe spillway with cantilever outlet; (c) inverted siphon.

used where water in an irrigation canal must be conveyed under a natural or artificial drainage channel. Inverted siphons must withstand hydraulic pressures much higher than those encountered in other pipe spillways and therefore require special attention to structural design.

9.10 Function and Limitations of Pipe Spillways

The pipe spillway used as a culvert has the simple function of providing for passage of water under an embankment. When combined with a riser or drop inlet, the pipe spillway serves to lower water through a considerable change in elevation and to dissipate the energy of the falling water. Drop-inlet pipe spillways are thus frequently used as gully control structures. This application is usually made where water may pond behind the inlet to provide temporary storage. The hydraulic capacity of pipe spillways is related to the square root of the head; hence they are relatively low-capacity structures. This characteristic is desirable where discharge from the structure is to be restricted.

9.11 Design Features of Pipe Spillways

Culverts. Culvert capacity may be controlled by either the inlet section or the conduit. The headwater elevation may be above or below the top of the inlet section. Several possible flow conditions are represented in Figure 9-11. Solution of a culvert problem requires determination of the type of flow that will occur under given headwater and tailwater conditions. Consider a culvert as shown in Figures 9-11a and 9-11b. Pipe flow (i.e., where the conduit controls capacity) will usually occur where the slope of the culvert is less than the neutral slope and entrance capacity is not limiting. The neutral slope s_n is

$$s_n = \frac{H_f}{L} = K_c \frac{v^2}{2g} \tag{9.2}$$

where H_f = friction loss in conduit of length L (L),
L = length of conduit (L),
K_c = conduit friction loss coefficient (Tables C.2 and C.3) (L^{-1}),
v = velocity of flow (LT^{-1}),
g = gravitational acceleration (LT^{-2}).

Inlet losses may be so great in some situations that pipe flow will not occur even though the slope of the culvert is shallower than neutral slope. The capacity of the culvert under conditions of full pipe flow is given by

$$Q = \frac{A\sqrt{2gH}}{\sqrt{1 + K_e + K_b + K_c L}} \tag{9.3}$$

where Q = discharge capacity (L^3T^{-1}),
A = conduit cross-sectional area (L^2),
H = head causing flow (L),
K_e = entrance loss coefficient,
K_b = bend loss coefficient,
K_c = conduit loss coefficient (L^{-1}).

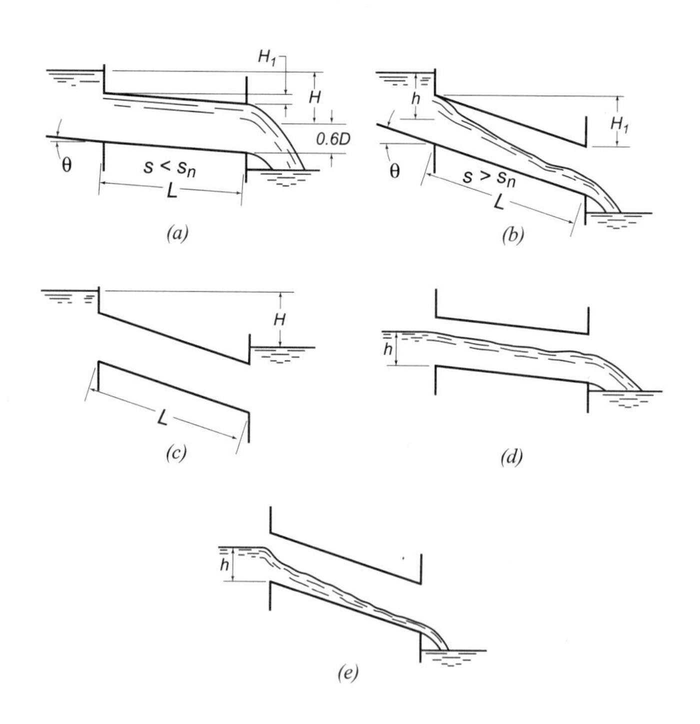

Figure 9–11

Possible conditions of flow through culverts. (a) Full: free outfall, pipe flow. (b) Part full: free outfall, orifice flow. (c) Full: outfall submerged, pipe flow. (d) Inlet not submerged: conduit controls, open channel flow. (e) Inlet not submerged: inlet controls, weir flow. (Modified from Mavis, 1943)

For full pipe flow, H is taken as the difference between the headwater elevation and the point 0.6 times the culvert diameter above the downstream invert (Figure 9-11a). Values of K_b, K_c, and K_e are given in Appendix C. Since most culverts do not have bends, K_b can usually be omitted.

If the inlet is submerged, the slope of the conduit is greater than neutral slope, and the outlet is not submerged, the flow will be controlled by the inlet section on short-length culverts, and orifice flow will control. Discharge capacity is then given by (Chapter 6)

$$Q = AC\sqrt{2gH} \tag{9.4}$$

where Q = discharge capacity (L^3T^{-1}),
A = conduit cross-sectional area (L^2),
H = head-causing flow (L),
C = orifice discharge coefficient.

The discharge coefficient C for a sharp-edged orifice is 0.6. For more detailed values and for other orifices, consult Brater & King (1996) or other hydraulics books. Examples 9.1 and 9.2 illustrate the discussions above.

In situations where the headwater elevation does not reach the elevation of the top of the inlet section, there is again the possibility of control of flow by either the conduit or the inlet section. In Figures 9-11d and 9-11e, the conduit controls if the slope of the conduit is too flat to carry the maximum possible inlet flow at the required depth. This depth is equal to the headwater depth above the inlet invert minus the static head loss resulting from entrance losses and acceleration. Conditions of control at the entrance section occur when the slope of the conduit is greater than that required to move the possible flow through the inlet. For conditions of control by the entrance section, solution for circular culverts may be made from Figure 9-12. This figure also applies when the inlet is submerged. Examples 9.3 and 9.4 demonstrate the solution for conditions of an unsubmerged inlet.

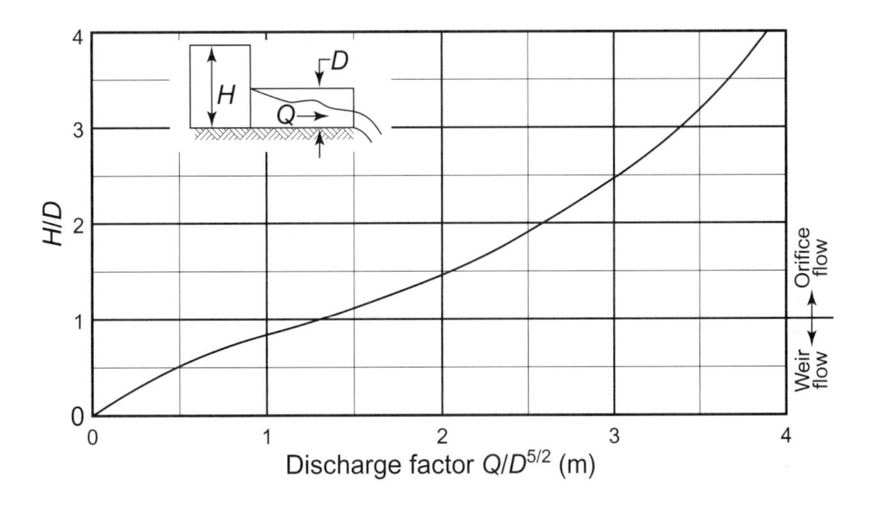

Figure 9–12

Stage-discharge relationship for control by a square-edge entrance inlet to a circular pipe. (Redrawn from Mavis, 1943)

Values for the Manning roughness coefficient n for conduits may be found in Appendix B.

Example 9.1

Determine the capacity of a 762-mm diameter corrugated metal (ring) culvert that is 20 m long with a square-edged entrance. The elevation of the inlet invert is 127.92 m, and the elevation of the outlet invert is 127.71 m. The headwater elevation is 129.54 m, and the tailwater elevation is 126.80 m.

Solution. Calculate discharges for both pipe control and orifice control conditions and compare them to see which type of control prevails. First, assume pipe flow controls and apply Equation 9.3. From Appendix B, find the roughness coefficient n for the culvert to be 0.025. From Figure C-1, find $K_e = 0.5$, and from Table C-2, $K_c = 0.112$. The head H is from the headwater elevation to the point that is 0.6 D above the outlet invert, i.e., 129.54 m – (127.71 m + 0.6 × 0.762 m) = 1.37 m. Equation 9.3 gives

$$Q = \frac{\pi(0.762\text{ m})^2}{4}\left(\frac{2 \times 9.81\text{ m/s}^2 \times 1.37\text{ m}}{1 + 0.5 + 0.112\text{ m}^{-1} \times 20\text{ m}}\right)^{1/2} = 1.22\text{ m}^3/\text{s}$$

Then assume that orifice flow controls and use Equation 9.4 to calculate the discharge.

$$Q = \frac{\pi(0.762\text{ m})^2}{4}(0.6)\sqrt{2(9.81\text{ m/s}^2)(1.24\text{ m})} = 1.35\text{ m}^3/\text{s}$$

The lesser of the two discharges represents the controlling condition, which in this case is pipe flow. As a further check, calculate the neutral slope for pipe flow using Equation 9.2.

$$S_n = \frac{0.112\text{ m}^{-1}}{2(9.81\text{ m/s}^2)}\left(\frac{1.22\text{ m}^3/\text{s}}{\frac{\pi}{4}(0.762\text{ m})^2}\right)^2 = 0.043$$

Compare this value to the actual slope which is (127.92 m – 127.71 m)/20 m = 0.011. The small value for the actual slope compared to the neutral slope indicates that pipe flow will control.

Example 9.2

Determine the capacity of the 762-mm culvert of Example 9.1 if the elevation of the outlet invert is 125.15 m, and the tailwater elevation is 124.34 m.

Solution. Assuming pipe flow controls, calculate the discharge with H = 129.54 m – (125.15 m + 0.46 m) = 3.93 m. Note that 0.6 D = 0.46 m.

$$Q = \frac{\pi(0.762\ \text{m})^2}{4}\left(\frac{2 \times 9.81\ \text{m/s}^2 \times 3.94\ \text{m}}{1 + 0.5 + 0.112\text{m}^{-1} \times 20\text{m}}\right)^{1/2} = 2.07\ \text{m}^3/\text{s}$$

$$\textit{Neutral slope:}\quad s_n = \frac{0.112\ \text{m}^{-1}}{2(9.81\ \text{m/s}^2)}\left(\frac{2.07\ \text{m}^3/\text{s}}{\frac{\pi}{4}(0.762\ \text{m})^2}\right)^2 = 0.118$$

$$\textit{Actual slope:}\quad s = (127.92 - 125.15)/20 = 0.139$$

Since the slope of the culvert is greater then neutral slope, pipe flow will not control. Entrance conditions will control, and the discharge will be that given by the orifice flow formula, Equation 9.4. From Example 9.1, Q = 1.35 m^3/s.

Figure 9-12 can be used in a graphical approach. With head H = 129.54 – 127.92 = 1.62 m, calculate H/D = 1.62/0.762 = 2.13, and read $Q/D^{5/2}$ = 2.66 from the curve. Solving for Q gives a discharge of 1.35 m^3/s, which is the same as that computed with Equation 9.4.

Example 9.3

Determine the capacity of a 1.52-m diameter concrete culvert that is 30 m long. The culvert entrance is square-edged. The elevation of the inlet invert is 57.76 m, and the elevation of the outlet invert is 55.60 m. The headwater elevation is 58.80 m, and the tailwater elevation is 52.40 m.

Solution. First, assume that pipe flow controls. Neglect for the moment the loss of static head at the culvert entrance resulting from acceleration of the flow into the entrance of the culvert. Under these assumptions, the depth of flow in the culvert would be 1.04 m (58.80 m – 57.76 m). The Manning formula can be used to calculate the discharge with a = 1.32 m^2, n = 0.015, R = 0.447 m, and s = 0.071, giving the result Q = 13.74 m^3/s. Compare this value to the discharge if the inlet controls. Using Figure 9-12, H/D = 0.68 and Q = 1.98 m^3/s. Since the inlet-controlled discharge is the smaller of the two, inlet conditions control the flow and the capacity is 1.98 m^3/s.

Example 9.4 Determine the capacity of a culvert as in Example 9.3, but having the outlet invert at an elevation of 57.73 m.

Solution. As in Example 9.3, the maximum possible discharge if inlet conditions control is 1.98 m^3/s; however, the slope of the culvert is much flatter in this case and the flow may be limited by the capacity of the conduit.

The depth of flow in the conduit can be determined by an iterative approach with the following steps: (1) assume a flow depth in the conduit; (2) use the Manning formula to calculate flow velocity and discharge; (3) using the calculated velocity from step 2, compute the velocity head and subtract that from the depth of water above the inlet invert (this assumes that the velocity head of the water in the section approaching the inlet is negligible); (4) compare the flow depth calculated in step 3 to the assumed depth in step 1; (5) adjust the assumed depth and repeat steps 2 through 4 until the assumed and computed depths are in close agreement. The flow velocity and discharge computed for that depth will be the correct values.

The surface of the water approaching the inlet is 58.80 m – 57.73 m = 1.07 m above the inlet invert. The slope of the conduit, s, is (57.76 m – 57.73 m)/30 m = 0.001. (1) Assume a flow depth of 0.8 m in the conduit. (2) This gives A = 0.97 m^2 (using Equation C.1), R = 0.39 m, v = 1.13 m/s, and Q = 1.09 m^3/s. (3) The corresponding velocity head is 0.065 m. Subtracting that from the initial water depth leaves 1.005 m. (4) This is much greater than the value assumed in step 1. (5) Assume a new flow depth of 0.99 m and recalculate, as follows.

(2′) a = 1.27 m, R = 0.44 m, v = 1.22 m/s, Q = 1.54 m^3/s. (3′) The velocity head is 0.076 m, giving a flow depth of 0.99 m. (4′) The assumed depth and calculated depth agree. The conduit-controlled discharge is 1.54 m^3/s, which is less than the 1.98 m^3/s discharge under inlet control. Therefore, the conduit conditions control the flow and the discharge is 1.54 m^3/s.

Culvert calculations can be more complicated than these examples. Many culverts are not long enough for uniform flow to fully develop if the culvert is not flowing full. Software packages are available that can perform more detailed analyses of the nonuniform flow profiles found in hydraulically short structures.

Drop Inlets. The discharge characteristics of a drop-inlet pipe spillway (Figure 9-13) are determined by the component of the system that controls the flow rate. At low heads, the crest of the riser controls the flow (as a weir) and discharge is proportional to $h^{3/2}$. Equation 9.1 should be used to calculate the discharge for these conditions. As the head increases, the capacity of the weir will eventually equal the capacity of the conduit (pipe flow) or the conduit inlet section (orifice flow). The flow will then be proportional to the square root of either the total head loss through the structure or the head on the conduit inlet, depending on whether pipe flow or orifice flow controls the discharge.

Hood Inlets. For mechanical spillways on ponds and similar small structures, the hood inlet provides a relatively simple and inexpensive alternative to the drop inlet. The hood inlet, when provided with a suitable antivortex device, will cause the pipe

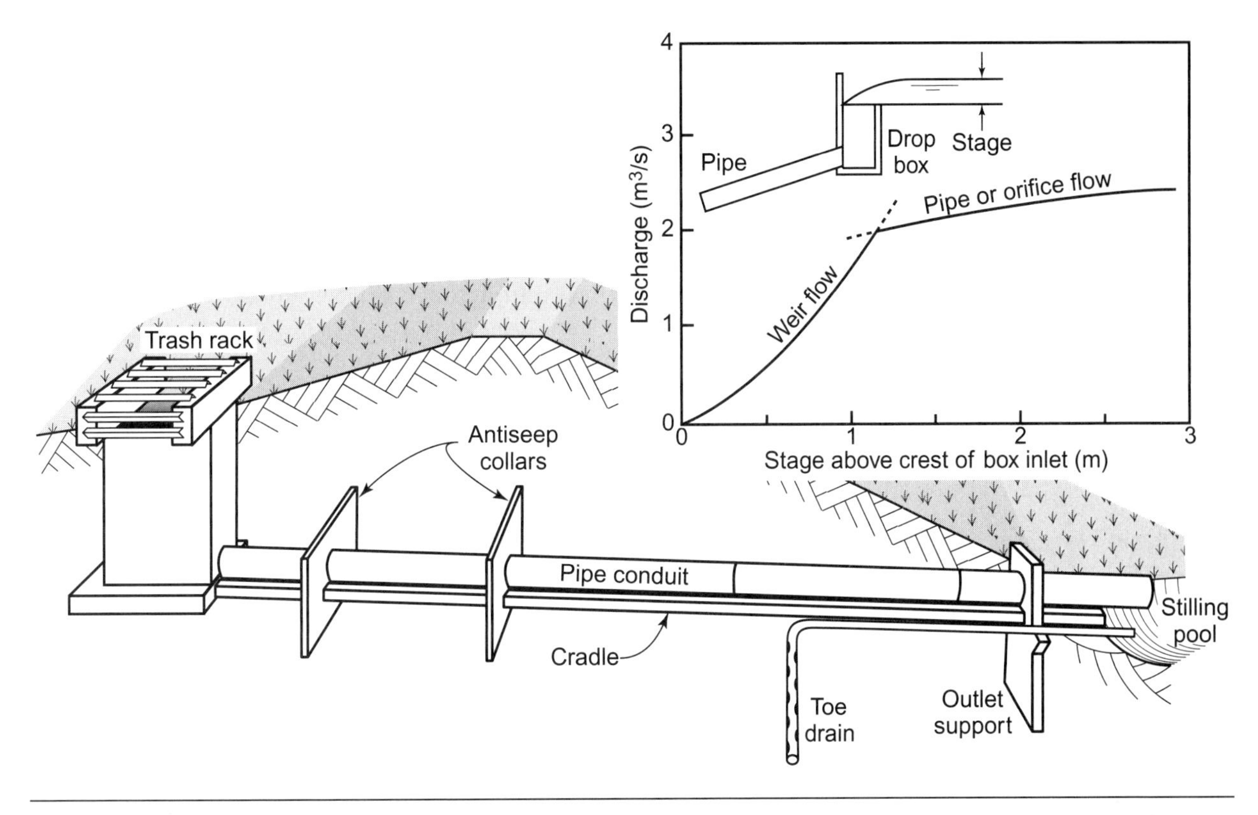

Figure 9–13 Drop inlet mechanical spillway. Risers are often fitted with a drain valve near the bottom to allow dewatering of the impoundment for maintenance.

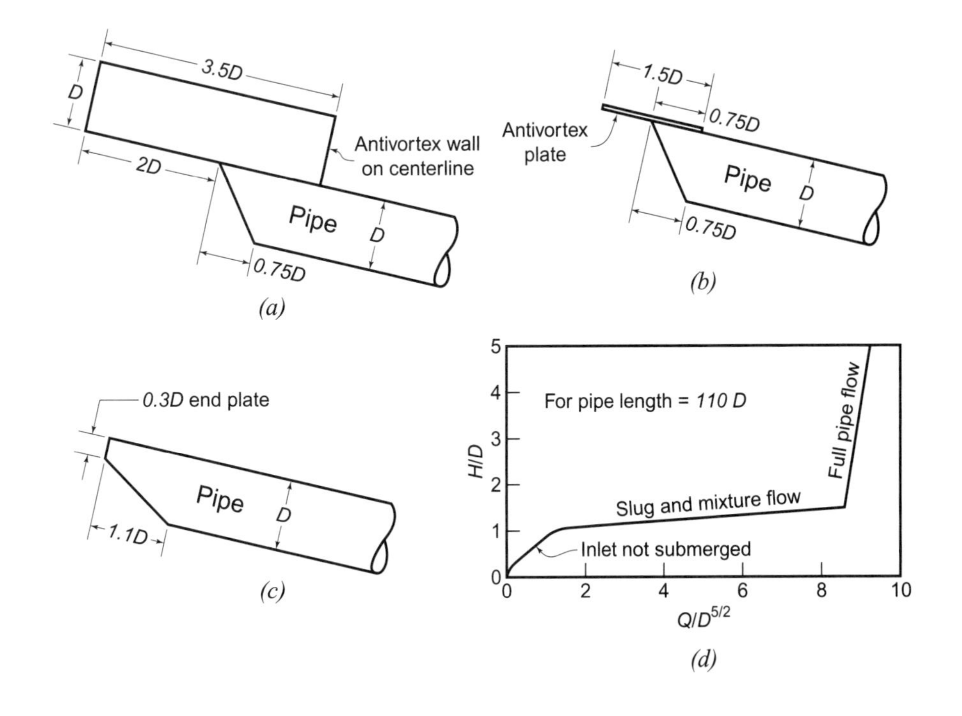

Figure 9–14

Types of antivortex pipe inlets. (a) Splitter-type antivortex wall. (b) Circular or square antivortex top plate. (c) End antivortex plate. (d) Hood inlet discharge-head relationships for (a), (b), and (c). (Blaisdell & Donnelly, 1958; Beasley et al., 1960)

to prime and flow full for spillway slopes up to 30 percent. Hood inlets shown in Figures 9-14a and 9-14b were developed by Blaisdell & Donnelly (1958). Beasley et al. (1960) reported that a hood inlet with an endplate as shown in Figure 9-14c gave satisfactory performance, although the entrance loss was somewhat higher than with the other two. The discharge characteristics of these three inlets are shown in Figure 9-14d for a pipe of length 110 *D*. For *H*/*D* less than 1, weir flow occurs. Up to *H*/*D* of about 1.4, the flow is erratic. Above *H*/*D* of 1.4 the vortex is eliminated and pipe flow controls.

The design capacity is determined from the pipe flow equation as described for culverts. For design of hood inlet spillways, a value of 1.0 is recommended for the entrance loss coefficient K_e. Although the approach conditions have little effect on spillway performance, the presence of the face of the dam will somewhat reduce the entrance-loss coefficients. High velocities near the hood inlet may erode the dam, though such a hole is small and becomes stabilized in a short time.

The hood drop inlet shown in Figure C-1 (Appendix C) is a hood inlet on the pipe of a drop-inlet pipe spillway like that shown in Figure 9-13. Since the hood inlet must have a head of 1.1 *D* to prime, the pipe spillway on large structures will require a specific and significant rise in the reservoir level before the spillway will flow at design capacity. The hood drop inlet will thus reduce the height and cost of the dam compared with that for a hood inlet.

Outlet Protection. For small culverts or drop-inlet pipe spillways, the cantilever-type outlet is usually satisfactory. The straight apron outlet may be used in some instances. Large drop-inlet pipe spillways may be provided with the SAF stilling basin discussed in Section 9.7.

Earthen Embankments

Earthen embankments are widely used as dikes, levees, and dams. They are important in water supply and flood control. Unlike the hard structures discussed in the preceding sections, earthen embankments are subject to degradation by erosion, sloughing, and other natural processes. Proper design, construction, and maintenance will yield a stable and reliable structure.

The basic design principles to be described here apply equally to all earthen embankments whose total height above ground level does not exceed 15 m.

9.12 Types of Earth Embankments

The discussions in this chapter are limited to the rolled-fill type of earthen embankments in which the earthen material is spread in uniform layers and then compacted at optimum water content to achieve maximum density. The selection and design of earthen embankments for water control depends on (1) the properties of the foundation, that is, stability, depth to impervious strata, relative permeability, and drainage conditions, and (2) the nature and availability of the construction materials.

There are three major types of earth fill. (1) The simple embankment type is constructed of relatively homogeneous material and is either keyed into an impervious foundation stratum, as shown in Figure 9-15a, or constructed with an upstream blanket of impervious material, as shown in Figure 9-15b. This type is limited to low fills and to sites having sufficient volumes of satisfactory fill materials available. (2) The core or zoned type of design includes a central section of highly impermeable materials extending from an impermeable stratum in the foundation to above the water line (Figure 9-15c). An upstream blanket is sometimes used in conjunction with this design. Core type construction reduces the percentage of high-grade fill materials needed. (3) The diaphragm type uses a thin wall of plastic, butyl, concrete, steel, or wood to form a barrier against seepage through the fill. A full-diaphragm cutoff extends from above the water line down to and sealed into an impervious foundation stratum as shown in Figure 9-15d. A partial diaphragm (Figure 9-15e) does not extend through this full range and is sometimes referred to as a cutoff wall. A diaphragm, particularly when constructed of rigid materials, has the disadvantage of being unavailable for inspection or repair if broken or cracked due to settlement in the foundation or the fill. The use of flexible diaphragms of plastic or butyl rubber has partially overcome this problem.

9.13 General Requirements for Earthen Embankments

Six basic requirements must be met to ensure an effective reservoir for water storage. (1) Topographic conditions at the site must allow economical construction; cost is a function of fill length and height, which determine volume of the structure. (2) Soil materials must be available to provide a stable and impervious fill. (3) Storage embankments must have adequate mechanical and flood spillway facilities to maintain a uniform water depth during normal conditions and to safely manage flood runoff. (4) Large storage embankments should be equipped with a bottom drain to facilitate maintenance and fish management. (5) Appropriate safety equipment must be installed around drop-inlets and other hazardous portions of the structure. (6) All design specifications must be followed during construction and a sound program of maintenance instituted to protect against damage by wave action,

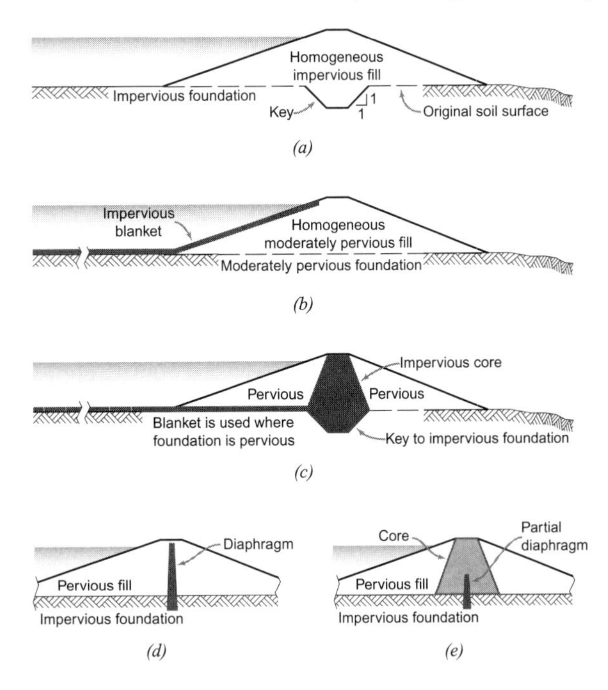

Figure 9–15

Earthen embankment designs. (a) Simple embankment using "key" construction. (b) Simple embankment with an impervious blanket for seepage control. (c) Embankment using a central core, key, and blanket of impervious material. (d) Embankment with full diaphragm for seepage control. (e) Embankment with core and partial diaphragm.

erosion, burrowing animals, livestock, farm equipment, and careless recreational use. All of these are necessary to ensure safety of the structure and prevent damage to downstream property.

9.14 Foundation and Earthen Fill Requirements

Earthen dams and embankments may be built on a wide range of foundation materials provided they are properly considered in the design process. A site investigation must be performed to determine the foundation conditions. For small structures, the investigation may be limited to soil pits or auger borings. For larger structures, the investigation should be more thorough to reliably determine the nature of the underlying soil and geologic conditions. A full discussion of exploration procedures, soil tests, classification, and interpretation of the data may be found in references such as USBR (1987).

Geological and foundation conditions can be broadly classified as follows (USBR, 1987):

Rock foundations. Competent rock foundations, those free of significant defects, have relatively high sheer strengths, and are resistant to erosion and percolation. Removal of weathered/disintegrated rock and grouting of fractures are often necessary. Weaker rocks (e.g., clay shales, weathered basalt, etc.) may present significant problems.

Gravel foundations. If these are well compacted, they may be suitable for earth fill structures. Special precautions may be needed to control seepage through the highly permeable materials.

Silt or Fine Sand foundations. These materials present problems of nonuniform settlement, potential soil collapse on saturation, uplift forces, piping, excessive seepage, and erodibility.

Clay foundations. These can support earth fill structures but require relatively flat embankment slopes because of relatively low foundation sheer strengths. Clay foundations can also consolidate significantly under the load of the fill.

Nonuniform foundations. It is not always possible to find a site with reasonably uniform materials of any of the types described above. Structures on such sites may require special design features that should be selected by experienced engineers.

Design of an earthen embankment should make the most economical use of the materials available on or near the site. For example, if satisfactory core materials must be hauled for some distance, the hauling cost should be compared with the cost of a diaphragm system.

The type of cross section depends on both the foundation conditions and the available fill materials. Where depth to an impervious foundation is not too great and where supplies of quality core materials can be found, designs shown in Figures 9-15c and e can be used. The combination core and blanket design shown in Figure 9-15c is adapted to sites having extremely deep pervious foundations. Other designs may be developed to use diaphragms alone or in combination with cores and other construction features to meet specific conditions.

Fill materials that will compact to maximum density are generally not over 20 percent gravel, 20–50 percent sand, not over 30 percent silt, and 15–25 percent clay. Soils having high shrink–swell characteristics and ungraded soils, when their use cannot be avoided, should be placed in the downstream interior of the embankment. There they are subject to less change in water content and, because of the overburden weight, will have less volume change than if used elsewhere in the embankment. Fills with higher percentages of graded sandy materials resist changes in water content, temperature, and internal stresses. Organic soils should never be used as fill.

In levee construction, the selection of materials is usually limited to that found adjacent to the structure. As these change along the route of the levee, it may be necessary to adjust the cross-sectional design of the embankment.

9.15 Seepage Through the Embankment

All earthen fill materials are porous and, even with optimal compaction, allow some seepage in response to differences in hydraulic potentials. The seepage will saturate portions of the embankment. One objective of design is to limit the saturation of the downstream portion. Saturated fill is less stable than unsaturated fill. If the zone of saturation includes part of the downstream face of the embankment, there will be danger of sloughing and piping. Saturation at the surface will also inhibit growth of protective vegetation.

The seepage line is analogous to a water table; there is no hydrostatic pressure above it and there is hydrostatic pressure below it. Its location depends on the permeability of the various fill materials and foundation, the ground water potentials at the site, the type and extent of any core or diaphragm within the embankment, and the type and placement of drains in the downstream portion of the structure.

A dam of homogeneous fill located on an impervious foundation will have a seepage line that intersects the downstream face above the base of the dam. Its location depends on the geometry of the cross section and is independent of the permeability as long as the fill is homogeneous. The seepage line will intersect the downstream face at a point about $h/3$ above the base for sideslopes flatter than 1:1. A toe drain or underdrain in or near the downstream edge of the embankment can intercept seepage before it reaches the face of the embankment, thereby keeping the entire downstream face unsaturated. The seepage discharge through an embankment of homogeneous fill on an impervious foundation (Figure 9-16a) can be approximated within ±20 percent as (Harr, 1962)

$$q = K(\sqrt{d^2 + h^2} - d) \tag{9.5}$$

where q = discharge per unit length of the embankment (L^2T^{-1}),

K = saturated hydraulic conductivity of the fill (LT^{-1}),

h = head of water (L),

d = adjusted flow length through the embankment (L).

If there is no drain, take the downstream end of the adjusted flow length to be below the midpoint of the seepage face.

In a zoned dam, the seepage line will drop sharply through the core, leaving more of the downstream zone unsaturated as shown in Figure 9-16b. The resistance to flow in the pervious materials is negligible compared to that in the core. The discharge can be conservatively estimated by considering only the core. For analysis of

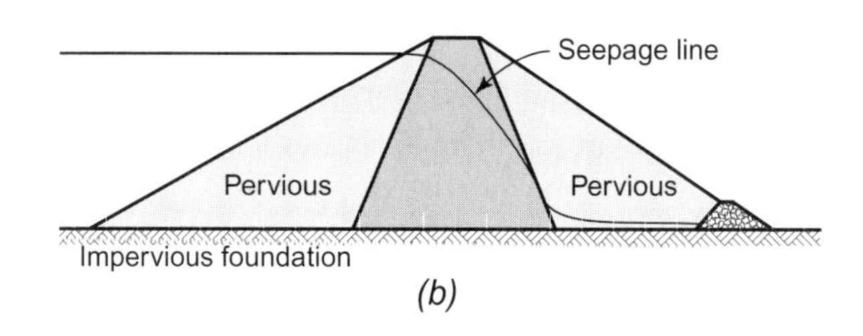

Figure 9–16
Seepage line through embankments on impervious foundations. (a) Homogeneous embankment with a trapezoidal rock-filled toe drain. (b) Embankment with a core and a rock-filled toe drain.

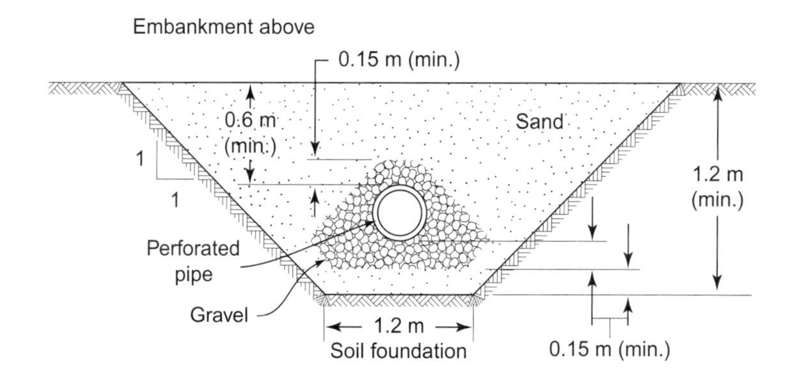

Figure 9–17
Typical filter and perforated pipe toe drain. (Redrawn and revised from USBR, 1987)

more complex designs, see references such as Harr (1962) and USBR (1987). Figure 9-17 shows a type of toe drain that uses a filter bed of sand and gravel to collect seepage into a perforated pipe that discharges below the downstream face of the dam. For filter design methods, see Chapter 14 or USBR (1987).

Example 9.5

Estimate the seepage through the dam in Figure 9-16a given: upstream water depth of 13 m, dam height of 15 m, top width of 6 m, upstream slope of 4:1, downstream slope of 3:1, and hydraulic conductivity of 0.038 m/d. There is no toe drain.

Solution. Compute d as $0.3 \times 4 \times 13 \text{ m} + 4 \times 2 \text{ m} + 6 \text{ m} + 45 \text{ m} - 3 \times \frac{13 \text{ m}}{3} \times \frac{1}{2} = 61.8 \text{ m}$. Then, using Equation 9.5 with h = 12 m and K = 0.038 m/d, find

$$q = 0.038 \text{ m/d}(\sqrt{(68.1 \text{ m})^2 + (13 \text{ m})^2} - 68.1 \text{ m}) = 0.047 \text{ m}^3\text{/d/m}$$

For each meter of length of the dam, about 47 L/d would seep through to the downstream face. This illustrates the need for seepage collection drains in embankments used for permanent or long-term water retention.

In the design of levees, maximum flood stage is substituted for the water surface elevation. Determination of the seepage line is important where sustained high water may be expected and where construction materials are particularly permeable. For short-duration floods, levees will rarely become sufficiently saturated to create a hazardous condition.

For impoundments where underlying materials are pervious, it is often necessary to line the entire area with a clay blanket. Blankets should not be applied to the upstream face of a dam where frequent, total, or rapid drawdown is anticipated because of the danger of slumping due to the internal hydrostatic pressure in the saturated portion of the dam. The effectiveness of a blanket can also be reduced by cracking during drying. The thickness of the blanket, for the structures discussed in this chapter, should be about 10 percent of the depth of water above the blanket, with a minimum of 1 m. A blanket on the upstream face of a small dam should extend out from the toe of the dam for 8 to 10 times the depth of the water.

If the soil materials on the bottom of an impoundment contain approximately 70 percent sharp, well-graded sand, 20–25 percent clay, and sufficient silt to provide good gradation of particle size, an adequate seal can be obtained by ripping the soil to a depth of about 0.3 m and then recompacting at optimum water content with a sheepsfoot roller until maximum density is attained. If proper soils are not available on the site, they must be selected from adjacent borrow pits, mixed, spread, and compacted.

Fractured rock presents a failure hazard under an impoundment. Under most conditions, there should be at least 0.6 m of soil between the compacted blanket and the underlying rock. Soils with the particle size gradation outlined above have sufficient structural resistance to the pressure of the water to prevent slippage and flow of the soil mantle into underlying rock crevices. High clay-content soils do not have this resistance and therefore require a greater thickness.

Where satisfactory construction materials are unavailable, swelling clays, such as bentonite, can be incorporated into the soil to obtain a seal. Bentonites have poor structural strength by themselves, so they should only be used with soils containing at least 10–15 percent sand. Polyphosphates and other chemical additives that tend to deflocculate the soil can reduce permeability for the purpose of obtaining a seal. Such additives must be carefully chosen for compatibility with the chemical characteristics of the soils.

Where special site or construction conditions dictate minimum seepage losses, lining an impoundment with a geotextile may be justified. Special installation procedures are needed to properly anchor the liner and protect it from damage.

9.16 Site Preparation

Exposed rock foundations must be examined for joints, fractures, and permeable strata that might contribute to seepage and eventual failure through piping. The usual treatment is to pressure-grout the foundation to a depth equal to the design head of water above the rock surface using techniques described in USBR (1987).

Where the foundation is porous material, seepage can be reduced by construction of a cutoff trench that extends to bedrock or an impervious stratum. In general, the bottom width of the trench should be 3 to 6 m. Sideslopes of the trench should be no steeper than 1:1. The depth of the cutoff trench depends on the depth to the impermeable material.

Where a foundation of fine-textured, impermeable material exists, the site should be prepared by removing all vegetation, topsoil with high organic matter content, and soil that has been loosened by frost action or that has been recently deposited. If no cutoff trench is needed, other measures should be taken to bond the embankment to the foundation, such as the use of a shallow (0.6 to 1 m) cutoff trench as shown in Figure 9-15c.

If suitable core materials are not available and the depth to an impervious foundation is too great for a cutoff trench, cutoff or core walls of steel sheetpile or reinforced concrete may be installed. The ultimate effectiveness of these depends on the care taken in installation and on properly keying them into the impervious materials. Cement-bound cutoff curtains and chemical grouting have been successfully used where their cost is justifiable.

9.17 Drainage

As mentioned in Section 9.15, a toe drain or underdrain may be needed to keep the seepage line sufficiently below the downstream face of the embankment. The capacity of such drains should be at least twice the estimated maximum seepage discharge through the embankment.

Where a saturated pervious foundation underlies an impervious top layer, consideration must be given to the possibility of piping resulting from uplift pressures exceeding the pressure of the impervious mantle. Interceptor ditches can be used if the mantle is thin enough; otherwise, pressure-relief wells can be used to intercept seepage and reduce the uplift pressures.

Drains should be designed (1) to provide adequate capacity, (2) to act as filters to prevent the movement of soil particles, and (3) to have sufficient weight to overbalance the high upward seepage pressures that might lead to downstream blowouts or related piping failures. Filter design criteria are provided in Chapter 14.

Adequate drainage is equally important in levees that may be subject to prolonged flood conditions that could saturate the structure. A pipe drain or ditch near the land-side toe at a sufficient depth to affect the lower end of the seepage line is standard. Care must be taken to assure an adequate and safe outlet for these drains.

9.18 Sideslopes and Berms

Minimum safe sideslopes depend on the height of the structure, the shearing resistance of the construction materials, and the duration of inundation. On structures less than 15 m high with average materials, the sideslopes should be no steeper than 3:1 on the upstream face and 2:1 on the downstream face. Coarse materials that do not compact well may require sideslopes of 3:1 or 4:1 to ensure stability. For more precise designs or for structures exceeding 15 m in height, a thorough soil analysis is needed to determine horizontal shear strength.

Levees on major waterways subject to prolonged flood peaks may need sideslopes as flat as 7:1. Such flat slopes are needed because the materials used in levees are often unstable and are not compacted in construction.

Most conservation levee construction is limited to protecting agricultural lands on the upper tributaries where flood peaks are of short duration. Sideslopes of 2:1 may be adequate if they are properly protected at critical points. Levees that protect land from backflow after pump drainage, as in certain muck areas, tidal marshes, and swampy areas, should be designed and constructed by the same criteria as those for dams.

9.19 Top Width

The minimum top width depends on the height of the structure and its use. For dams up to 4.3 m in height, the top width should be at least 2.4 m. If there will be a road (paved or unpaved) across the top, the minimum width should be 3.6 m to provide a 0.6-m shoulder. Road runoff must be controlled to prevent erosion of the sideslopes. Recommended minimum top widths for dams exceeding 4.3 m can be calculated as (SCS, 1985)

$$W = 0.2H + 2.1 \tag{9.6}$$

where W = top width in m,
H = maximum height of the embankment in m.

For levees, top widths vary from 1 to 6 m or more. For levees subject to short-duration flood peaks, 1 m is adequate. For levees subject to prolonged flood peaks, the general rule for the top width is $W = 2\sqrt{H}$.

9.20 Freeboard

Net freeboard is the distance between the maximum design high-water level or flood peak and the top of the structure (after settlement). Reference is sometimes made to gross freeboard, or surcharge, which is the vertical distance between the crest of the mechanical spillway and the top of the embankment.

The net freeboard should be sufficient to prevent waves or spray from overtopping the embankment or from reaching that portion of the fill that may have been weakened by frost action (which rarely exceeds 0.15 m).

Wave height for moderate-size water bodies can be estimated by

$$h = 0.014D_f^{1/2} \tag{9.7}$$

where h = wave height (trough to crest) under maximum wind velocity in m,
D_f = fetch or exposure in m.

All freeboards should be based on the water level at the maximum flow heights for which the embankments and spillways are designed. For small ponds and other detention reservoirs with small mechanical spillways, the design flood height will include a depth of flow in the emergency spillway of not more than 0.3 m. The flood storage depth, which is the difference in elevation between the mechanical and emergency spillway crests, can be accurately determined by flood-routing procedures. For a small storage reservoir with a 200- to 250-mm-diameter mechanical spillway, this depth can be estimated from Figure 9-18. The following example, as illustrated in Figure 9-19, shows the procedure for estimating freeboard.

Example 9.6

Determine the net and gross freeboard for a pond having 0.6-ha water surface area and an exposure length of 180 m. Assume the frost depth is 0.15 m, the 25-year design peak runoff rate is 4.0 m^3/s, and the flow depth in the flood spillway is 0.3 m.

Solution. From Equation 9.7, the wave height is calculated as

$$h = 0.014\sqrt{180} = 0.19 \text{ m}$$

From Figure 9-18, read the flood storage depth as 0.6 m. Then,

$$\text{Net freeboard} = 0.15 \text{ m} + 0.19 \text{ m} = 0.34 \text{ m}$$

$$\text{Gross freeboard} = 0.34 \text{ m} + 0.3 \text{ m} + 0.6 \text{ m} = 1.24 \text{ m}$$

Additional freeboard should be added as a safety factor where lives and high-value property would be endangered by an embankment failure. On large dams the net freeboard is 50 percent more than the wave height and is increased for higher winds. Levee freeboard is designed on the same basis as for dams.

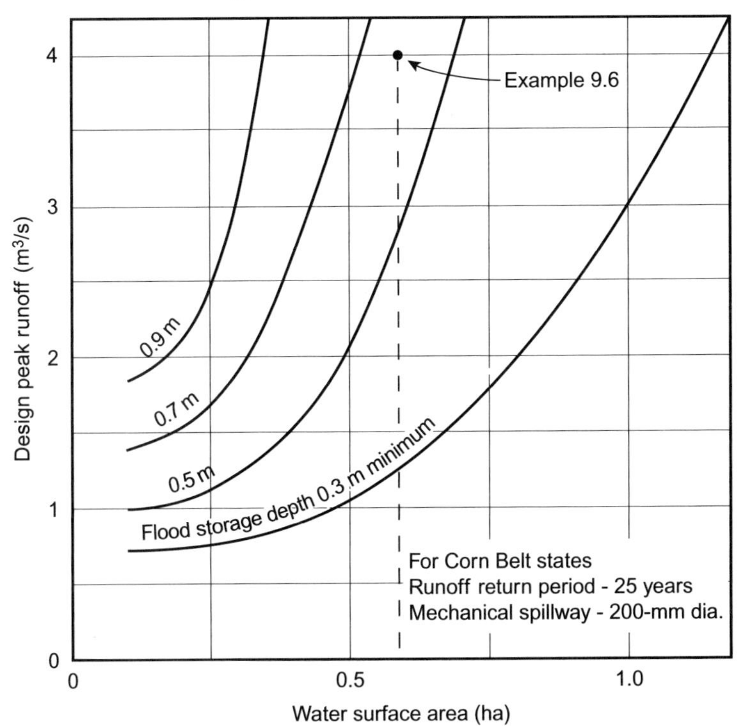

Figure 9–18

Flood storage depth for small storage reservoirs.

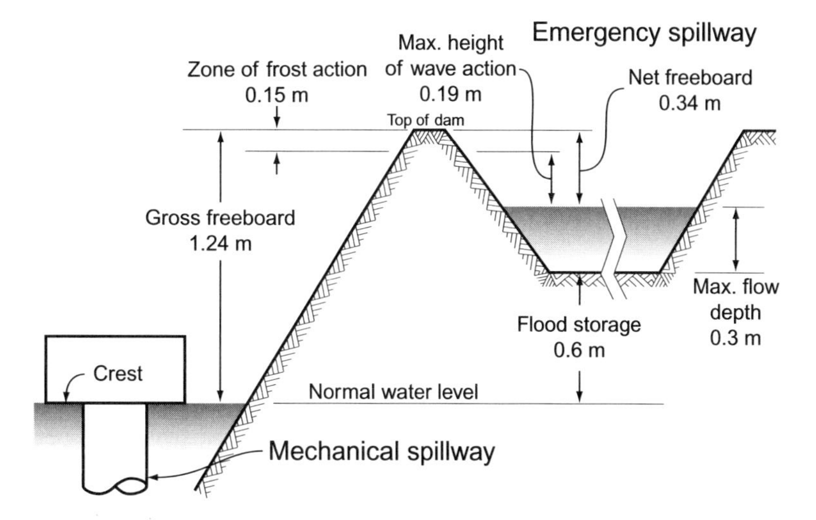

Figure 9–19
Net and gross freeboards for Example 9.6.

9.21 Compaction and Settlement

The volume relationships of soil may be expressed by the formula

$$V = V_s + V_e \tag{9.8}$$

where V = total in-place volume (L^3),
V_s = volume of solid particles (L^3),
V_e = volume of voids, either air or water (L^3).

The void ratio e is defined as

$$e = V_e/V_s \tag{9.9}$$

The dry bulk density of the soil is defined as the oven-dry mass per unit volume of soil in place.

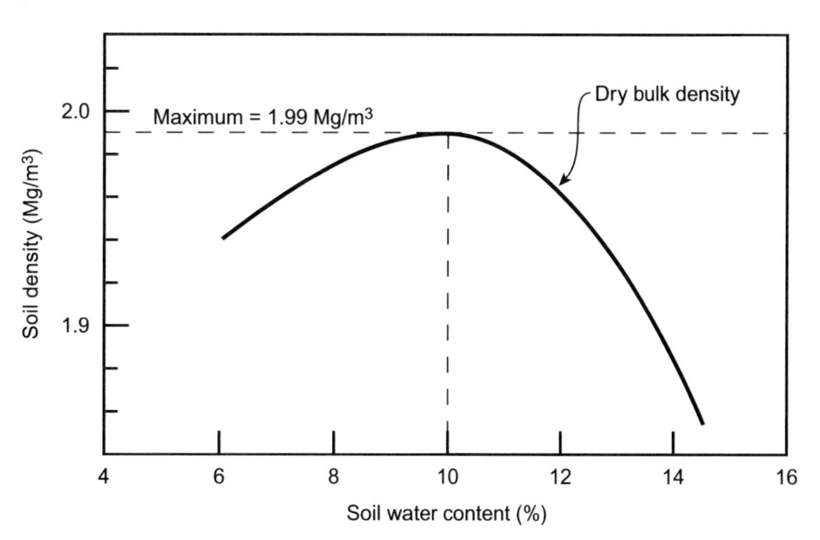

Figure 9–20
A typical Proctor soil density curve.

The degree to which a soil can be compacted by a given amount of applied energy depends on the water content of the soil. The Proctor density test is a standard procedure used to determine the optimum water content, which is the water content at which maximum compaction will occur. Disturbed soil is placed in a container and compacted, layer by layer, by dropping a weight from a specified height for a specified number of times. The density and water content are determined after each test. By adding water and repeating the test several times, a Proctor density curve can be drawn as shown in Figure 9-20. As water content increases, the density curve increases to a maximum and then decreases. The Proctor density is the maximum density, which defines the optimum water content for compaction. This method is widely used in engineering construction.

Compacted earthen fill is normally denser than it was while in place in the borrow area. Equations 9.8 and 9.9 can be used to estimate the volume of borrow material needed to produce a given volume of compacted fill.

Example 9.7

An earthen dam requires 1200 m^3 of compacted fill. If the desired void ratio of the fill is 0.6 and the void ratio of the material in the borrow area is 1.1, what is the volume of material that must be removed from the borrow area?

Solution. The volume of solids in the borrow material is the same as the volume of solids in the fill, but the volume of voids will be smaller in the fill. From Equation 9.9, $V_e = 0.6V_s$ for the fill. Substituting in Equation 9.8,

$$V = 0.6V_s + V_s = 1.6V_s = 1200 \text{ m}^3$$

$$V_s = 750 \text{ m}^3$$

For the borrow area, $V_e = 1.1V_s$. Substituting in Equation 9.8,

$$V = 1.1V_s + V_s = 2.1V_s = 2.1 \times 750 \text{ m}^3 = 1575 \text{ m}^3$$

Rolled-fill embankments that have been properly placed in thin layers (lifts) and compacted at optimum water content and that are on unyielding foundations will not settle more than about 1 percent of the total fill height. On deep plastic foundations, settlement may reach 6 percent. Compaction on small structures such as farm ponds is generally not as thorough as on larger structures; therefore, the fill is usually constructed to about 105 to 110 percent of the designed settled fill height.

In levee construction a minimum allowance for settlement of 20 to 25 percent is usually added because placement by dragline or conveyor will not give the fill a high degree of consolidation.

9.22 Wave Protection

If the water side of a dam or levee is exposed to considerable wave action or current, it should be protected by a covering of erosion-resistant material, such as riprap, or by energy dissipaters such as anchored floating logs. For levees and small ponds that are not subject to excessive wave action, a dense sod of a suitable grass will usually provide adequate slope protection.

Erosion-resistant materials such as riprap or various types of concrete blocks or slabs should be carefully placed on a bedding layer of graded gravel or crushed stone 0.2 to 0.5 m thick, depending on quality. This layer keeps the underlying soil from being washed out by wave action, which would allow the top material to gradually settle into the embankment.

9.23 Construction

Construction according to plan and adjustment to soil conditions while the work is in progress are essential for the success of earthen structures. All trees, stumps, and major roots should be removed from the site. Sod and topsoil should be removed and stockpiled for later spreading over the project area to help establish good grass cover. If necessary, the fill should be wetted to optimum water content for compaction. Proper water content makes it possible to force some of the fill material into the original surface, thus eliminating any dividing plane between the fill and the foundation.

After the site is cleared, the cutoff trench is excavated. Compacted fill should consist only of carefully selected materials that are placed in thin layers and compacted at optimum water content. Layers for pervious soils should not exceed 0.25 m; more plastic and cohesive soils should not exceed 0.15 m in thickness. Layers should be nearly horizontal, with slopes of 20:1 to 40:1 away from the center of the structure to provide good surface drainage should rainfall occur during construction.

The design degree of compaction is specified as the ratio of the embankment density to a standard density for the soil. For earthen dam construction, the degree of compaction should be 85 to 100 percent of the Proctor density.

High clay-content soils should be carefully compacted, with the soils slightly drier than the lower plastic limit to prevent formation of shear planes called slickensides. Pervious materials, such as sands and gravels, consolidate under the natural loading of the embankment; however, additional compaction aids in increasing shear strength and in limiting the embankment settlement. Compaction of these noncohesive materials is best achieved when they are nearing saturation.

Special care should be taken when compacting materials close to core walls, antiseep collars, pipes, conduits, and so on. All such structures should be constructed so that they are wider at the bottom than at the top so that settlement of the soil will create a tighter contact between the two materials. Thin layers of soil, at water contents equal to the remainder of the fill, should be tamped into place next to all structures. Hand-operated pneumatic or motor tampers are well suited for this. Heavy equipment should take varying routes over fill to prevent excessive compaction along travel paths.

9.24 Protection and Maintenance

The entire impoundment area should be fenced to prevent damage to embankments, spillways, and banks. Where this is not practicable, at least the spillways and dam should be protected.

To minimize sediment deposition, the entire watershed should be protected by adequate erosion control practices. Buffer or filter strips of dense close-growing vegetation should be maintained around the water's edge.

A properly designed and constructed earthen dam should require minimal maintenance; however, a program of regular inspection and maintenance should be established. Particular attention should be given to surface erosion, development of seepage areas on the downstream face or below the toe of the dam, development of sand boils and other evidence of piping, erosion by wave action, and damage by animal or human activity. Early detection and repair of such conditions will prevent their growth into dangerous and expensive problems. Under no conditions should trees be permitted to grow on or near the embankment.

Spillways

Most ponds and reservoirs are protected from overtopping by vegetated flood or emergency spillways. This vegetation would not survive if base flow entering the impoundment and frequent minor flood-flows kept the spillways unduly wet. Various types of mechanical spillways may be used to protect flood spillways and to carry low flows.

9.25 Mechanical Spillways

A mechanical spillway has its inlet set at the elevation at which the water surface is to be maintained (permanent pool elevation) and extends through the embankment to a safe outlet below the dam. The mechanical spillway should be constructed of a durable material that has a long life expectancy and is not subject to damage by settling, loads, or impacts. The portion passing through the embankment should be placed on a firm, impermeable foundation, preferably undisturbed soil, and should be located to the side of the original channel. If a rigid conduit, such as concrete, cast iron, or clay tile, is placed on a rock foundation, it must be cradled in concrete. Flexible conduits, such as plastic or corrugated metal pipe, may be cradled in compacted impermeable soil material. Where a conduit is laid on a foundation or passes through a fill that is subject to more than nominal settlement, consideration should be given to laying the conduit with a camber to compensate for such settlement.

All structures passing through the embankment should be covered by thin, well-compacted layers of high-grade material. Where the fill surrounding the pipe is of low quality, concrete or metal antiseep collars that extend out at least 0.6 m in all directions should be used (Figure 9-13). The number of collars should be sufficient to increase the length of the creep distance by at least 10 percent. The creep distance is the length along the pipe within the dam, measured from the inlet to the filter drain or the point of exit on the backslope of the dam. The increase in creep distance for an antiseep collar is measured from the pipe out to the edge of the collar and back to the pipe. Where concrete or mortar joints are used, it is advisable to complete backfilling and tamping before it has set, thus avoiding the danger of cracking the concrete. At least 1.2 m of protective fill should be carefully placed over the structure before heavy earth-moving equipment is allowed to pass over it.

A mechanical spillway similar to the one shown in Figure 9-13 can serve as a drain for cleaning, repairing, or restocking the impoundment. All valve assemblies should be equipped to operate from above the maximum water surface elevation. The inlet should be protected with a screening device to prevent floating debris, turtles, and other foreign objects from clogging the pipe. The device should be designed to be self-cleaning. Outlets should extend well beyond the toe of the dam and be protected against scouring.

A mechanical spillway may be designed to carry a small percentage of the peak flow. For small watersheds and for flood spillways lined with concrete or other stable material, a mechanical spillway may not be required. Where considerable flood storage volume is available, flood-routing procedures should be used to determine the appropriate pipe size. The mechanical spillway should always be large enough to carry the base flow so that the flood spillway will be dry and stable. Approximate pipe sizes for small impoundments and watersheds are given in Table 9-1.

9.26 Flood or Emergency Spillways

All impoundments must be equipped with an emergency spillway that will safely conduct flood flows in excess of the temporary storage capacity of the structure. Where a natural spillway, such as a depression in the rim of the impoundment area, does not exist, a trapezoidal channel should be constructed in undisturbed earth. The flood spillway consists of an approach channel, a level control section, and an exit (grassed) section.

After estimation of the peak rates and amounts of runoff to be handled, the weir formula and vegetated waterway design procedures may be used to design the flood spillway (Chapters 5, 6, and 8).

Example 9.8

The design capacity of a flood spillway for a dam is 3.4 m^3/s. The exit section was designed as a grassed waterway with a bottom width of 12 m. Assuming a maximum depth of flow of 0.3 m, what width is required for the control section so that it does not restrict the flow, that is, increase the design level in the reservoir?

Solution. Equation 9.1 with $C = 1.8$ will give a satisfactory estimate for the width of the control section.

$$L = 3.4/(1.8 \times 0.3^{3/2}) = 11.7 \text{ m}$$

The grassed approach channel should be flared out on each side at least 0.2 m per meter of length as measured along the center line of the channel upstream from the control section.

Example 9.8 assumes that the flow passes through critical depth at the downstream end of the control section. For this to occur, the slope in the exit section must be greater than the critical slope; otherwise, the exit section would control the flow. The exit section must be able to carry the design flow at velocities that do not exceed those recommended for grassed waterway use given in Chapter 8. Outlets should extend well beyond the dam or into an adjacent waterway and should be

TABLE 9-1 Mechanical Spillway Pipe Diameters for Small Impoundments

Peak Runoff Rate for 25-Year Return Period (m^3/s)	*Water Surface Area at Normal Water Level (ha)*				
	0.2	*0.4*	*0.8*	*1.2*	*2.0*
			Diameter (mm)		
≤ 0.5	200	150	150	—	—
0.5–1.0	200	200	150	150	—
1.0–1.5	250	250	200	200	150
1.5–2.0	250	250	250	250	200
2.0–2.5	—	300	300	300	250

Note: Suggested sizes of corrugated metal pipe primarily for ground water or seepage flow. For smooth wall pipes, use the next smaller diameter.

protected against scouring. The slope of the approach channel should be at least 2 percent to provide drainage and reduce inlet losses.

The design capacity for flood spillways on small impoundments is normally based on a return period of 25 years from the contributing watershed. For larger structures where high-value property or human lives are endangered, a return period of 100 years or more should be considered. If the capacity of the mechanical spillway is large compared with the flood flow, the design flow for the flood spillway may be reduced.

Flood Routing

The design of storage/detention structures and their appurtenances requires analysis of the dynamic response of the system to a design storm. Flood routing is the process of determining the time-varying flows, stages, and storage for the system in response to an inflow hydrograph.

9.27 Principles of Flood Routing

Flood routing involves time-varying inflows, storage within a stream reach or impoundment, and outflows through the spillways or downstream reach. The objective is to find a combination of dam height and spillway sizes that will safely provide the desired permanent storage and flood wave attenuation. The basic inflow–storage–outflow relationship is the continuity equation for unsteady flow

$$i\,dt - o\,dt = s\,dt \tag{9.10}$$

where i = inflow rate, instantaneous (L^3T^{-1}),
o = outflow rate, instantaneous (L^3T^{-1}),
s = storage rate, instantaneous (L^3T^{-1}),
dt = time differential (T).

Since the time period of interest for flood routing through small impoundments is typically no more than a day, seepage and evaporation losses during that period will be small compared to the flood flows and may be neglected. For small impoundments, backwater effects may be assumed to be negligible (level-pool flood routing). For large reservoirs, backwater effects may increase the reservoir storage.

Equation 9.10 can be solved by numerical integration procedures if the inflow hydrograph, stage–storage relationship, and stage–discharge relationship were all available as analytical functions. A simpler solution approximates Equation 9.10 by using small time intervals. The difference form of the equation is

$$\frac{i_1 + i_2}{2}\Delta t - \frac{o_1 + o_2}{2}\Delta t = S_2 - S_1 \qquad \textbf{9.11}$$

where S = storage volume (L^3),

Δt = time interval (T),

1,2 = subscripts denoting the beginning and end of the time interval.

The value of Δt influences the quality and efficiency of the process; too-large values will give poor results and too-small values will give good results but waste time.

9.28 Elements of Flood Routing

Any flood-routing procedure requires three basic inputs and generates one output. The inputs are (1) the inflow hydrograph, (2) the stage–storage relationship for the stream reach or structure, and (3) the stage–discharge relationship for the stream reach or spillway system. The output is the outflow hydrograph.

Inflow Hydrograph. The inflow hydrograph is the same as a runoff hydrograph, which is discussed in Chapter 5.

Stage–Storage Relationship. This gives the volume of water stored within the channel reach or structure as a function of water surface elevation. For flood routing through impoundments, the elevations of interest are those above the level of the mechanical spillway. For routing through the stream reach, elevations are those above an arbitrary datum. The stage–storage relationship must be determined empirically from the topography of the stream reach or structure.

Stage–Discharge Relationship. This gives the outflow rate from the stream reach or spillway system as a function of water surface elevation. For stream routing through successive reaches, the discharge from the upstream reach will be the inflow into the next reach downstream.

Outflow Hydrograph. This is the product of the flood-routing procedure. It shows the value of the discharge from the stream reach or structure as a function of time.

9.29 Flood-Routing Procedure for Small Impoundments

Numerous software packages are available for flood routing through stream and storage systems. The Hydrologic Engineering Center (HEC) of the U.S. Army Corps of Engineers develops and distributes hydrologic analysis packages, including their Reservoir System Simulation program (HEC-ResSIM).

In general, the height of embankment and sizes of spillways cannot be determined directly. The flood routing analysis must be done with estimated values. Using those results, the engineer can then modify the inputs iteratively until satisfactory outcomes are obtained. Consideration must be given to feasibility and costs of construction as well as hydraulic performance.

For relatively simple flood-routing problems, a semigraphical method may be used. The principles of flood routing will be illustrated by the following discussion and example.

For most drop-inlet pipe spillway structures, the discharge rate when the pipe first flows full will be only slightly less than the maximum discharge through the structure. If peak runoff, total runoff, drainage area, and available storage are known, the discharge through the pipe, and consequently the approximate pipe size, can be estimated with an equation developed by Culp (1948):

$$Q_0 = Q\left[1.25 - \left(\frac{1500V}{RA} + 0.06\right)^{1/2}\right] \tag{9.12}$$

where Q_0 = outflow rate when the pipe first flows full (m^3/s),
Q = peak inflow rate in m^3/s,
V = available storage in ha-m,
R = runoff depth in mm,
A = drainage area in ha.

Example 9.9

Design a small reservoir for a site that has a contributing watershed of 50 ha. The total runoff for a 50-year return period is 90 mm, and the peak runoff rate is 5.4 m^3/s. A depth of 2.40 m for the pond is available below an elevation of 29.26 m, which is the position of the crest of the mechanical spillway. The stage–storage relationship for the structure above the 29.26 m elevation is shown in Figure 9-21. A box inlet spillway and circular concrete pipe are used for the mechanical spillway. The maximum allowable water surface elevation is 30.98 m, giving 1.72 m depth for flood storage. Using flood-routing procedures, determine the size of the outlet structure, the actual maximum water elevation, the elevation of the flood spillway crest, and the required height of the dam. Allow a net freeboard of 0.6 m and a flow depth of 0.3 m in the flood spillway.

Solution. For a flood storage depth of 1.72 m, the available storage volume is 1.92 ha-m (5.34 m^3/s-h). Using Equation 9.12, estimate the outflow rate when the pipe first flows full.

$$Q_0 = 5.4\left[1.25 - \left(\frac{1500 \times 1.92}{90 \times 50} + 0.06\right)^{1/2}\right] = 2.23\ m^3/s$$

Assume a 1 m × 1 m box inlet (crest length 3 m) and a 762-mm outlet pipe. (The area of the box inlet should be about twice the area of the pipe.) Use the weir formula with C = 1.7 and the pipe flow formula with K_e = 1.0, n = 0.014, and L = 35 m to compute the spillway discharge curves as shown in Figure 9-21. The intersection

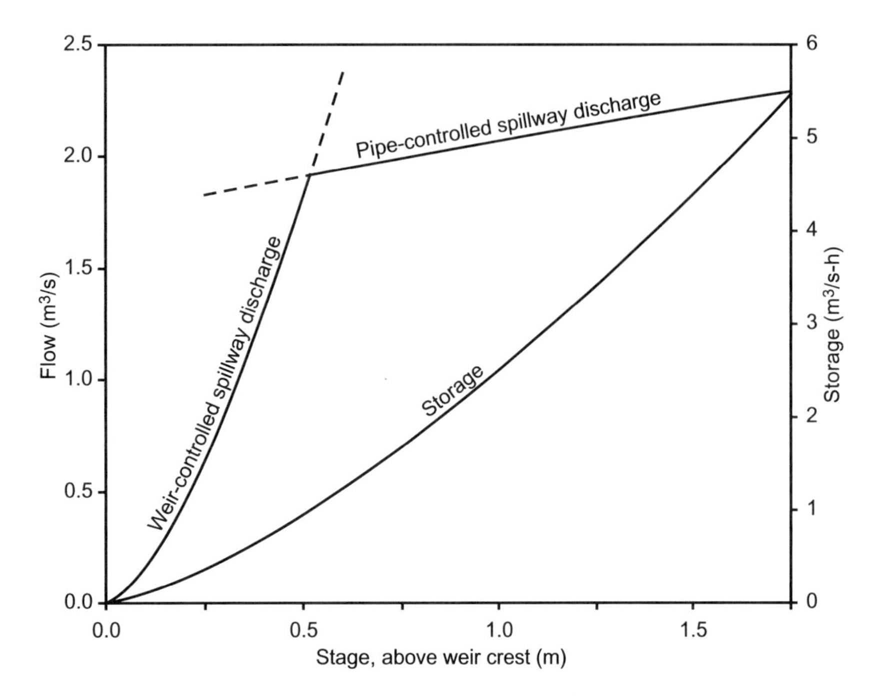

Figure 9–21

Reservoir storage and outflow curves for Example 9.9.

of the weir and pipe curves gives the value for Q_o as 1.93 m^3/s, which is within 15 percent of the estimate using Equation 9.12. Use the dimensionless hydrograph (Chapter 5) to synthesize the inflow hydrograph in Figure 9-22.

To facilitate calculations, Equation 9.11 is rearranged so that all the unknowns at the beginning of a time interval are on the left side of the equation.

$$\left(\frac{S}{\Delta t} + \frac{o}{2}\right)_2 = \left(\frac{S}{\Delta t} + \frac{o}{2}\right)_1 + \frac{i_1 + i_2}{2} - o_1 \qquad \textbf{9.13}$$

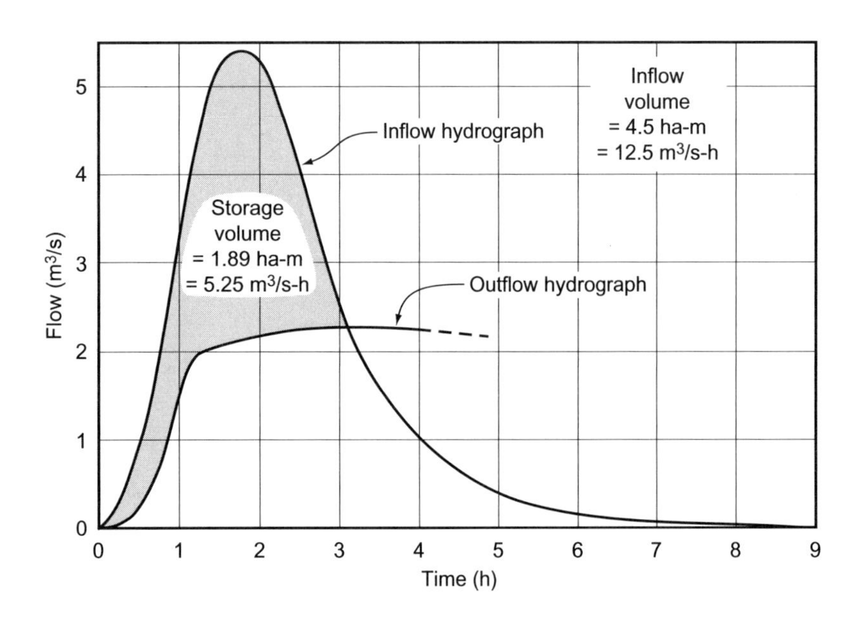

Figure 9–22

Inflow and outflow hydrographs for Example 9.9.

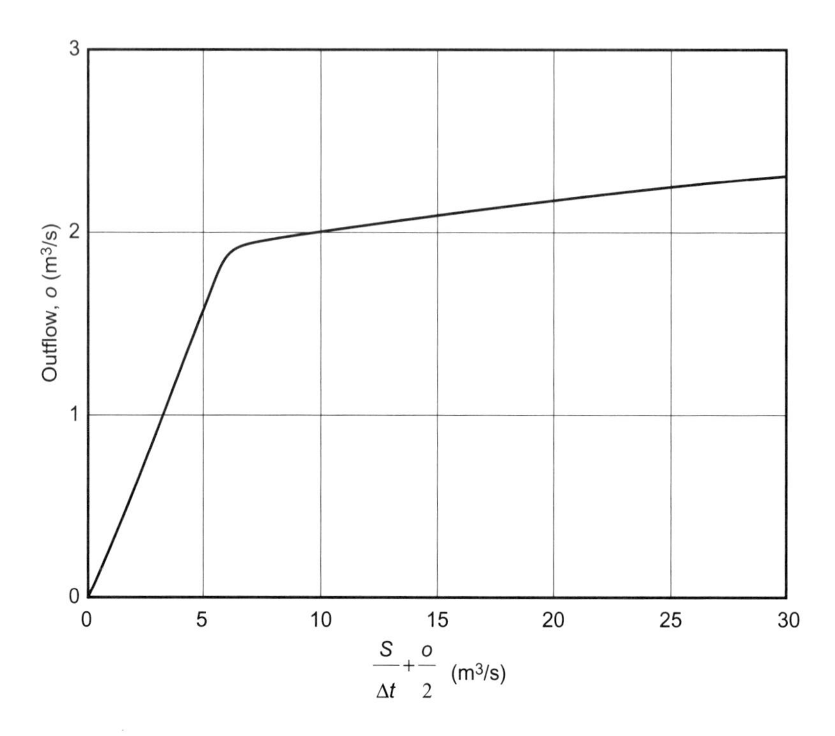

Figure 9–23
Routing curve for Example 9.9.

For any time step, the value for $(S/\Delta t + o/2)_2$ can be calculated from the values known at the beginning of the time step. $S/\Delta t + o/2$ is uniquely related to outflow, o, by their common dependence on stage. That relationship, the routing curve, can be developed as shown in Table 9-2, where o is the lesser of the weir-controlled

TABLE 9-2 Computations for Outflow versus $(S/\Delta t+o/2)$ using $\Delta t = 0.2$ h

Stage (m)	*Outflow, o (m^3/s)*	*Storage, S (m^3/s-h)*	*$S/\Delta t+o/2$ (m^3/s)*
0.00	0.00	0.00	0.00
0.10	0.16	0.10	0.58
0.20	0.46	0.26	1.54
0.30	0.84	0.46	2.74
0.40	1.29	0.69	4.11
0.50	1.80	0.95	5.64
0.60	1.95	1.22	7.09
0.70	1.98	1.52	8.58
0.80	2.01	1.83	10.15
0.90	2.04	2.16	11.81
1.00	2.07	2.50	13.54
1.10	2.10	2.86	15.34
1.20	2.13	3.23	17.20
1.30	2.16	3.61	19.13
1.40	2.19	4.00	21.12
1.50	2.22	4.41	23.16
1.60	2.25	4.83	25.26
1.70	2.28	5.25	27.41

TABLE 9–3 Flood-Routing Computations for Example 9.9

Time (h)	i (m^3/s)	$(i_1+i_2)/2$ (m^3/s)	$S/\Delta t+o/2$ (m^3/s)	o (m^3/s)	Stage (m)
0.0	0.00	0.00	0.00	0.00	0.00
0.2	0.20	0.10	0.10	0.03	0.03
0.4	0.65	0.43	0.50	0.12	0.08
0.6	1.24	0.95	1.32	0.40	0.18
0.8	2.20	1.72	2.64	0.76	0.28
1.0	3.24	2.72	4.60	1.50	0.44
1.2	4.25	3.75	6.85	1.94	0.58
1.4	5.00	4.63	9.53	2.00	0.75
1.6	5.35	5.18	12.71	2.06	0.95
1.8	5.40[a]	5.38	16.02	2.12	1.15
2.0	5.25	5.33	19.23	2.15	1.25
2.2	4.86	5.06	22.13	2.21	1.45
2.4	4.38	4.62	24.54	2.22	1.50
2.6	3.82	4.10	26.42	2.26	1.65
2.8	3.18	3.50	27.66	2.27	1.68
3.0	2.48	2.83	28.22	2.28[b]	1.70
3.2	2.04	2.26	28.20	2.28	1.70
3.4	1.74	1.89	27.81	2.27	1.68
3.6	1.45	1.60	27.14	2.27	1.68
3.8	1.23	1.34	26.21	2.26	1.65
4.0	1.05	1.14	25.09	2.24	1.58

[a]Maximum inflow.

[b]Maximum outflow. Corresponds to max. stage and max. storage.

flow or the pipe-controlled flow (Figure 9-21). The routing curve is plotted in Figure 9-23. Once $(S/\Delta t + o/2)_2$ is calculated, that value is used with Table 9-2 or Figure 9-23 to determine the value of o_2. Table 9-2 or Figure 9-21 is used to get the stage for that time step. The process is continued time step by time step at least until the outflow rate exceeds the inflow rate. From that point on, both the stage and outflow diminish.

The flood storage volume is represented graphically by the area between the inflow and outflow hydrographs in Figure 9-22 and should agree with the storage at the maximum stage.

The maximum stage, in Table 9-3, was 1.70 m, which was just within the specified limit. Had the critical stage been exceeded, a higher-capacity spillway would be needed. The elevation for the crest of the flood spillway is 29.26 m + 1.70 m = 30.96 m. The maximum settled height of the dam is then 2.40 + 1.70 + 0.3 + 0.6 = 5.00 m.

In this example, the entire flood wave was routed through the mechanical spillway. To include a flood spillway in the routing, the stage–discharge relationship would have to be modified to include the flows in both the mechanical and flood spillways.

Sediment Control Structures

The first objective in any sediment management system should be to minimize detachment and therefore the possibility of transport. Various methods for managing upland areas are presented in Chapters 7 and 8. Measures for minimizing sediment production from streams are discussed in Chapter 10. Chapter 20 deals with erosion by wind, which can be a significant sediment source in some areas. The structures presented earlier in this chapter may be incorporated into systems that either reduce detachment or encourage deposition. The underlying principle in all cases is the management of the energy of flowing water.

Figure 9-1 illustrates the use of hard structures for gully stabilization. By dissipating much of the energy of the water in drop boxes, chutes, and stilling basins, the energy of flow in the reaches between structures can be kept below critical values. Impoundments with hard spillway structures can both attenuate flood waves and dissipate energy.

In many situations, such as mining, construction, and agricultural operations, it is not possible to protect the soil surface at all times. Rainfall onto disturbed soils will detach large amounts of sediment. The focus then shifts to the capture of sediment before it leaves the site and creates problems elsewhere. Sediment-trapping devices and basins have been designed for service lives ranging from a few weeks to many years. These range from small, semipervious barriers such as bales of straw or silt fences to earthen impoundments with special outflow controls.

9.30 Principles of Sediment Capture

The transport capacity of flowing water depends on its velocity. Higher velocity flows, with higher energies, are able to transport larger particles and larger amounts of sediment than slower flows. The main objectives in sediment capture are to (1) reduce the energy of the flow to the point that sediment can settle out of the water column and (2) detain the water long enough for that settlement to take place. The structure must also have sufficient capacity to store sediment accumulated during its required service life or a design cleanout interval.

The rate at which a spherical particle falls through nonturbulent water can be estimated with Stokes's law

$$u = \frac{d^2 g}{18\mu}(\rho_s - \rho_w) \qquad \textbf{9.14}$$

where u = velocity downward (LT^{-1}),
d = particle diameter (L),
g = gravitational acceleration (LT^{-2}),
μ = absolute viscosity of water ($LM^{-1}T^{-1}$),

ρ_s = density of particle (ML^{-3}),
ρ_w = density of water (ML^{-3}).

From Equation 9.14, it is apparent that larger particles, such as sand, settle more rapidly than smaller silt or clay particles. As water flows through a detention structure, the amount of its sediment load that will be deposited depends on the sizes and densities of the particles, the depth of the water column, and the residence time of the water in the structure. Flocculation of particles into larger clumps can enhance settling. Flocculants are often used in water and wastewater treatment plants, but cost and environmental concerns make them impractical for general use. Assuming a water temperature of 20°C (μ =1.002 cp and ρ_w = 0.998 g/mL) and particles with the density of quartz (2.65 g/cm^3), Equation 9.14 simplifies to

$$u = 89.9d^2 \tag{9.15}$$

where u = velocity in cm/s,
d = particle diameter in mm.

Example 9.10

Calculate the time required for particles of fine sand (0.2 mm), silt (0.01 mm), and clay (0.001 mm) to settle out of a 1-m column of water. Assume that all particles are spherical and have the density of quartz.

Solution. Using Equation 9.15, calculate the settling velocities and times for each particle size.

Particle	*Diameter (mm)*	*Velocity (cm/s)*	*Time to Settle 1 m*
Fine sand	0.2	3.59	27.8 s
Silt	0.01	0.00899	3.09 h
Clay	0.001	0.0000899	12.9 d

The settling times shown above suggest that capture of sand is relatively easy, silt is difficult, and clay is nearly impossible.

9.31 Measures of Performance

The performance of a sediment detention basin or similar structure may be characterized by its sediment-trapping efficiency. Trapping efficiency is defined as the percentage of the mass of sediment in the inflow that is retained in the structure. Another measure of performance is the concentration of sediment in the effluent from the structure.

Haan et al. (1994) list the major factors controlling sediment transport through a detention basin as

- Physical characteristics of the sediment
- Hydraulic characteristics of the basin

- Inflow sedimentgraph
- Inflow hydrograph
- Basin geometry
- Chemistry of the water and sediment

Due to the many and highly variable factors that affect performance, design of detention structures to specific standards is exceedingly complex. For example, a detention basin that traps 75 percent of the sediment leaving a construction site during the early phases of construction when the site is greatly disturbed is likely to have much lower efficiency later in the project when most of the area has been stabilized. The cause is the nature of the sediment load. Erosion on the disturbed site is likely to produce a relatively high percentage of sand, which is easily captured. As the site is stabilized and revegetated, sand is likely to be a very small fraction of the sediment load. As the dominant particle sizes in the inflow tend toward silt and clay, capture becomes more difficult and trapping efficiency drops. Since trapping efficiency makes no allowance for the changes in the nature of the sediment load, the falling efficiency would give the impression that the structure is failing somehow. Haan et al. (1994) discuss such factors and their impact on design.

9.32 Detention Basin Characteristics

To maximize the effectiveness of a sediment detention structure, it should be designed to minimize turbulence, minimize dead areas, and maximize residence time. Sedimentation basins may be designed to maintain a permanent pool, or to empty completely between runoff events. In either case, the structure must provide volumes for

1. Storage of accumulated sediment
2. Detention storage to give the necessary residence time
3. Flood storage

Most detention basins are earthen structures like those discussed is Sections 9.12–24. The difference is in the type and placement of the outlet structures. Several outlet designs are in common usage. One type of design uses a perforated riser, as shown in Figures 9-24a and b. While this meters the outflow over the range of storage depths, water from the entire column can enter the outlet, including the more sediment-laden water near the bottom of the basin. Millen et al. (1997) evaluated an alternative method that uses a Faircloth skimmer, essentially a floating outlet, to discharge water from only the top of the water column (Figure 9-24c). They found that the skimmer performed significantly better than the perforated riser. Another option, often used for temporary structures because of its ease of construction, is the rock filter outlet (Figure 9-24d). If the rock filter extends to the top of the embankment and has a notch formed in the top (Figure 9-24e), it can also serve as a flood spillway.

Sedimentation basins may be either dry (Figure 9-24a) or wet, i.e., having a permanent pool (Figure 9-24b). If a dry basin is used, the first runoff that enters the basin may flow all the way to the outlet and begin to discharge, carrying its sediment load with it. If a permanent pool is used, the first runoff entering the basin will be slowed by the standing water and drop at least some of its sediment before reaching the outlet.

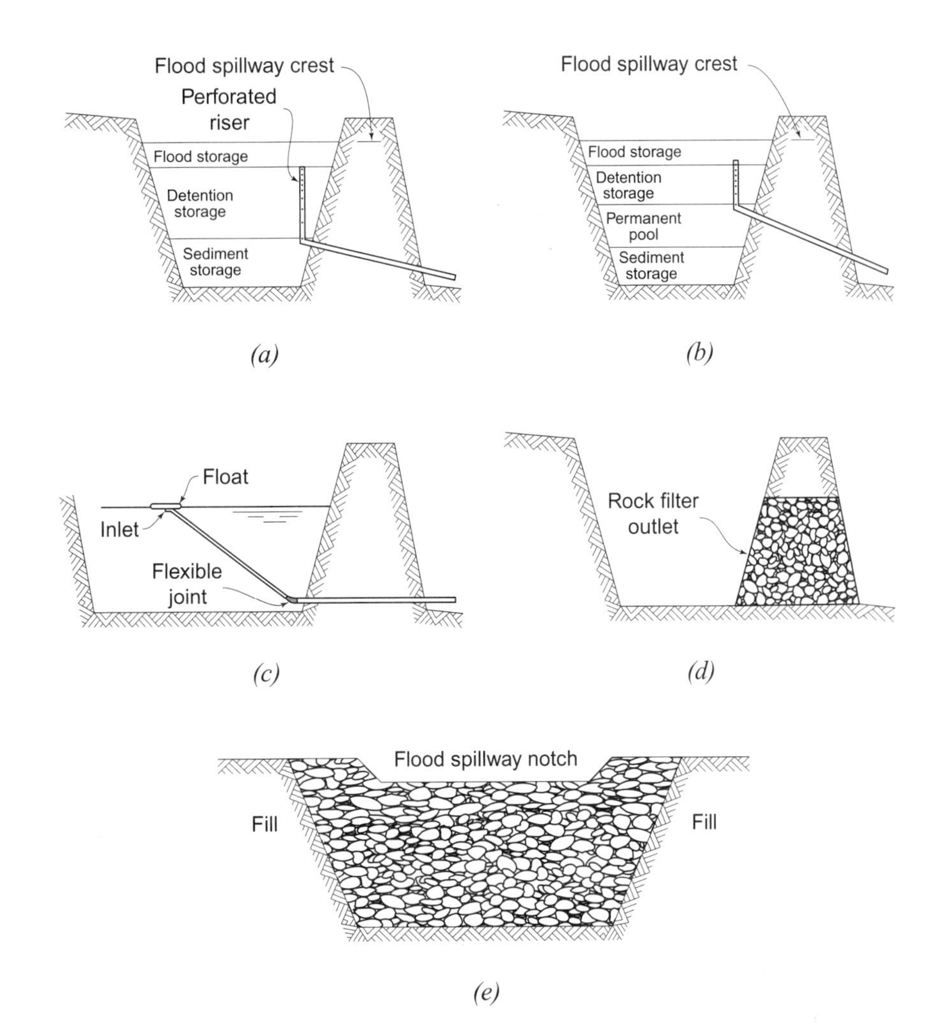

Figure 9–24

Detention basin types and outlets. (a) Dry basin with perforated riser. (b) Wet basin with perforated riser. (c) Basin with skimmer outlet. (d) Basin with rock filter outlet. (e) Rock filter outlet with notch for flood spillway.

The geometry of the basin influences its effectiveness. In general, basins should be long and narrow in the direction of flow. Griffin et al. (1985) recommended that the length-to-width ratio be at least 2:1 to minimize dead zones. Baffles, usually made with geotextiles, can be inserted into a basin to create a longer flow path and suppress turbulence. These can be helpful where site geometry makes it difficult to shape the basin itself with an appropriate length–width ratio.

9.33 Basin Sizing

Sediment storage volume may be specified by local regulations as a certain volume per unit of contributing area. The expected sediment load over the service life or cleanout interval, trapping efficiency, and bulk density of the settled sediment are used to calculate the sediment storage volume.

The storage volume may also be specified by local regulations. If hydrologic design is to be used, the basin should be sized to detain the runoff from a design storm, typically having a return period of 2 to 10 years (Haan et al., 1994). The flood storage volume and size of the flood spillway will depend on the design storm selected for that purpose. For small structures, the flood spillway is typically designed to pass the peak inflow rate.

9.34 Regulatory Compliance

Many states have developed erosion and sediment control handbooks that specify approved designs and practices. Regulations may require that only those practices outlined in the handbooks be used. Although strict adherence to the guidelines may protect the engineer legally, the result may be over- or underdesign, since site conditions may differ considerably from those for which the standards were developed. Every effort should be made to provide economical designs that will perform adequately while complying with regulations.

Permanent structures must often be designed for both sediment and flood control. Regulations may require that postdevelopment peak runoff rates not exceed predevelopment rates. To satisfy this requirement, both pre- and postdevelopment hydrographs must be estimated (Chapter 5). Using flood routing procedures (Section 9.29), the postdevelopment hydrograph is routed through the structure. The flood storage and flood spillway are then sized to reduce the peak outflow to match the predevelopment rate.

Internet Resources

U.S. Army Corps of Engineers, Hydrologic Engineering Center (HEC)
http://www.hec.usace.army.mil

USDA Natural Resources Conservation Service (NRCS)
http://www.nrcs.usda.gov
http://www.nrcs.usda.gov/technical/ENG/efh.html

U.S. Bureau of Reclamation (USBR)
http://www.usbr.gov

References

Beasley, R. P., L. D. Meyer, & E. T. Smerdon. (1960). Canopy inlet for closed conduits. *Agricultural Engineering, 41*, 226–228.

Blaisdell, F. W. (1959). *The SAF Stilling Basin: A Structure to Dissipate the Destructive Energy in High-Velocity Flow from Spillways*. Agriculture Handbook No. 156, USDA Agricultural Research Service in cooperation with the Minnesota Agriculture Experimental Station and St. Anthony Falls Hydraulic Laboratory of the University of Minnesota. Washington, DC: U.S. Government Printing Office.

Blaisdell, F. W., & C. A. Donnelly. (1951). *Hydraulic Design of the Box Inlet Drop Spillway*. USDA SCS-TP-106. Washington, DC: USDA Soil Conservation Service.

———. (1958). *Hydraulics of Closed Conduit Spillways, Pt X: The Hood Inlet*. University of Minnesota, St. Anthony Falls Hydraulic Lab. Tech. Paper No. 8, Ser. B.

———. (1966). *Hydraulic Design of the Box-Inlet Drop Spillway*. Agriculture Handbook No. 301. USDA Agriculture Research Service in cooperation with the Minnesota Agriculture Experimental Station and St. Anthony Falls Hydraulic Laboratory of the University of Minnesota. Washington, DC: U.S. Government Printing Office.

Brater, E. R., & H. W. King. (1996). *Handbook of Hydraulics*, 7th ed. New York: McGraw-Hill.

Culp, H. L. (1948). The effect of spillway storage on the design of upstream reservoirs. *Agricultural Engineering, 29*, 344–346.

Griffin, M. L., B. J. Barfield, & R. C. Warner. (1985). Laboratory studies of dead storage in sediment ponds. *Transactions of the ASAE, 28*(3), 799–804.

Haan, C. T., B. J. Barfield, & J. C. Hayes. (1994). *Design Hydrology and Sedimentology for Small Catchments*. San Diego: Academic Press.

Harr, M. E. (1962). *Groundwater and Seepage*. New York: McGraw-Hill.

Mavis, F. T. (1943). *The Hydraulics of Culverts*. Pennsylvania Engineering Experimental Station Bulletin 56. College Station: Pennsylvania State University.

Millen, J. A., A. R. Jarrett, & J. W. Faircloth. (1997). Experimental evaluation of sedimentation basin performance for alternative dewatering systems. *Transactions of the ASAE, 40*(4), 1087–1095.

Morris, B. T., & D. C. Johnson. (1942). Hydraulic design of drop structures for gully control. *Proceedings of the ASCE, 68*, 17–48.

U.S. Bureau of Reclamation (USBR). (1987). *Design of Small Dams*, 3rd ed. Washington, DC: U.S. Government Printing Office.

U.S. Soil Conservation Service (SCS). (1984). *Engineering Field Manual*. Chapter 6: Structures. Washington, DC: Author.

———. (1985). *Earth Dams and Reservoirs*. Technical Release No. 60. Washington, DC: Author.

Wooley, J. C., M. W. Clark, & R. P. Beasley. (1941). *The Missouri Soil Saving Dam*. Agricultural Experiment Station Bulletin 434. Columbia: University of Missouri.

Problems

9.1 What is the maximum capacity of a straight-drop spillway having a crest length of 3 m and a depth of flow of 0.9 m?

9.2 Determine the design dimensions for the drop spillway in Problem 9.1, using the Morris & Johnson outlet, if the drop in elevation is 1.5 m. The waterway is 4.6 m wide.

9.3 Determine the crest length for a straight-inlet drop spillway to carry 7.5 m^3/s if the depth of flow is not to exceed 0.9 m. What should be the dimensions of a square-box inlet for the same conditions?

9.4 Determine the design dimensions for a 1.5 × 1.5-meter box-inlet drop spillway to carry 7.0 m^3/s if the end sill is 3 m in length.

9.5 What is the capacity of a formless flume 1.8 m wide when the flow depth is 0.6 m?

9.6 Determine the discharge of a 30-m long, 0.9- × 0.9-m concrete box culvert having a square entrance, $n = 0.013$, $s = 0.003$, and elevations of 10.97 m at the center of the conduit at the outlet, 12.89 m for the headwater, and 12.34 m for the tailwater.

9.7 Calculate and plot the head–discharge relationship up to 1.5-m head for a drop-inlet culvert having 0.9- × 0.9-m square inlet attached to a 457-mm concrete pipe 50 m in length. Assume that pipe flow controls in the conduit, the tailwater height is below the center of the pipe at the outlet, the radius of curvature of the pipe entrance is 0.1 m, and the difference in elevation between the crest and center of the pipe at the outlet is 5.5 m.

9.8 Determine the capacity of a 610-mm-diameter culvert that is 30 m long. $K_e = 0.25$, $K_c = 0.05$, $s = 0.032$, headwater is 2.16 m above the inlet invert, and tailwater is 0.40 m above the outlet invert. What is the neutral slope assuming full pipe flow?

9.9 Determine the discharge of a circular-plate hood-inlet pipe spillway through a dam if the conduit is a 305-mm-diameter corrugated metal pipe, 30 m long on a slope of 16 percent. The depth of water above the inlet invert is 0.94 m.

9.10 If the critical frost depth is 0.15 m, maximum exposure of the water surface is 400 m, depth of flow in the flood spillway is 0.3 m, and flood storage depth is 1.2 m, determine the net freeboard and the gross freeboard.

9.11 An earthen dam is to be constructed having a volume after settlement of 1450 m^3. The desired wet Proctor density of the dam is to be 1910 kg/m^3 with an optimum water content of 20 percent. If the void ratio of the dam is 0.75, how many cubic meters of borrowed soil with a void ratio of 1.15 are required? What is the dry mass of the soil in the dam in Mg?

9.12 Immediately after construction of an earthen embankment, it had a void ratio of 0.40. What is the settlement of the embankment, expressed as a percentage of the height before consolidation, if the void ratio after its consolidation is 0.30?

9.13 Compute the seepage rate per unit length through a homogeneous earthen dam resting on an impervious foundation if the top width is 4.0 meters, the height of the dam is 12 meters, the upstream sideslope is 3:1, the downstream sideslope is 2:1, the hydraulic conductivity of the fill is 3.3×10^{-8} m/s, and the net freeboard is 1.0 m.

9.14 Use flood-routing analysis to determine the maximum water level for a flood control reservoir that has a watershed area of 50 ha. The runoff depth for a 50-year return period storm is 90 mm, and a peak runoff rate is 5.4 m^3/s. The outlet structure is a box-inlet spillway, 0.9 m square (one side is the headwall) with a weir coefficient of 3.0 and a 610-mm diameter concrete outlet pipe. The pipe is 34 m long, with $K_c = 1.0$ and $n = 0.014$. The elevation of the crest of the box inlet is 26.8 m, and elevation of the center of the pipe at its outlet is 25.6 m. The cumulative storage available at each 0.6-meter stage above the crest is 0.06, 0.17, 0.41, 0.76, 1.32, 2.10, and 3.22 ha-m.

9.15 Design the outlet structure (box inlet and pipe size) for the reservoir in Problem 9.14 using the inflow hydrograph given below. The runoff depth is 80.8 mm, and the peak runoff rate is 6.12 m^3/s. The maximum allowable elevation of the flood spillway crest is 30.2 m.

Time (min)	*Q (m^3/s)*	*Time (min)*	*Q (m^3/s)*
0	0	87	5.86
10	0.31	97	5.21
23	0.91	110	4.42
32	2.15	129	3.06
42	3.43	151	2.01
52	4.70	171	1.47
58	5.52	193	0.99
64	5.95	217	0.68
74	6.12	270	0.25
		322	0

9.16 Calculate the settling velocities of spherical particles having the density of quartz and diameters of 1, 3, 10, 30, and 100 μm. Generate a graph that shows the length of flow path through a sedimentation basin needed for particles of each size to settle out of flow that is 1.0 m deep with flow velocities of 0.001, 0.01, and 0.1 m/s.

9.17 At maximum detention storage, the water surface elevation in a wet sedimentation basin is 82.5 m. The permanent pool elevation is 81.5 m. The water surface at 82.5 m is 25 m wide and 100 m long. The interior sideslopes are 3:1. (a) What is the detention volume of the structure? (b) Regulations require that the detention volume be drained within 72 hours. The outflow device is a skimmer that uses an orifice to control the flow. Assuming an orifice coefficient of 0.6, what size orifice should be specified if there is a constant head of 0.4 m on the orifice?

CHAPTER 10

Channel Stabilization and Restoration

A dominant characteristic of most inland cities of the world is a river. To have access to freshwater, many coastal cities were formed near the outlet of a river. Historically, rivers have represented a source of water, transportation route, food supply, and means for waste removal. As cities grew, the river systems became impacted by human activities. These impacts include pollution, sedimentation, channelization, and impoundments. Nearly every major river system is now controlled to such an extent that it can be represented by a series of reservoirs. Tributaries of these large rivers have also been affected by the placement of impoundments to control flow and flooding.

The end of the twentieth century saw a shift from analyzing and designing stream systems strictly for hydraulic function to the design of systems with a balance between hydraulic, environmental, and ecosystem functions. The Kissimmee River in Florida is an example of a river that was channeled from 160 km of meandering river to a series of five impoundments connected by a large drainage canal (Brookes & Shields, 1996). The channelization caused a dramatic change in the ecosystems of the region, with significant losses in aquatic and waterfowl species. Procedures are now underway to restore the river to its natural condition. Because of the relatively new science of stream restoration and stabilization, few design procedures are available (FISRWG, 1998; Doll et al., 2003).

The stream channel is the focal point for characterizing the hydrology of a watershed. Over time, most natural channels meander back and forth across a well-developed floodplain. In addition to the lateral movement, the stream channel can move vertically. A foundation for the ecological design of natural stream corridors, even those that contain stable channels, is that they are constantly changing.

The stream corridor describes the linkage between the stream channel and the floodplain. The characteristics of the stream corridor vary from the headwaters to the mouth. The changes include stream slope, stream width, stream depth, bed material, riparian vegetation, and aquatic life. Some rather dramatic changes in the stream corridor can occur during a flood event, but most changes in the stream corridor occur over long periods of time.

10.1 Watershed Classification

In the 1970s, the U.S. Geological Survey (USGS) developed a system for classifying and numbering drainage basins in the United States (Figure 10-1). The original system included four levels of classification with each level utilizing a 2-digit code. The four levels of the hydrologic unit codes (HUC) were region, subregion, accounting unit, and cataloging unit. There are 21 regions that represent the drainage of major river systems. There are 222 subregions in the United States, representing a smaller river system, reach of river, or major tributary. The 222 subregions are divided into 352 hydrologic accounting units (also called basins). The fourth level of classification includes 2149 cataloging units (also called sub-basins) representing drainage basins or distinct hydrologic features. The smallest USGS unit or 8-digit HUC contains approximately 181 000 ha. Since the area encompassed by the 8-digit codes was still quite large, these sub-basins were further subdivided by the NRCS into 11-digit watersheds (fifth level) and 14-digit subwatersheds (sixth level). There are approximately 22 000 watersheds ranging in size from 16 000 to 100 000 ha and 160 000 subwatersheds ranging in size from 1200 to 16 000 ha.

As an example of the coding system, the watershed that includes Lexington, Kentucky, is cataloged in unit 05100205, which is the Lower Kentucky sub-basin. Just to the east of Lexington is the small watershed of Clear Creek. Its 14-digit code is 05100205-020-080. In the coding system, the Clear Creek subwatershed is a part of the Lower Kentucky sub-basin and the Ohio River region (05).

10.2 Drainage Networks

Excess water either from surface runoff or baseflow from ground water makes its way to a particular channel or series of channels. The drainage network is the pattern of rivers, streams, and minor rills that define the surface transport of water in a watershed. In most cases, the organization of the drainage network fits a simple pattern that can be described by stream order, number, and length (Horton, 1945).

Figure 10–1
Two-digit hydrologic unit codes of the United States.

Stream order is a numbering system that describes the position of a particular stream in the hierarchy of tributaries in a watershed. First-order streams are streams with no definite tributaries. When two first-order streams meet, a second-order stream is formed. In the Horton system, the second-order stream extends headward to the tip of the longest tributary it drains (Figure 10-2b). A third-order stream is created when two second-order streams combine. As with the second-order stream, the longest tributary of the third-order stream is determined and reordered in the Horton ordering system. This process of ordering the stream channels continues to the outlet of concern. With the Horton ordering system, the longest tributary in the watershed has the same order as the outlet channel. The primary disadvantage of the Horton system is the need to reorder or relabel the longest tributary whenever a higher-order stream is formed. Strahler (1952) modified the Horton ordering system by not extending higher-order streams headward to the tip of the longest tributary (Figure 10-2a).

Horton (1945) analyzed many watersheds and determined that the number, length, and drainage area of streams were functions of stream order. The bifurcation ratio is the slope of the line formed by a semilog plot of stream number to stream order. It relates the ratio of the number of stream segments of a given order to the number of stream segments of the next highest order. The bifurcation ratio ranges from 2 to 4, but is generally between 3.2 and 3.5. For a fourth-order stream, there

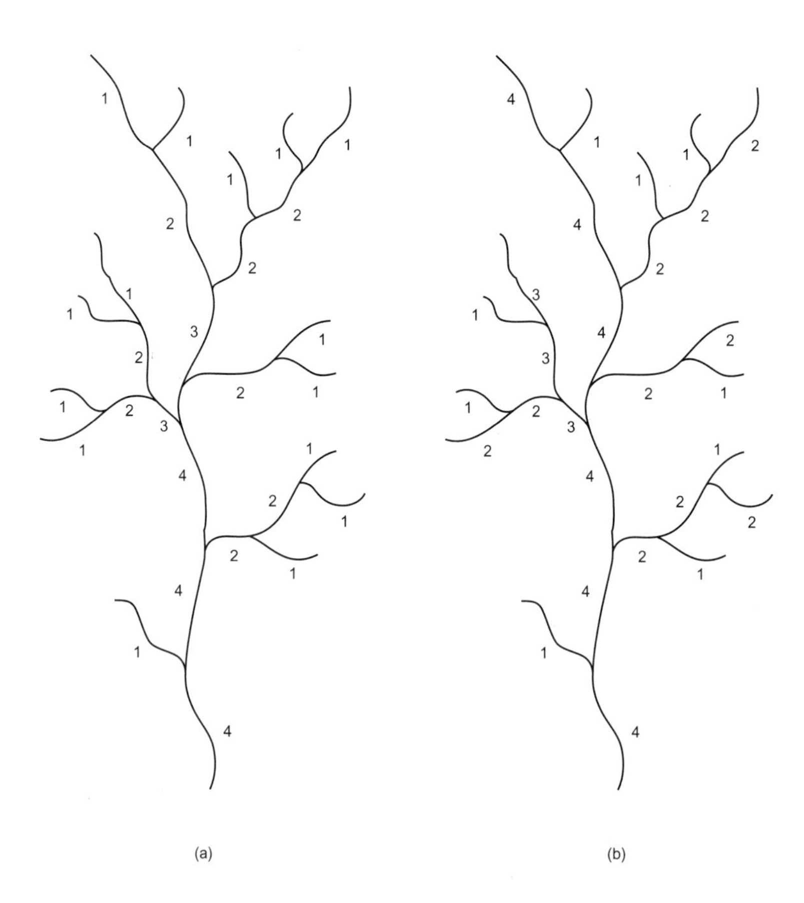

Figure 10–2
Illustration of the Strahler (a) and Horton (b) stream-ordering systems for watersheds.

are normally between 3 and 4 third-order streams. For a third-order stream, there are normally 3 to 4 second-order streams and so on.

In addition to the relationship between number of streams and stream order, a semilog plot of either stream length or drainage area versus stream order results in a straight line. These relationships are important if extrapolations need to be made for higher- or lower-order streams. Stream length is related to drainage area (Knighton, 1998):

$$L = 1.27A_d^{0.6} \tag{10.1}$$

where L = stream length (km),

A_d = drainage area (km^2).

One difficulty in classifying drainage networks is caused by the scale of the map being used for the classification. Detailed topographic maps are required to identify the smallest tributaries (first-order streams) in a large drainage basin. In many cases, a low-resolution topographic map is available. The bifurcation ratio and Equation 10.1 can be used to estimate the number and length of lower-order streams if a site visit indicates stream orders were miscalculated.

Example 10.1

A topographic map was used to determine the number and average area of first-through third-order streams in a watershed in eastern Kentucky (table below). A visit to the area indicates that the first-order streams are actually second-order streams. Using the data already obtained, estimate the number of first-order streams and the drainage area and length of these streams.

Order	*Number*	*Average Drainage Area (km²)*
1	32	0.6
2	10	2.8
3	3	11.2
4	1	48.1

Solution.

(1) The first step is to realize that the stream order is shifted by 1. First-order streams in the table are actually second-order streams and so on.

(2) An average bifurcation ratio can be obtained by averaging the individual ratios between each stream order. (3 + 3.33 + 3.2)/3 = 3.18)

(3) If there are 32 second-order streams, then there are 32(3.18) = 102 first-order streams in the basin.

(4) The data can be placed in a spreadsheet and plotted in a semilog plot. A trend-line can be used to estimate the slope and intercept of the exponential relationship described by the semilog plot of drainage area D_a and stream order O:

$$D_a = 0.034e^{(1.45O)}$$

Using this relationship, the average drainage area for a first-order stream is 0.145 km^2.

(5) The stream length can be estimated by Equation 10.1.

$$L = 1.27\,D_a^{0.6} = 0.4\,\text{km}$$

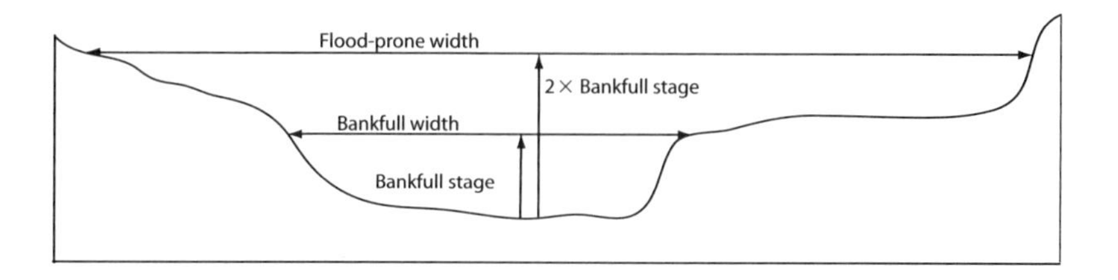

Figure 10–3

Illustration of bankfull depth and width. The flood-prone width is used to calculate the entrenchment ratio.

10.3 Channel Geometry

Most precipitation events that produce runoff increase stream flow. In addition to water, the runoff process may cause erosion and an addition of sediment to the stream. Infrequent large storms can produce large quantities of stream flow and sediment discharge, but some of the flow occurs on the floodplain adjacent to the channel. The excess water flowing on the floodplain does not affect channel formation. Bankfull discharge represents the breakpoint between channel formation processes and floodplain processes (Figure 10-3). Bankfull discharge has the greatest effect on channel formation and maintenance over the long term.

Bankfull stage corresponds to the discharge at which overbank flow is incipient. Bankfull stage is easily determined for streams with well-defined floodplains (Figure 10-3). For channels without a well-defined floodplain, changes in bank vegetation, staining of rocks, and the tops of depositional features such as point bars may be useful in identifying bankfull stage (Rosgen, 1996).

Where streamflow data exist, the bankfull discharge can be estimated by studying recurrence interval data. Since the discharge is responsible for developing and maintaining the channel dimensions, the bankfull or channel-forming discharge must occur frequently enough to allow the proper conditions to develop. Comparisons of bankfull discharge and associated recurrence interval data have shown that recurrence intervals of 1 to 3 years are essential for channel formation. Rosgen (1996) found that recurrence intervals of 1.5 years for rural watersheds and 1.2 years for highly developed, urban watersheds corresponded to field-determined bankfull discharge. Doll et al. (2002) found return intervals of 1.3 years for urban streams and 1.4 years for rural streams in North Carolina. Williams (1978) analyzed data from 233 sites and found a good correlation between bankfull area A_b (m^2), channel slope S (m/m), and bankfull discharge Q_b (m^3/s):

$$Q_b = 4.0A_b^{1.21}S^{0.28} \quad \textbf{10.2}$$

Doll et al. (2002) studied the relationship between A_b and Q_b as a function of watershed drainage area D_a (km^2) in the Piedmont region of North Carolina. Excellent correlations ($r^2 > 0.87$) were found for these relationships:

$$A_b = 3.02D_a^{0.65} \text{ and } Q_b = 4.77D_a^{0.63} \text{ for urban streams}$$

$$\textbf{10.3}$$

$$A_b = 1.08D_a^{0.67} \text{ and } Q_b = 1.32D_a^{0.71} \text{ for rural streams}$$

A comparison of urban streams to rural streams indicated that the slope of the relationship (exponent) was not statistically different based on level of urbanization. A shift in the relationship was noted. The shift indicates that as watersheds become increasingly urbanized, the size of the stream and channel forming discharge both increase.

Other terms such as *dominant discharge* or *effective discharge* have been used synonymously with *bankfull discharge*. Dominant discharge is the streamflow that is largely responsible for forming the geometry of the channel and is a more appropriate term than *bankfull discharge* for incised channels that may not have a distinct floodplain. Effective discharge is the discharge where the product of cumulative sediment transport and streamflow frequency of occurrence is maximized. Small streamflows occur frequently but carry small sediment loads. Large streamflows occur infrequently but can transport large sediment loads. Within these two extremes exists a discharge, the effective discharge, that occurs frequently enough to carry a combined total of sediment that is a maximum. Comparisons of effective, dominant, and bankfull discharges indicate that the more easily field-determined value of bankfull discharge is a good estimate for the channel-forming discharge.

Stream width w, stream depth d, and average water velocity v at bankfull discharge are important indicators of channel form. Leopold & Maddock (1953) showed that these factors were related according to simple power functions.

$$\begin{aligned} w &= aQ_b^b \\ d &= cQ_b^f \\ v &= kQ_b^m \end{aligned} \qquad \textbf{10.4}$$

Since discharge is the product of average velocity and cross-sectional area, the product of a, c, and k must equal 1 and the sum of b, f, and m must equal 1. Parameters for Equation 10.4 should be determined from measurements at a stable riffle on streams in a particular area. The resulting curves are called regional curves (Leopold et al., 1992). Natural stream reaches that are stable with little aggradation or degradation occurring should be chosen for measurement. If the regional curves are being developed for a stream restoration project, the data for the regional curves should extend across a range of flows that encompass the bankfull discharge expected at the project site. Knighton (1998) summarized the work of numerous studies and listed ranges for a (2.55–5.23), b (0.45–0.52), c (0.26–0.69), and f (0.31–0.46) for rivers and canals.

Depending on channel slope, natural streams either form a single channel that meanders across the floodplain or a system of channels that form a braided pattern within the floodplain. Leopold and Wolman (1957) developed a threshold between meandering streams and braided streams related to bankfull discharge.

$$S = 0.0125Q_b^{-0.44} \qquad \textbf{10.5}$$

where S = threshold channel slope (L/L),

Q_b = bankfull discharge (m^3/s).

Channel slopes greater than the threshold slope will promote the formation of braided streams. Slopes less than the threshold will result in meandering streams (Figure 10-4). A simple model of the meander pattern is the equation for a sine-generated curve (Knighton, 1998):

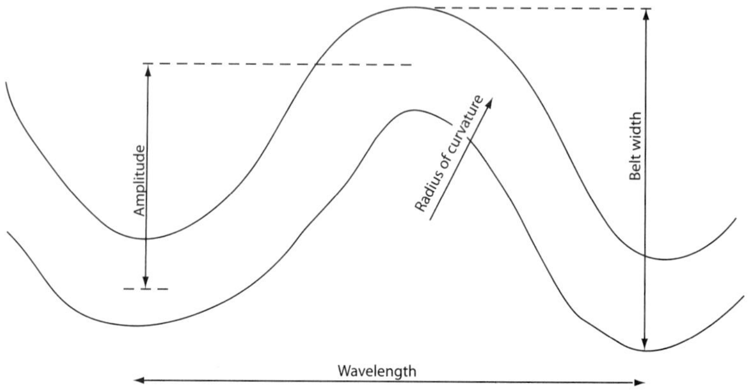

Figure 10–4
Characteristic pattern and descriptors for a meandering stream.

$$\theta = \omega \sin\left(\frac{2\pi x}{KL_m}\right) \qquad \textbf{10.6}$$

where θ = stream direction relative to the down-valley axis (radians),
ω = crossover angle between the stream segment and the down-valley axis,
K = sinuosity (L/L),
L_m = meander wavelength (L),
x = distance along the stream (L).

Note that the sine-generated curve allows for up-valley directions in flow to occur that are not possible with the typical sine wave curve (Ferguson, 1973). Langbein and Leopold (1966) used the sine-generated curve to develop a relationship between the radius of curvature R_c, channel sinuosity K, and wavelength L_m:

$$R_c = \frac{L_m K^{1.5}}{13(K - 1)^{0.5}} \qquad \textbf{10.7}$$

Sinuosity is the ratio of stream length to valley length. Streams with significant meandering would have high values of sinuosity.

Chang (1988) completed a theoretical analysis of nonbraided channels of various width-to-depth ratios and found the following relationship for the ratio of radius of curvature to bankfull width:

$$\frac{R_c}{w} = 2.2 + 0.15\left(\frac{w}{d} - 4\right) \qquad \textbf{10.8}$$

Williams (1986) determined a mean R_c/w value of 2.43 in an analysis of 79 alluvial stream sites with a sinuosity greater than 1.2. Most of the values fell within the range of 2 to 3, meaning that the radius of curvature is approximately 2 to 3 times the bankfull width. For 438 stream sites around the world with sinuosity greater than 1.2, Copeland et al. (2001) determined that meander wavelengths were 11 to 13 times the bankfull width.

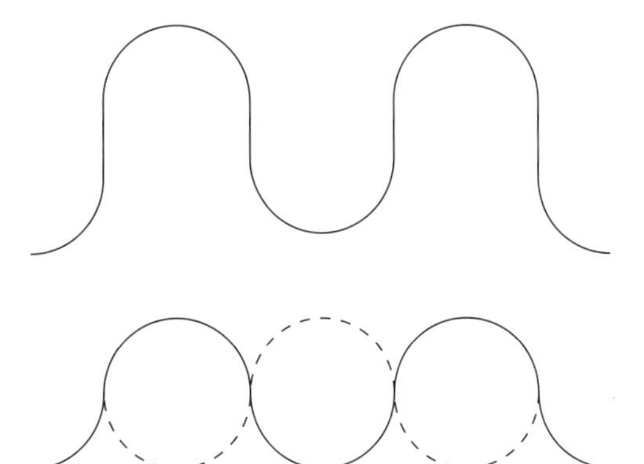

Figure 10–5
Meander patterns predicted using a sine-generated curve (top) and successive applications of circular arcs (bottom).

Another model for the meander pattern of a stream is the use of successive circular arcs (Figure 10-5). The sine-generated curve and the circular arcs produce regular, systematic meander patterns that rarely occur in nature. The primary advantage of using these types of patterns in restoration work is the smooth, rounded curvature that is produced. This rounded curvature of the bends minimizes the energy expended by the stream in making turns.

Example 10.2

Regional curves for stable natural streams in an area have been developed. The parameters for the regional curves (Equation 10.4) are $a = 3.18$, $b = 0.48$, $c = 0.52$, and $f = 0.45$. A concrete-lined channel in an urban area is to be converted to a naturalized area. Estimate the dimensions of a stable channel for the area if the 1.2-yr recurrence interval storm for the urban area is 3.5 m^3/s. Estimate the meander pattern for the stream.

Solution.

(1) Substitute the a, b, c, and f parameters and bankfull discharge of 3.5 m^3/s into Equation 10.4 to obtain the stable width and depth at bankfull conditions:

$$w = aQ_b^b = 3.18(3.5)^{0.48} = 5.8\,\text{m}$$

$$d = cQ_b^f = 0.52(3.5)^{0.45} = 0.9\,\text{m}$$

(2) A meandering stream must have a slope less than the threshold slope described by Equation 10.5.

$$S = 0.0125Q_b^{-0.44} = 0.0125(3.5)^{-0.44} = 0.007\,\text{m/m}$$

(3) The radius of curvature for the stream meanders can be estimated from Equation 10.8.

$$\frac{R_c}{w} = 2.2 + 0.15\left(\frac{w}{d} - 4\right) = 2.2 + 0.15\left(\frac{5.8}{0.9} - 4\right) = 2.57$$

The radius of curvature is 2.57(5.8) = 15 m. The meander wavelength should be approximately 12(5.8) = 70 m.

Natural channels will not have uniform dimensions and meander patterns. These calculations serve as a starting point and should be used in conjunction with natural variability found in stable natural streams in the area.

10.4 Stream Classification

A wide variety of stream forms exist on the landscape ranging from steep high-energy streams in mountainous areas to low-gradient streams of the coastal plain. There have been many efforts to classify streams into categories that allow a stream in one location to be referenced against or compared to a stream in another location. Figure 10-6 illustrates a modification of the Rosgen (1996) classification system that can be used to identify and classify streams into eight types. The primary advantage of the modified system is the capability of uniquely identifying a stream type from measurements of width-to-depth ratio and entrenchment ratio.

The width-to-depth ratio is defined as the ratio of the bankfull width to the mean bankfull depth in a stable riffle. The width-to-depth ratio provides an indication of the energy level or stability of the stream. The entrenchment ratio describes the vertical containment of the river or stream in the valley floor. It is defined as the flood-prone width divided by the bankfull width (Figure 10-3).

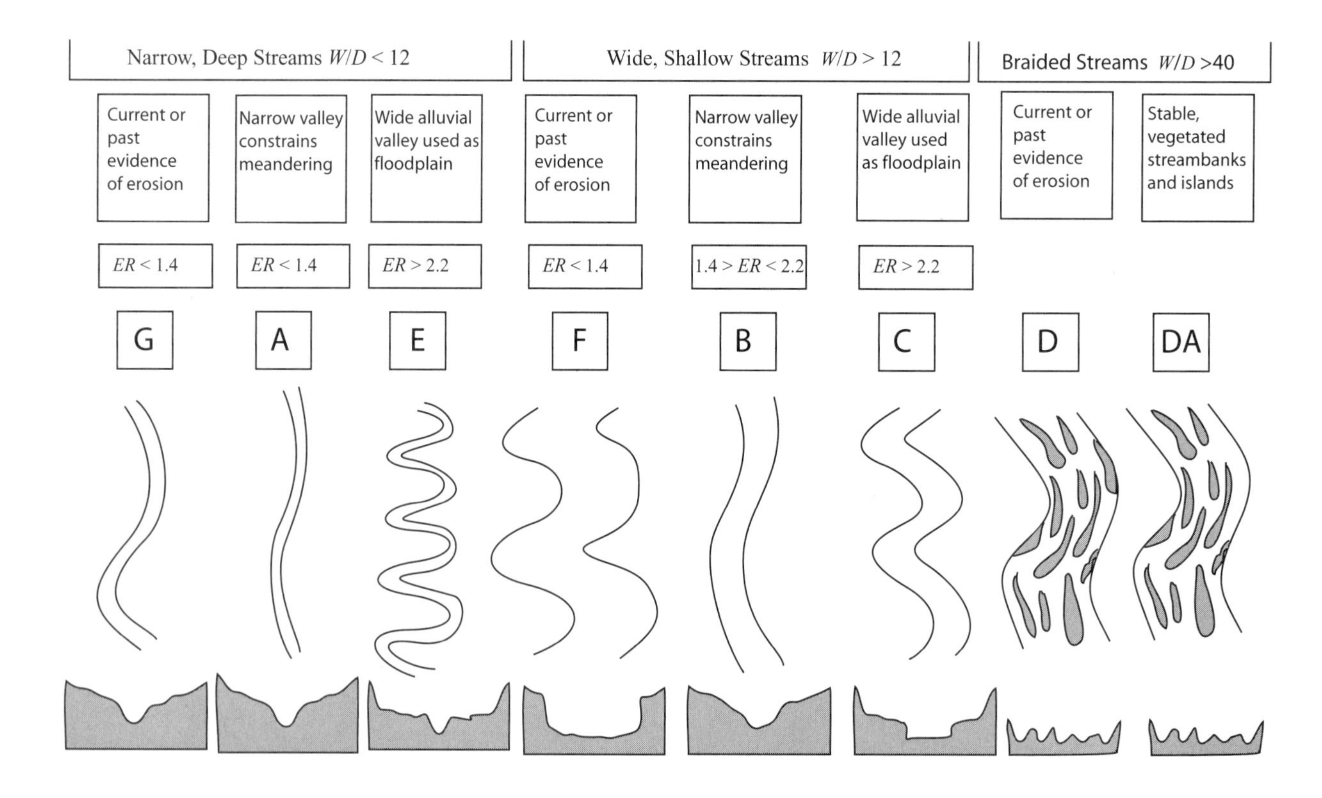

Figure 10–6 Schematic of the stream classification system.

The flood-prone width is defined as the width of water that exists when the water level is equal to twice the depth of the maximum bankfull depth in a stable riffle.

Additional parameters can be used to further classify a stream. These parameters include channel slope, sinuosity, and meander width ratio. The meander width ratio (belt width/bankfull width) describes how much of the floodplain is used for the stream corridor. The belt width describes the amplitude of the meanders in a stream (Figure 10-4). Meander width ratio is an important parameter for stream restoration. Stream types are further classified based on the bed material of the stream. These classifications range from 1 to 6 for bedrock, boulder, cobble, gravel, sand, and silt-clay bed materials, respectively.

The most common method of characterizing the bed material is with the pebble count as described by Wolman (1954). The method is appropriate for small streams that can be safely waded. Samples are collected and diameters are recorded. Diameter ranges are silt/clay at 0–0.062 mm; sand at 0.062–0.13, 0.13–0.25, 0.25–0.5, 0.5–1, 1–2 mm; gravel at 2–4, 4–6, 6–8, 8–11, 11–16, 16–22, 22–32, 32–45, 45–64 mm; cobbles at 64–90, 90–128, 128–180, 180–256 mm; boulders at 256–362, 362–512, 512–1024, 1024–2048 mm; and bedrock.

Prior to conducting the pebble count, key features of the stream should be noted, such as riffles, pools, runs, or glides within a stream length of at least 20 bankfull widths. Pools are deep, low-gradient areas within the stream that show little surface movement. Riffles are higher-gradient transition zones that exhibit riffles on the surface caused by fast-moving flow. Glides are transition zones between pools and downstream riffles that represent shallow, smooth flow. A run is a transition from a riffle to a downstream pool. Runs often occur as deeper, faster-flowing sections of a wider portion of a stream.

If a stream section contains 30 percent riffles/runs and 70 percent pools/glides, then pebble counts should assure that approximately 30 percent of the samples are collected from the riffles and 70 percent are collected from the pools. Ideally the reach should be broken up into 10 segments according to the appropriate features. At each segment 10 samples should be taken at uniform distances across the channel, starting at the bankfull elevation on one bank and extending to bankfull elevation on the other bank. A zigzag pattern is used to move up the stream from one side to the other. Randomly select a particle, and measure its length (longest axis), width (intermediate axis), and thickness (shortest axis). Determine the pebble count diameter (intermediate axis) class, and record a count for that class. Take an increment toward the opposite bank, randomly select a particle at this location, and repeat the process. One method of randomly selecting a particle is to reach down without looking, place an index finger next to the big toe of your foot, and select the first particle the index finger touches. For large particles, the particle may be left in place and the pebble count range for the smaller of the two exposed axes should be recorded.

A grain-size distribution curve is usually obtained by plotting the cumulative percent finer against the grain size. The D_{50} grain-size diameter is the diameter at which 50 percent of the particles are larger than that size and 50 percent are smaller. A frequency distribution can also be plotted to understand the variability in pebble sizes in a stream.

Narrow, Deep Streams (W/D < 12)

Entrenched. The G streams typically exhibit gully characteristics. The low entrenchment means that the stream channel must be able to withstand significant stream energies caused by flood flows. These energy levels cause gully streams to be unstable, with grade control problems and high streambank and streambed erosion rates. Because of these high erosion rates, gully streams typically have high bedload and suspended sediment transport. In the Rosgen system, gully streams and A streams share the same entrenchment ratio and are distinguished by a difference in slope. Slopes greater than 4 percent are classified as A streams, and slopes less than 4 percent are classified as G streams.

Moderately Entrenched. Narrow, deep streams with an entrenchment ratio between 1.4 and 2.2 are classified as A streams (Figure 10-6). The primary distinguishing factor for these streams is the narrow valley that constrains floodplain development. Unlike the G streams where flood flows occur primarily in the channel, A streams are not incised to the same degree and overbank flow does occur. The A streams generally occur in mountainous regions or areas with moderately steep valleys. Depending on the steepness of the stream, a cascading, step/pool, or plane/bed channel may form (Montgomery & Buffington, 1997). In very steep regions, streambeds generally consist of bedrock or boulders because the stream energy easily erodes smaller materials and limits the amount of deposition. The narrow valley limits the sinuosity (<1.2) and meander width (<1.5) of the stream.

Not Entrenched. The E streams are evolutionary in that they represent endpoints in stream development with highly sinuous channels and stable well-vegetated streambanks. The meander width ratio may be as high as 20 to 40 (Rosgen, 1996). Type E streams exist in broad valleys with alluvial materials contained in the floodplains. Since these streams are well connected to the floodplain, they are very efficient and stable. The E stream generally has a consistent series of riffle/pool sequences. Removal of streambank vegetation or alteration of the bankfull discharge can cause the stream system to become unstable and shift the stream to another stream type. The E stream can be distinguished from the C stream by the low width-to-depth ratio, wider beltwidth, and the high sinuosity. Many channelized streams in the eastern United States are E channels that have been modified to reduce the sinuosity.

Wide, Shallow Streams (W/D > 12)

Entrenched. As with the G stream, the low entrenchment in F streams forces the channel to absorb high stream energies during flood flows, making these streams generally unstable with high streambank erosion rates. Unlike the G streams where downcutting is the primary form of degradation, lateral bank erosion is the dominant process. These streams meander laterally across the valley. In many cases, these streams are reestablishing a functional floodplain by eroding away the streambanks and laterally extending the channel across the valley floor. A type C or type E stream often forms inside the F stream channel with the F stream channel forming the floodplain for the new channel form. The meander width ratio for F streams ranges from 2 to 10, with an average of 5.3 (Rosgen, 1996).

Moderately Entrenched. As with A streams, the primary determining factor for B streams is the narrow valley that constrains the development of a wide floodplain. The combination of channel slope and entrenchment of the stream causes B-type streams to form rapids with scour pools around large boulders. Spacing between pools is approximately three to five bankfull widths, which decreases as slope increases. Meander width ratios are low (2 to 8, with an average of 3.7) because of the narrow valley. The moderate slope may allow some sinuosity to develop (>1.2) as constrained by the valley.

Not Entrenched. Type C streams have well-developed floodplains and exhibit entrenchment ratios greater than 2.2. These streams are normally located in valleys formed from previous alluvial deposition. Slope is less than 2 percent, which promotes greater sinuosity (>1.2) and the larger width-to-depth ratios (>12). The meandering C streams exist in broad valleys and have well-defined point bars and riffle/pool sequences. These riffle/pool sequences occur at intervals of 5 to 7 bankfull widths. The point bars that form because of the larger width-to-depth ratio indicates the aggradation/degradation potential of these streams. Since these channels are formed in alluvial valleys with easily eroded banks, they can be easily destabilized by alterations in the watershed. The meander width ratio ranges from 4 to 20, with an average of 11.4, indicating the broad floodplain that is developed for these stream types.

Braided Streams

The D and DA streams consist of multiple channels typically found in deltas, depositional fans, wetlands, and glacial outwash valleys. Width-to-depth ratios for these streams are greater than 40. Because of a general lack of vegetation, bank erosion rates can be very high for D streams, which promotes lateral movement of the stream. Sediment supply is available from upstream sources, making aggradation a dominant process with significant channel adjustments during and after runoff events. Since the braided stream covers most of the valley width, meander width ratios are extremely low (1 to 2, with an average of 1.1). A more stable braided stream with well-vegetated banks is classified as a DA stream. As a result of low velocities, the DA streambanks are more stable than the D streambank. The DA systems can be viewed as a series of broad wetland features linked together.

10.5 Fluvial Processes

A stream channel has the primary purpose of carrying water and sediment from a watershed. The channel shape is generally parabolic but can vary greatly depending on the bed materials and flow regime (pool, riffle, run, etc.) present. The channel dimensions are determined from four basic factors. These are streamflow Q_w, channel slope S, sediment discharge Q_s, and sediment particle size D_{50}. At equilibrium, the carrying capacity of the stream defined as a combination of streamflow and stream slope must be balanced against sediment transport, defined as a combination of the sediment discharge and size. Lane (1955) showed this relationship as

$$Q_w S \propto Q_s D_{50} \qquad \textbf{10.9}$$

Channel equilibrium occurs when all factors are in balance. If any of the factors increase or decrease, there must be a proportional increase or decrease in another factor to maintain a balance. For example, if streamflow is increased and slope remains unchanged, then sediment discharge or particle size must be increased to maintain balance. Generally this results in degradation of the streambank.

Flowing water in a stream exerts a shear force on the streambed. The hydraulic shear stress per unit area in the flow direction can be estimated by

$$\tau_f = \gamma RS \tag{10.10}$$

where τ_f = hydraulic shear stress ($MT^{-2}L^{-1}$),
γ = specific weight of water ($MT^{-2}L^{-2}$),
R = hydraulic radius (L),
S = channel slope (L/L).

For most natural channels, the stream width is generally large compared to the stream depth. Under these conditions the hydraulic radius R can be approximated by the stream depth d.

Scour is the removal of bed material by the action of the flowing water. Particles at the interface between the flowing water in a stream and the streambed exert a downward force equal to the immersed weight of the particle. For a particle to be transported downstream, stream velocity must be great enough to move the particle. The bed or particle shear stress at incipient motion is (Knighton, 1998)

$$\tau_c = \eta g(\rho_s - \rho)\frac{\pi}{6} D \tan\phi = kg(\rho_s - \rho)D \tag{10.11}$$

where τ_c = critical bed shear stress ($MT^{-2}L^{-1}$),
k = Shields factor,
g = gravity (LT^{-2}),
ρ_s and ρ = particle and water densities (ML^{-3}),
D = particle diameter (L),
η = degree of packing,
ϕ = packing angle.

The Shields factor k ranges from a minimum of 0.03 for bed grain size less than 0.7 mm to a maximum of 0.06. A value of 0.045 is a conservative approximation for k (Knighton, 1998). At incipient motion, the critical shear stress of the bed material (Equation 10.11) should equal the average shear stress of the flowing water (Equation 10.10). The relationship between particle diameter and average shear stress is important for stream restoration work because streambed materials must be stable. If a value of 0.06 is used for the Shields factor, then the hydraulic shear stress in N/m^2 is approximately equal to the particle diameter in millimeters. Channel dimensions, channel slope, and streambed particle size should be evaluated to assure that scour does not occur.

Example 10.3

Determine the necessary mean particle diameter for the streambed specified in Example 10.2 to assure that degradation of the stream channel does not occur. Assume that the newly constructed channel has a slope of 0.6 percent.

Solution.

(1) Compute the hydraulic shear stress of the flowing water in the stream from Equation 10.10.

$$\tau_f = \gamma R s = 9806\left(\frac{5.8(0.9)}{2(0.9) + 5.8}\right)0.006 = 40.4\,\text{N/m}^2$$

(2) At incipient motion, the critical shear stress must be equal to the hydraulic shear stress of the stream. Assuming a particle density of 2650 kg/m^3, solve for the minimum diameter from Equation 10.11.

$$D = \frac{\tau_c}{kg(\rho_s - \rho)} = \frac{40.4}{0.045(9.81)(2650 - 1000)} = 0.055\,\text{m} = 55\,\text{mm}$$

Mean particle diameter must be larger than 55 mm for the streambed to remain stable.

The energy of the flowing water in a fluvial system carries sediment from the upper reaches of a watershed to the downstream depositional areas. The total sediment load can be separated into three components. These are wash load, suspended load, and bed load.

Wash load represents the fine-grained fraction of the total sediment load. Settling velocities of the wash load fraction are so small that these particles can be carried in flow with extremely low velocities. Particle diameters are generally less than 0.062 mm (Knighton, 1998). The nearly continuous transport of the wash load fraction in fluvial systems causes this fraction to be a significant portion of total sediment transport. Since the wash load fraction is generally independent of the flow characteristics of the fluvial system, it depends on the rate of supply of these fine particles from erosional zones in the watershed.

Suspended load are materials carried by the turbulent forces of the flowing water. As the depth of water increases, the average shear force in the stream (Equation 10.10) and the velocity increase. These increases allow more sediment to be lifted into motion as a result of the turbulent nature of the flow. As shown in Lane's relationship (Equation 10.9), there is a relationship between the flow of water and the associated sediment discharge.

One of the most important sediment components of streamflow is the bed load. The term *bed load* applies to the particles that slide or roll along the bottom of the stream channel. The size distribution of these particles is important in describing channel characteristics and is usually determined with the Wolman Pebble Count described earlier. As indicated from the discussion concerning average shear force of the flowing water and the critical bed shear force of the streambed, the bedload represents the armor that resists erosion.

10.6 Restoration Process

In many cases, stream restoration represents a conversion from a man-made or degraded condition to a more natural, stable form. The existing condition is often the result of previous work either to the stream itself or to areas drained by the stream. A typical practice in the past was to straighten meandering streams to increase the passage of flood flows. This caused many problems. Since the hydraulic gradient was increased as a result of the straightening, higher-velocity water and possibly higher flows in the channel destabilized the area downstream of the straightened area. The straightened channels generally had uniform depths of flow, which limited the diversity of aquatic species present in the stream. Since straight channels are inherently unstable, protective measures were required for the streambank to maintain channel stability.

Alterations in a watershed as a result of urbanization, increased agricultural activity, or other land use changes can affect the volumes of sediment and water that enter a stream. Urbanization has the primary effect of increasing the bankfull discharge resulting from the additional impervious surfaces such as roads, buildings, and parking lots. Agricultural activity can include overgrazing with subsequent degradation of the streambanks or increased sediment loads from a change in management from a pasture to a row crop. As noted earlier (Equation 10.9), any change in the flow or sediment concentration of a stream will cause an alteration of channel form.

One of the first steps in stream restoration is an identification of the existing condition of the watershed and stream (Doll et al., 2003). Current and historical land uses should be identified and analyzed with respect to the probable effect land uses have had on channel stability. Land use changes might include vegetation removal, soil compaction, urbanization, overgrazing, roads, retention structures, and impoundments. For instance, if streamflow is increased from urbanization, the stream channel dimensions will increase to accommodate the additional flow (Equation 10.4). Moderate changes in channel width and depth are not always problematic, but meander length, radius of curvature, and belt width are all related to the bankfull width. The change in meander geometry usually causes channel instability and the ultimate requirement to control the stream.

After the current watershed conditions have been determined, the next step is to determine the most stable form in terms of dimension, pattern, and profile for the stream. This step may include the determination of the most appropriate stream type for the stream in question. The most appropriate stream type may not be an altered form of the present stream, but might be a form that represents the future potential of the stream based on the watershed characteristics.

Part of the responsibility in determining existing conditions is to develop regional curves for natural channels in the region. The selected reference reach should be of the same stream type as that proposed for the stream restoration site. Ideally, the reference reach should be located near the stream restoration site. Reference reaches are stream segments that have stable stream dimensions. While gathering information to develop the regional curves, data should be obtained regarding streambed materials (pebble count), meander geometries, belt width, riffle/pool or step/pool sequences, and vegetation.

Because of its significance in the determination of all other terms, the bankfull discharge has to be determined for the restoration site. An estimate of the bankfull dimensions of the existing channel may be used in the restoration, but a hydrologic analysis of the watershed should be conducted to determine existing and proposed flow in the watershed for 1- to 2-yr recurrence interval storms. Width-to-depth ratios for the selected stream type can be used to specify bankfull width. From bankfull width, other dimensions such as meander wavelength, radius of curvature, belt width, sinuosity, slope, and spacing of pools can be determined. Bed materials of the proposed stream must be of a size to resist average shear forces of the streamflow. Flows greater than the bankfull discharge are expected to cover the floodplain area. The floodplain should be large enough to handle the desired return interval storm/flood.

Streambank Stabilization

10.7 Vegetative Control

Streambanks need protection during and after the restoration process. Natural vegetation of existing streams may be unable to provide adequate protection requiring additional stablization. In many cases, trees that had once stabilized the bank have been cut, streambank vegetation has been removed by tillage or overgrazing, or the channel has been artificially straightened, leading to instability of an entire stream reach. Reducing these disturbances adjacent to the stream may be adequate for natural revegetation to stabilize the stream reach.

One of the most common vegetative control methods is to insert large willow cuttings or similar vegetation, into the streambank. Generally these cuttings range from 1 m to 3 m in length, and are up to 75 mm in diameter. The cuttings quickly root in the damp soil, and a dense cover of willow generally follows. These cuttings are usually taken from nearby streamside stands and thus minimize the risk of introducing undesirable plants. More expensive treatments include planting shrubs appropriate for the area, or installing straw wattles or similar structures containing seeds, including bundles of branches called facines.

10.8 In-stream Structures

The restoration process may require the use of in-stream structures to create riffle/pool sequences, step/pool sequences, and improved aquatic habitat. Rosgen (1996), FISRWG (1998), and Doll et al. (2003) describe many habitat improvement structures, including check dams, boulder placement, channel constrictors, covers and shelters, migration barriers, gravel traps, and bank covers. Some structures are used for both channel stability and habitat improvement (Figure 10-7). These dual-use structures include boulder placement, weirs or vanes, native material revetment, and log vanes.

The weir or cross vane directs flow toward the center of the stream and away from the bank, which reduces the shear forces at the bank and reduces bank erosion. The weir also creates a pool, increases sediment transport capacity, and controls grade to prevent downcutting. A weir is generally used to transition from a restored reach to downstream reaches. The partial vane or J hook vane is used to redirect the velocity distribution in a stream. The J hook can be effective in stabilizing streambanks and dissipating energy near the streambank. The J hook is effective for helping to create meander bends. The W rock weir serves as a multiapplication of the cross vane by redirecting flows toward the center of a stream and is generally used on streams greater than 12 m in width. The W form creates a diversity of velocities across the stream channel while providing in-stream cover for habitat development. The W weir may be used upstream of a bridge pier to direct flow energy away from the pier. The cross vane, J hook, and W weir are oriented upstream at an angle of 20–30 degrees with the streambank. The structures are highest near the bank, to provide streambank protection. In most cases, the rocks that make up

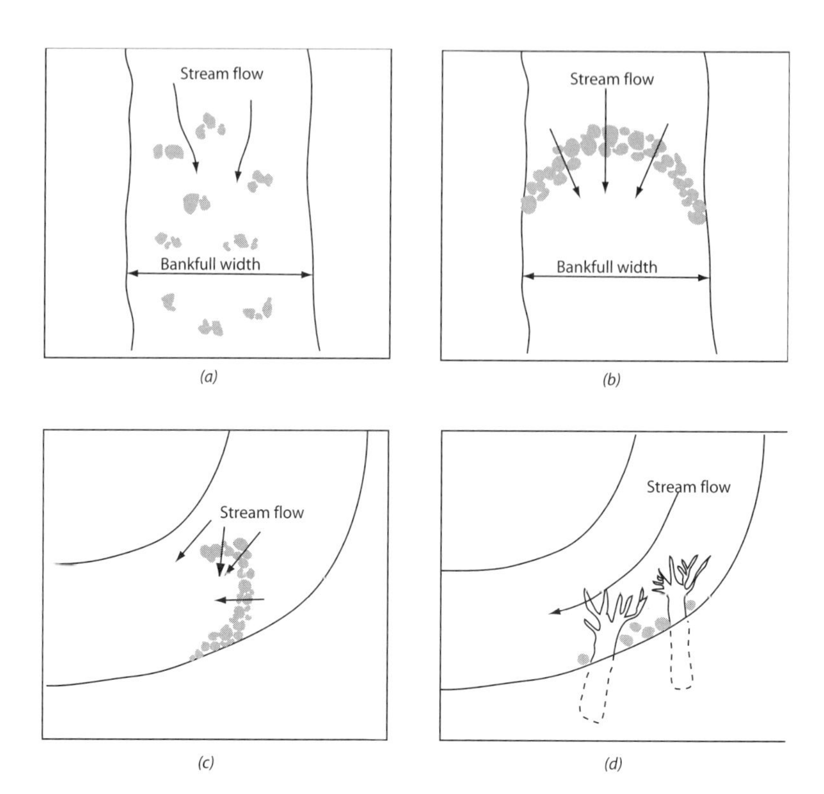

Figure 10–7

Typical structures used for both channel stability and habitat improvement during stream restoration. Structures include (a) boulder placement, (b) weirs or vanes, (c) partial vanes or J hooks, and (d) tree revetment. All structures are designed to provide maximum protection at the streambank by guiding flow toward the middle of the stream.

these structures are quite large, with sizes heavier than 1 to 2 tonnes. Footer rocks should be placed above and below the structure to help hold it in place.

Native material revetment through the placement of root wads, transplant of vegetation, and placement of log vanes can be used to protect the streambank from erosion, provide in-stream cover for habitat, and increase the diversity of habitats along the stream corridor (Figure 10-7). They need to be anchored to ensure they are not lost during high flows. These logs can survive from 10 to 20 years in some streams, until natural vegetation is able to recover.

Newbury and Gaboury (1993) discussed the construction of riffles in Canadian streams to mimic the natural spawning rifflebed profile. Riffles generally occur at the inflection point between adjacent meander arcs, causing riffle spacing to be approximately five to seven times the bankfull width. The riffle is similar to an elongated cross vane. It has a broad V-shaped crest across the width of the channel to guide flow toward the center of the riffle. At the stream midpoint, the riffle is 0.6 m high. A 4:1 rock slope is placed on the upstream side of the crest, and a 20:1 slope is used to form the riffle section. A shear force analysis (Equations 10.10 and 10.11) is required to size the rock for the riffle area.

10.9 Road Crossings

Road crossings represent the single greatest challenge to natural stream function. It is natural for a stream to meander across the floodplain over time, however, the road crossing represents a permanent location for the streambed. Ideally, the road crossing will occur at the inflection point between meander arcs. Cross vanes and J-hook structures should be used to guide the flow toward the road crossing.

The road crossing creates additional problems related to the sediment and debris carrying capacity of the stream. Culvert failures often occur due to blockage of the inlet by debris. By having the culvert span the bankfull width of the natural channel, many of these failures can be prevented. In addition to assuring that bankfull discharge is always passed through the road crossing, additional openings can be installed through the crossing to pass flood flows onto the floodplain.

A constriction created by the culvert or bridge may cause water to pond. The decrease in turbulence or velocity of the stream allows coarser sediments to settle out of the stream and be deposited on the upstream side. Smooth-walled concrete or metal culverts typically used to transport water through the crossing increase the velocity of water. On the downstream side of the road crossing, high-velocity water deprived of sediment can erode the streambank and scour the streambed. Stream protection such as cross vanes or riffles may be required at or near the outlet to limit degradation and form a transition to the natural channel.

A poorly designed road crossing may affect the natural movement of aquatic species in the stream because of poor transitions to the stream or too-high velocities created in the culvert. If aquatic movement is an important characteristic for the road crossing, transition zones will be required on the upstream and downstream sides of the crossing to maintain appropriate flow characteristics through the crossing area.

Internet Resources

Hydrologic unit maps
- http://water.usgs.gov/GIS/huc.html
- http://cfpub.epa.gov/surf/locate/index.cfm

Stream restoration software
- http://www.ohiodnr.com/so. Key search term: stream morphology

Stream Corridor Restoration manual (FISRWG, 1998)
- http://www.usda.gov/stream_restoration/

North Carolina Stream Restoration Manual
- http://www.bae.ncsu.edu. Key search term: stream restoration

References

Brookes, A., & F. D. Shields Jr. (1996). *River Channel Restoration: Guiding Principles for Sustainable Projects.* New York: Wiley.

Chang, H. H. (1988). *Fluvial Processes in River Engineering.* New York: Wiley.

Copeland, R. R., D. N. McComas, C. R. Thorne, P. J. Soar, M. M. Jonas, & J. B. Fripp. (2001). *Hydraulic Design of Stream Restoration Projects.* Technical Rep. No. ERDC/CHL TR-01-28. Vicksburg, MS: U.S. Army Engineer Research and Development Center.

Doll, B. A., D. E. Wise-Fredrick, C. M. Buckner, S. D. Wilkerson, W. A. Harman, R. E. Smith, & J. Spooner. (2002). Hydraulic geometry relationships for urban streams throughout the Piedmont of North Carolina. *Journal of the American Water Resources Association, 38*(3), 641–651.

Doll, B. A., G. L. Grabow, K. R. Hall, J. Halley, W. A. Harman, G. D. Jennings, & D. E. Wise. (2003). *Stream Restoration: A Natural Channel Design Handbook.* Raleigh: North Carolina State University.

Federal Interagency Stream Restoration Working Group (FISRWG). (1998). *Stream Corridor Restoration: Principles, Processes, and Practices.* By the Federal Interagency Stream

Restoration Working Group (FISRWG) [15 federal agencies of the U.S. government]. GPO Item No. 0120-A; SuDocs No. A 57.6/2:EN 3/PT.653. Washington, DC: U.S. Government Printing Office.

Ferguson, R. I. (1973). Regular meander path models. *Water Resources Research, 9*(5), 1079–1086.

Horton, R. E. (1945). Erosional development of streams and their drainage basins: hydrophysical approach to quantitative morphology. *Bulletin of the Geological Society of America, 56*, 275–370.

Knighton, D. (1998). *Fluvial Forms and Processes: A New Perspective*. New York: Oxford University Press.

Lane, E. W. (1955). Design of stable channels. *Transactions of the American Society of Civil Engineers, 120*, 1234–1260.

Langbein, W. B., & L. B. Leopold. (1966). *River Meanders—Theory of Minimum Variance.* United States Geological Survey Professional Paper 422H.

Leopold, L. B., & T. Maddock. (1953). *The Hydraulic Geometry of Stream Channels and Some Physiographic Implications.* United States Geological Survey Professional Paper 252.

Leopold, L. B., & M. G. Wolman. (1957). *River Channel Patterns-Braided, Meandering, and Straight.* United States Geological Survey Professional Paper 282B 39–85.

Leopold, L. B., M. G. Wolman & J. G. Miller. (1992). *Fluvial Processes in Geomorphology*. New York: Dover Publications.

Montgomery, D. R., & J. M. Buffington (1997). *Channel-reach morphology in mountain drainage basins.* GSA Bulletin, 109(5): 596-611. Boulder, CO: Geological Society of America.

Newbury, R., & M. Gaboury. (1993). Exploration and rehabilitation of hydraulic habitats in streams using principles in fluvial behavior. *Freshwater Biology, 29*, 195–210.

Rosgen, D. L. (1996). *Applied River Morphology*. CO: Wildland Hydrology.

Strahler, A. N. (1952). Hypsometric (area-altitude) analysis of erosional topography. *Bulletin of the Geological Society of America, 63*, 1117–1142.

Williams, G. P. (1978). Bankfull discharge of rivers. *Water Resources Research, 14*, 1141–1158.

———. (1986). River meanders and channel size. *Journal of Hydrology, 88*, 147–164.

Wolman, M. G. (1954). A method of sampling coarse river-bed material. *Transactions of the American Geophysical Union, 35*, 951–956.

Problems

10.1 Using the available Internet resources determine the 14-digit hydrologic unit code for your home watershed.

10.2 Determine the stream order, number, and average length for streams in your locality using available topographic maps. Compare results to another watershed in the area.

10.3 Compute the required diameter of rock material to form a riffle in a stream with a width-to-depth ratio of 8.3. The bankfull depth is 1.3 m, and the slope of the stream is 0.01 m/m. What would be the diameter required if the channel slope was 0.001 m/m?

CHAPTER

11 Water Supply

Precipitation is the primary source of renewable freshwater supply for all agricultural, industrial, and domestic uses. Large-scale desalinization of brackish or salty waters may eventually result in reasonable supplies of water for high-value uses in some locations, but desalinization is energy-intensive and expensive. Developed water supplies in the United States use only 4 percent of precipitation, which is only 13 percent of the residual precipitation after allowing for all evaporation and transpiration. There is actually ample water for our needs, though it is often not available at the desired time and place. The development of water resources involves storage and conveyance systems to deliver water from the time and place of natural occurrence to the time and place of beneficial use. This chapter emphasizes the development of water resources for agricultural use while briefly discussing other uses and/or benefits.

Reasons for Developing a Water Supply

11.1 Irrigation

Irrigation is the largest user of water worldwide. In humid regions, it may be a relatively minor user, but in semiarid and arid regions, irrigation exceeds all other uses except in large metropolitan areas. The quality of waters used for irrigation can vary widely, particularly in dissolved solids. Crop tolerances to salts determine the minimum quality water that may be used and the amount of excess water that must be applied to flush accumulated salts from the root zone (Chapter 15). As competition for water resources increases, use of lower-quality waters, including wastewater, for irrigation becomes more economically attractive. Widespread implementation of wastewater reuse systems will be slow, however, since they require separate piping systems for distribution.

11.2 Potable: Human or Livestock

Potable (i.e., fit for drinking) water supplies are important for both human and animal consumption. A few sources (most typically ground water) are sufficiently pure to be used with little or no treatment. Most supplies, however, will require some level of treatment to remove contaminants and pathogens prior to consumption (Chapter 2). As a rule, surface waters require disinfection for human use. Drinking water for livestock can be of lesser quality. Surface supplies are commonly used for watering livestock without treatment, but they are vulnerable to direct contamination. Livestock access to surface waters must be restricted to prevent contamination and the spread of pathogens.

11.3 Recreation

Recreational use is usually a secondary consideration in development of surface water supplies. Fishing, boating, and water sports, with complementary businesses, can be important contributors to economic development. Waterfront properties are highly desirable and usually have premium valuations.

11.4 Wildlife Habitat

Water supplies may be managed to sustain or improve habitat for wildlife. Wildlife and environmental issues are some of the most highly charged. In several cases in the United States, large water control projects have been altered or reversed to alleviate degradation to ecosystems. A prominent example is the dechannelization of the Kissimmee River in Florida. The slow-flowing river between Lake Kissimmee and Lake Okeechobee had been channelized (1962–1971) for flood control by the U.S. Army Corps of Engineers. The project dramatically altered the hydrology of the river floodplain. The changing habitat resulted in shifts in populations of many plants, fish, birds, and animals. The Kissimmee River Restoration Project, authorized in 1992, is restoring the river to its previous flow conditions. As habitat is restored and water quality improves, flora and fauna are recovering.

Because of low snowpack in the summer of 2001, water from the Klamath River in Oregon that was normally allotted to irrigation was withheld to provide flows considered important for the survival of salmon. The impacts of water allocation shifts on agricultural, fishing, and environmental interests in the region led to acrimonious court battles. Such conflicts will become more frequent as competition for water resources increases.

11.5 Process Water (Manufacturing, Food Processing, Waste Handling)

Industrial processes are major water users. These include mining, manufacturing, food processing, and many others. The majority of domestic water use is for nondrinking purposes such as washing and waste disposal. Some of these needs can be satisfied with lower-quality waters, such as untreated surface waters or treated wastewaters. Use of such waters will require new infrastructure.

Characteristics of a Water Supply

A potential water supply may be characterized according to its quality, time of availability, storage volume, and rate at which water can be extracted. Water quality requirements (Chapter 2) vary widely with the intended use of the water. If the water is intended for human consumption, it must meet strict standards for physical, biological, radiological, and chemical contaminants. For livestock, those standards are somewhat lower. The quality required for irrigation water depends on the crop (Chapter 15) and can depend on the time remaining until harvest, particularly for fruits and vegetables intended for the fresh market. Irrigation of forage crops with wastewater may also have restrictions based on time-to-harvest or time-to-grazing. Consult local regulatory agencies for specifics.

Timing can be a crucial aspect of water supply development. Precipitation is ultimately the most reliable source of fresh water, but it is intermittent and highly variable between seasons and between years (Chapter 3). Topography, vegetation, and geology influence the translation of precipitation into stream flow and ground water flow. Major rivers with large watersheds tend to have persistent baseflows and provide reliable supplies for large users. Flows from smaller and steeper watersheds tend to be more variable. Flows in small streams often cease entirely in dry periods.

About 97 percent of all available fresh water is ground water. Ground water supplies are less seasonal than surface waters. Aquifers act as underground reservoirs whose storage capacity attenuates the variability of precipitation. Ground water levels fluctuate, however, and can drop significantly in drought periods. Where ground water drops below a pump intake, one might (1) deepen the well and/or (2) lower the pump or intake. In some situations, none of these will be feasible.

The potential yield (rate at which water can be obtained from a source) is an important supply characteristic. Where the minimum flow in the stream is inadequate to satisfy demand by immediate extraction, surface impoundments (reservoirs, lakes, ponds) can provide carryover storage. Ground water, although it can be found almost anywhere, is highly variable in terms of ease of access, quantity, and quality. The potential yield of a well depends on the characteristics of the aquifer: porosity, hydraulic conductivity of the primary matrix, fracturing, and thickness of the water-bearing strata. Ground water supplies in some areas provide high yields and excellent quality, whereas ground water in other areas is very limited and/or may have quality problems such as extreme hardness, dissolved salts, iron, etc., that may limit its usefulness.

Demand Assessment

Prior to developing a water supply, an assessment of the demand is needed. The demand assessment should estimate timing, volume, and flow rates required. Constraints on quality must also be known before the suitability of a supply can be determined.

In most cases, demands for irrigation will be highly seasonal. Irrigation requirements for most field crops ramp up during the early growth stages, peak during late vegetative growth stages, and drop off as the crop matures (Chapters 4 and 15).

Water requirements for livestock are less seasonal. Animals will require more water during hot periods, but their basic requirements are fairly constant. Demands for personal uses (washing, bathing, drinking) are relatively steady, but domestic/urban demands may include significant landscape irrigation that will peak during summer months. Municipal water demands show strong diurnal fluctuations as well as weekday–weekend shifts between industrial and domestic uses and locations. Processing demands vary with the type of industry and production schedules. The variability of demand rates can often be handled by incorporating storage components into the supply system.

Storage elements (reservoirs, ponds, tanks) can supply short-term discharges in excess of the rates available from the water sources (wells, streams, etc.). Large reservoirs can provide year-to-year carryover to reduce the effects of short-term droughts. Sizing a reservoir requires a hydrologic analysis of the supply and estimation of losses by evaporation and seepage. This chapter will deal only with relatively small surface impoundments.

Surface Impoundments

11.6 Types of Impoundments

Three types of small impoundments are in common use: (1) dugout ponds that are supplied by ground water; (2) on-stream ponds fed by continuous or intermittent flow of surface runoff, streams, or springs, and (3) off-stream ponds.

Dugout Impoundments. Dugout ponds are limited to areas having gentle slopes (less than 4 percent) and a high water table (typically within 1 meter of the surface). Subsoils with high hydraulic conductivities are desirable to facilitate recharge. Design is based on the required storage capacity, water table depth, and stability of the sideslope materials. If surface water flows are diverted from the pond, an overflow spillway will probably not be necessary.

On-Stream Impoundments. The on-stream impoundment depends on surface runoff for replenishment. Storage capacity must consider both the demand requirements and the probability of a reliable supply of runoff. Where heavy usage is expected, the design capacity of the impoundment must be sufficient to supply several years' needs to ensure time for replenishment in the event of one or more years of low runoff (Chapter 5). Spring- or stream-fed impoundments are formed by either excavating a basin or damming a stream valley or depression below the source. A dam across a natural depression can capture diffuse surface water, i.e., where no defined channel exists.

The maximum surface elevation of a spring-fed pond should be below the spring outlet to avoid the hazard of diverting the spring flow by increasing the head on its natural outlet. If the spring discharge is sufficient to meet the demand requirements, surface water should be diverted around the impoundment to reduce the sediment load and required spillway size.

Off-Stream Impoundments. The off-stream or bypass impoundment is typically constructed adjacent to a flowing stream, and is supplied by an intake, either a pipe or an open channel, between the stream and the impoundment. Sedimentation can be minimized by installing controls on the intake to prevent the entry of floodwater. A pump may be used to transfer water from the stream to the impoundment. Reasonable provision should be made to protect the impoundment from damage during stream flooding.

11.7 Site Selection

The location for an impoundment should be chosen to minimize the size and cost of structures as well as minimize the amount of pumping and piping required to deliver the water to its intended point of use. The site should have suitable soils for construction of embankments. Underlying materials having low permeability are desirable to minimize seepage losses. Channel slopes above the dam normally range from 3 to 6 percent to provide adequate depth and water surface area. Slopes in this range provide a good tradeoff between depth and surface area.

Choice of the site for an impoundment also depends on the contributing watershed. It must be sufficient to supply the necessary runoff, but should not be so large as to require routing excessive flows through the spillways. Channels carrying high sediment loads should be avoided if possible, because the useful life of the impoundment would be dramatically shortened by sediment accumulation. Construction of impoundments is discussed in Chapter 9.

11.8 Storage Requirements

The storage requirement for an impoundment depends first on the direct water needs. In addition, losses to evaporation and seepage through the dam, bottom, and sides must be considered. Allowances must be made for sediment storage and carryover from year to year. A minimum depth of 2 to 4 meters is needed for fish to survive where the water surface freezes. Deeper waters also provide a greater range of temperatures and light conditions that help support diverse biota. If the structure is to provide flood control, the design must also include flood storage (Chapter 9).

The watershed area needed to produce a certain volume of runoff can be estimated from Figure 11-1. Water needs for domestic uses, livestock, spraying, irrigation, and fire protection may be estimated from Table 11-1. The evaporation from an impoundment can be estimated with methods presented in Chapter 4. Evaporative losses can be minimized by choosing a site that will have a smaller surface area and greater depth. Since seepage losses depend strongly on soil properties and construction techniques, they are difficult to predict (Chapter 9). Studies have shown that impoundments with large ratios of storage volume to watershed area provide a reliable supply even in dry years. Within reason, the impoundment should be designed for maximum capacity. The minimum storage requirement is usually determined by estimating the total annual needs and allowing 40 to 60 percent of the total storage volume for seepage, evaporation, and other nonconsumptive requirements.

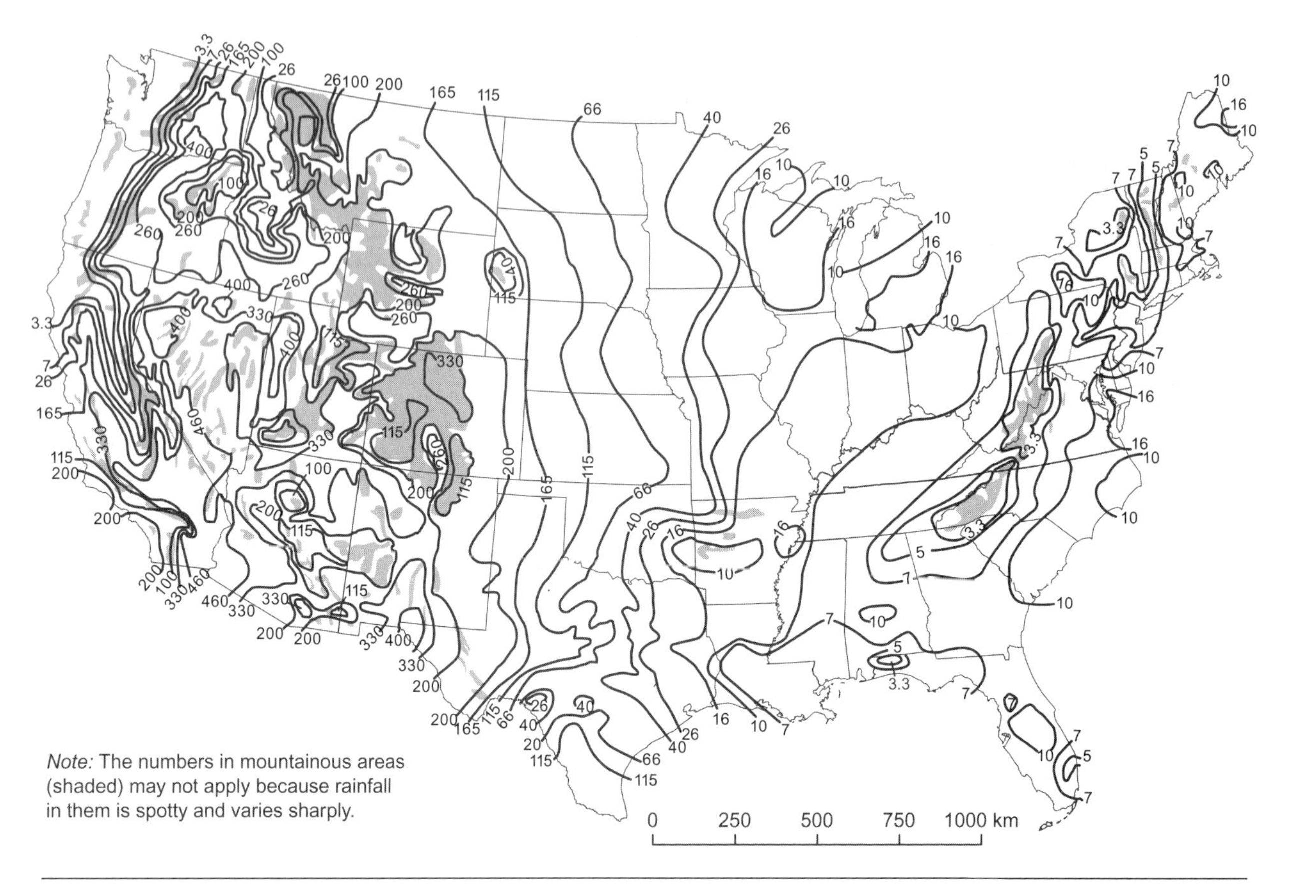

Figure 11–1 Watershed area, in hectares, required to yield one hectare-meter of runoff per year. (Adapted from NRCS, 1997)

TABLE 11–1 Approximate Water Requirements for Selected Agricultural Uses

	Average Use	
Type of Use	***L/day***	***ha-m/yr***
Household, all purposes, per person	250	0.0091
Steer or dry cow, per 450 kg	60	0.0022
Milk cow, per 450 kg (drinking and barn needs)	130	0.0047
Horse or mule, per 450 kg	60	0.0022
Turkeys, per 100 birds	50	0.0018
Chickens, per 100 birds	30	0.0011
Swine, per 45 kg	10	0.0004
Sheep, per 45 kg	4	0.00015
Orchard spraying	4 L tree^{-1}application^{-1}(yr of tree age)$^{-1}$	
Irrigation (humid regions)	0.3–0.45 m per growing season	
Irrigation (arid regions)	0.3–1.5 m per growing season	

Notes: Livestock water consumption varies with climate. Consult local sources for more precise values. Consult local building codes for requirements for on-site fire protection water supplies.

Example 11.1 Determine the annual water requirement, storage volume needed, and watershed area required for an impoundment in central Indiana. The supply must serve a family of four, 60 steers, 8000 chickens, allow 0.03 ha-m/year for fire protection, and supply irrigation for 4 ha of vegetable crops. Seepage and evaporative losses are estimated as 55 percent of the storage capacity.

Solution. Estimate the annual direct use from Table 11-1.

Type of Use	*Units*	*Annual Water Use Rate*	*Annual Water Use*
Household use	4 people	0.0091 ha-m/year/person	0.036 ha-m/year
Steers	60 steers	0.0022 ha-m/year/steer	0.132 ha-m/year
Chickens	8000 birds	0.0011 ha-m/year/100 birds	0.088 ha-m/year
Irrigation (humid)	4 ha	0.4 ha-m/year/ha	1.60 ha-m/year
Fire protection			0.03 ha-m/year
Total			1.89 ha-m/year

$$\text{Storage volume} = \frac{\text{use volume}}{1 - \text{loss fraction}} = \frac{1.89\,\text{ha-m}}{1 - 0.55} = 4.20\,\text{ha-m}$$

Using Figure 11-1, find the watershed area required in central Indiana to yield 1 ha-m/year of runoff to be 13 ha.

$$\text{Watershed area} = 4.20\,\text{ha-m}\frac{13\,\text{ha}}{\text{ha-m}} = 55\,\text{ha}$$

It would be desirable to increase the storage, if possible, to provide for year-to-year carryover and maintain a minimum pool level for recreational or wildlife uses.

As the numbers in Example 11.1 suggest, private surface impoundments for irrigation water supply may be impractical, especially for large irrigated areas. Development of surface waters for irrigation is usually done by public agencies. Large-scale examples include the Columbia Basin Project in Washington, the Central Arizona Project, and the Central Valley Project in California. Although there are many benefits of such projects, there are also many disagreements over management strategies and environmental impacts.

Surface storage can also be used where the water supply is from ground water. For example, many golf courses must irrigate heavily during short periods overnight. If they use ground water, they can pump water into ponds during the day so that high flows can be supplied to the irrigation system when needed. The storage ponds are often aesthetically integrated into the course as water hazards.

Ground Water

Ground water varies greatly in accessibility and quality. In some areas, ground water exists within a few meters of the soil surface, whereas in others it is many tens of meters down. While all ground water contains some dissolved solids, concentrations range from near zero up to brines that are more concentrated than seawater.

11.9 Terminology

Ground water hydrology has special terminology. All water below the earth's surface is called underground water. Typically, there is some zone from the surface downward that is not completely saturated, i.e., the void spaces in the soil and rock (porous media) are not completely filled with water. That is the unsaturated zone, which is also called the vadose zone. At some depth, water completely fills the void spaces. From that depth on, it is called the saturated zone.

Both the vadose and saturated zones occur in the zone of rock fracture, where rock formations fracture in response to stresses from deformations of the earth's crust (folding and faulting). At much greater depths, in the zone of rock flowage, the rock is under such pressure that it deforms plastically. Although water exists in the rock at such depths, it is unavailable for use. The vast majority of ground water extraction is from the upper few hundred meters. The various zones and terms are shown in Figure 11-2.

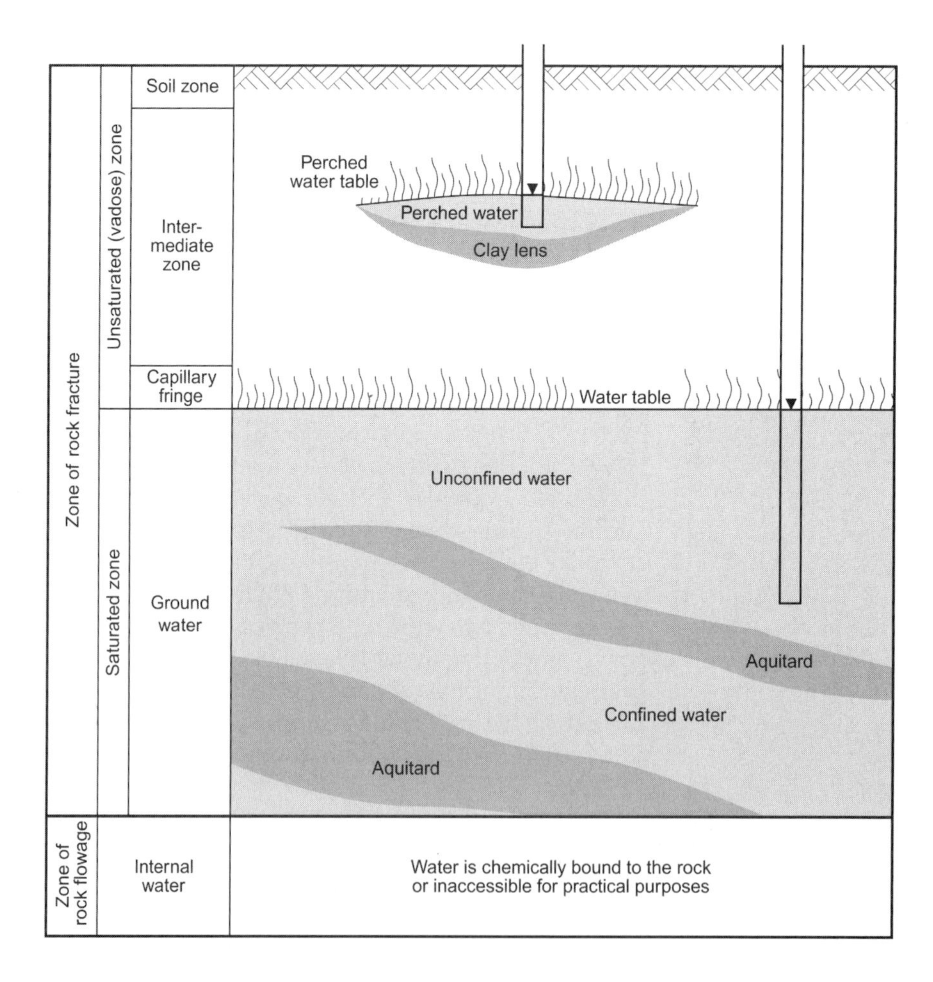

Figure 11–2

Classification of near-surface zones of the earth's crust and underground waters.

A water table (also called a phreatic surface) is the upper surface of a saturated zone. If a hole were bored into the earth (e.g., a well), water would enter that hole to the level of the water table. At the water table, the pressure in the underground water is equal to atmospheric pressure.

Just above the water table is a zone where the porous media are virtually saturated. That zone is the capillary fringe and ranges in depth from a few millimeters in coarse sands to 2–3 meters in clays. Because water wets almost all media, the surfaces of the solids are covered by a film of water. Water is drawn up into pores until the surface tension effects are balanced by the pull of gravity. The height of capillary rise can be estimated using Poiseuille's formula (see any soil physics text) and the characteristic pore size of the medium. The capillary fringe can be an important consideration in the design of drainage systems (Chapter 14), but is generally ignored in ground water development. Water in the capillary fringe is not readily extractable.

The void or pore space occurs in many configurations, depending on the type of formation—unconsolidated or consolidated, sedimentary, metamorphic, or igneous—and its history since deposition. Primary porosity is the void space between particles. Sediments may be of similar particle sizes (well sorted) or a mixture of sizes (poorly sorted) as shown in Figures 11-3a and b. Precipitation of cementing agents (primarily silica, iron oxide, or calcite) partially fills the pores (Figure 11-3c). The term *secondary porosity* refers to voids or void systems that are much larger than the primary particles. Massive formations such as granite may have negligible primary porosity, but have systems of fractures or joints through which water can move (Figure 11-3d). Dissolution of carbonate rock by (weakly acidic) ground water increases porosity, mainly along cracks and joints (Figure 11-3e). In the extreme, it produces caverns.

The actual configuration of the subsurface can be quite complicated. Rock formations often occur in alternating strata of differing porosities and hydraulic conductivities. Near the surface there may be various soil layers ranging from coarse sands to tight clays. Although water moves downward easily through the more permeable materials (such as sands), it can be impeded by the less permeable materials (such as clays) to the point that it ponds above the restrictive layer(s), creating localized saturated zones that have unsaturated media below them. Those are called perched waters and they have perched water tables. A relatively impermeable layer that restricts vertical flow is called an aquitard (or, if very restrictive, aquiclude).

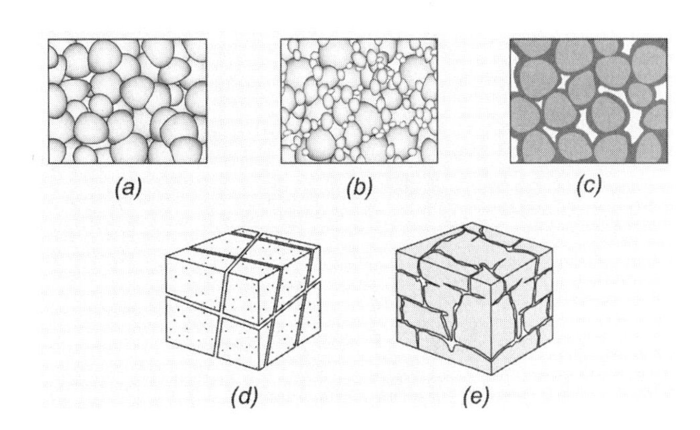

Figure 11–3

Several types of porous media. (a) Well-sorted sand. (b) Poorly sorted sand. (c) Well-sorted sand with precipitates that reduce porosity. (d) Rock with fractures. (e) Rock with dissolution channels.

11.10 Aquifers: Unconfined and Confined

Ground water development requires creating access to an aquifer (Latin for "water bearing"). An aquifer is a saturated stratum of relatively high permeability. Aquifers occur in two major types—unconfined and confined. An unconfined aquifer has no aquitard above it. Water percolating from the land surface can freely join the saturated zone. The upper boundary of an unconfined aquifer is a water table, which will fluctuate up or down in response to precipitation, evapotranspiration, and lateral flows. If an aquifer has an aquitard as its upper boundary (e.g., sand overlain by clay), it is confined. In some terrains, there may be several confined aquifers separated by aquitards. In parts of the Atlantic Coastal Plain in the eastern United States, for example, hydrogeologists have identified six or more distinct aquifers in vertical succession overlying the basement rock.

The designation of a stratum as an aquifer or aquitard is relative to its neighboring strata. Media that would be considered aquifers in some locations could be considered aquitards in others, and vice versa. Potential water yields range over about three orders of magnitude for similarly constructed wells, depending on the aquifer characteristics. Bear (1979) lists media having hydraulic conductivity of 10^{-5} m/s or higher as good aquifers, media with hydraulic conductivity of 10^{-9}–10^{-5} m/s as poor aquifers. Media with hydraulic conductivity less than 10^{-9} m/s will not yield sufficient water to be considered for development.

Just like the surface of the earth, strata below the surface are rarely level. Confined aquifers act as huge, flattened pipes conducting water laterally from recharge areas to eventual discharge areas that may be many kilometers distant. (Typical flow velocities are on the order of a meter per day.) Where confined aquifers are inclined, the hydraulic head (sum of elevation and pressure heads) in some areas may exceed the elevation head at the upper boundary of the aquifer. In a well bored into such an aquifer, the water will rise above the top of the aquifer formation. This is an artesian condition. Where the hydraulic head in the aquifer is sufficient to lift water above the ground surface, flowing artesian wells may occur. (Artesian wells take their name from the Artois region in northern France, where flowing wells were common.) Flowing artesian wells can also occur where a well in a discharge area taps water at a higher potential than the ground surface (see Freeze & Cherry, 1979, for detailed explanation).

In an unconfined aquifer, the water table is the loci of points where the hydraulic pressure is equal to atmospheric pressure. An analog exists for confined aquifers and is called the potentiometric surface (or piezometric surface). It is not physically present in the profile, but represents the surface to which water would rise in wells penetrating the confined aquifer. Figure 11-4 shows an unconfined aquifer overlying a confined aquifer. Note that the water table and potentiometric surfaces do not generally coincide.

Since aquitards are restrictive but not completely impermeable, there can be some hydraulic communication between aquifers. If the strata separating aquifers are discontinuous or so thin that the hydraulic communication is not negligible, the aquifers are called leaky aquifers (though it is actually the aquitards that are leaky). The analysis of leaky aquifer systems is beyond the scope of this text.

11.11 Well Hydraulics

The behavior of most wells can be adequately predicted by assuming that the aquifer is homogeneous, of uniform depth/thickness, and extends far beyond the influence of the well. The flow field is then axially symmetric about the well.

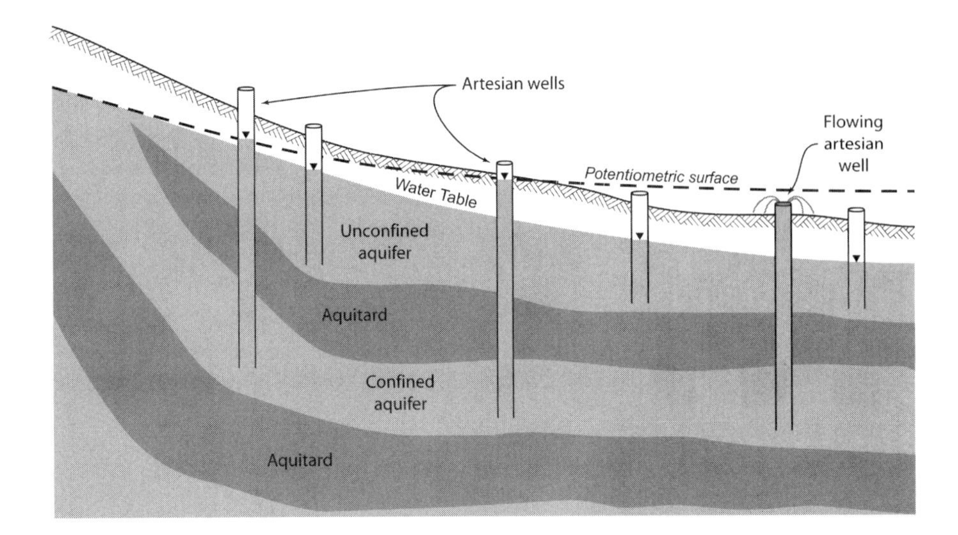

Figure 11–4
Common aquifer structure in gently sloping sedimentary systems.

The phreatic or potentiometric surface is assumed to be level prior to pumping. As pumping progresses, a cone of depression develops around the well. The shape of the cone of depression approaches a steady-state configuration as the time of pumping becomes large. A cross section through the well and cone of depression shows the drawdown curve (Figures 11-5 and 11-6). The vertical difference between the static water level and the water level during pumping is called drawdown. (The most common usage of the term is with respect to the water level difference inside the well.) Note that the drawdown just outside the well is slightly less than the drawdown inside the well. This is due to head losses as water passes through the well screen (a special section of casing with narrow holes).

For a well that fully penetrates an extensive, unconfined aquifer (Figure 11-5), the steady-state well discharge is given by

$$Q = \frac{\pi K(h_2^2 - h_1^2)}{\log_e(r_2/r_1)} \tag{11.1}$$

where Q = discharge (L^3/T),
K = hydraulic conductivity (L/T),
r_1, r_2 = radial distances from the center of the well (L),
h_1, h_2 = height of the water level at r_1, r_2 above the bottom of the aquifer (L).

The steady-state discharge for a well that fully penetrates an extensive, confined aquifer (Figure 11-6) is

$$Q = \frac{2\pi Kd(h_2 - h_1)}{\log_e(r_2/r_1)} \tag{11.2}$$

where d = thickness of the confined aquifer (L),
r_1, r_2 = radial distances from the center of the well (L),
h_1, h_2 = height of the potentiometric surface at r_1, r_2 above the bottom of the aquifer (L).

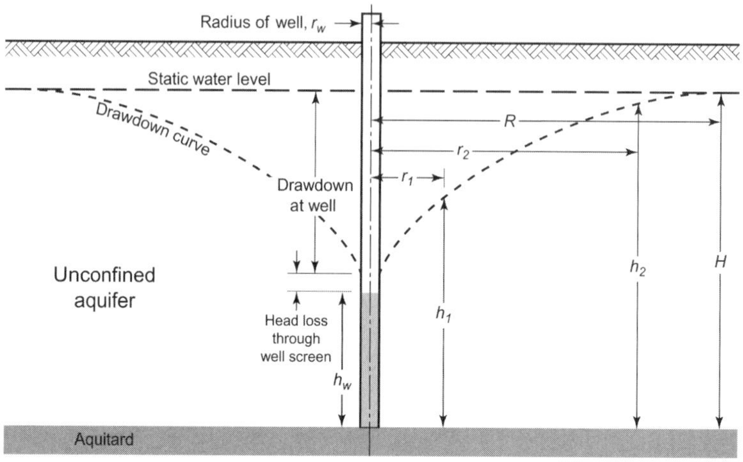

Figure 11–5

Cross section of a well in a homogenous unconfined aquifer.

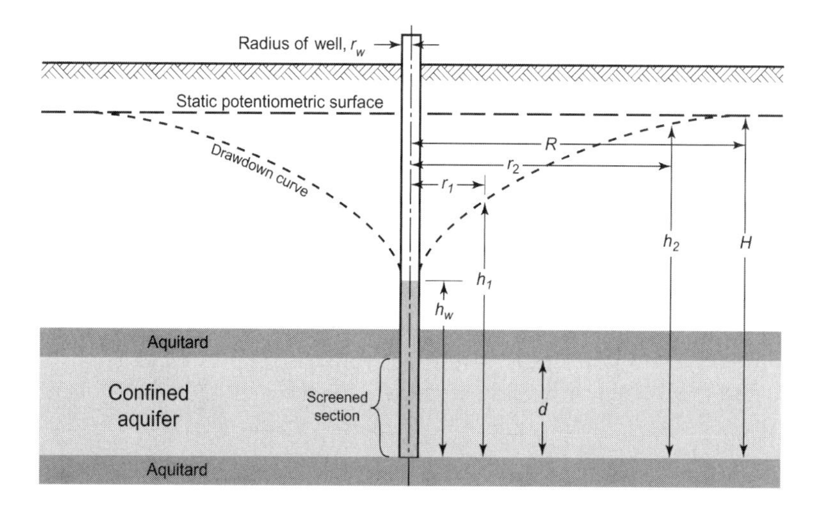

Figure 11–6

Cross section of a well in a homogeneous confined aquifer.

In practice, the values of r_1 and r_2 are chosen to address various objectives. If only the stratigraphy is known (from installation of a well), the hydraulic conductivity can be estimated by installing piezometers at two distances from the well and conducting a pumping test. Table 11-2 gives ranges of hydraulic conductivity for selected porous media. Pumping test data can also be used to calculate a radius of influence, which is the distance at which there is negligible change in the phreatic or potentiometric surface between static and pumped conditions. The radius of influence is only an approximate value because it actually depends on the well discharge. Since only the logarithm of the radius appears in Equations 11.1 and 11.2, errors in estimating the radius of influence have little effect on the discharge estimate. Bear (1979) gives several empirical and semiempirical formulas for estimating the radius of influence. The Siechardt formula (Chertousov, 1962) is commonly used.

$$R = 3000 s_w K^{1/2} \tag{11.3}$$

TABLE 11-2 Approximate Characteristics of Various Natural Porous Media

Media	*Porosity (%)*	*Hydraulic Conductivity (m/s)*
Unconsolidated		
Gravel	25–40	$1 \times 10^{-3} - 1$
Coarse sand	30–45	$1 \times 10^{-4} - 1 \times 10^{-2}$
Sand, mixture	20–35	$5 \times 10^{-5} - 1 \times 10^{-4}$
Fine sand	25–50	$1 \times 10^{-5} - 5 \times 10^{-4}$
Silt	35–50	$1 \times 10^{-7} - 5 \times 10^{-6}$
Clay	40–70	1×10^{-8} and lower
Unweathered marine clay		$5 \times 10^{-13} - 1 \times 10^{-9}$
Consolidated		
Karst limestone	5–50	$8 \times 10^{-7} - 1 \times 10^{-2}$
Fractured igneous and metamorphic rocks	0–10	$5 \times 10^{-9} - 2 \times 10^{-4}$
Limestone and dolomite	0–20	$3 \times 10^{-10} - 2 \times 10^{-6}$
Sandstone	5–30	$8 \times 10^{-11} - 2 \times 10^{-6}$
Shale	0–10	$1 \times 10^{-3} - 1 \times 10^{-9}$
Unfractured igneous and metamorphic rocks	0–5	$<10^{-13} - 2 \times 10^{-10}$

Sources: Harr (1962), Freeze & Cherry (1979), McWhorter & Sunada (1977).

where R = radius of influence, in m,
s_w = drawdown in the well, in m,
K = hydraulic conductivity of the aquifer medium, in m/s.

Example 11.2

Estimate the discharge from a 0.1-m diameter well fully penetrating an unconfined aquifer given the following:

Depth to restrictive layer	14 m
Depth to static water table	3 m
Drawdown at the well	4 m
Drawdown at a radius of 80 m	0.2 m
Aquifer medium	medium sand hydraulic conductivity = 6×10^{-5} m/s

Solution. For a well in an unconfined aquifer, use Equation 11.1. Radius r_2 is 80 m and radius r_1 is 0.05 m (the radius of the well). At r_2, h_2 is (14 m – 3 m) – 0.2 m = 10.8 m and at r_1, h_1 is (14 m – 3 m) – 4 m = 7 m. Then,

$$Q = \frac{\pi K(h_2^2 - h_1^2)}{\log_e(r_2/r_1)} = \frac{\pi(6 \times 10^{-5}\,\text{m/s})[(10.8\text{ m})^2 - (7\text{ m})^2]}{\log_e(80\text{ m}/0.05\text{ m})}$$
$$= 1.7 \times 10^{-3}\,\text{m}^3/s = 1.7\text{ L/s}$$

Example 11.3

Estimate the discharge from a 0.3-m diameter well fully penetrating a confined aquifer given the following:

Aquifer thickness	26 m
Height of static potentiometric surface above the top of the aquifer	85 m
Drawdown at 20-m radius	35 m
Drawdown at 100-m radius	5 m
Aquifer medium	mixed sand hydraulic conductivity = 2×10^{-5} m/s

Solution. For a well in a confined aquifer, use Equation 11.2. At 100 m from the well, h_2 = 85 m - 5 m = 80 m. At 20 m from the well, h_1 = 85 m - 35 m = 50 m. Then,

$$Q = \frac{2\pi Kd(h_2 - h_1)}{\log_e(r_2/r_1)} = \frac{2\pi(2 \times 10^{-5}\,\text{m/s})(26\,\text{m})(80\,\text{m} - 50\,\text{m})}{\log_e(100\,\text{m}/20\,\text{m})}$$
$$= 0.061\,\text{m}^3/\text{s} = 61\,\text{L/s}$$

Yields of this magnitude could supply many medium-sized irrigation systems, large animal operations, or small public water systems.

Example 11.4

Estimate the hydraulic conductivity of an unconfined aquifer from the following pump test data:

Depth to restrictive layer	21 m
Depth to static water table	4 m
Drawdown at 10-m radius	8 m
Drawdown at 50-m radius	1.2 m
Well discharge	6 L/s

Solution. Use Equation 11.1 for an unconfined aquifer. Enter the appropriate values, and solve for K.

$$K = \frac{Q\log_e(r_2/r_1)}{\pi(h_2^2 - h_1^2)} = \frac{(0.006\,\text{m}^3/\text{s})\log_e(50\,\text{m}/10\,\text{m})}{\pi[(15.8\,\text{m})^2 - (9\,\text{m})^2]} = 1.8 \times 10^{-5}\,\text{m/s}$$

This value can now be used to calculate drawdowns at other pumping rates or yields at other drawdowns.

As the equations show, there is a direct relationship between the amount of drawdown and the well yield. Results of the equations must be evaluated critically. If the drawdown in an unconfined system approaches the thickness of the aquifer, the calculated yield is not likely to be valid. For a confined aquifer, if the drawdown is so great that the cone of depression penetrates the top of the aquifer itself, the aquifer will begin to dewater, i.e., the upper part of the aquifer near the well will become unsaturated, altering the hydraulic behavior. The well will still produce, but the flow analysis is much more complicated.

Equations 11.1 and 11.2 assumed steady-state conditions in the aquifers. Although such conditions are never found in nature, the approximation is good for relatively short-term pumping. Where a well will be pumped for extended periods, e.g., for irrigation or municipal supplies, pumping tests should be run for as much as 30 days to evaluate the capacity of a well. The equations also assume that the wells fully penetrate the aquifers. If the well is screened in only part of the aquifer, there will be a vertical component added to the flow field, requiring a more complex analysis (see, e.g., Bear, 1979).

11.12 Well Construction

There are several primary methods of well construction. Wells may be dug, driven, jetted, bored, or drilled. Until the last century, dug wells were most common. They were generally large diameter (>1 m) and were often lined with rock or brick. They could be constructed in any type of formation with proper tools and persistence. Dug wells were usually shallow (<10 m), though some were much deeper. Most were open at the surface for access, which left them especially vulnerable to contamination.

Driven wells are generally small in diameter (30–70 mm) and shallow. A combination screen and drive point is pounded into the ground. Joints of pipe (typically 1.5 m long) are added and the point advanced until a water-producing zone is reached. Installation by hand can reach up to 10-m depths; mechanically driven hammers can reach depths of 15 m or more. This method only works in relatively rock-free, unconsolidated formations.

Jetting can only be used in unconsolidated formations. Water is pumped through the casing and emitted through a special jetting valve and/or the screen. For dense clays, shell beds, or partially cemented formations, a chisel bit can be added to the end of the casing. Mechanical agitation by raising and lowering the casing can help cut the hole, while the water carries cuttings to the surface. If the water can escape into the formation, circulation along the length of the casing will cease and the formation will "lock" the casing in place. Jetting is most common with small-diameter, relatively shallow wells. It can be used for well diameters of 50–300 mm and reach depths of 30 m.

Bored wells are constructed with augers. Three types in common usage are bucket augers, solid-stem augers, and hollow-stem augers. Well diameters are typically 50–760 mm. The larger wells are sometimes lined with concrete casings. Typical depths are less than 15 m, but can be as much as 30 m. Suitable formations are unconsolidated, with minimal cobbles or boulders.

Drilled wells can be several hundred meters deep and range in diameter from 100 to 600 mm. Drilling includes a variety of methods using different equipment. Cable-tool percussion drilling (developed in China 4000 years ago) advances the borehole by repeatedly dropping a weighted bit. Cuttings are periodically removed by bailing. Cable-tool rigs have largely been replaced by rotary drilling rigs. Rotary rigs use rotating bits at the lower end of the string of drill rods to break up the formation. Continuously circulating fluid (air, water, or clay slurry) cools the bit and carries cuttings to the surface. In hard rock formations, another variant, the down-the-hole hammer, may be used. These are air-powered percussion tools with tungsten carbide inserts. The hammering of the tool on the rock pulverizes it and the cuttings are then carried to the surface by the exhaust air.

Drilled wells can be constructed in anything from unconsolidated sediments to hard rock. A variety of bits and tooling has been developed for work in different formations. Glacial till with its mixture of everything from clay to large boulders presents some of the most challenging conditions (Heath, 1980; Driscoll, 1986).

Most wells need some type of casing to stabilize the formation and prevent intrusion of contaminants or undesirable waters. Steel is the most common material for casings. Plastic casing can be used in many applications, especially with highly corrosive waters where steel casing would be short-lived. Plastic should be avoided where it may be subject to uneven loading. Concrete casing has been used with larger-diameter wells, but the difficulty of sealing joints between sections makes it a poor choice. Local regulations usually specify acceptable casing materials.

Well screens can be as simple as a section of casing with holes punched through it, but manufactured screens with precisely sized openings give better results. Screens are selected to permit 50 to 70 percent of the particles in the formation to pass through, leaving a matrix of larger particles with larger pores behind after development. To help stabilize fine-textured formations and increase the effective size of the well, gravel is sometimes packed around the outside of the screen (Figure 11-7). Gravel packs are common practice where high yields are required, such as for irrigation or municipal wells. A gravel pack is a cost-effective alternative to using very-large-diameter casing.

Well Development. Following installation of the casing and screen(s), a well should be developed to flush the finer particles from the formation near the screen. This improves the yield and reduces the sediment entering the well under normal use.

Most wells can be developed by pumping for an extended period at rates higher than the design pumping rate. Surging alternates normal pumping with forcing water to flow back into the formation. The changing flows help to dislodge the fines and remove them from the formation. Where yields are inadequate, more extreme measures may be used. In hydrofracturing, the well is sealed with packers and water is injected into the formation at very high pressures (3.5—14 MPa) to increase the fracturing of the rock near the borehole. Another method is to detonate explosive charges in the borehole to increase the intensity of fracturing. In limestone or dolomite formations, acid can be injected to dissolve some of the formation, enlarging cracks and channels. Hydrofracturing, explosives, and acids are dangerous and should be handled only by experienced professionals (Driscoll, 1986).

Once development is completed, a pump should be selected that will provide a maximum discharge that is somewhat less than that used during development. During normal operation the hydraulic gradients in the formation will then be lower than during development, so very little additional sediment will be dislodged.

Grouting. Most well construction methods create a borehole that is larger than the casing that is eventually installed. The annular space between the casing and the formation should be filled, from the soil surface down to the top of the well screen, with an impermeable material to prevent the vertical flow of water along the outside of the casing. Improperly grouted or ungrouted wells are especially vulnerable to contamination from surface or near-surface sources such as pathogens or

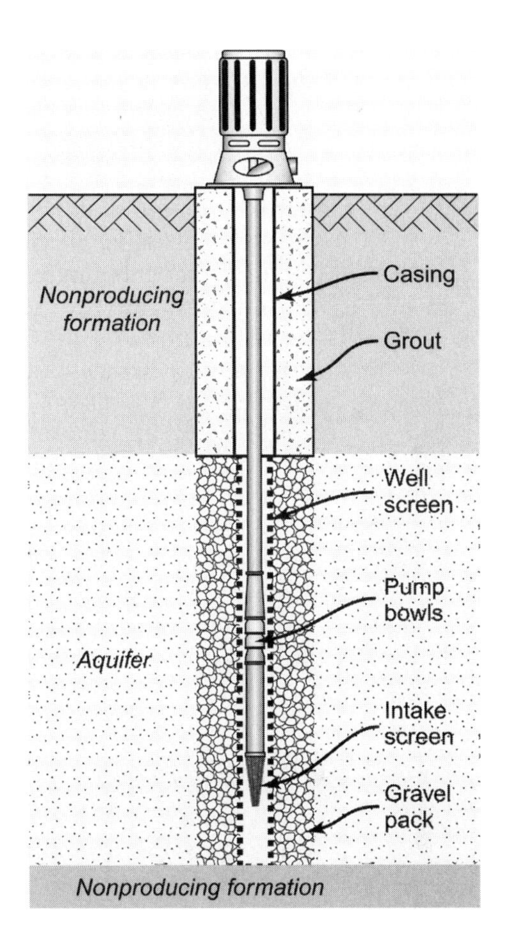

Figure 11–7
Cross section through a gravel-packed well.

agricultural chemicals. Local regulations will specify acceptable grout materials, such as a neat (nonshrinking) concrete or bentonite.

Construction of driven or jetted wells does not create a continuous annular space that can be grouted (although an initial borehole is sometimes used at the surface). The disturbed soil around the casing could allow vertical flows, so the wellhead should be finished properly with a concrete pad to keep surface waters away from the casing. Wellhead standards are usually specified by local regulations.

11.13 Effects of Ground Water Development

Ground water, though renewable, is a limited resource. Recharge rates vary with local climate and geology. Extractions in excess of recharge rates can alter local hydrology by lowering ground water levels, changing gaining streams (where ground water discharges into the stream) into losing streams (where surface water is lost to the ground water). The reduction in stream flow can impact both wildlife and downstream users of the resource. Wetlands can dry up as water tables drop. Population shifts in flora and fauna can result. Some of these effects have been observed in the High Plains of the central United States where decades of irrigation dropped the water levels in the Ogallala Aquifer by tens of meters. More efficient irrigation methods and conservation efforts have reduced the rates of decline since the 1980s.

A dramatic and very serious result of excessive pumping can be land subsidence. As water pressure drops in the aquifers and confining beds, the weight of the overburden must be carried more and more by the solids in the formation. Clay particles in aquitards readily rearrange in response to the increased mechanical stresses, becoming more compact. As this progresses, the land surface drops. Drainage patterns can shift; pavement, buildings, and pipes may shift and break. Subsidence of almost nine meters has occurred in the Los Baños-Kettelman City area of California. Parts of the Houston-Galveston area in Texas have subsided almost three meters. About four meters of subsidence has occurred at Stanfield and Eloy in Arizona (Poland, 1985). Subsidence is irreversible. Even if ground water levels rise again, the land surface will not rebound.

Whereas an isolated well has a particular yield, if a number of wells tapping the same aquifer all pump together, it is possible that the combined extraction will exceed the capacity of the aquifer. This can happen at local and regional scales. Government agencies may have authority to regulate extractions in areas of limited ground water supply. Parts of eastern North Carolina, for example, have been designated "capacity use" areas. Planning and cooperation are needed to maximize utility while meeting the needs of all parties.

Water Rights

Water law varies from place to place, but the following distinctions are generally recognized: (1) diffuse surface waters, (2) surface water in well-defined channels, (3) water in well-defined aquifers, and (4) percolating underground waters. Diffuse surface waters and percolating underground waters are normally regulated by common or civil law rather than legislation. In some western states, however, diffuse surface waters are treated the same as waters in well-defined channels. Most states consider diffuse surface waters the property of the landowner. Especially in the eastern states, the law for diffuse surface waters has been concerned with the damage caused by such water and the assignment of liability. Percolating underground water is defined as that subsurface water that flows in small pores or filters through the soil in such a way that its course or direction cannot be easily determined. Ruling in *Frazier v. Brown, 12 Ohio St. 294, 311* (1861), the Ohio Supreme Court declared,

> Because the existence, origin, movement and course of such waters, and the causes which govern and direct their movements, are so secret, occult and concealed, that an attempt to administer any set of legal rules in respect to them would be involved in hopeless uncertainty, and would be, therefore, practically impossible.

The court finally overturned this ruling in 1984. In a concurring opinion, Justice Holmes wrote,

> . . . a primary goal of water law should be that the legal system conforms to hydrologic fact. Scientific knowledge in the field of hydrology has advanced in the past decade to the point that water tables and sources are more readily discoverable. This knowledge can establish the cause and effect relationship of the tapping of underground water to the existing water level. Thus, liability can now be fairly adjudicated with these advances which were sorely lacking when this court decided *Frazier* more than a century ago.

Unfortunately, the 1861 ruling is still frequently cited in legal arguments.

Two basic doctrines exist regarding water rights of landowners, namely, riparian and appropriation. They are recognized either separately or as a combination of both doctrines in different states. A comparison of the main features of these two doctrines is shown in Table 11-3.

11.14 Riparian Doctrine

Riparian doctrine is largely derived from old English water law, which was borrowed in part from Roman civil law. Riparian land is that which is contiguous to a stream or other body of surface water. The landowner has the right of access to and use of the water, and this right is not lost by nonuse. Strict riparian doctrine allows the riparian landowner to use the water for domestic use and livestock, as long as the stream is not diminished in either quality or quantity. In practice, this strict interpretation is unworkable. The riparian doctrine in its U.S. version recognizes the right of a riparian landowner to make reasonable use of the stream's flow, provided the water is used on riparian land. The stream may be diminished in quality and/or quantity as long as the use is reasonable, which includes domestic and livestock uses. This variant of riparian doctrine is known as the Reasonable Use doctrine. Where conflicts arise, what is deemed reasonable must be decided by the courts for each case. Under this principle, for example, the California Supreme Court established that no proprietor can legally absorb all the water of the stream so as to allow none to flow down to a neighbor.

For percolating ground water, most states recognize absolute ownership and do not restrict the use of percolating water. This principle is known as common-

TABLE 11-3 Comparison of Water Rights Doctrines[a]

Characteristic	*Riparian Doctrine*	*Doctrine of Prior Appropriation*
Acquisition of water right	By ownership of riparian land	By permit from state (state ownership)
Quantity of water	Reasonable use	Restricted to that allowed by permit
Types of use allowed	Domestic, livestock, etc., but not precisely defined	Beneficial use required
Water rights lost by nonuse	No	Yes, but continued use not always required
Location where water may be used	On riparian land, with some exceptions	Anywhere, unless specified in permit

[a] Generally applicable only for surface water in well-defined channels and for water in well-defined aquifers. Some state laws on ground water deviate from the above.

law rule of capture. In California, the riparian doctrine is further modified by establishing correlative rights. Under this doctrine, the landowner's use of ground water must be reasonable in consideration of the similar rights of others and must be correlated with uses by others in times of shortage.

In states that do not have statutory laws governing water rights, the riparian doctrine is based on previous court decisions. Many of the eastern states have modified the riparian doctrines by regulating use through the issuance of permits for specific amounts of water. Special restrictions may be placed on extractions in "capacity use" areas, i.e., where potential extractions would exceed recharge, leading to eventual depletion of the resource.

11.15 Doctrine of Prior Appropriation

This doctrine is based on the priority of development and use; that is, the first to develop and put water to beneficial use has the right to continue that use. The doctrine asserts that whereas no one owns the water in a stream, any person, corporation, or municipality has the right to use the water for beneficial purposes. Ownership of riparian land is not required.

The doctrine of prior appropriation was developed in the western United States to deal with situations where the scarcity of water made riparian doctrine unworkable. The right to use the water (called a "priority") is acquired by filing a claim in accordance with the laws of the state. The first user ("senior appropriator") is protected against claims by later users ("junior appropriators"). Administrative details vary among the states. Water rights may be lost by abandonment or forfeiture.

Although the prior appropriation doctrine gives broad rights to the appropriators, extractions may be limited by in-stream flow requirements. The term *in-stream flow* refers to the amount of water necessary to sustain in-stream values and uses such as fish and wildlife habitat and migration, recreation, navigation, power generation, waste assimilation, riparian and floodplain wetlands, and channel geomorphology. Some states have statutes that alter the priorities during times of water shortage to give preference to more critical uses (such as domestic) irrespective of the normal rank of priorities.

11.16 Water Rights Law by States

Water rights doctrines vary from state to state; however, because of increased demand for water and the need for better utilization, water law is continually evolving. Newer laws have generally given more authority to state agencies to protect the public values associated with water resources, such as recreation and wildlife habitat, by establishing water duties and other stringent criteria for use. For discussion and examples, see ASAE (1986).

Internet Resources

US Bureau of Reclamation
http://www.usbr.gov

USDA Natural Resources Conservation Service
http://www.nrcs.usda.gov

US Geological Survey
http://www.usgs.gov

MidWest Plan Service
htt://www.mwpshq.org

References

American Society of Agricultural Engineers (ASAE). (1986). *Water Resources Law, Proceedings of the National Symposium on Water Resources Law.* St. Joseph, MI: Author.

Bear, J. (1979). *Hydraulics of Groundwater.* New York: McGraw-Hill.

Chertousov, M. D. (1962). *Hydraulics* (in Russian). Moscow: Gosenergouzdat.

Driscoll, F. G. (1986). *Groundwater and Wells,* 2nd ed. Johnson Filtration Systems Inc., St. Paul, MN 55112.

Freeze, R. A., & J. A. Cherry. (1979). *Groundwater.* Englewood Cliffs, NJ: Prentice-Hall.

Harr, M. E. (1962). *Groundwater and Seepage.* New York: McGraw-Hill.

Heath, R. C. (1980). *Basic Elements of Ground-Water Hydrology with Reference to Conditions in North Carolina.* U.S. Geological Survey Water Resources Investigations Open-File Report 80–44. Denver, CO: USGS Open-File Services Section.

McWhorter, D. B., & D. K. Sunada. (1977). *Groundwater Hydrology and Hydraulics.* Fort Collins, CO: Water Resources Publications.

Poland, J. F., ed. (1985). *Guidebook to Studies of Land Subsidence Due to Ground Water Withdrawal.* UNESCO Publishing, ISBN 92-3-102213-X.

Problems

11.1 Calculate the annual water requirement, storage volume needed, and watershed area required for an impoundment in central Arkansas. The supply must serve a poultry grower with 200 000 chickens. Seepage and evaporation losses are 45 percent of the storage capacity.

11.2 An irrigation reservoir has a storage capacity of 14 ha-m. If the seasonal irrigation requirement for the crop is 540 mm and the losses to seepage and evaporation are 56 percent of the stored water, how many hectares can be irrigated?

11.3 Determine the discharge of a 200-mm diameter well that fully penetrates an unconfined aquifer. The bottom of the aquifer is 22 meters below the ground surface. Water levels in observation wells at distances of 10 and 40 m from the well are 12.8 and 5.7 m, respectively, below the ground surface.

11.4 Determine the discharge of a 600-mm diameter well that fully penetrates a confined aquifer. The top of the aquifer is 28 meters below the ground surface. The aquifer is 16 m thick. The water level in an observation well at a distance of 100 m is 7.6 m below the surface. The water level in the well is 24.0 meters below the surface.

11.5 Estimate the radius of influence for the conditions given in Example 11.2.

11.6 Derive the equation for the discharge of a well that completely penetrates an unconfined aquifer.

11.7 Derive the equation for the discharge of a well that completely penetrates a confined aquifer.

CHAPTER

12 Wetlands

Wetlands have had a profound effect on the world ecosystem. Much of our fossil fuel supplies were formed in the swampy environments of the Carboniferous Period approximately 300 million years ago. Most of our world's population centers were developed on or adjacent to what are now classified as wetland areas. People were drawn to these areas because of easy access to food, transportation, and clean water. Examples include Mexico City, Mexico; Cairo, Egypt; Paris, France; Tokyo, Japan; Christchurch, New Zealand; and in the United States, Chicago, New Orleans, and Washington, D.C.

Wetlands are sources, sinks, and transformers of a multitude of chemical, biological, and genetic materials, making them important ecosystems. They have been described as the "kidneys of the landscape" because of their ability to cleanse polluted waters. The capabilities of natural wetlands have been widely recognized and, taking a lesson from nature, artificial or constructed wetlands are being designed to treat a wide range of waste streams.

Approximately 6 percent of the land surface of the world (8.6 million km^2) is wetland (Mitsch & Gosselink, 2000). During the nineteenth and twentieth centuries, large portions of the world's wetlands were artificially drained for agricultural or urban uses. Over 50 percent of U.S. wetlands were drained during this period (Dahl, 1990). However, since approximately 1970, the widespread destruction of wetland ecosystems has been curtailed. The Convention on Wetlands of International Importance, also referred to as the Ramsar Convention, held in 1971, developed a global treaty to protect wetlands as habitats for migratory birds (Blasko, 1997). The treaty requests countries to designate and protect their wetland resources. As of 2003, 136 nations have agreed to the Ramsar Convention by protecting 1288 wetland sites, totaling 108.9 million hectares.

Wetland Definition

The definition of a wetland would seem to be a fairly simple task; however, no widely accepted definition exists. All definitions refer to the combined influences of hydrology, soils, and vegetation. The identification or delineation of a wetland requires the evaluation of each of these components. Generally, a difference in elevation of 0.1 m can separate an area from being classified as a wetland versus an upland site (Figure 12-1) (Lyon, 1993). Unfortunately these transitional areas can be classified as a wetland under one set of criteria and upland under another (Skaggs et al., 1994).

Much emphasis has been placed on the development of criteria for delineating wetlands (Lyon, 1993; Mitsch & Gosselink, 2000). According to the U.S. Army Corps of Engineers, Section 404 of the 1977 Clean Water Act Amendments, wetlands are defined thus (Mitsch & Gosselink, 2000):

> The term "wetlands" means those areas that are inundated or saturated by surface or groundwater at a frequency and duration sufficient to support, and under normal circumstances do support, a prevalence of vegetation typically adapted for life in saturated soil conditions. Wetlands generally include swamps, marshes, bogs, and similar areas.

While developing a classification system for wetlands, the U.S. Fish and Wildlife Service adopted the following more detailed description (Cowardin et al., 1979):

> Wetlands must have one or more of the following three attributes: (1) at least periodically, the land supports predominantly hydrophytes; (2) the substrate is predominantly undrained hydric soil; and (3) the substrate is nonsoil and is saturated with water or covered by shallow water at some time during the growing season.

Hydrophytes are water-loving plants found where water is at or near the surface. Hydric soils are saturated, flooded, or ponded long enough during the growing season to develop anaerobic conditions near the surface. Nonsoils are the rocky areas along shorelines near oceans and lakes.

The Swampbuster Provision of the 1985 Food Security Act enhanced the Fish and Wildlife Service definition to produce the following:

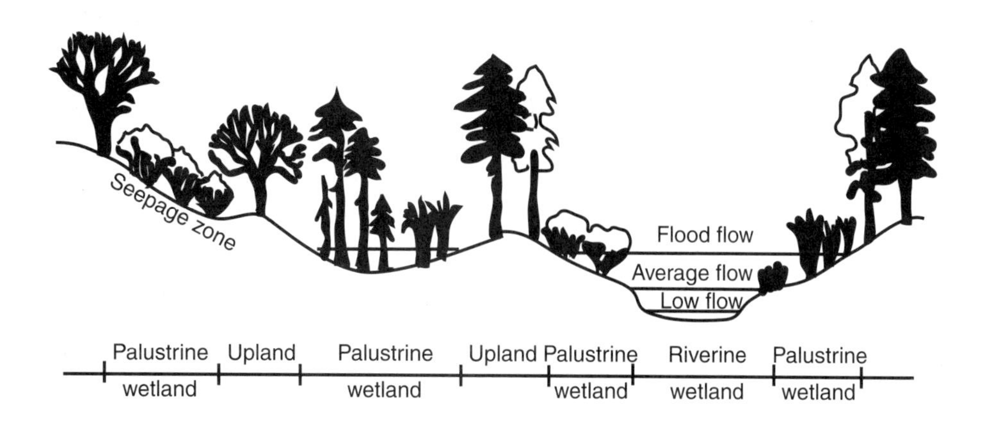

Figure 12–1

Cross section illustrating the small difference in elevation between uplands and wetlands.

> The term "wetland" except when such term is part of the term "converted wetland" means land that (A) has a predominance of hydric soils; (B) is inundated or saturated by surface or ground water at a frequency and duration sufficient to support a prevalence of hydrophytic vegetation typically adapted for life in saturated soil conditions; and (C) under normal circumstances does support a prevalence of such vegetation.

A converted wetland is an area that was drained, dredged, filled, leveled, or otherwise manipulated before the Swampbuster Provision was enacted (December 23, 1985) for the purpose of, or to have the effect of, making the production of an agricultural commodity possible. Under the Swampbuster Provision, there are no restrictions on either drainage maintenance or additional drainage on these prior converted wetlands, which are estimated to total more than 20 million ha. Wetlands falling under the description of Section 404 of the 1977 Clean Water Act definition or the Swampbuster Provision are called *jurisdictional wetlands* and represent legally defined wetlands in the United States.

The international definition developed at the Ramsar Convention, extends the definition of wetlands to include (Blasko, 1997):

> . . . areas of marsh, fen, peatland or water, whether natural or artificial, permanent or temporary, with water that is static or flowing, fresh, brackish or salt, including areas of marine water the depth of which at low tide does not exceed six meters.

The international definition includes a wide variety of habitats such as rivers, lakes, and coastal areas that are not typically described as wetlands by other definitions. The logic for extension to areas with water depths up to 6 m was used to include all habitats of migratory birds.

Wetland Classification

Landscape position, along with hydrology and vegetation, are used to classify the various types of wetlands. Wetlands are found on landscapes ranging from alpine slopes to ocean coastlines and in temperature regimes from the polar regions to the tropics. In 1974, the U.S. Fish and Wildlife Service began an inventory of wetland resources and, in the process, developed a wetland classification system that classifies natural wetlands into five major systems (Cowardin et al., 1979). These are

- Marine (coastal wetlands, including coastal lagoons, rocky shores, and coral reefs);
- Estuarine (including deltas, tidal marshes, and mangrove swamps);
- Lacustrine (wetlands associated with lakes);
- Riverine (wetlands along rivers and streams); and
- Palustrine (meaning "marshy"—marshes, swamps, fens, and bogs).

Outside of Alaska (43 percent of surface area is wetland), most of the wetlands in the United States are in the upper Midwest and eastern coastal regions (Figure 12-2). In addition to these natural wetland systems, there are numerous human-influenced wetland environments, including farm ponds and reservoirs, paddy fields, gravel pits, and constructed wetlands.

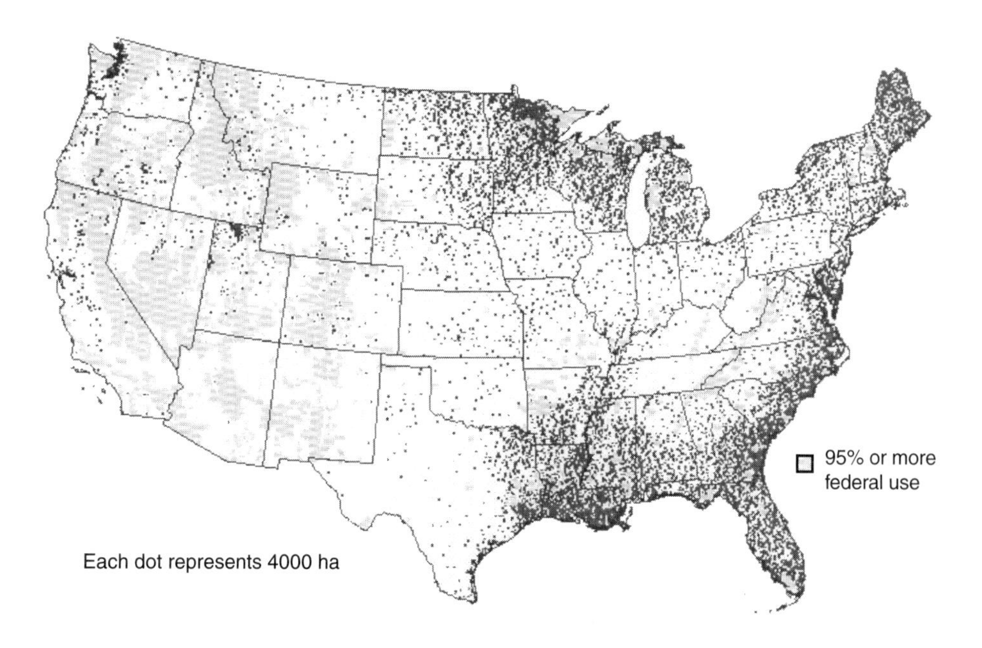

Figure 12–2
Wetland distribution within the United States. (NRCS website, 1997 data)

12.1 Marine Wetlands

The marine wetland system extends from the outer edge of the continental shelf (water depths less than 6 m) to shoreline features that are impacted by the high-energy waves produced by the ocean. The most recognizable aspects of marine wetlands are the salinity content greater than 30 ppt (parts per thousand) and the action of waves, which greatly reduces the presence of vegetation. Marine wetlands are nourished only by ocean water and the minerals and nutrients contained therein. Subsystems of marine wetlands include subtidal (where the substrate is continuously submerged) and intertidal (where the substrate is alternately exposed and flooded by tides). Typical examples include the rocky cliffs and islands adjacent to the ocean, ocean beaches, coral reefs, and tide-influenced lagoons. Aquatic beds of algae, sea lettuce, and kelp are also classified as marine wetlands. Examples include Monterey Bay and Bolinas Lagoon sites on the northern coast of California and the 62 000-ha Baie du Mont St. Michel site on the western coast of France.

12.2 Estuarine Wetlands

Estuarine wetlands occur at the interface between saline and freshwater environments with salinity ranging between 0.5 and 30 ppt. The absence of high-energy wave action distinguishes the estuarine system from the marine system and allows salt marshes to develop in the middle to high latitudes and mangrove swamps to develop in tropical and subtropical regions (Mitsch & Gosselink, 2000). Mangroves are salt-tolerant trees that have specially adapted roots and leaves that enable them to occupy the saline coastal waters where most plants cannot survive. Many estuarine wetlands are located in areas where freshwater rivers merge into the oceans. As with the marine system, estuarine wetlands are further classified as subtidal and intertidal subsystems. Estuarine systems described as salt marshes can be found along the Atlantic and Gulf coasts of the United States, including the 45 000-ha Chesapeake Bay Estuarine Complex and the 51 000-ha Delaware Bay. Mangrove swamps are located along the southern tip of Florida and the tropical coasts of Africa, Central and South America, and Asia.

12.3 Lacustrine Wetlands

The lacustrine system, as delimited by Cowardin et al. (1979), includes deep freshwater habitats and wetlands with a total area greater than 8 ha situated on landscape depressions or dammed rivers. These wetlands lack trees, shrubs, persistent emergent plants, emergent mosses, or lichens with greater than 30 percent areal coverage. In effect, lacustrine wetlands are essentially shallow lakes with limited "rooted" vegetation that are classified with regard to water depth. Areas with water depths between 2 m and 6 m are considered limnetic, and areas between the shoreline and a depth of 2 m are considered littoral. Distinctions between lacustrine and palustrine wetlands are made based on the total area and vegetation characteristics. Lacustrine wetlands are the second most prevalent type of wetland in the United States.

12.4 Riverine Wetlands

Riverine systems include rivers and streams characterized by water flowing from an upstream position to a downstream position through a defined channel. These systems extend from near-ocean locations with ocean-derived salt contents less than 0.5 ppt to the river source. Additional terminal points of riverine systems are where a channel enters or leaves a lake. Other than the streambank, riverine systems do not include the land adjacent to the stream that may contain wetland-supported trees, shrubs, persistent emergent plants, emergent mosses, or lichens. The Cowardin classification system labels these wetland areas adjacent to the stream as palustrine wetlands.

There are four subsystems of riverine wetlands. Riverine systems close to the oceans with low gradients and fluctuating flow velocity influenced by the tides are classified as tidal. Continuously flowing streams without tidal influence are either lower perennial or upper perennial depending on having either a low gradient or high gradient. Rivers and streams that do not flow for part of the year are classified as intermittent systems.

12.5 Palustrine Wetlands

The palustrine system includes all nontidal wetlands dominated by trees, shrubs, persistent emergent plants, emergent mosses, or lichens and similar tidal wetlands that have less than 0.5 ppt ocean-derived salt contents. The dominant feature of palustrine wetlands is the presence of persistent wetland vegetation. The palustrine system can include nonvegetated wetlands if they are less than 8 ha in area, lack a wave-formed or bedrock shoreline, water depth is less than 2 m at the deepest point, and salinity due to ocean-derived salts is less than 0.5 ppt. Most inland wetlands are classified as palustrine and include marshes, swamps, bogs, fens, tundra, potholes, and floodplains.

Wetland Functions

As described earlier, wetlands are sources, sinks, and transformers of a multitude of chemical, biological, and genetic materials. The two primary processes that occur in wetland environments are sedimentation and nutrient cycling. The unique characteristics of a wetland allows for the hydrology, soils, and vegetation to interact to create an ideal environment for chemical and biological changes to take place.

12.6 Hydrology

The presence of water is a key ingredient for delineating a wetland. A method to track the water flows within a wetland is the water budget consisting of a mass balance of inflows and outflows of water:

$$P + SWI + GWI = ET + SWO + GWO + \Delta S \qquad \textbf{12.1}$$

where P = precipitation (L),
SWI = surface water inflow (L),
GWI = ground water inflow (L),
ET = evapotranspiration (L),
SWO = surface water outflow (L),
GWO = ground water outflow (L),
ΔS = change in storage (L).

The relative importance of each component varies both spatially and temporally. A particular lacustrine wetland may be influenced by surface inflows and outflows, whereas a nearby lacustrine wetland may be more influenced by ground water inflows. All the components interact to define the hydrology of an individual wetland and must be included in the mass balance. Factors that influence the water budget are landscape position, vegetation, soil, and climate.

An identifiable aspect of the water storage in a wetland is the water depth. During some periods, inflows are greater than outflows and water becomes ponded on the soil surface. During other times, outflows may be greater than inflows and water may disappear from the surface as happens during low tide or during extended dry periods. The water depth may actually fall just below the soil surface forming a shallow water table. The water table is defined as the upper surface of a saturated zone below the soil surface where the water is at atmospheric pressure (ASAE, 2001). The water table remains close enough to the soil surface that the soil profile is nearly saturated or can become saturated for a sufficient time to allow hydrophytic (water-tolerant) vegetation to survive.

The daily, monthly, or seasonal pattern of water level changes of a few wetland systems are shown in Figure 12-3. These patterns describe the hydroperiod and represent the hydrologic signature of the wetland (Mitsch & Gosselink, 2000). Coastal wetlands exhibit relatively uniform hydroperiods influenced by the regular tidal fluctuations. Inland wetlands may be influenced by climatic factors such as dry or wet years or seasonal effects related to evapotranspiration (ET) demands. Riverine wetlands or palustrine wetlands near rivers often show hydroperiods associated

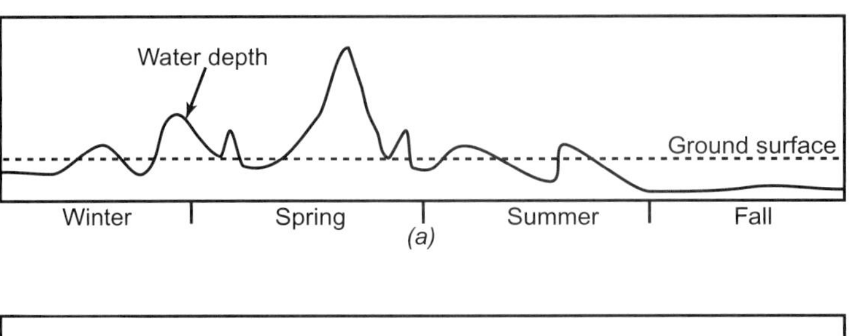

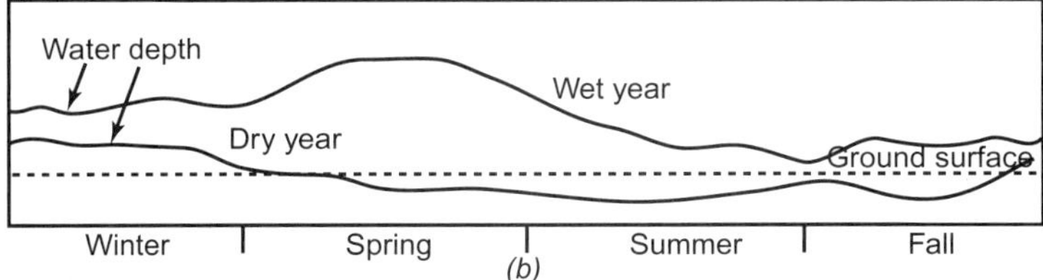

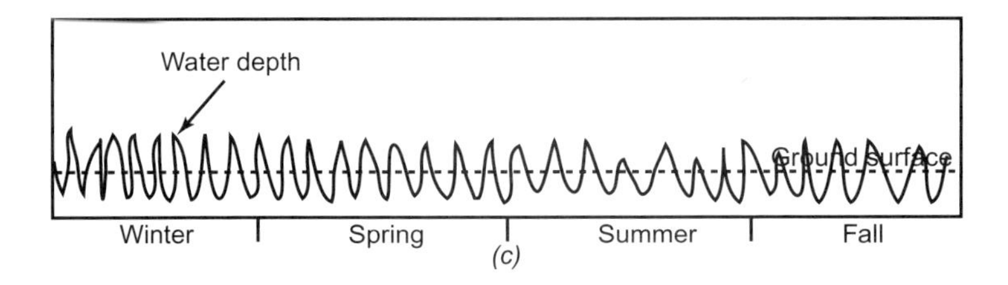

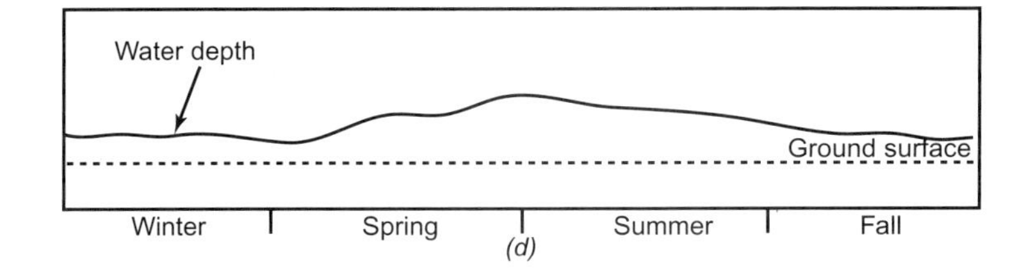

Figure 12–3
Typical hydroperiods for (a) riparian, (b) palustrine, (c) estuarine, and (d) lacustrine wetlands.

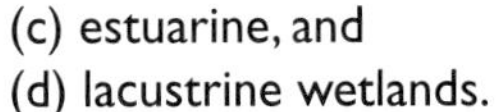

with seasonal floods. For wetlands that encounter random changes in water level, the amount of time the wetland is flooded is called the flood duration and the average number of flood events within a given period is called the flood frequency. The wetland hydroperiod can last for weeks, months, or even years. A depth–duration curve (Figure 12-3) can be constructed for any wetland to summarize the water depth and hydroperiod.

Another key hydrologic parameter for wetlands is the residence or retention time R_t, which describes the average amount of time water is retained in the wetland. An estimate of R_t is

$$R_t = \frac{\text{wetland volume}}{\Sigma \text{inflows}} \tag{12.2}$$

where the wetland volume is a function of the surface area of the wetland and the water depth. Note that this calculation generally overestimates the actual residence time as plug flow is assumed. In reality, short-circuiting can occur in most natural wetlands, allowing water to bypass regions within the wetland. Constructed wetlands are designed with baffles or berms to control the flow pattern and to reduce the amount of short-circuiting that can occur.

Example 12.1

The following data were obtained for a coastal marsh. Precipitation was 965 mm, and evapotranspiration was 680 mm for the year. Surface inflows to the marsh totaled 87 300 m^3 and surface outflows were 101 700 m^3. Ground water inflows were 13 270 m^3. Determine the change in storage for this 1.3-ha wetland. If the depth of water in the wetland was 0.3 m at the start of the year, what was the depth at the end of the year? Estimate the residence time.

Solution. In order to compute a mass balance, all quantities must be converted to the same units. Precipitation and *ET* can be converted to units of volume.

$$P = 965 \text{ mm} \times 1.3 \text{ ha} = 12\,545 \text{ m}^3$$

$$ET = 680 \text{ mm} \times 1.3 \text{ ha} = 8840 \text{ m}^3$$

Now apply Equation 12.1.

$$\text{Inflows} = \text{Outflows} + \Delta\text{Storage}$$

$$12\,545 + 87\,300 + 13\,270 = 8840 + 101\,700 + \Delta\text{Storage}$$

$$\Delta\text{Storage} = 2575 \text{ m}^3 = 0.2 \text{ m added depth over the 1.3-ha wetland}$$

Depth at the end of the year was

$$0.3 \text{ m} + 0.2 \text{ m} = 0.5 \text{ m}$$

Use the average depth over the year and Equation 12.2 to estimate the residence time.

$$R_t = \frac{0.5(0.3 + 0.5)13\,000}{(12\,545 + 13\,270 + 87\,300)} = 0.023 \text{ yr} = 8.4 \text{ day}$$

Sediments, nutrients, trace metals, microorganisms, and organic materials are transported to wetlands by ground water and surface water. Wetlands can trap, precipitate, transform, recycle, and export many of these constituents, causing the quality of outflow water from a wetland to differ markedly from that of the entering water (Mitsch & Gosselink, 2000).

The hydroperiod and residence time significantly affect the capability of a wetland to alter water quality. Short, pulsing hydroperiods resulting from tidal influences force the wetland area to cycle through periods of wetting and drying. As the water level rises, sediment and other solutes are transported into the wetland where they may settle out; however, the short residence time limits the amount of nutrient transfer that can occur. Long hydroperiods found in the Everglades or swamps in the southeastern United States allow for significant sedimentation and nutrient cycling to occur as water moves slowly through these systems.

Water-quality enhancement from wetlands can affect an entire drainage basin. Water chemistry in basins that contain a large proportion of wetlands is usually different from that in basins with fewer wetlands. Basins with more wetlands tend to have water with lower specific conductance and lower concentrations of chloride, lead, inorganic nitrogen, suspended solids, and phosphorus (total and dissolved). Wetlands may change water chemistry sequentially; that is, upstream wetlands may serve as the source of materials that are transformed in downstream wetlands. Estuaries and tidal rivers depend on the flow of freshwater, sediments, nutrients, and other constituents from upstream sources.

12.7 Soil

Soils that are subject to flooding or ponding of water for extended periods will exhibit hydric or waterlogged characteristics. An important aspect of a saturated soil profile is the lack of oxygen (anaerobic). As water fills the pores, air is driven out. Air, most importantly oxygen, cannot be replaced because it diffuses very slowly through water (Lyon, 1993). The resulting anaerobic conditions that exist greatly affect the chemical and biological transformations that can occur in wetlands. Depending on the microbial population and the availability of organic material, the change from aerobic to anaerobic conditions can occur on the order of a few hours to a few days (Mitsch & Gosselink, 2000). After the soil profile becomes anaerobic, microbial activity and organic decomposition are decreased, which leads to the accumulation of organic matter in wetlands.

The redox potential or oxidation-reduction potential is the measure of the tendency of a wetland soil solution to exchange electrons. Since there are no free electrons in the soil solution, every oxidation reaction is accompanied by a simultaneous reduction reaction. Oxidation occurs during the uptake of oxygen or the removal of hydrogen. In essence, it is the giving up of electrons. Reduction is the release of oxygen, the gaining of hydrogen, or the gaining of electrons (Mitsch & Gosselink, 2000). Once oxygen is removed from the system, reduction of other compounds occurs in the order nitrogen, manganese, iron, sulfur, and carbon.

A platinum electrode can be constructed and used to measure the redox potential relative to a hydrogen electrode in units of millivolts (mV) (Mitsch & Gosselink, 2000). When dissolved oxygen is available, the redox potential is in the range of 400 to 700 mV. As dissolved oxygen is depleted and other compounds in the soil subsequently release oxygen (electrons), the redox potential becomes lower and eventually negative. The redox potential provides information about compounds that are or have undergone change in the wetland environment (Table 12-1). If a redox potential of -150 mV is measured in a wetland, then it can be assumed that no dissolved oxygen is present and that nitrate, manganese, and iron have all been reduced. Since sulfur is reduced at -150 mV, there is probably a slight odor of hydrogen sulfide.

Nitrogen. Nitrogen occurs in many forms in a wetland depending on the oxidation-reduction state. The transformation from one form to another involves several microbial processes, including ammonification, nitrification, and denitrification.

Ammonification or N mineralization is the biological transformation (decomposition and degradation) of organic nitrogen to the ammonium ion (NH_4^+) that occurs in aerobic (oxidized) and anaerobic (reduced) environments. Although ammonification occurs more rapidly in aerobic environments, ammonium is more

TABLE 12–1 Typical Redox Potential Values for Chemical Reactions in a Wetland

Chemical Reaction	*Redox Potential (mV)*
O_2 available	400 to 700
NO_3^- converted to N_2	250
Mn^{4+} converted to Mn^{2+}	225
Fe^{3+} converted to Fe^{2+}	100 to −100
SO_4^{2-} converted to H_2S	−100 to −200
CO_2 converted to CH_4	< −200

likely to accumulate in anaerobic environments because nitrification is inhibited (Kadlec & Knight, 1996). Once the ammonium ion is formed it can be absorbed by plants, adsorbed onto negatively charged soil particles, or nitrified into nitrate.

Nitrification is the two-step process of converting the ammonium ion into nitrate. Since nitrification is the process of removing hydrogen ions and adding oxygen ions, it must occur in an aerobic environment. *Nitrosonomas* bacteria conduct the first step of the process by slowly converting ammonium to nitrite (NO_2^-). *Nitrobacter* bacteria rapidly convert the nitrite to nitrate with addition of more oxygen. Since the nitrate ion is negatively charged, it cannot be immobilized by attaching to soil particles and is highly mobile in solution. The mobility of nitrate makes it a potential ground and surface water contaminant.

Denitrification represents a significant path for nitrogen removal in a wetland since it occurs in anaerobic environments. In the absence of free oxygen, several microorganisms such as *Bacillus, Enterobacter, Micrococcus, Pseudonomas,* and *Spirillum* bacteria use oxygen from the nitrate ion for metabolism. These organisms are called denitrifying bacteria. Nitrate is transformed to nitrogen gas (N_2), nitrous oxide (N_2O), or nitric oxide (NO) in the process (Kadlec & Knight, 1996). Denitrification, however, can be inhibited if a carbon source is not available for the bacteria. The denitrifying bacteria increase in population based on the quantity of nitrate and carbon available. Bacterial populations are greater in treatment wetlands because of the consistent and higher inputs of organic materials in these wetlands than generally occur in natural wetlands.

Phosphorus. Phosphorus retention is an important attribute in natural and constructed wetlands (Mitsch & Gosselink, 2000). The retention occurs via sedimentation, sorption, and uptake. If residence time is sufficiently long, sedimentation will occur whereby phosphorus sorbed to soil particles can settle out of the water and be deposited in the wetland. Riparian wetlands or other wetland environments that are exposed to short-term high flows can actually be phosphorus sources if erosional forces remove significant quantities of phosphorus-rich sediment that had previously been deposited in the wetland. Phosphates also form complexes with iron, calcium, and aluminum. Since phosphorus is an important nutrient for plant growth, plant uptake represents an important removal mechanism in wetlands. Unfortunately, the phosphorus is retained in the plant tissues and can be returned to the soil profile when the plant dies, causing the soil–plant system to be saturated with phosphorus with no net decrease. Biomass removal or harvest may be an important maintenance requirement for some treatment wetlands to remove the phosphorus and keep the wetland functional.

12.8 Vegetation

The types of plants growing in a wetland depend to a large degree on the hydroperiod or amount of time the wetland is inundated with water (Table 12-2). For cases with continual flooding of water 1 m or deeper, free-floating plants will dominate. In cases where the water level is less than a meter in depth or the area is occasionally flooded, emergent herbaceous plants and woody species become more dominant.

TABLE 12–2 Typical Species Found in Wetlands

Common Name	***Scientific Name***
Submerged Aquatic Species (0.1–3 m water depth)	
Water mosses	*Fontinalis* spp.
Brazilian waterweed	*Egeria densa*
Elodea	*Elodea* spp.
Hydrilla	*Hydrilla verticillata*
Eelgrasses	*Vallisneria* spp.
Naiads	*Najas* spp.
Water milfols	*Myriophyllum* spp.
Bladderworts	*Utricularia* spp.
Fanwort	*Cabomba caroliniana*
Floating Aquatic Species	
Mosquito fern	*Azolla caroliniana*
Water fern	*Salvinia rotundifolia*
Water lettuce	*Pistia stratiotes*
Duckweeds	*Lemna* spp.
Giant duckweeds	*Spirodela* spp.
Watermeals	*Wolffia* spp.
Bog mats	*Wolffiella* spp.
Water hyacinth	*Eichhornia crassipes*
Floating Rooted Aquatic (0.1–3 m water depth)	
Frog's bit	*Limnobium spongia*
Pondweeds	*Potamogeton* spp.
Water shield	*Brasenia schreberi*
Floating hearts	*Nymphoides* spp.
Lotuses	*Nelumbo* spp.
Spatterdocks	*Nuphar* spp.
Water lilies	*Nymphaea* spp.
Pennyworts	*Hydrocotyle* spp.
Woody Shrubs and Trees (continuous inundation)	
Cypresses	*Taxodium* spp.
Water-willow	*Decodon verticillatus*
Tupelo gums	*Nyssa* spp.
Buttonbush	*Cephalantus occidentalis*
Willows	*Salix* spp.

Common Name	*Scientific Name*
Red mangrove	*Rhizophora mangle*
Loblolly bay	*Gordonia lasianthus*
Woody Shrubs and Trees (less than continuous inundation)	
Pond pine	*Pinus serotina*
Cabbage palm	*Sabal palmetto*
Maples	*Acer spp.*
Hollies	*Ilex ssp.*
Black mangrove	*Avicennia germinans*
Hazel alder	*Alnus serrulata*
River birch	*Betula nigra*
Swamp dogwood	*Cornus foemina*
Titi	*Cyrilla racemiflora*
Fetterbush	*Lyonia lucida*
Oaks	*Quercus spp.*
St. John's wort	*Hypericum spp.*
Emergent, Herbaceous Plants (continuous inundation)	
Sphagnum mosses	*Sphagnum spp.*
Arrowheads	*Sagittaria spp.*
Wild taro	*Colocasia esculenta*
Spoon flowers	*Peltandra spp.*
Canna lilies	*Canna spp.*
Sedges	*Carex spp.*
Sawgrass	*Cladium jamaicense*
Spikerushes	*Eleocharis spp.*
Beak rush	*Rhynchospora spp.*
Bulrush	*Scirpus spp.*
Watergrass	*Hydrocloa*
Maidencane	*Panicum hemitomon*
Torpedograss	*Panicum repens*
Common reed	*Phragmites spp.*
Wild rice	*Zizania aquatica*
Southern wild rice	*Zizaniopsis milacea*
Blue flag iris	*Iris spp.*
Rushes	*Juncus spp.*
Arrowroot	*Thalia geniculata*
Pickerelweeds	*Pontederia spp.*
Bur reed	*Sparganium americanum*
Cattails	*Typha spp.*
Alligator weed	*Alternanthera philoxeroides*
Water primroses	*Ludwigia spp.*
Smartweeds	*Polygonum spp.*
Lizard's tail	*Saururus cernuus*
Reed canary grass	*Phalaris arundinacea*
Mannagrass	*Glyceria spp.*

(continued)

Common Name	*Scientific Name*
Emergent, Herbaceous Plants (less than continuous inundation)	
Swamp fern	*Blechnum serrulatum*
Chain ferns	*Woodwardia spp.*
Royal ferns	*Osmunda spp.*
Giant cane	*Arundinaria gigantea*
Knotgrass	*Paspalum distichum*
Cordgrasses	*Spartina spp.*

Although most plants require aerobic soil for normal root development, wetland plants are adapted to saturated soil conditions. One of these adaptations is the development of aerenchymous tissues that transport air to the roots (Kadlec & Knight, 1996). Air, including oxygen, can move out of the roots to develop a small, aerated zone around each root. Other plants develop adventitious roots that extend from the plant stem into the water. Adventitious roots have the capability of absorbing dissolved oxygen and nutrients from the water profile. The hydroperiod, water depth, water chemistry, and dissolved oxygen content of the water create the environmental conditions that allow individual species to thrive in wetlands.

Wetland Creation and/or Restoration Techniques

As stated earlier, significant wetland areas have been removed from the environment over time through drainage and filling. Restoration of these impaired/abolished wetlands may be easier than the process of creating a wetland in an upland area. Since ditches or subsurface drains have been used to drain many of the wetlands in the midwestern and coastal plain regions of the United States, the simple "plugging" of the ditches or drains is thought to be a viable method of restoring wetland functions. Unfortunately, agricultural or silvicultural activities, in addition to land-forming or -leveling procedures have removed most if not all of the original wetland topography or surface roughness. In areas with high soil organic carbon content, land subsidence may have occurred after the area was drained. The combination of a lack of surface roughness and inundation promotes a monoculture of wetland plants and may encourage a different population of vegetation to inhabit the area than existed prior to drainage.

Since the restoration of wetlands on prior drained sites is a relatively new practice, no design criteria are available concerning the amount of surface roughness necessary for restoring the appropriate hydrologic function. Saturation will slowly restore the anaerobic environment necessary for wetland function, but some surface roughness should be created. If possible, minimum and maximum water levels should be measured in nearby functional wetlands. Topographic "islands" and "depressions" can be created with a bulldozer prior to blocking the drain outlets. The development of surface roughness is an important part of the restoration process for promoting diverse communities of plants and wildlife.

The surface roughness can be used to promote mixing in the restored or created wetland. It is advisable to construct berms or barriers to force inflows to spread out over the wetland area. Serpentine flow patterns tend to maximize the residence time of a wetland.

In areas with minimal topographic relief, the blocking of a drainage ditch could flood many square kilometers with significant loss of prime farmland, commercial land, or housing sites. In Ohio, researchers have promoted a Wetland Reservoir Subirrigation System that allows drainage water to flow through a wetland before being stored in a small reservoir. Water from the reservoir is subsequently used during the growing season to irrigate a crop. In North Carolina, researchers have removed soil from the sides of a drainage ditch to create an instream wetland. In both cases, soil is removed to create a depression for the wetland. The advantages of these types of systems are the ability to maintain productive agricultural lands and to remediate water quality problems without having to flood large areas.

The restoration of the wetland hydrology to a site and a little patience (3 to 5 years) are enough to promote the formation of a wetland (Hammer, 1997; Mitsch & Gosselink, 2000). The seed bank not present in the soil will be transported over time to the site via wildlife. However, the development of a forested swamp, for instance, will take many more years (50–75) to develop than a simple marsh because the growth processes are much different for an oak or cypress tree compared to a cattail or bulrush. The types and diversity of plants colonizing a wetland will depend to a great extent on the water depth and hydroperiod of the wetland.

For those cases where quick establishment is important, seeding and planting techniques may be required. Water levels must be carefully managed to produce a wet, but not saturated, soil surface, because few seeds will germinate while completely submerged. Planting times and methods depend on the species and materials, such as seeds, tubers, rootstocks, plants, cuttings, or seedlings. Herbaceous vegetation and seeds are generally planted in the spring, whereas tubers, rootstalks, and seedlings are generally planted in the fall (Hammer, 1997). Ideally, plant materials can be retrieved from a nearby functional wetland and used as a plant and seed source for the new wetland. After planting, the hydrologic conditions must be carefully controlled to assure proper plant growth, because flooding too quickly can cause the death of many species.

Constructed Wetlands

Constructed wetlands are typically built to treat wastewater. Sources are as diverse as milkhouse wastes, animal wastes, agricultural runoff, home and municipal sewage, and industrial waste streams. These sources can have high biochemical oxygen demand (BOD), total suspended solids (TSS), nitrogen, phosphorus, fecal coliform bacteria, etc. Initial filtration or primary treatment techniques may be required to treat the waste stream prior to passing it through the constructed wetland. Unlike natural wetlands, constructed wetlands are heavily controlled by inlet and outlet conditions. Flows are monitored and cells are created to maximize the retention time. Detailed design procedures based on climate and waste characteristics should be followed (Kadlec & Knight, 1996).

Constructed wetlands (Figure 12-4) are classified as surface flow (SF) or subsurface flow (SSF) wetlands. Subsurface flow wetlands utilize large gravel as the porous media and as a result of the anoxic conditions are generally more efficient than

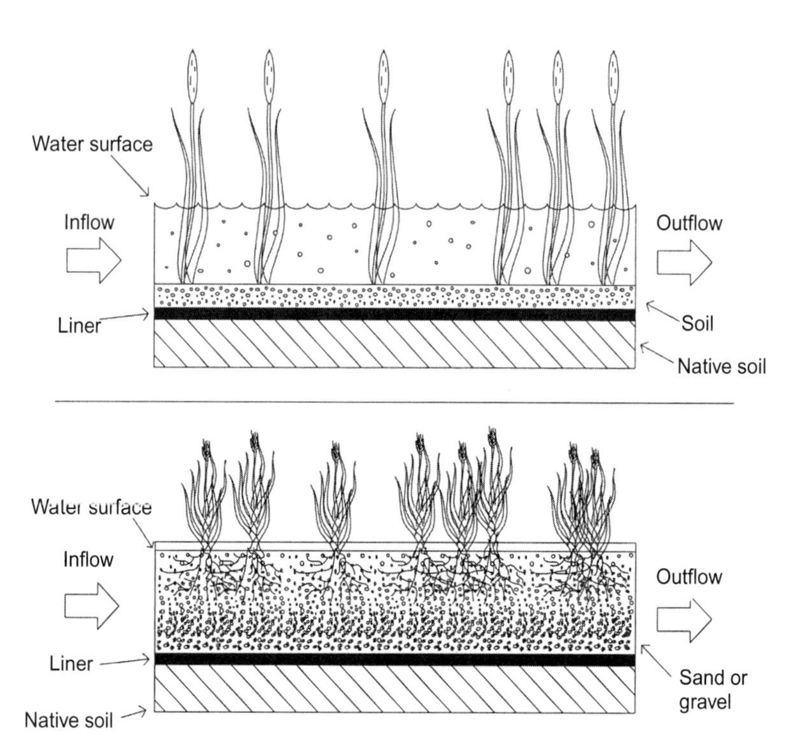

Figure 12–4
Cross sections of a surface flow wetland (upper) and subsurface flow wetland (lower).

surface flow wetlands. The primary disadvantage of subsurface flow wetlands is the inability to treat waste streams high in total solids because of clogging. Sizing of SF or SSF wetlands is based on either the assurance of a specified retention time or on the capacity of the wetland to make a specified reduction in a particular constituent in the waste stream.

Intuitively, the amount of time a waste stream is held in a wetland influences the quality of water exiting the wetland. Total suspended solids and BOD have been successfully reduced in SF wetlands with retention times of 7 to 10 days and in SSF wetlands after 2 to 4 days (Kadlec & Knight, 1996). Wetland area and water depth are the controlling factors for determining retention time. Equation 12.2 can be rewritten in terms of area, depth, and media porosity as

$$R_t = \frac{\eta A b}{Q} \tag{12.3}$$

where R_t = retention time (T),

η = porosity (L^3/L^3),

A = area (L^2),

b = depth (L),

Q = inflow rate (L^3/T).

For SF wetlands most of the flow is above the soil surface, with typical depths of 0.15–0.45 m. A portion of the flow is below the soil surface or around plants in the wetland giving a typical porosity in the range of 0.9 to 0.95. Either increasing the wetland area or increasing the depth of flow can increase retention time. However, at full inundation deeper water provides no significant increase in pollutant removal (Kadlec & Knight, 1996). For this reason, the term "hydraulic loading rate" q is often used in surface flow wetlands, which is the inflow rate Q divided by the area A.

Pollutant removal in subsurface flow wetlands is controlled by the development of biofilms on the porous media in addition to the anoxic environment. SSF wetlands have depths of 0.3 to 0.6 m and porosities of 0.3 to 0.5 m^3/m^3.

Another approach used to size treatment wetlands is the mass balance design model or Field Test Method (Kadlec & Knight, 1996). In this model, the wetland area is a function of the desired inlet and outlet pollutant concentrations along with a rate constant describing the pollutant decay relationship. For a particular pollutant, the equation is

$$A = -\frac{Q}{k}\log_e\left(\frac{C_2 - C^*}{C_1 - C^*}\right) \tag{12.4}$$

where A = area (L^2),
Q = inflow (L^3/T),
k = rate constant (L/T),
C_1 = inflow concentration (M/L^3),
C_2 = outflow concentration (M/L^3),
C^* = background concentration (M/L^3).

The background concentration (C^*) represents the concentration of the pollutant that resides in the wetland before inflow is introduced. Values for the rate constant have been derived by Kadlec & Knight (1996) for treatment wetlands, and by CH2M Hill & Payne Engineering (1997) for animal waste treatment wetlands (Table 12-3). Because the rate constant varies with temperature and maturity of the wetland, site-specific or regional rate constants should be developed and used in the design process.

TABLE 12–3 Rate Constants for the Mass Balance Design Model (m/yr)

Pollutant	*SF Wetland*	*SSF Wetland*	*Animal Waste Treatment Wetland*
BOD	34	180	22
Total nitrogen	22	27	14
Nitrate nitrogen	35	50	Not reported
Ammonia nitrogen	18	34	10
Total phosphorus	12	12	8
Fecal coliform	75	95	Not reported
Total suspended solids	1000	Not recommended	21

Example 12.2 Compute the wetland area required to reduce ammonia concentrations of an inflow stream (47 450 m^3/yr) by 50 percent. The inflow concentration to the surface flow wetland was 200 mg/L and the background concentration of ammonia in the wetland was 25 mg/L. Estimate the retention time.

Solution. The outflow concentration is 50 percent of 200 mg/L or 100 mg/L. The rate constant from Table 12-3 for ammonia and SF wetlands is 18 m/yr.

$$A = -\frac{47\,450}{18}\ln\left(\frac{100 - 25}{200 - 25}\right) = 2233 \text{ m}^2$$

If a wetland depth of 0.4 m and a porosity of 0.9 are assumed, the retention time can be computed (Equation 12.3)

$$R_t = \frac{0.9(2233)0.4}{(47\,450/365)} - 6.2 \text{ days}$$

Internet Resources

Listing of wetland sites agreeing to the Ramsar Convention
http://www.ramsar.org

An indication of the wetlands systems in each state of the United States can be found at
http://www.nrcs.usda.gov/technical/NRI/1997/summary_report/table16.html
http://www.nrcs.usda.gov/technical/land/wetlands.html
http://www.nrcs.usda.gov/programs/wrp/

References

American Society of Agricultural Engineers (ASAE). (2001). *Soil and Water Terminology.* S526.1. ASAE Standards. St. Joseph, MI: Author.

Blasko, D. (1997). *The Ramsar Convention Manual: A Guide to the Convention on Wetlands,* 2nd ed. Gland, Switzerland: Ramsar Convention Bureau.

CH2M Hill & Payne Engineering. (1997). *Constructed Wetlands for Livestock Wastewater Management: Literature Review, Database, and Research Synthesis.* Prepared for the EPA's Gulf of Mexico Program, Alabama Soil and Water Conservation Committee, Montgomery, AL.

Cowardin, L. M., V. Carter, F. Golet, & E. T. LaRoe. (1979). *Classification of Wetlands and Deepwater Habitats of the United States.* FWS/OBS-79/31. Office of Biological Services, U.S. Fish and Wildlife Service.

Dahl, T. E. (1990). *Wetlands losses in the United States 1780's to 1980's.* U.S. Department of the Interior, Fish and Wildlife Service, Washington, DC. Jamestown, ND: Northern Prairie Wildlife Research Center Home Page. Online at http://www.npwrc.usgs.gov/resource/othrdata/wetloss/wetloss.htm.

Hammer, D. A. (1997). *Creating Freshwater Wetlands,* 2nd ed. Boca Raton, FL: Lewis Publishers.

Kadlec, R. H., & R. L. Knight. (1996). *Treatment Wetlands.* Boca Raton, FL: Lewis Publishers.

Lyon, J. G. (1993). *Practical Handbook for Wetland Identification and Delineation.* Boca Raton, FL: Lewis Publishers.

Mitsch, W. J., & J. G. Gosselink. (2000). *Wetlands,* 3rd ed. New York: Wiley.

Skaggs, R. W., D. Amatya, R. O. Evans, & J. E. Parsons. (1994). Characterization and evaluation of proposed hydrologic criteria for wetlands. *Journal of Soil and Water Conservation, 49*(5), 501–510.

Problems

12.1 A subsurface flow wetland with a porosity of 0.43 m^3/m^3 will be used in a waste treatment process to treat a waste stream of 28 m^3/d. How large of a treatment wetland is required if the flow depth is 0.5 m and the required retention time is 6 days?

12.2 Nitrate nitrogen will be reduced from a concentration of 48 mg/L to a concentration of 8 mg/L in a surface flow wetland. The inflow stream is 33 400 m^3/yr. The background concentration in the wetland is 4 mg/L. Compute the required size of the surface flow wetland.

CHAPTER 13

Drainage Principles and Surface Drainage

Variable climatic patterns and soil characteristics combine to cause many areas to exhibit either excess or deficit soil water conditions. Few areas can maintain optimal growing conditions without some additional drainage or irrigation. In fact, drainage and irrigation practices have been used for thousands of years to create or enhance productive lands. Some notable examples of large drainage projects are the polders in Holland and the fens in England, which are lowlands reclaimed from the sea. Much of the Corn Belt or upper Midwest in the United States was drained and made productive following the passage of the Swamp Lands Acts of 1849 and 1850. The drainage in the Midwest was instigated by medical professionals, who realized that it reduced malaria before the mosquito was the known cause. Good soil drainage has long been recognized as essential for permanent irrigated agriculture to alleviate salinity problems in areas such as the western United States, Egypt, and India. With the current trend of world population increase, additional land may be needed for production of food and fiber.

Drainage is the removal of excess water from the soil surface (surface drainage) or from the soil profile (subsurface drainage). In most cases, surface drainage represents the easiest and least expensive drainage method. Excess water can be directed to shallow drains or ditches by using the natural topography of the area to remove the water. Some of the most recognizable surface drains are the ditches that drain roads and highways.

Many areas contain a naturally high water table. Although good surface drainage can be used to remove excess surface water, subsurface drainage consisting of a series of drainage ditches and possibly buried perforated pipe may be necessary to lower the water table to a desired position. Poorly drained lands that require subsurface drainage are usually topographically situated so that when drained they may be farmed with little or no erosion hazard. Many soils having poor natural drainage are, when properly drained, rated among the most productive soils in the world.

The benefits from drainage in a humid area growing corn are illustrated in Table 13-1. Subsurface drainage gave more than twice the increase in yields compared with surface drainage; however, surface drainage resulted in a higher benefit–cost ratio because of the much lower investment cost than for subsurface

TABLE 13–1 Drainage Benefits and Costs for Corn in a Silty Clay Soil in Northern Ohio (1990 prices)

	Drainage System		
	Surface Only	*Subsurface Only*	*Surface Plus Subsurface*
Corn yields (kg/ha) (13-year average)	5750	7250	7600
Increased yields (undrained 4400 kg/ha)	1350	2850	3200
Benefits at $0.12/kg	$162	$342	$384
Annual costs/ha[a]	$47	$182	$229
Benefit–cost ratio	3.5	1.9	1.7
Benefits – costs	$115	$160	$155

[a] Estimates do not include fertilizer and other production costs as these would be similar for all drainage systems. Costs were computed assuming 5 percent per year depreciation (20-yr economic life), 10 percent interest on average investment, and $2.00/ha per year for maintenance of subsurface drains; and 10 percent annual interest on investment and $10.00/ha per year for maintenance of surface drainage, only. Initial investment was $370/ha for surface drains and $1800/ha for subsurface drains (12-m spacing, 1 m deep).

Source: Schwab et al. (1993).

drainage. For this reason, enhancements to surface drainage should be made prior to any large investments in subsurface drainage for correcting problems of excess water. Costs were based on 1990 prices, and will vary greatly from field to field, especially for surface drainage. Costs also vary from region to region.

13.1 Drainage Benefits

Soils that require drainage are generally wet during the periods when most planting and harvesting operations occur. Any excess precipitation, combined with the low ET rates during these periods, tends to cause wet soil conditions and poor trafficability. Trafficability is the ability of the soil profile to support wheel traffic with minimal compaction. The inability to till or plant causes planting dates to be delayed, which has a significant effect on the yield of many crops. Late planting can cause more susceptibility of the crop to frost damage later in the season, risk poor rainfall/temperature patterns, and increase weed and insect problems.

Excess water affects plant growth by reducing soil aeration, because plants are not adversely affected even in total water culture if air is provided. Although for design purposes the water table is a convenient term for reference, it is not always satisfactory for explaining the water relationships with respect to the plant root environment. Saturation may occur for some distance above the water table because of the capillary fringe, especially in a clay soil. A better criterion for drainage depth would be the "upper surface of saturation" or a depth based on the oxygen diffusion rate of the soil. Several investigators have found that roots do not generally penetrate to a static water table, but stop growing a short distance above it.

Drainage requirements for optimum plant growth are determined largely by the volume and content of the soil air. These needs depend on type of crop, soil, availability of plant nutrients, climatic conditions, biological activity, and management practices. The rooting depth and the tolerance to excess water are the most important crop characteristics. For example, rice is tolerant because it has special internal mechanisms for obtaining oxygen from the air, whereas tomato is sensitive to excess water. Plants have widely varying oxygen requirements. Evans & Fausey (1999) summarized many laboratory and small plot studies that relate crop yield to excess water conditions.

Soil and air temperatures affect oxygen diffusion rates as well as soil organisms and the biological processes in the plant. Increasing the temperature usually causes a decrease in oxygen and an increase in carbon dioxide concentration because of the increase in respiration rates of roots and organisms. Normally, during the dormant season, plants are not affected by flooding because at low temperatures the biological processes are slow. Drainage influences soil temperature because of the differences in specific heat, thermal conductivity, and evaporation between wet and dry soils. The specific heat of dry soil particles is about 0.8 kJ kg^{-1} $°C^{-1}$ as compared to 4.2 kJ kg^{-1} $°C^{-1}$ for water. Because of the added heat capacity of water, wet soils warm slower than dry soils. The thermal conductivity of dry soil is one third to one half that of water. Evaporation from the surface requires solar energy that would otherwise be available to warm the soil.

Flooding of plant roots causes a rapid reduction in transpiration, reduced absorption of oxygen and other plant nutrients with a corresponding increase in the carbon dioxide content of the soil, a disturbance of microbiological activity, and a reduced mass of soil from which nutrients can be obtained. Changes in the oxygen and carbon dioxide balance may affect the growth of disease organisms. With prolonged flooding and reduced biological activity, soil structure can be destroyed, which in turn reduces aeration. Heavy-textured soils swell and shrink during wetting and drying cycles, having a beneficial effect.

13.2 Environmental Impacts of Drainage

In recent years the major impacts of drainage have been related to the loss of wetlands (Chapter 12) for wildlife habitat and the potential for nonpoint pollution from drainage discharge water. Effects of drainage on water quality of discharge water are of concern in both arid and humid regions. Compared with surface runoff in humid areas, subsurface drainage water generally contains less sediment and smaller amounts of phosphorus, potassium, and pesticides, and has a higher pH (Gilliam et al., 1999). Nitrogen is normally greater in subsurface water because it moves through the soil with the water. Water table management (WTM) techniques (Chapter 14) have been shown to reduce nitrogen discharges from drained lands. In irrigated areas, sodium and trace elements, such as selenium, boron, molybdenum, and arsenic have long been known to have adverse effects on water quality.

The primary concern with the drainage of wetlands is the loss of habitat for wildlife. Federal legislation generally supports the concept that there should be no net loss of wetlands. The greatest controversy is not that wetlands should be drained, but in what constitutes a wetland and how a landowner is to be compensated for wetlands that should be preserved.

The hydrologic impact of subsurface drainage on flood reduction is often not realized. In Ohio, studies showed that peak runoff rates from small, flat plots (0.2 ha) were reduced 7 percent by installing subsurface drains and that the number of floods was reduced by 46 percent. Similar results were obtained from adjacent 32-ha watersheds in North Carolina. These reductions in peak flow are attributed to the increased capacity of the soil to store water, but the effect may not always be consistent for all soil and hydrologic conditions.

Surface Drains

The type of surface drainage selected for individual field areas depends largely on the topography, soil characteristics, crops, and availability of suitable outlets. Surface drains are shallow and must be suitable for mechanized operations on various types of topography, such as pothole areas, flat fields, and gently sloping land. Pothole areas are frequently found in glaciated regions where the topography is relatively flat and where geologic erosion has not had time to develop natural outlets. Flat or level land having impermeable subsoils with shallow topsoil often requires surface drainage because subsurface drains are not practical or economical. Claypan, hardpan, and tight alluvial soils are examples. In irrigated areas, surface drains are needed to remove excess water from surface irrigation. If irrigation is done properly, there should be little surface runoff.

Shallow surface drains cannot remove ground water and cannot give the benefits inherent to good subsurface drainage; however, surface drains are required even though subsurface drains are installed. Since most field crops are able to withstand inundation for only a short period without damage, it is desirable to remove surface water within 12 to 24 h. The quick removal of surface water via surface drains reduces the amount of water that has to be removed by the subsurface drainage system.

Surface drains include (1) bedding, in which tillage equipment is used to form a series of low, narrow ridges separated by parallel, shallow drains called deadfurrows; and (2) field drains, which are shallow, graded channels that are typically formed with construction equipment and are wider spaced than deadfurrows. The bedding system and field drains generally drain to drainage ditches that are much deeper and cannot be crossed with farm machinery.

13.3 Bedding

The design and layout of a bedding system involves the proper spacing of deadfurrows, depth of bed, and grade in the channel (Figure 13-1). The depth and width of beds depend on land slope, drainage characteristics of the soil, and cropping system. Bed widths recommended for the Corn Belt region of the United States vary from 7 to 11 m for very slow internal drainage, 13 to 16 m for slow internal drainage, and 18 to 28 m for fair internal drainage. The length of the beds may vary from 90 to 300 m. In the bedded area the direction of farming operations should be parallel or normal to the deadfurrows. Tillage practices should be parallel with the deadfurrows.

13.4 Random Field Drains

These drains are best suited to the drainage of scattered depressions or potholes (Figure 13-2) where the depth of cut is less than 1 m. In cross section they have a flat "V" or parabolic shape. The design of field drains is similar to the design of grass waterways (Chapter 8).

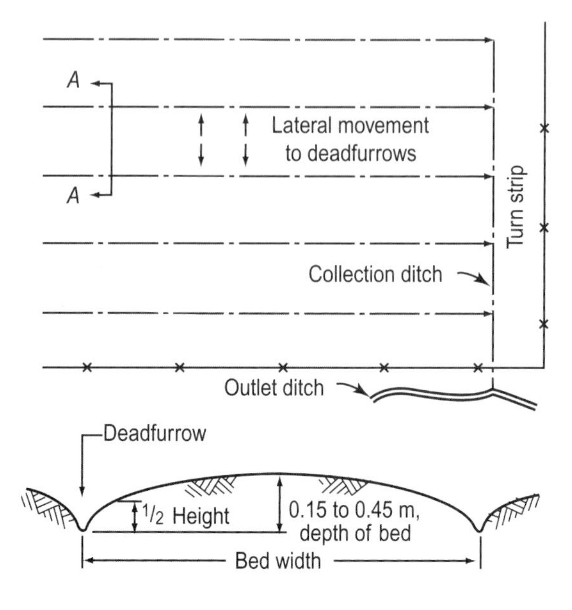

Figure 13–1
Bedding system of surface drainage.

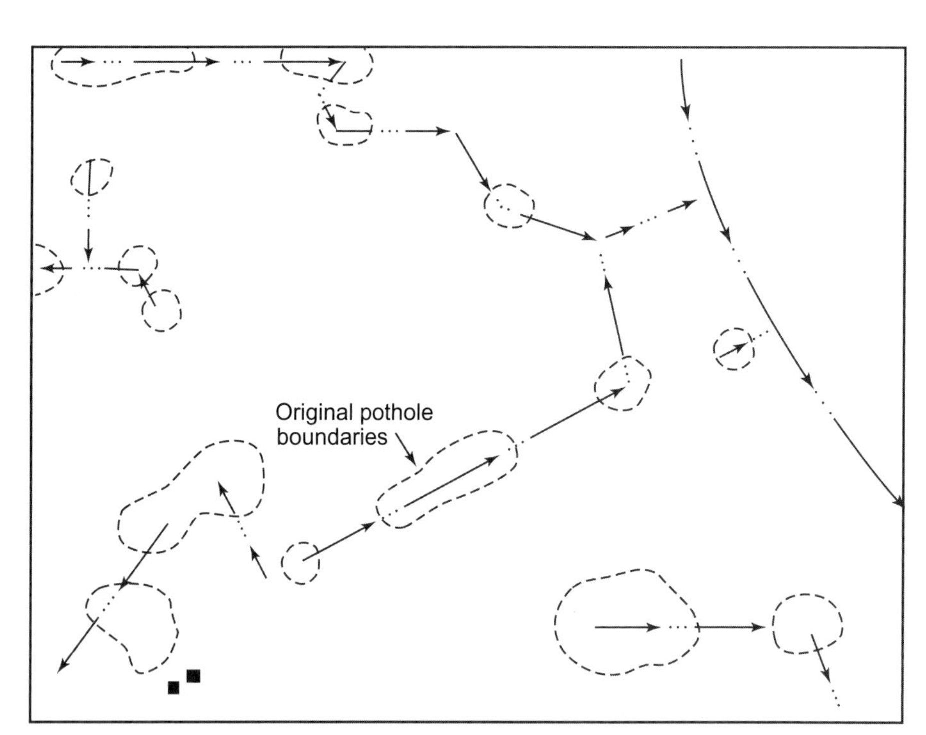

Figure 13–2
A random field drain system in central Iowa.

Where farming operations cross the channel, the sideslopes should be 10:1 or flatter (ASAE, 2003b). Minimum sideslopes of 4:1 are possible if the field is farmed parallel to the field drain. The depth is determined primarily by the topography of the area, outlet conditions, and capacity of the channel. The grade in the channel should be such that the velocity does not cause erosion or sedimentation. The capacity of the drain is usually not considered for areas less than 2 ha provided the minimum design specifications are met; however, where the area is larger than 2 ha, the capacity should be based on a 10-year return period storm, making allowances for minimum infiltration and interception losses.

The layout of a typical random field drain system is shown in Figure 13-2. Normally, the channel should follow a route that provides minimum cut and least interference with farming operations. Where possible, several potholes should be drained with a single drain. The outlet for such a system may be a natural stream, constructed drainage ditch, or protected slope if no suitable ditch is available. Where the outlet is a broad, flat slope, the water should be permitted to spread out on the land below. This type of outlet is practical only if the drainage area is small.

13.5 Parallel Field Drain System

Parallel field drains are similar to bedding (Figure 13-1) except that the channels are spaced farther apart and may have a greater capacity than the deadfurrows. This system is well adapted to flat, poorly drained soils with numerous small depressions that must be filled by land grading.

The design and layout are similar to those for bedding except that drains need not be equally spaced and the water may move in only one direction. The size of the drain may be varied, depending on grade, soil, and drainage area. The depth should be a minimum of 0.2 m and have a minimum cross-sectional area of 0.5 m^2. For trapezoidal cross sections the bottom width should be 2.4 m (ASAE, 2003b). Sideslopes of 8:1 or flatter facilitate crossing with farm machinery. As in bedding, plowing operations must be parallel to the channels, but planting, cultivating, and harvesting are normally perpendicular to them.

Drainage Ditches

For a given watershed area, the required capacity of a drainage ditch is considerably different than the design capacity of a vegetated waterway (Chapter 8). Drainage ditches generally have flatter bed slopes, lower velocities, steeper sideslopes, and a greater depth of channel flow than do vegetated waterways. Vegetated waterways are designed to carry peak runoff, whereas drainage ditches are designed to remove water much more slowly, but still rapidly enough to prevent serious damage to crops on adjacent lands.

Drainage ditches are arbitrarily designed as mains, submains, laterals, and field ditches. The drainage plan should incorporate, as needed, levees, pump installations, field drains, and pipe drains into a coordinated system.

13.6 Parallel Lateral Ditches

The parallel lateral ditch system is similar to the parallel field drain system except that the ditches are deeper to lower the water table and also provide surface drainage. These ditches cannot be crossed with farm machinery. The minimum depth for lateral ditches is 0.3 m deep with sideslopes steeper than 6:1. These ditches are sometimes used to drain peat and muck soils to obtain initial subsidence prior to subsurface drainage. For the same depth to water table, ditches provide the same degree of subsurface drainage as pipe drains.

The design specifications for lateral ditches are given in ASAE (2003b). Maximum recommended spacing for lateral ditches is 200 m in sandy soils, 100 m in mineral soils, and 60 m in organic soils. If the primary purpose of the lateral ditch system is for water table management, steady and unsteady state design equations (Chapter 14) should be used to determine the proper spacing. Since these ditches are too deep to cross with farm machinery, farming operations must be parallel to the ditches.

Discharge of Surface Drains and Drainage Ditches

The rate of conveyance of water by surface drains and drainage ditches is influenced by (1) rainfall rate; (2) size of the drainage area; (3) runoff characteristics including slope, soil, and vegetation; (4) potential productivity of the soil; (5) crops; (6) degree of protection warranted; and (7) frequency and height of flood waters from rivers and streams. Although the degree of protection is important in design, it is one of the most difficult factors to evaluate because costs must be balanced against anticipated flood damage. More frequent flooding is permissible for agricultural and forested land than for home or building sites.

Drains should be capable of removing all excess water from the soil surface within 24 h after rainfall ceases. More rapid removal may be necessary for high-value crops or high-value sites (sports fields, parking lots, etc.). Empirical curves have been developed from numerous field measurements of drainage rates (ASAE, 2003a). The curves are applicable for drainage areas having slopes less than 4.7 m/km and are of the form

$$Q = CA^{5/6} \tag{13.1}$$

where Q = design discharge (m^3/s),
C = coefficient for location and level of drainage (Table 13-2),
A = drainage area (km^2).

TABLE 13–2 Coefficients Relating to Desired Level of Drainage Needed for Regions in the United States

Drainage Desired	*Coefficient*
Corn Belt and northeast cultivated land	0.48
Corn Belt and northeast pasture	0.33
Coastal plain cultivated land	0.59
Coastal plain pasture	0.39
Coastal plain woodland	0.13
Coastal plain and Delta riceland	0.30
Delta cultivated	0.52
Southwestern rangeland	0.20
Red River valley (Minnesota and North Dakota) cultivated	0.26

The design discharge should be computed for each component of the surface drainage system. Local data relating drainage area to discharge may be available for some areas. If the design discharge is to be estimated below the junction of two ditches draining watersheds with different C values, an empirical method called the 20–40 rule should be used as described in the following cases:

(1) If one ditch drains an area less than 20 percent of the total watershed area, base the design discharge on the total area and the C value of the larger area.

(2) If one ditch drains 40–50 percent of the total watershed area, the total design discharge is the sum of the discharges computed with C values for each area.

(3) If one ditch drains 20–40 percent of the area, interpolate between the values obtained from Case 1 and Case 2.

Example 13.1

Determine the combined discharge from a 4.5-km² watershed draining a coastal plain woodland and a 2.5-km² watershed consisting primarily of coastal plain farmland.

Solution. First determine which case is appropriate for combining watershed discharges. The total watershed area is 4.5 + 2.5 = 7 km². The smaller watershed consists of approximately 36 percent of the area (2.5/7 = 0.36). Since this falls between 20 and 40 percent, base the solution on Case 3, which requires the data from Case 1 and Case 2.

Case 1: The C value for the 4.5-km² coastal plain woodland is 0.13.

$$Q = CA^{5/6} = 0.13(7)^{5/6} = 0.66 \text{ m}^3/\text{s}$$

Case 2: The C value for the 2.5 km² coastal plain farmland (cultivated) is 0.59. The total flow is the sum of the flow from the woodland and the cultivated areas.

$$Q = Q_W + Q_C = 0.13(4.5)^{5/6} + 0.59(2.5)^{5/6} = 0.456 + 1.266 = 1.72 \text{ m}^3/\text{s}$$

Case 3: This is an interpolation from Case 1 (20 percent) and Case 2 (40 percent) for the 36 percent area in the watershed.

$$Q = 0.66 + \frac{36 - 20}{40 - 20}(1.72 - 0.66) = 1.51 \text{ m}^3/\text{s}$$

Installation

13.7 Construction and Layout

Since the location of surface drains and ditches requires experience and good judgment combined with a careful study of local conditions, only a few general rules can be given: (1) follow the general direction of natural drainageways, particularly with mains and submains; (2) provide straight channels with gradual curves, especially for large ditches; (3) locate drains along property lines and utility rights-of-way, if practical; (4) make use of natural or existing ditches as much as possible; (5) use the available grade to best advantage, particularly on flat land; and (6) avoid unstable soils and other natural conditions that increase construction and maintenance costs.

The selection of equipment and procedures for construction of surface drains varies with the depth of cut and quantity and distribution of excavated soil. For depths of cut up to 0.75 m, graders, scrapers, and heavier excavators are suitable (Figure 8-11). Construction may require spreading the spoil from both sides of the drain to prevent ponding behind the spoil. For deep cuts over 0.75 m, crawler tractors equipped with push- or pull-back blades and scrapers may be used to fill the pothole area or other depressions near the point of excavation. See Chapter 8 concerning other construction techniques such as the track hoe excavator.

The field layout of random surface drain is shown in Figure 13-3. Grade stakes are usually set every 15 m along the centerline. Slope stakes are placed about 2 m outside the edge of the cut so as not to be removed by construction equipment.

Example 13.2 Determine the volume of cut for the random field drain shown in Figure 13-3.

Solution. The following cuts were determined by instrument survey and were computed for a channel grade of 0.15 percent and a sideslope of 10:1.

Sta. (m)	*Cut (m)[a]*	*One Half Top Width (m)*	*Cross-Sectional Area (m^2)*	*Average Cross-Sectional Area (m^2)*	*Distance (m)*	*Volume of Cut (m^3)*
0 + 00	0	0	0			
				0.27	15	4.05
0 + 15	0.23	2.3	0.53			
				1.07	15	16.05
0 + 30	0.40	4.0	1.60			
				3.62	15	54.30
0 + 45	0.75	7.5	5.63			
				3.74	15	56.10
0 + 60	0.43	4.3	1.85			
				0.93	15	13.95
0 + 75	0	0	0			
					Total Volume	144.45 m^3

[a] If the spoil is not placed in the pothole area, all cuts should be increased by 0.15 m.

13.8 Grades

There is frequently little choice in the selection of grades for open ditches, because the grade is determined largely by the outlet elevation, elevation and distance to the lowest point to be drained, and depth of ditches. Where open ditches drain flat land, the grade should be as steep as possible, provided maximum permissible velocities are not exceeded (Chapters 6 and 8).

The depth at all points along the channel should be sufficient to adequately drain the area. A minimum depth of 1 to 2 m is required where subsurface drains outlet into the ditch. In peat and muck soils the ditch should be made deeper to allow for subsidence. Because of reduced velocities, sedimentation and vegetation

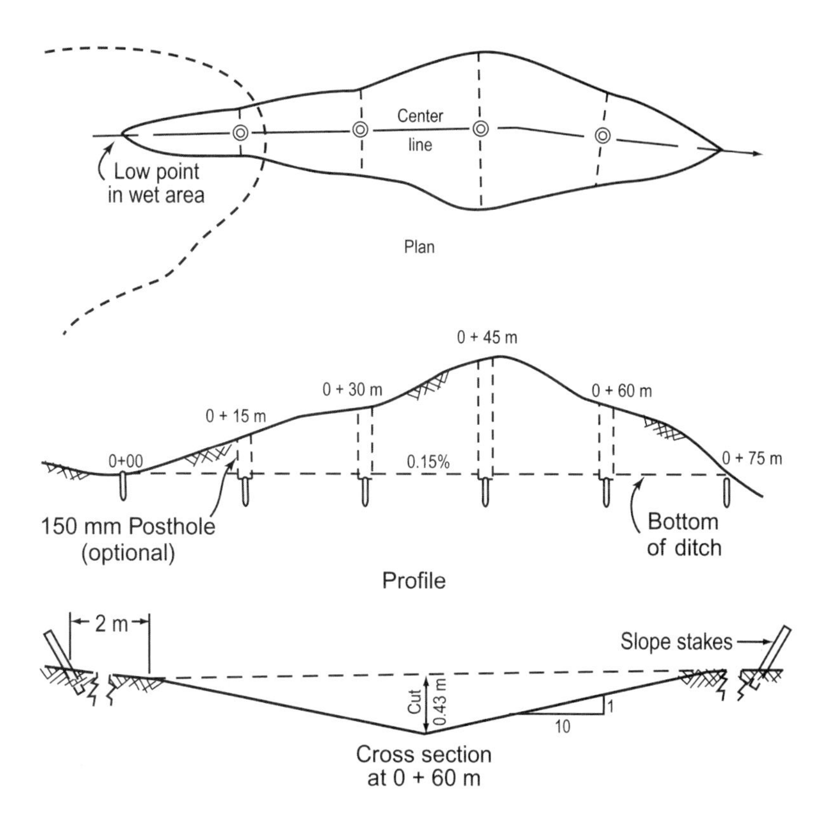

Figure 13–3
Construction layout for a random field drain.

growth may be a problem. In some instances an allowance is made for accumulation of sediment, depending on channel velocities and soil conditions.

13.9 Controlled Drainage Structures

Controlled drainage with open ditches requires structures in the channel to maintain the water at a level to optimize plant growth. Several types of control structures are the burlap bag dam, timber or sheet piling cutoff, reinforced concrete structure, and a combination of culvert and control gate. The elevation views of some of these structures are illustrated in Figure 13-4. Because the structures shown are designed for organic soils, the cutoff wall is larger than required for less permeable soils.

All control dams should be provided with an erosion-controlling apron constructed below the dam as for a drop spillway (Chapter 9). Burlap bag (concrete-filled) dams with a facing of timber are suitable for low heads and small drainage areas (Figure 13-4). Reinforced concrete dams are better suited for handling large quantities of drainage flow. Where seepage is a problem, sheet piling may be required both under and to the sides of the structure. Combination culvert and control gate structures with crestboards at the upper end of the conduit may provide an economical and practical dam. Crestboards are placed in the openings when the water level is to be increased or removed when drainage is required.

13.10 Legal Aspects

Mutual Drainage Enterprises. Many states have laws that provide for the organization of mutual drainage enterprises, also called drainage associations.

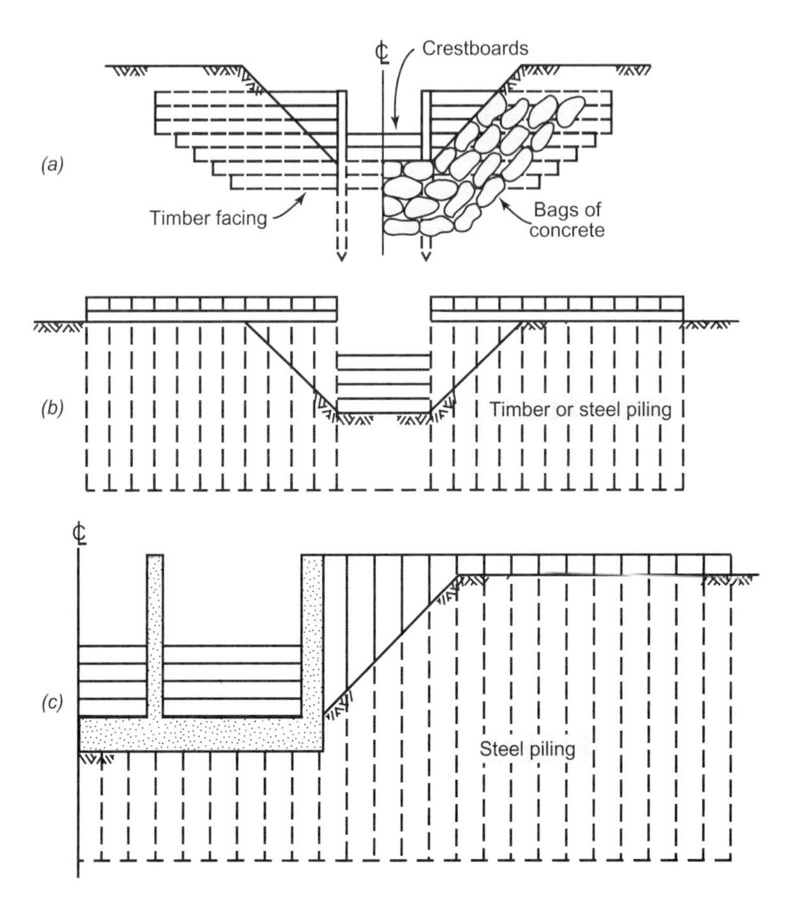

Figure 13–4
Controlled drainage structures for organic soils.

To establish such an enterprise the landowners involved must be fully in accord with the plan of operation and with the apportionment of the cost. After the agreement has been drawn up and signed, it must be properly recorded in the drainage record of the county or other political subdivision. The local court may be asked to name the district officials, sometimes called commissioners, who are responsible for the functioning of the district, or they may be named in the agreement. The principal advantage of the mutual district is that less time is required for establishment and the costs are held to a minimum. Because it may be difficult for several landowners to come to an agreement, particularly on the division of the costs, districts are difficult to organize where the number of landowners is large or where considerable area is involved. Although a mutual enterprise cannot assess taxes, it may petition to become an organized district in some states to overcome this disadvantage.

Organized Drainage Enterprises. An organized drainage enterprise, often called a county ditch or a drainage district, is a local unit of government established under state laws for the purpose of constructing and maintaining satisfactory outlets for the removal of excess surface and subsurface water. It is different from a mutual enterprise in that minority landowners can be compelled to go along with the project. For further details on drainage enterprises consult the laws or publications of the state involved. The method of organization, powers of district officials, and methods of assessing benefits, damages, and financing vary greatly from state to state.

Land Grading

Although *land leveling* is the term generally associated with surface irrigation, *land grading* is synonymous but somewhat more descriptive. For most conditions, a sloping plane surface rather than a level surface is desired.

13.11 Factors Influencing Design

Slopes, cuts, and fills are influenced by the soil, topography, climate, crops to be grown, and method of irrigation or drainage. The major problem with land grading is the effect of removing the topsoil on plant growth. Reduced growth may occur on the fill areas, although the exposure of subsoil in the cuts is usually a more serious problem. Topsoil can be stockpiled and placed over the cut areas, where the cost can be justified.

A uniform design slope is more important for surface irrigation than for drainage. Having a variable slope for drainage is not usually objectionable, provided flow velocities are not erosive. The topography limits the length and degree of slope as well as the location of slope changes. For flood irrigation, the land slope in both directions may be restrictive, whereas for furrow irrigation the length-of-run and furrow grade are the most critical. In semihumid to humid climates, land grading can be made compatible for both drainage and irrigation. For irrigation purposes, the largest flow occurs at the upper end of the slope; for drainage, rainfall enters along the entire slope length with the highest runoff at the lower end. These factors should be carefully considered in design.

Field elevation data can be obtained from a traditional survey (for example, ground elevations taken to the nearest 0.01 m on a 30-m square grid for horizontal control). Recent developments with laser equipment or GPS receivers allow survey data to be obtained more quickly. A person with a truck or all-terrain vehicle equipped with a laser or GPS receiver drives back and forth over the area, and elevations are recorded by computer at selected distance intervals (Carter, 1999). Care is taken to assure that elevations are taken at critical points, such as obvious high and low points, and the water surface in the supply ditch or in the drainage outlet. Aerial photography with stereo plotting is a survey method used for large areas (Carter, 1999). The data can be transferred to a computer for plotting. Computer software systems are available to determine the plane of best fit, a cut-fill map, earthwork volumes, and three-dimensional and profile views of a field. After obtaining the desired balance between cuts and fills, the cut volume and fill volume are computed.

13.12 Plane Method

Most computer programs use the least-squares method to determine the plane of best fit for a field (Scaloppi & Willardson, 1986). This method assumes that the entire area is to be graded to a true plane. The general equation for a plane surface is

$$E = E_o + S_x x + S_y y \tag{13.2}$$

where E = elevation at any point (L),
E_o = elevation at the origin (L),
S_x, S_y = slope in the x and y directions, respectively (L/L),
x, y = distance from the origin (L).

Any plane passing through the centroid will produce equal volumes of cut and fill. The least-squares plane by definition is that which gives the smallest sum of squared differences in elevation between the grid points and the plane. In most applications of the least-squares method, the grid points are assumed to represent a grid square. Since the least-squares plane minimizes the difference between the computed elevation and the elevation of the grid square, the amount of cuts and fills is minimized and the plane is called the plane of best fit.

Although the plane of best fit minimizes the cuts and fills, the design for a field may have to meet certain constraints. For instance, the design may specify a predetermined slope in one or both directions. If fixed design specifications such as edge slopes or control points must be explicitly incorporated in the design, then simplified methods described by Easa (1989a) may be used. In this method, the field elevations are specified for grid intersection points. First, the volume of soil in the field before grading V_g is computed.

$$V_g = \frac{uv}{4}(\Sigma e_c + 2\Sigma e_b + 4\Sigma e_i) \tag{13.3}$$

where V_g = volume of soil before grading (L³),
u = grid spacing in x direction (L),
v = grid spacing in y direction (L),
Σe_c = sum of corner elevations (L),
Σe_b = sum of intermediate boundary elevations (L),
Σe_i = sum of all interior elevations (L).

For a rectangular area, Easa (1989a) showed that for equal cut and fill volumes, the after-grading volume should be equivalent to the before-grading volume (V_g from Equation 13.3):

$$V_g = PE_o + QS_x + RS_y$$

$$P = WL$$

$$Q = \frac{WL^2}{2}$$

$$R = \frac{W^2L}{2}$$

where W is the total length in the y direction and L is the total length in the x direction. For the special case where slopes in the x and y directions are specified

$$E_o = \frac{V_g - (QS_x + RS_y)}{P} \tag{13.5}$$

Once E_o is determined, Equation 13.2 can be used to determine elevations of points across the site. Easa (1989b) presented additional formulas for triangular areas and multiple rectangular areas. In addition to having the slopes specified, the cases of one edge slope and a control point or two control points were also presented. Equations 13.2 through 13.5 are useful for many grading applications where outlet conditions or design slopes have to be met.

Example 13.3

The field shown in Figure 13-5 is to be graded to a slope of 0.3 percent in the x direction. There is to be no slope in the y direction. Determine the equation of the plane to meet these conditions. The grid spacing is 30 m in both directions.

Solution.

(1) The first step in the process is to compute the volume of soil under the ground surface using Equation 13.3 which requires $\sum e_c$, $\sum e_b$, and $\sum e_i$ to be computed.

$$\Sigma e_c = 8.22 + 8.62 + 8.88 + 7.98 = 33.7$$

$$\Sigma e_b = 8.25 + 8.10 + 8.36 + 8.42 + 8.69 + 8.74 + 8.82 + 8.71 + 8.34 + 8.22 + 8.00 + 8.10 = 100.75$$

$$\Sigma e_i = 8.23 + 8.30 + 8.60 + 8.88 + 8.27 + 8.53 + 8.51 + 8.70 = 68.02$$

(2) Apply Equation 13.3 with these data.

$$V_g = \frac{uv}{4}(\Sigma e_c + 2\Sigma e_b + 4\Sigma e_i)$$

$$= \frac{30 \times 30}{4}(33.7 + 2(100.75) + 4(68.02)) = 114\ 138\ \text{m}^3$$

(3) The values for P, Q, and R are needed for Equation 13.5.

13.4

$$P = WL = 90 \times 150 = 13\ 500$$

$$Q = \frac{WL^2}{2} = \frac{90 \times 150^2}{2} = 1\ 012\ 500$$

$$R = \frac{W^2L}{2} = \frac{90^2 \times 150}{2} = 607\ 500$$

(4) Equation 13.5 can be used to compute the elevation of the origin (E_o).

$$E_o = \frac{V_g - (QS_x + RS_y)}{P} = \frac{114\ 138 - 1\ 012\ 500(0.003)}{13\ 500} = 8.23\ \text{m}$$

(5) The equation of the plane, Equation 13.2, is then

$$E = E_o + S_x x + S_y y = 8.23 + 0.003x$$

(6) Elevations at each grid point are then computed to the nearest 0.01 m (shown at the base of Figure 13-5).

13.13 Profile Method

The profile method is appropriate for land grading for surface drainage applications. The resulting field grade will not be uniform for irrigation applications, but will provide a continuous grade to field drains (Sevenhuijsen, 1994). Ground profiles are plotted that are adapted to making changes in design slopes. A grade is established that will provide a desirable ratio of cuts to fills, as well as reduce haul distances to reasonable limits. The design slopes and grade lines are selected by trial and error so the desired cut–fill ratio can be obtained.

13.14 Earthwork Volumes

The average end area and the prismoidal formulas are suitable for making earthwork calculations, but these are time consuming. A more common procedure, called the four-point method by the SCS (1961), is sufficiently accurate for land grading. Volume of cuts for each grid square is

$$V_c = \frac{L^2(\Sigma C)^2}{4(\Sigma C + \Sigma F)} \qquad \textbf{13.6}$$

where V_c = volume of cut (L^3),
L = grid spacing (L),
C = cut on the grid corners (L),
F = fill on the grid corners (L).

For computing V_f the volume of fills, $(\Sigma C)^2$ in the numerator of Equation 13.6 is replaced by $(\Sigma F)^2$. Easa (1989a) presented a three-point method for estimating cuts and fills. The three-point method is applicable to triangular grids and rectangular grids with unequal intervals.

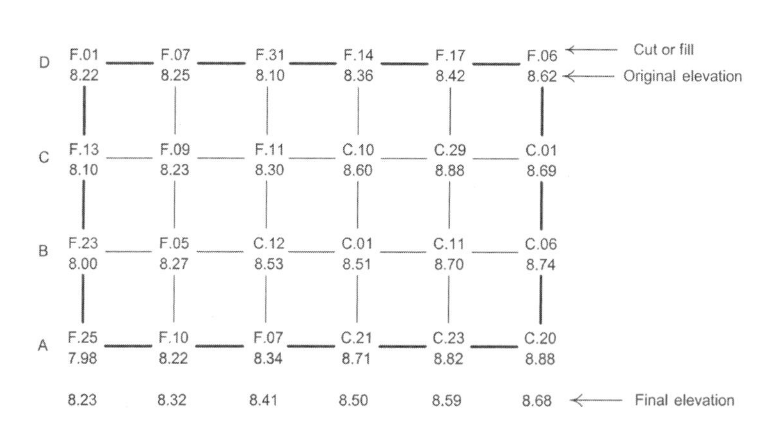

Figure 13–5

Plane method of land grading (all measurements are in meters). The grid spacing is 30 m in both directions. Final elevations are shown for the slope in the x direction. No slope was designed in the y direction.

Experience has shown that, in leveling, the cut–fill ratio should be greater than 1. Compaction from equipment in the cut area, which reduces the volume, and also compaction in the fill area, which increases the fill volume needed, are believed to be the principal reasons for this effect. The cut–fill ratio for volume is usually 1.3 to 1.6, but may range from 1.1 to 2.0. With the plane method of computing cuts and fills, a settlement correction for the whole field is more convenient to apply. The settlement allowance or the amount of lowering of the elevation may range from 0.003 to 0.01 m for compact soils and from 0.015 to 0.05 m for loose soils. A small change in elevation may cause a considerable change in the cut–fill ratio. If extra quantities of soil are needed outside the area to be leveled, such as for a roadway or depression, the plane surface can be lowered by the amount of earthwork required.

Example 13.4 Determine the volumes of cuts and fills and the cut–fill ratio for the field in Figure 13-5 using the four-point method. All elevations are in meters.

Solution. Compute the volumes of fills and cuts using Equation 13.6. Starting in the upper left corner of the figure

$$\Sigma V_f = \frac{30 \times 30}{4}\left[\frac{(0.01 + 0.07 + 0.09 + 0.13)^2}{(0.01 + 0.07 + 0.09 + 0.13)} + \frac{(0.13 + 0.09 + 0.05 + 0.23)^2}{(0.13 + 0.09 + 0.05 + 0.23)} + \cdots + 0\right] = 691 \text{ m}^3$$

$$\Sigma V_c = \frac{30 \times 30}{4}\left[0 + 0 + 0 + 0 + \frac{0.12^2}{(0.09 + 0.11 + 0.12 + 0.05)} + \cdots + \frac{(0.11 + 0.06 + 0.20 + 0.23)^2}{(0.11 + 0.06 + 0.20 + 0.23)}\right] = 691 \text{ m}^3$$

Note that the volume of cuts equals the volume of fills because of the methodology used to determine the plane. Lowering of the plane by 0.01 m will increase the volume of cuts (758 m^3) and decrease the volume of fills (653 m^3). The cut–fill ratio resulting from a lowering of the plane is 1.16.

Internet Resources

Land-leveling software
http://www.engineering.usu.edu. Key search term: land leveling

References

American Society of Agricultural Engineers (ASAE). (2003a). *Agricultural Drainage Outlets—Open Channels*. EP407.1 DEC99. St. Joseph, MI: Author.

———. (2003b). *Design and Construction of Surface Drainage Systems on Farms in Humid Areas*. EP302.4 JUN00. St. Joseph, MI: Author.

Anderson, C. L., A. D. Halderman, H. A. Paul, & E. Rapp. (1980). Land shaping requirements. In M. E. Jensen, ed., *Design and Operation of Farm Irrigation Systems.* ASAE Monograph 3, Chap. 8, pp. 2810–2314. St. Joseph, MI: ASAE.

Carter, C. E. (1999). Surface drainage. In R. W. Skaggs, & J. van Schilfgaarde, eds. *Agricultural Drainage.* Agronomy Monograph 38, pp. 1023–1048. Madison, WI: ASA, CSA, SSSA.

Easa, S. M. (1989a). Direct land grading of irrigation plane surfaces. *ASCE Journal of Irrigation and Drainage Engineering, 115*(2), 285–301.

Easa, S. M. (1989b). Three-point method for estimating cut and fill volumes of land grading. *ASCE Journal of Irrigation and Drainage Engineering, 115*(3), 505–511.

Evans, R. O., & N. R. Fausey. (1999). Effects of inadequate drainage on crop growth and yield In R. W. Skaggs, & J. van Schilfgaarde, eds. *Agricultural Drainage.* Agronomy Monograph 38, pp. 13–54. Madison, WI: ASA, CSA, SSSA.

Gilliam, J. W., J. L. Baker, & K. R. Reddy. (1999). Water quality effects of drainage in humid regions. In R. W. Skaggs, & J. van Schilfgaarde, eds. *Agricultural Drainage.* Agronomy Monograph 38, pp. 801–830. Madison, WI: ASA, CSA, SSSA.

Scaloppi, E. J., & L. S. Willardson. (1986). Practical land grading based on least squares. *ASCE Journal of Irrigation and Drainage Engineering, 112*(2), 99–109.

Schwab, G. O., D. D. Fangmeier, W. J. Elliot, & R. K. Frevert. (1993). *Soil and Water Conservation Engineering,* 4th ed. New York: Wiley.

Sevenhuijsen, R. J. (1994). Surface drainage systems In H. P. Ritzema, ed. *Drainage Principles and Applications,* 2nd ed. Institute for Land Reclamation and Improvement Publication 16, pp. 799–826. The Netherlands: Wageningen.

U.S. Soil Conservation Service (SCS). (1961). Land leveling. In *Irrigation, National Engineering Handbook,* Sect. 15, Chap. 12 (litho.). Washington, DC: U.S. Government Printing Office.

Problems

13.1 Compute the volume of soil to be excavated in cubic meters for a random field drain with 8:1 sideslopes if the cuts at consecutive 15-m stations are 0.15, 0.30, 0.73, 0.55, 0.24, and 0 m. The first station is in a depression where a cut of 0.15 m is needed for adequate drainage.

13.2 Determine the design discharge from 5.6 km^2 of land in the Corn Belt region. Compare the design discharge to the discharge from a similarly sized area in the coastal plain region.

13.3 Determine the equation of the plane for the data shown in Figure 13-5 if there will also be a slope of 0.2 percent in the y direction. Compute the volume of cuts and fills.

CHAPTER

14 Water Table Management

As described in Chapters 12 and 13, poor internal drainage, high water tables, and inadequate surface drainage combine to produce waterlogged soils. Although some of these areas may be classified as wetlands, significant land resources experience wet soil conditions for brief periods during the year that limit their maximum productivity but not to a degree that creates wetland conditions. Increases in world population and the need for food, clothing, and housing will require that additional land resources are available to meet our needs without negatively affecting the environment.

For thousands of years, naturally wet areas have been drained with ditches, canals, and, more recently, buried pipe to develop sites for recreation, construction, and agriculture. The practice of controlling or altering the position of the water table with ditches or subsurface pipes is called water table management (WTM).

Initially, WTM systems were designed for the purpose of removing excess water from poorly drained areas to allow farming activities to occur. Much of the farmland in the midwestern United States (Illinois, Indiana, Iowa, and Ohio) was made productive by drainage of the saturated soils in the region. In the Netherlands, a system of ditches and pumps is used to maintain large areas of drained land for agricultural production. Many other countries have benefited from the use of subsurface drainage to lower the water table in agriculturally productive, but poorly drained, areas. In some cases, WTM systems can be operated in a reverse mode by pumping water into the system to irrigate the crop (Skaggs, 1999b).

Water table control is also needed in irrigated areas such as the Nile Delta in Egypt or the San Joaquin Valley in California where poor natural drainage occurs. To control salinity in arid climates, excess water for leaching has to be added during an irrigation event. The excess water and solutes move downward through the profile and gradually accumulate if good drainage is not available. When the water table approaches the root zone, a WTM system is necessary to remove the excess water and salts (ASAE, 2001).

In addition to the use of WTM principles to alter the water table in agricultural fields, these principles have been used to manage excess water on golf courses, on athletic fields, and around the foundations of buildings. The design of water table management systems includes the determination of the type of system needed;

selection of a suitable outlet; determination of the proper depth and spacing of drains; preliminary layout and arrangement of the drains; determination of the length, grade, and size of drains; design of any special inlet, outlet, or control structures; specification of materials; consideration of any installation requirements; and consideration of water supply constraints. If enough inputs are collected, simulation models such as DRAINMOD have the capability to evaluate the effects of many system variables such as depth and spacing of drains on crop yield, water quality, and salinity (Skaggs, 1999a). An economic analysis can then be conducted to determine the most effective design. As described in Chapter 13, improvements in surface drainage should usually be considered when evaluating WTM systems.

Description of Systems

Water table management systems can be designed to function in a combination of ways depending on the level of control desired (Figure 14-1). Each level of control imposed on the system increases both the cost of the system and the labor requirement.

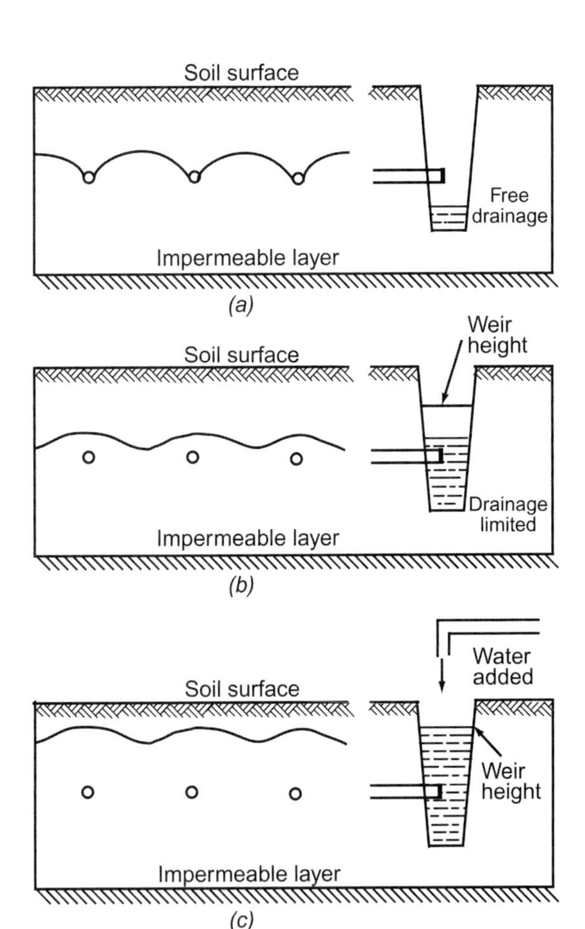

Figure 14–1
Illustrations of the three modes of water table management (a) conventional drainage, (b) controlled drainage, and (c) subirrigation.

14.1 Conventional

Conventional drainage represents the typical WTM system installed to lower the water table in poorly drained soil. The system operates by gravity flow within drains, which can be nonpressurized conduits located below the soil surface or a system of shallow ditches. Water moves through the soil profile to the drains that transport the water to the outlet. The system needs to be checked periodically to ensure that water is flowing freely. Although the shallow ditches may need to be reformed occasionally, the conventional drainage system requires the least management.

14.2 Controlled

The primary need for subsurface drainage is during the planting and harvesting periods when excess water may limit the ability to operate heavy equipment. Drainage may not be needed during other portions of the season. A control structure can be placed in the drainage outlet or along a submain to block the flow of water. The discharge can be controlled during the growing season and at other times (ASAE, 1999b).

During the growing season, the objective of controlled drainage is to limit the discharge of water from the site. Typically, evapotranspiration (ET) is greater than rainfall during the growing season and by limiting the discharge, water is stored in the soil profile. However, water is not pumped back into the drains, so the water table will continue to fall because of ET. The process works as long as sufficient rainfall occurs to replenish any water lost from ET. During periods of excess rainfall, the control mechanism is lowered and the drainage system is allowed to operate in the conventional mode. In some cases, the topography may require multiple control levels.

When crops are not present, the control structure can be used to limit off-season discharge of water and chemicals. Typically, the greatest losses of nitrogen from agricultural fields with improved drainage occur when rainfall exceeds ET (Gilliam et al., 1999). Limiting the amount of discharge reduces the potential for chemicals to leave the site. An added benefit is that the wetter soil conditions created by the controlled water table can enhance denitrification.

A controlled drainage system will require more management than a conventionally operated system. The operator must monitor rainfall conditions and be knowledgeable of the soil water status to decide if, when, and for how long to lower the water table (ASAE, 1999b).

14.3 Subirrigation

If a source of water is available to meet ET demands and the WTM system is designed correctly, water can be pumped into the drains to raise the water table (ASAE, 1999b). Essentially, this is a controlled drainage system with the added capability to provide water to the system. However, unlike the conventional or controlled drainage systems where the goal is to lower the water table, an additional goal for subirrigation is to maintain a water table at a prescribed depth below the soil surface. Because of this, subirrigation systems must be designed to meet ET conditions in addition to the excess water conditions.

Of the three systems, the subirrigation system is the most costly and requires the most management. The operator should check the water table depths over the drain and at the midpoint location between drains at least two times each week to ensure the system is operating correctly (ASAE, 1999b). Having the water table maintained at such a high position may cause some problems. Very little drainable pore space is available between the water table and the soil surface, which means that small rainfall events can completely saturate the profile. The operator has to monitor climatic conditions and begin lowering the water table prior to rainfall events. Since roots from most agricultural crops will not extend into the saturated soil, the shallow water tables created during subirrigation cause a shallow root zone to develop. If irrigation water is unavailable during the growing season and the water table falls, the crop will be more susceptible to drought stresses because of the shallower root zone.

System Layout

The primary reason for installing a WTM system is to alleviate excess water in humid environments. A site visit should be made to understand the level of drainage needed for the field. It is very important to note the topography, available surface drainage, potential outlets, and location of all utility lines. For WTM systems that will be designed for subirrigation, a convenient, plentiful source of good-quality water is needed, because it is difficult to justify the cost of piping water from long distances for typical agricultural crops. Sources of water may include a pond, stream, or well. A typical need is 6–8 mm/d to meet the ET demands (Chapter 4).

14.4 Random

A random layout of drains is used to remove excess water from localized depressions in a field (Figure 14-2a). Typically, surface drainage is adequate and allows most of the field to be easily farmed. Since most of the field is well drained, installing drains under the depressions can decrease expenses. The random drainage system is flexible as well as economical, because the drain lines follow natural draws or other depressions. The primary design considerations are to maintain adequate slope to the outlet and to ensure that adequate cover over the drain is maintained. A random layout is typically used as the drainage system under rolling topography, constructed grass waterways, and the greens of golf courses. Areas that cannot be drained with one drain can include additional laterals, which would be spaced using the principles of parallel systems. Generally, random systems are not suitable for controlled drainage or subirrigation because of the small land area affected.

Another type of random drain is the interceptor drain (or cutoff drain). Depending on the geologic history, hillslopes can represent the outcropping of various rock strata or soil horizons. Perched water tables often develop wet seeps along the hillslope at these intersection points. Drains can be placed across the flow of ground water and installed to intercept and collect subsurface flow before it resurfaces (Figure 14-2d). In some cases, these drains can be designed to collect and remove surface water.

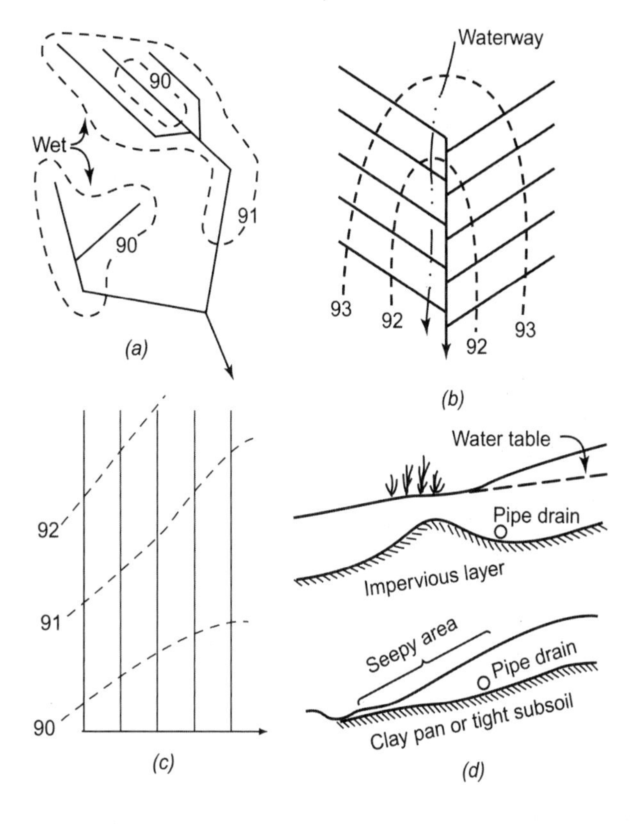

Figure 14–2
Common types of pipe drainage systems: (a) natural or random, (b and c) parallel drains, and (d) cutoff or interceptor drains.

14.5 Parallel

In most cases, the entire field is poorly drained and a system of parallel drains, called laterals, is installed throughout the entire field (Figure 14-2). The laterals are connected to a series of submains and mains that convey water to the outlet. The spacing between the parallel drains is important to the cost of the system and can result in a considerable savings if an optimal design is chosen. In permeable soils where drain spacing is wide, a system of open ditches can be used in lieu of pipes.

Depth and Spacing of Drains

In the design of WTM systems used solely for drainage, both the depth of the drain and the drain spacing must be specified. For good crop and soil management practices, the depth and spacing for normal conditions vary within the ranges given in Table 14-1. Drain depth depends mainly on the characteristics of the soil profile and equipment capabilities (Madramootoo, 1999). Placement of drains deeper than 2.5 m is difficult and costly with current equipment. Drains should be placed in the most permeable soil layer in the 0.9–1.2 m depth range. A minimum cover (depth of soil over the pipe) of 0.6 m should be maintained for tillage considerations and the inability of the pipe to handle the dynamic loads due to traffic (ASAE, 1998a). In nonagricultural areas that do not have appreciable dynamic loading, minimum covers may be reduced to 0.3 m (ASAE, 1998a).

A combination of factors influences the depth and spacing of laterals of a WTM system. These include hydraulic conductivity, drainable porosity, thickness of soil layers, depth to a restrictive layer, outlet conditions, and availability of surface drainage. It is important to visit the field site and identify the various soil layers, textures, and soil series in the field.

The single most important soil property affecting the design of WTM systems is hydraulic conductivity. Although the determination of hydraulic conductivity will be costly and time consuming, the optimum system cannot be designed without accurate knowledge of this value (Madramootoo, 1999). Since hydraulic conductivity is spatially variable, it is important to determine hydraulic conductivity at positions across the site. In the design of large systems where large differences in conductivity may occur, the spacing may be wider or narrower in some regions of the field. From a construction standpoint, however, it is best to maintain uniform drain spacing.

The design discharge of drainage systems is based on a parameter called the drainage coefficient, *DC*, which is defined as the depth of water to be removed in 24 hours by the drainage system. In humid areas where uniform, low-intensity rainfall events are common, the *DC* depends largely on a design rainfall event. In areas where high-intensity, short-duration storms frequently occur, it is difficult to correlate rainfall with the *DC*, because some of the rainfall is removed as surface runoff.

The choice of an appropriate *DC* is based primarily on local conditions, experience, and judgment. Typical *DC* values are presented in Table 14-2. Selection of an appropriate *DC* is important in the design process, as all drain spacing and drain diameters will be based on this value. The *DC* should be selected to remove excess water rapidly enough to prevent serious damage to the crop. Loss of crop will generally occur if the soil profile is allowed to remain saturated for more than 24 hours. For golf courses and athletic fields where it is important to achieve optimal conditions as quickly as possible after or during a rainfall event, a *DC* of 100–200 mm/d is appropriate. For these applications, the soil profile can be constructed to transmit water as quickly as possible to the drains.

A restrictive or impermeable soil layer generally occurs at some depth in the soil profile. The restrictive layer inhibits the vertical transport of water through the soil, and a perched water table forms. The depth to this restrictive layer is a factor in the design process. These depths are generally recorded during the initial soil investigations.

TABLE 14–1 Average Depth and Spacing for Pipe Drains

Soil	*Hydraulic Class*	*Conductivity (mm/h)*	*Spacing (m)*	*Depth (m)*
Clay	Very slow	0.5–1	3–6	0.9–1.1
Clay loam	Slow	1–5	5–10	0.9–1.1
Loam	Mod. slow	5–20	10–25	1.1–1.2
Fine, sandy loam	Moderate	20–65	25–50	1.2–1.4
Sandy loam	Mod. rapid	65–130	50–70	1.2–1.5
Peat and muck	Rapid	130–250	70–100	1.2–1.5

TABLE 14–2 Typical Drainage Coefficients for Drainage Systems

Soil	*No Direct Inlets (mm/d)*	*Blind Inlets (mm/d)*	*Open Inlets (mm/d)*
Mineral			
Field crops	10–13	13–19	13–25
High-value crops	13–19	19–25	25–38
Organic			
Field crops	13–19	19–25	25–38
High-value crops	19–38	38–51	51–102

Source: ASAE (1998a).

14.6 Steady-State Design

One method of designing a WTM system assumes that a uniform rainfall occurs on the soil surface over a long period of time. The WTM system is designed to remove the excess water while maintaining the water table below the crop root zone. Figure 14-3 depicts the water table profile expected between two parallel ditches assuming a constant rainfall rate of R. At section "A-A," the flow of water per unit length along the drain can be computed with Darcy's law (Chapter 5).

$$q_x = -Kh\frac{dh}{dx} \tag{14.1}$$

where q_x = flow rate per unit length along the drain (L^2/T),
K = saturated hydraulic conductivity (L/T),
h = height of the water table (L),
dh/dx = hydraulic gradient (L/L).

Under steady-state conditions, all water entering the soil profile between the midpoint location and section "A-A" must pass through the section while moving toward the drain (negative x direction). This rate of flow is

$$q_x = -R\left(\frac{S}{2} - x\right) \tag{14.2}$$

where R is the rainfall rate (L/T), S is the drain spacing (L), and x is the position of section "A-A." At steady state, all inflows must equal outflows; therefore

$$q_x = -Kh\frac{dh}{dx} = -R\left(\frac{S}{2} - x\right) \tag{14.3}$$

which can also be written

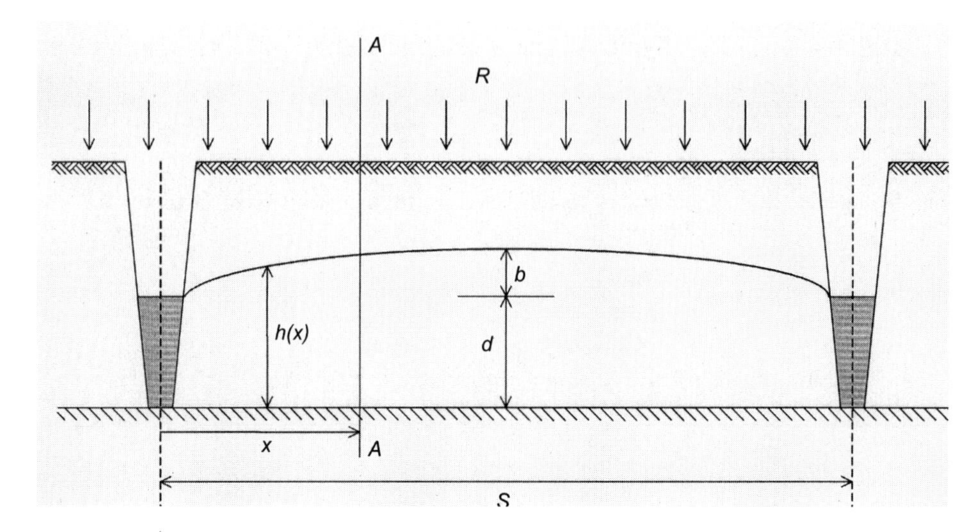

Figure 14–3
Cross section illustration of the design variables for steady-state drainage through parallel ditches.

$$Kh\,dh = R\left(\frac{S}{2} - x\right)dx \tag{14.4}$$

with the limits of integration for $x = 0$, $h = d$ and for $x = S/2$, $h = d + b$. Upon integration,

$$K\frac{h^2}{2}\Bigg|_{d}^{d+b} = \left(\frac{RSx}{2} - \frac{Rx^2}{2}\right)\Bigg|_{0}^{S/2} \tag{14.5}$$

Substitute these values and solve for R, which is also the drain flow rate under steady-state conditions.

$$R = q = \frac{4K[(d + b)^2 - d^2)]}{S^2} \tag{14.6}$$

This equation, which is often called the ellipse equation, was first developed by Colding in 1872. The equation can be reduced, which is generally attributed to Hooghoudt (van der Ploeg et al., 1999).

$$R = q = \frac{8Kdb + 4Kb^2}{S^2}$$

or

$$S = \left[\frac{8Kdb + 4Kb^2}{R}\right]^{1/2} \tag{14.7}$$

The primary limitation to Equation 14.7 is that flow is assumed to move horizontally toward a fully penetrating ditch according to the Dupuit-Forchheimer assumptions. For the case of flow to drainpipes, there is a vertical component to the flow or convergence that must be included. From a functional viewpoint, this is similar to reducing the thickness of the soil profile. These equations, developed by van der Molen & Wesseling (1991), can be used to compute the effective depth.

$$d_e = \frac{\pi S}{8\left(\log_e\left(\frac{S}{\pi r_e}\right) + F(x)\right)} \quad \text{for } x = \frac{2\pi d}{S} \qquad \textbf{14.8a}$$

$$F(x) = 2\sum_{n=1}^{\infty}\log_e[\coth(nx)]$$

$$= \sum_{i=1}^{\infty}\frac{4e^{-2ix}}{i\left(1 - e^{-2ix}\right)} \quad \left(n = 1,2,3,\ldots\right),\left(i = 1,3,5,\ldots\right) \qquad \textbf{14.8b}$$

For $x < 0.5$, $F(x)$ may be closely approximated by

$$F(x) = \frac{\pi^2}{4x} + \log_e\left(\frac{x}{2\pi}\right) \qquad \textbf{14.8c}$$

where r_e is the effective radius for the pipe, accounting for the small fraction of the pipe wall available for water entry. Table 14-3 lists effective radii for some typical drain materials. Equation 14.7 can now be written as

$$S = \left[\frac{8Kd_eb + 4Kb^2}{R}\right]^{1/2} \qquad \textbf{14.9}$$

Note that the drain spacing and effective depth are included in Equations 14.8 and 14.9. This means that an iterative solution must be completed.

In the design process for steady-state analysis, the drainage coefficient is considered to be the design rainfall rate. If, for instance, a *DC* of 13 mm/d is selected for the design, then this is equivalent to designing for a uniform rainfall of 13 mm/d. If the rainfall rate is greater than 13 mm/d, then the water table will rise and may reach the soil surface causing saturated conditions to occur. For most field crops, the water table should not rise to within 0.3 m of the soil surface. Water tables within 0.3 m of the surface limit the aeration in the root zone and cause significant crop stress to occur (Evans & Fausey, 1999). In areas where frequent

TABLE 14-3 Effective Radius of Drains, Considering the Openings for Water Entry

Drain	*Outside Diameter (mm)*	*r_e (mm)*
Corrugated plastic	89	3.5
Corrugated plastic	114	5.1
Corrugated plastic with synthetic envelope or filter	114	40.0
Corrugated plastic	140	10.3
Corrugated plastic	165	14.7
Clay with 1.6-mm joints	127	3.0
Clay with 3.2-mm joints	127	4.8
Corrugated plastic surrounded by a gravel envelope with a square cross section of 2*a* length on each side	2*a*	1.177*a*

high-intensity storms occur during the growing season, a transient analysis should be conducted to assure that the water table falls quickly enough to maintain a well-aerated root zone.

Example 14.1

The field shown in Figure 14-4 can be split into two distinct zones for drainage. The western portion has a deeper profile (total soil thickness of 2.8 m) and higher hydraulic conductivity ($k = 1.0$ m/d). The eastern portion has a shallower soil profile (total soil thickness of 1.8 m) and lower hydraulic conductivity ($k = 0.75$ m/d). What is the required drain spacing to adequately drain the field, assuming steady-state drainage conditions exist? Field crops are to be grown and there are no surface inlets.

Solution. The first step is to choose an appropriate drainage coefficient from Table 14-2. A drainage coefficient of 12 mm/d will be used for mineral soil growing field crops with no surface inlets. To minimize installation costs and maintain sufficient cover over the drain, select a drain depth of 1 m. The depth from the drain to the restrictive layer is 1.8 m (2.8 – 1m) for the western portion (Soil 1).

(1) The drain spacing can be computed from the Hooghoudt equation. If a 0.3 m depth to the water table at the midpoint is provided to allow aeration of plant roots, then the value for b is 0.7 m. Start the iterative procedure by assuming $d_e = d = 1.8$ m.

$$S = \left[\frac{8Kd_eb + 4Kb^2}{R}\right]^{1/2} = \left[\frac{8(1)(1.8)(0.7) + 4(1)(0.7)^2}{0.012}\right]^{1/2} = 31.7 \text{ m}$$

(2) Next, check the estimate for d_e.

$$x = \frac{2\pi d}{S} = \frac{2\pi(1.8)}{31.7} = 0.36 \text{ m}$$

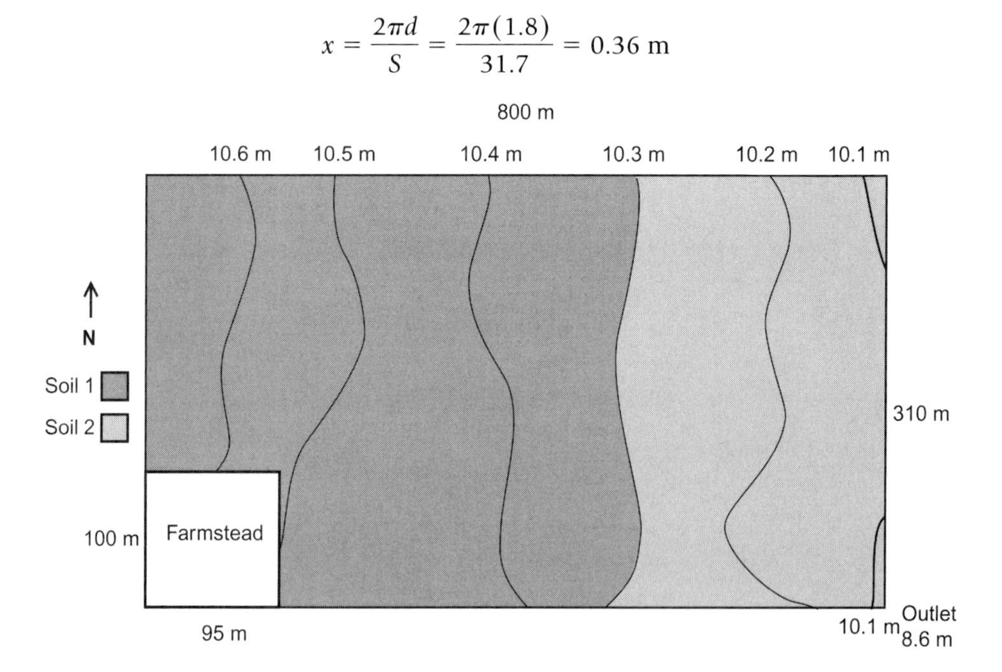

Figure 14–4

Field consisting of two soil types that requires drainage. The outlet is located in the lower right corner of the field.

$$F(x) = \frac{\pi^2}{4x} + \log_e\left(\frac{x}{2\pi}\right) = \frac{\pi^2}{4(0.36)} + \log_e\left(\frac{0.36}{2\pi}\right) = 4.0$$

(3) With an effective radius, r_e, of 5.1 mm from Table 14-3 the effective depth can be computed.

$$d_e = \frac{\pi S}{8\left[\log_e\left(\frac{S}{\pi r_e}\right) + F(x)\right]} = \frac{\pi(31.7)}{8\left[\log_e\left(\frac{31.7}{\pi(0.0051)}\right) + 4.0\right]} = 1.07$$

(4) Now recalculate the drain spacing.

$$S = \left(\frac{8Kd_e b + 4Kb^2}{R}\right)^{1/2} = \left(\frac{8(1)(1.07)(0.7) + 4(1)(0.7)^2}{0.012}\right)^{1/2} = 25.7\text{ m}$$

The next estimate of d_e is 0.98 m, which is similar to the last calculation. The corresponding drain spacing is 24.9 m. Note that the solution converges rapidly and can usually be accomplished in less than three iterations. The d_e and drain spacing for Soil 2 are 0.55 m and 17.8 m, respectively.

14.7 Transient Method

In humid regions where high-intensity, short-duration storms occur, there will be instances when the rainfall rate will far exceed the typical drainage coefficient values used in the steady state analysis. For instance, a 24-h 5-yr storm for Ames, Iowa, is about 100 mm/day (Chapter 3). A system designed for the design storm would be cost prohibitive. For storms of this magnitude, a portion of the rainfall event may be removed with surface drainage if the intensity exceeds infiltration. The water table may rise to the soil surface requiring the drainage system to be designed for a quick fall of the water table.

Mathematical analysis of the falling water table case is far more difficult than for the steady state case. The governing equation for lateral saturated flow is the Boussinesq equation:

$$f\frac{\partial h}{\partial t} = k\frac{\partial}{\partial x}\left[h\frac{\partial h}{\partial x}\right] \qquad \textbf{14.10}$$

The drainable porosity f (L^3/L^3) is the fraction of pore volume that empties as the water table recedes.

Figure 14-5 describes the important design variables in the transient solution for drainage. In the transient case, the drainage coefficient is related to the water table drop over a period of a day. The drainage rate is then the amount of water released as the water table drops

$$DC = f(b_0 - b) \qquad \textbf{14.11}$$

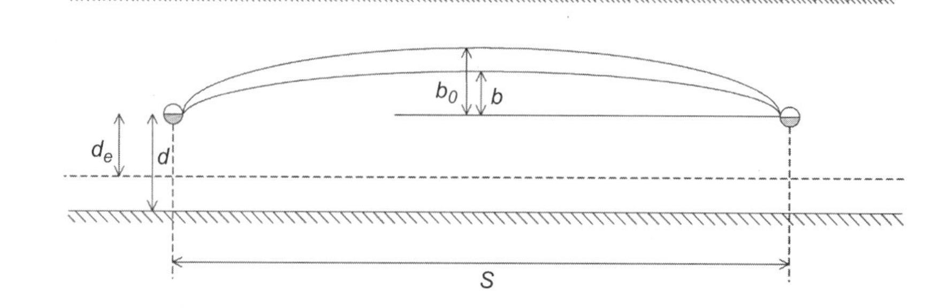

Figure 14–5
Cross section illustration of the design variables for transient drainage through parallel subsurface drains.

where $(b_o - b)$ is the prescribed drop in the water table in 24 hours. Note that the drainable porosity is a smaller fraction than the porosity of the soil because only the pores that drain as the water table falls from b_o to b are included.

van Schilfgaarde (1963) developed a solution to the Boussinesq equation for the falling water table case.

$$S = \left[\frac{9Ktd_e}{f \log_e\left(\frac{b_o[2d_e + b]}{b[2d_e + b_o]} \right)} \right]^{1/2} \quad \textbf{14.12}$$

where the effective depth to the impermeable layer (d_e) has been included to account for convergence losses to the drain.

Example 14.2

For the field identified in Example 14.1, determine the necessary drain spacing to lower the water table from the soil surface to a position 0.3 m below the soil surface in 1 day. Laboratory measurements indicate that the drainable porosity for the upper 0.3 m of the soil profile is 0.04 m^3/m^3.

Solution.

(1) b_o is the distance of the water table above the drain at the beginning of drawdown or 1 m. The value of b is then 1.0 m – 0.3 m = 0.7 m.

$$S = \left[\frac{9Ktd_e}{f \log_e\left(\frac{b_o[2d_e + b]}{b[2d_e + b_o]} \right)} \right]^{1/2} = \left[\frac{9(1)(1)(1.8)}{0.04 \log_e\left(\frac{1[2(1.8) + 0.7]}{0.7[2(1.8) + 1]} \right)} \right]^{1/2} = 37.4 \text{ m}$$

(2) A new estimate of d_e needs to be determined.

$$x = \frac{2\pi d}{S} = \frac{2\pi(1.8)}{37.4} = 0.302$$

$$F(x) = \frac{\pi^2}{4x} + \log_e\left(\frac{x}{2\pi}\right) = \frac{\pi^2}{4(0.302)} + \log_e\left(\frac{0.302}{2\pi}\right) = 5.1$$

(3) From Table 14-3 an effective radius for a 114-mm-OD (outer diameter) corrugated plastic pipe is 5.1 mm. The effective depth to the restrictive layer is

$$d_e = \frac{\pi S}{8\left[\log_e\left(\frac{S}{\pi r_e}\right) + F(x)\right]} = \frac{\pi(37.4)}{8\left[\log_e\left(\frac{37.4}{\pi(0.0051)}\right) + 5.1\right]} = 1.14 \text{ m}$$

(4) Repeat the calculation for drain spacing using the new d_e.

$$S = \left[\frac{9(1)(1)(1.14)}{0.04\log_e\left(\frac{1[2(1.14) + 0.7]}{0.7[2(1.14) + 1]}\right)}\right]^{1/2} = 31.4 \text{ m}$$

Additional iterations give $d_e = 1.05$ m and a drain spacing of 30.5 m. The corresponding d_e and drain spacing for Soil 2 are 0.59 m and 21.8 m, respectively.

Depth and Spacing of Drains for Subirrigation

14.8 Design Parameters

Figure 14-6 shows the water table profile that is expected for subirrigation using either a system of parallel ditches or subsurface pipes. The water level is raised in the outlet structure to a position above the height of the drain. Water flows through the WTM system and into the soil. If there is no resistance to flow, then the water level over the drain should equal the water level in the control structure. Unfortunately, there is resistance of the soil around the subsurface pipes that causes the water level over the drain to always be less than the water level in the control structure. The operator will generally raise the water level in the control structure to offset this resistance.

Note that the water level is maximum over the drain and minimum at the midpoint between drains as water moves from the drain towards the midpoint. The design difference between these depths depends on the crop being irrigated. A value of 0.15 m is typically used to provide a nearly uniform supply of water to field crops. The system is designed for the maximum daily evaporative demand of the crop. For the humid regions of the United States, this evaporative demand is 6–8 mm/d (Chapter 4). If losses to deep seepage or perimeter effects are large, the design rate may need to be increased by 10–15 percent.

14.9 Steady-State Methods

A steady-state analysis is typically used to design subirrigation systems. Ernst (1975) developed the following equation that relates drain spacing to the evaporative demand.

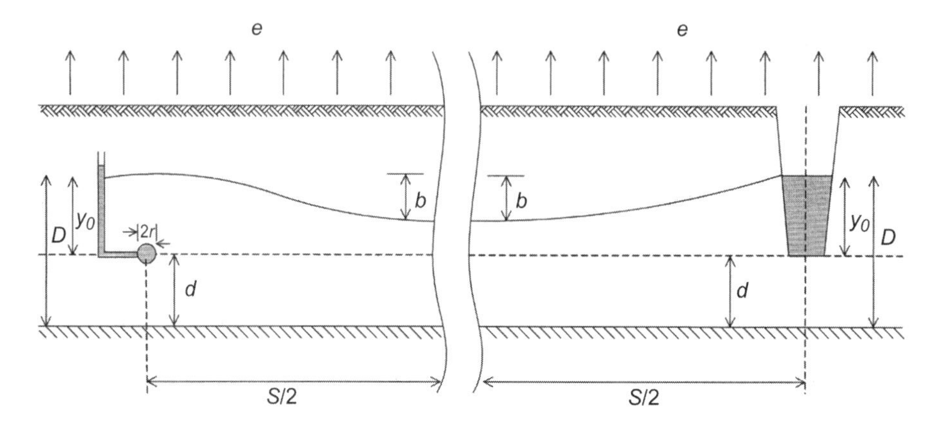

Figure 14–6
Cross section illustration of the design variables for subirrigation through subsurface drains (left) or parallel ditches (right).

$$S = \left[\frac{4Kbh_o\left(2 + \frac{b}{D}\right)}{e}\right]^{1/2} \tag{14.13}$$

where e is the evaporative demand of the crop, $D = y_o + d$, $h_o = y_o + d_e$ and y_o is the pressure head of water in the drain. These values are shown in Figure 14-6.

Note that the Ernst equation and the Hooghoudt equation are similar. Generally, the evaporative demand, e, will be less than the DC, which suggests a wider drain spacing for subirrigation. Unfortunately, the most important condition that determines the drain spacing in this case is the value of b. For the drainage case, the value of b is on the order of 0.7 to 1 m, whereas for the subirrigation case, the value of b is on the order of 0.15 m or less. The combination of these two factors means that a subirrigation system usually has a narrower drain spacing than a conventional drainage system.

Example 14.3

Determine the necessary drain spacing to subirrigate the field described in Example 14.1. The water table should be 0.3 m below the soil surface and should be no greater than 0.45 m below the soil surface at any point. Design for an evaporative demand of 8 mm/d.

Solution.

(1) The drain spacing is computed from the Ernst equation. Initially assume that $d_e = d$.

$$y_o = 1.0 - 0.3 = 0.7 \text{ m}$$

$$h_o = y_o + d_e = 0.7 + 1.8 = 2.5 \text{ m}$$

$$D = y_o + d = 0.7 + 1.8 = 2.5 \text{ m}$$

$$S = \left[\frac{4Kbh_o\left(2 + \frac{b}{D}\right)}{e}\right]^{1/2} = \left[\frac{(4)(1)(0.15)(2.5)\left(2 + \frac{0.15}{2.5}\right)}{0.008}\right]^{1/2} = 19.7 \text{ m}$$

(2) A new estimate of the effective depth is computed.

$$x = \frac{2\pi d}{S} = \frac{2\pi(1.8)}{19.7} = 0.58$$

(3) Since $x > 0.5$, Equation 14.8b must be used to compute $F(x)$.

$$F(x) = \sum_{i=1}^{\infty} \frac{4e^{-2ix}}{i(1 - e^{-2ix})} \qquad (i = 1, 3, 5, \ldots)$$

$$= \frac{4e^{-2(0.58)}}{(1 - e^{-2(0.58)})} + \frac{4e^{-2(3)(0.58)}}{(3)(1 - e^{-2(3)(0.58)})} + \frac{4e^{-2(5)(0.58)}}{5(1 - e^{-2(5)(0.58)})} = 1.9$$

(4) As in the previous examples, a value of 5.1 mm is used for the effective radius of the drain to compute the depth to the restrictive layer.

$$d_e = \frac{\pi S}{8\left[\log_e\left(\frac{S}{\pi r_e}\right) + F(x)\right]} = \frac{\pi(19.7)}{8\left[\log_e\left(\frac{19.7}{\pi(0.0051)}\right) + 1.9\right]} = 0.86 \text{ m}$$

Additional iteration gives $d_e = 0.73$ m and a drain spacing of 14.9 m. The corresponding d_e and drain spacing for Soil 2 are 0.48 m and 11.8 m, respectively.

The following summarizes the drain spacing computed for each of the example problems.

Soil	***Steady-State Drainage Hooghoudt Example 14.1***	***Transient Drainage van Schilfgaarde Example 14.2***	***Subirrigation Ernst Example 14.3***
	—	Spacing (m)	—
Soil 1	24.7	30.5	14.9
Soil 2	17.8	21.8	11.8

For these examples, the transient drainage analysis gave a slightly wider spacing. The low drainable porosity caused the water table to drop quickly. The drain spacing for subirrigation will always be narrower than that required for conventional drainage.

System Design

14.10 Layout

An adequate outlet is needed for the drainage network because the entire WTM system is designed to transport excess water to that location. The outlet must have sufficient capacity. The outlet may be a drainage channel, an existing main, or a stream channel. If the outlet is an open channel, the drainage discharge pipe should be at least 0.3 m above the normal low water flow in the channel (ASAE, 1998a). If an existing pipe is to serve as the outlet, it should be checked for capacity. If it cannot carry the design discharge of the current and proposed systems, then either an additional outlet pipe should be added or a new pipe should be installed to discharge water from both systems. If a sufficient outlet for gravity drainage is not available, then a drainage pump may be needed to remove the excess water (Chapter 19 and ASAE, 1999a).

Once the outlet location is chosen, the next step is to determine the layout of the system in the field. Existing utilities or utility easements need to be considered, as they may alter the layout. Most areas have a utility protection service that will help in locating the utilities.

There are some key characteristics of an efficiently designed system. First, the laterals should be as long as possible (Figure 14-7). The laterals are the smallest drains in the system and are typically installed with a large machine (drain plow or trencher). If there is a mild slope in the field, place the laterals parallel to the contour lines. This allows any surface runoff to be directed toward the drains and maximizes the capability of the drainage system. Figure 14-8 shows laterals placed roughly parallel to the contours to allow a more uniform distribution of water in subirrigation systems and the capture of surface flows. Laterals do not need to extend to the boundaries. Maintain a one half to one quarter drain spacing border around the field.

In addition, the least expensive system will generally have the shortest possible lengths of submains and mains. The submains and mains are the most costly drains in the system. The minimum diameter for a main is 125 mm (ASAE, 1998a). Connections and fittings are an extra expense and should be minimized.

14.11 Design Flow Rate

The design flow rate for the system is related to the drainage coefficient used to compute the drain spacing. Multiplication of the *DC* times the area drained gives a flow rate.

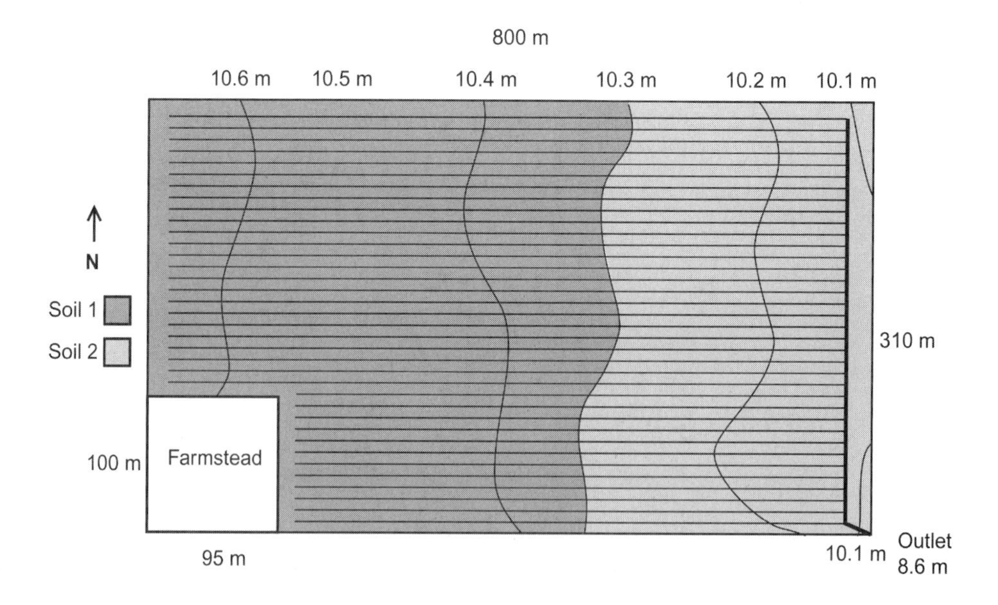

Figure 14–7
Layout of parallel drains that minimizes the length of the main.

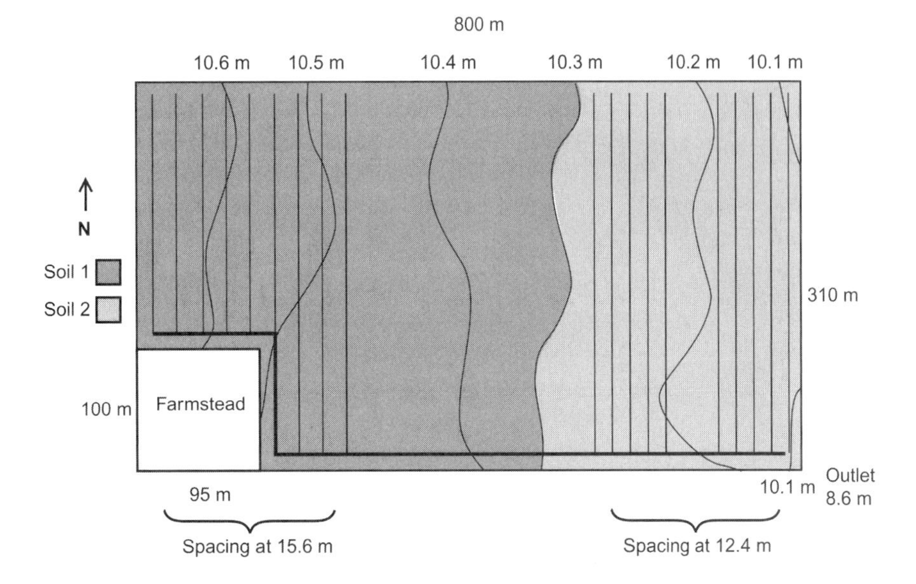

Figure 14–8

Layout of parallel drains that allows drains to be spaced according to the requirement of the soils.

$$Q = DC \times A_d \qquad (14.14)$$

where Q = flow rate (L^3/T),
DC = drainage coefficient (L/T),
A_d = area drained (L^2).

14.12 Grades

Maximum grades are limiting only where pipes are designed for near-maximum capacity or where pipes are placed in unstable soil. It is important to avoid pressure flow in the drainage system, which means that the pipes should always flow partially full. Pressure flow causes water to move out of the pipe and may cause washouts or remove soil from around the drain at high velocities. A pressure relief structure may be required where grade is changed from steep to mild. Subirrigation systems are operated under pressure, however, the water elevation is always less than the soil surface and water velocity in this instance is low.

For nearly level areas, the drain should be as steep as possible while maintaining adequate depth at all locations to reduce the size of mains and submains. As a general guide, the desirable minimum working grade is 0.1 percent. The grade may be reduced to the values presented in Table 14-4. The primary concern with shallow grades is sediment accumulation in the drains. Envelopes may be necessary to limit the transport of sediment into the drains (ASAE, 1998a, 2001). In addition, sediment traps and cleanout systems may be necessary. Where minimum grades are used, a shorter life for the system is expected because of the effects of sedimentation.

TABLE 14–4 Minimum Recommended Grade for Drains (percent)

Inside Diameter of Pipe (mm)	*Pipes NOT Subjected to Fine Sand or Silt*		*Pipes Subjected to Fine Sand or Silt*	
	Clay/Concrete	*Corrugated*	*Clay/Concrete*	*Corrugated*
75	0.08	0.10	0.60	0.81
100	0.05	0.07	0.41	0.55
125	0.04	0.05	0.30	0.41
150	0.03	0.04	0.24	0.32

Source: ASAE (1998a).

14.13 Pipe Size

Pipe drain capacity is the product of the nominal cross-sectional area of the pipe and the velocity of flow. Since the flow is not pressurized, the Manning equation (Chapter 6) can be used to determine the drain diameter.

A nomograph (Figure 14-9) has been developed that relates drain discharge and grade for selecting an appropriate-diameter drain. To use the chart, locate the

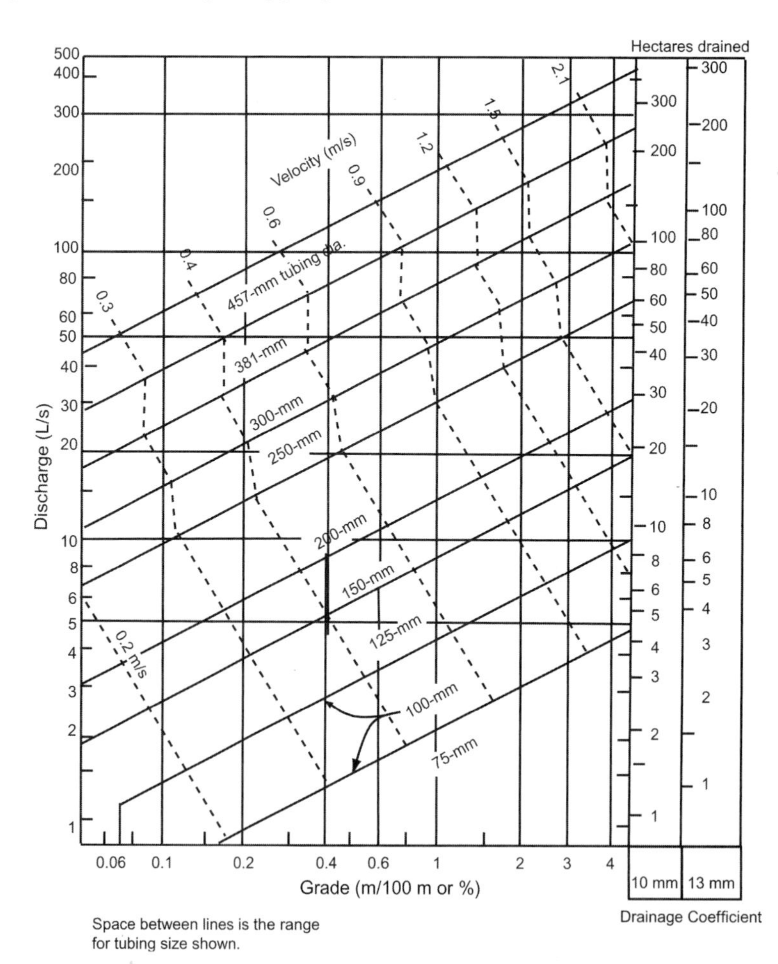

Figure 14–9
Nomograph for determining the size of pipe drains.

design discharge and grade on the vertical and horizontal axes, respectively. Horizontal and vertical lines are drawn from these points to a point of intersection in the chart. The diameter indicated in the space between diagonal solid lines is the required pipe diameter. The diagonal solid lines represent a full pipe flow condition for that drain diameter and grade. The dashed diagonal lines indicate the velocity of flow in the pipe.

Spreadsheets can be constructed to solve the Manning equation using length of lateral, drain spacing DC, and roughness coefficients. For a pipe flowing full, the hydraulic radius is one fourth the pipe diameter. After substitution in the Manning equation and rearranging, the drain diameter can be computed as

$$d = 51.7(DC \times A_d \times n)^{3/8} s^{-3/16} \qquad \textbf{14.15}$$

where d = pipe diameter for a pipe flowing full (mm),
DC = drainage coefficient (mm/d),
A_d = drainage area entering the drain (ha),
n = Manning roughness coefficient,
s = drain slope (m/m).

Example 14.4

Lay out a subirrigation system, and size all laterals and the main for the field in Examples 14.1–14.3.

Solution. The required drain spacing to meet the irrigation needs was computed in Example 14.3. Another constraint of the subirrigation system is the need to maintain a uniform water table elevation over as large an area as possible. To illustrate the calculations, a parallel system with a main extending down the center of the field will be used (Figure 14-10).

(1) Start by sizing the laterals at the furthest point from the outlet. Remember that the laterals do not need to extend to the boundaries. The laterals near the farmstead have lengths of

$$\frac{310}{2} - \frac{14.9}{2} = 147.6 \text{ m on one side of the main}$$

$$\frac{310}{2} - \frac{14.9}{2} - 100 = 47.6 \text{ m on the other side}$$

(2) Although the drains do not extend to the boundaries, they do drain the entire area. The drainage coefficient used for the determination of drain spacing was 12 mm/d. The flow for the longer laterals is

$$Q = DC \times \text{length} \times \text{spacing} = 0.012(155)(14.9) = 27.7 \text{ m}^3\text{/d} \ \ (0.32 \text{ L/s})$$

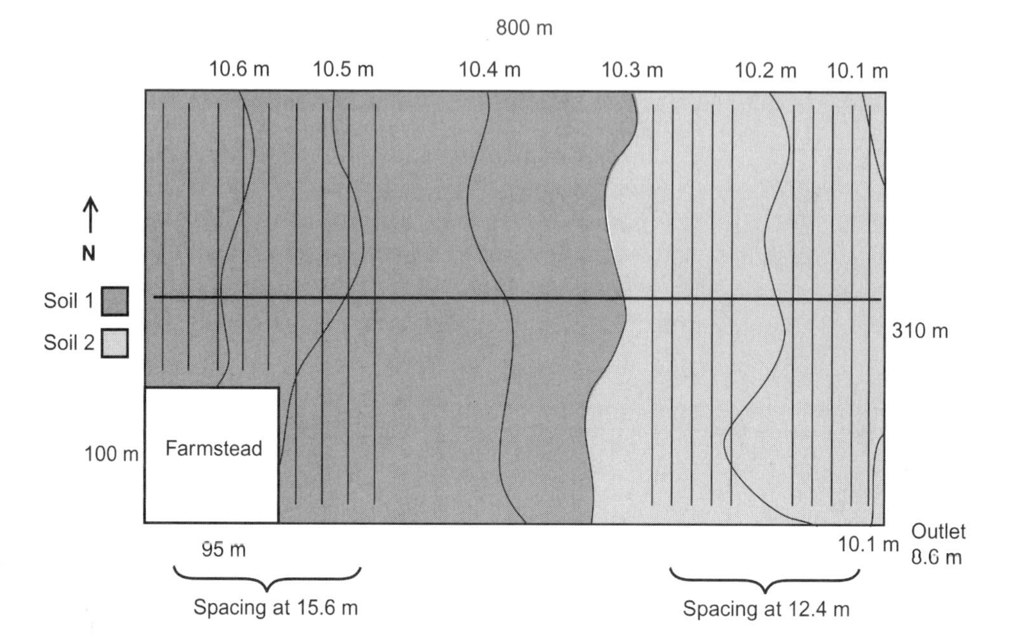

Figure 14–10
Layout of subirrigation system for Example 14.4.

(3) For subirrigation purposes, use the minimum grade possible, so design for a 0.1 percent grade. A 100-mm diameter pipe will easily carry the flow of 0.32 L/s (Figure 14-9). A 100-mm drain can also be used for the shorter section. The reader may want to verify that a 100-mm diameter drain will be adequate for all laterals shown in the figure.

(4) Now the main line can be sized. At the end furthest from the outlet, each junction supplies 0.012(210)(14.9) = 37.5 m^3/d (0.43 L/s) of drainage water. At a grade of 0.1 percent, a 125-mm drain can discharge 2.75 L/s of water. How many junctions can enter the pipe before a larger diameter pipe is needed?

$$\frac{\text{Lateral}}{0.43 \text{ L/s}} \times \frac{2.75 \text{ L/s}}{125\text{-mm drain}} = 6 \frac{\text{laterals}}{125\text{-mm drain}}$$

After 6 laterals, a larger-diameter pipe is needed. A 150-mm diameter pipe will carry 5 L/s when flowing full. There are already 6 junctions or 6(0.43) = 2.58 L/s of water in the main.

(5) Note that the laterals at the seventh junction remove water from 155 m in both directions and 0.64 L/s is added at each of these junctions. Because 5.0 L/s – 2.58 L/s = 2.42 L/s, 3 more junctions can be added.

(6) The next larger pipe is 200 mm, which can carry 10 L/s. The process is repeated until the outlet is reached. In the lower portion of the field, the water added at each junction is

$$0.012(11.8)(310) = 43.9 \text{ m}^3/\text{d} = 0.51 \text{ L/s}$$

(7) At the outlet, an area 310 m × 800 m minus the farmstead (100 m × 95 m) will be drained with a drainage coefficient of 0.012 m/d. The discharge will be 33.1 L/s and a 300-mm diameter pipe is required.

14.14 Drain Envelopes

Since subsurface drains are placed in close contact with the soil using very small grades, any sediment that moves into the drains can potentially clog the system. Problem soils tend to be very fine sandy and silty soils with low clay contents (Stuyt & Willardson, 1999). Risk for clogging depends on the texture, organic matter content, soil cations (calcium and sodium), and soil water content. Soils that tend to clog drains benefit from the addition of a high-permeability envelope of materials or fabric (sometimes improperly referred to as a filter) around the drain to improve the efficiency of the drainage system. A properly designed envelope will allow very small particles to pass through and into the drain system for removal. The primary purpose of the envelope is to stabilize the soil particles around the drain. An advantage of the envelope is a larger effective radius of the drain (Table 14-3).

Natural envelope materials include coarse sand, fine gravel, crushed stone, and, in some cases, topsoil. Natural envelope materials are added around the drainpipe as the pipe is being installed. The primary design criteria relate the D_{15} of the envelope to the D_{15} of the soil or the D_{85} of the soil (Stuyt & Willardson, 1999).

$$\frac{D_{15}(\text{filter})}{D_{15}(\text{soil})} \geq 5 \tag{14.16a}$$

$$\frac{D_{15}(\text{filter})}{D_{85}(\text{soil})} \leq 5 \tag{14.16b}$$

$$\frac{D_{85}(\text{filter})}{\text{Maximum size opening of the pipe}} \geq 2 \tag{14.16c}$$

ASAE (1998a) uses 4 as the criterion for Equation 14.16a, and 7 as the criterion for Equation 14.16b. Added criteria for the filter material are that the material cannot be less than 0.5 mm in diameter, less than 5 percent material can pass a No. 200 sieve, and the maximum diameter shall be less than 35 mm (ASAE, 1998a).

Synthetic envelope materials include fiberglass and plastic geotextile fabrics. The fabrics differ widely in their density, weave, fiber size, thickness, and uniformity. Fabrics are generally wrapped around the drainpipe during pipe manufacture. The ability of a synthetic material to be used as an envelope is expressed as a ratio of a characteristic pore size of the fabric to the particle size of the soil (Stuyt & Willardson, 1999). A widely used criterion is the O_{90} where 90 percent of the fabric pores have a smaller diameter:

$$\frac{O_{90}}{D_{90}}(\text{soil}) \leq \begin{cases} 2.5 \text{ for thin fabrics} \\ 5 \text{ for thick (at least 5 mm) fabrics} \end{cases} \tag{14.17a}$$

Soil bridging across the fabric pores can be a problem. To prevent soil bridging and clogging of the envelope

$$\frac{O_{90}}{D_{90}(\text{soil})} \geq 1.25 \tag{14.17b}$$

Accessories

Accessories for WTM systems include surface inlets, blind inlets, sedimentation basins, and control structures.

14.15 Surface Inlets

As shown in Figure 14-11, a surface inlet, sometimes called an open inlet structure, is used to remove surface water from potholes, road ditches, or other depressions. Whenever practical, surface water should be removed with surface drains rather than surface inlets to subsurface drains. Surface inlets increase the volume of water that enters the drainage system and can add debris and sediment to the drainage system. As indicated by Table 14-2, the addition of surface inlets in a drainage system increases the design drainage coefficient.

Surface inlets should be placed at the lowest point along fence rows or in land that is in permanent vegetation. Where the inlet is in a cultivated field, the area immediately around the intake should be kept in grass to minimize sediment entry. The surface inlet should be constructed of nonperforated pipe and extended at least 2 m as a short lateral on either side of the main line. At the ground surface, a concrete collar should extend around the intake to prevent the growth of vegetation and to hold the riser pipe in place. On top of the riser a beehive cover or other suitable grate is necessary to prevent trash and vermin from entering the WTM system.

14.16 Blind Inlet or French Drain

Where the quantity of surface water to be removed is small or the amount of sediment is too great to permit surface inlets to be installed, blind inlets may at least temporarily improve drainage. Although these inlets often do not function satisfactorily for more than a few years, they are economical to install and do not interfere with surface operations.

As shown in Figure 14-12, a blind inlet is constructed by backfilling the trench with various gradations of material. The coarsest material is placed immediately

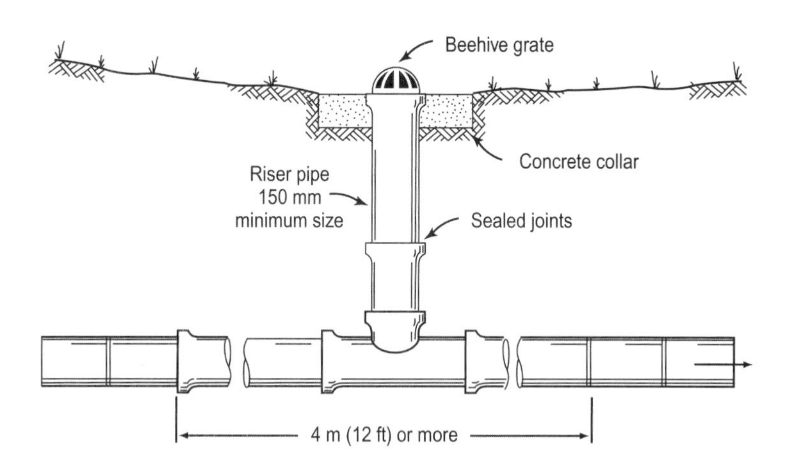

Figure 14–11
Surface inlet for a pipe drain.

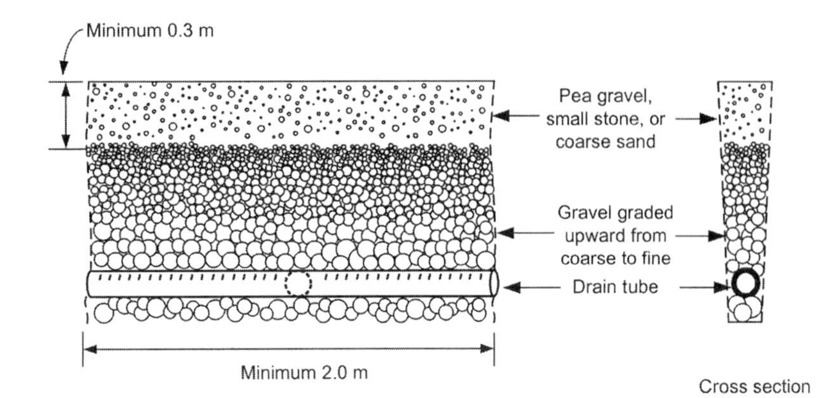

Figure 14–12
Blind inlet or French drain.

over the drain and the size is gradually decreased toward the surface. These inlets seldom meet the design criteria for drain envelopes and for that reason are not permanently effective. Because the soil surface has a tendency to seal, the area should be kept in grass or permanent vegetation.

14.17 Sediment Traps

Soils containing large quantities of fine sand frequently cause sedimentation, because the particles enter the drains through the openings. A sediment trap is any type of structure that provides for sediment accumulation, thus reducing deposition in the drain.

A sediment trap may be desirable where the slope downstream is greatly reduced, where several laterals join the main at one point, or where surface water enters. The structure shown in Figure 14-13 has a turned-down elbow on the outlet

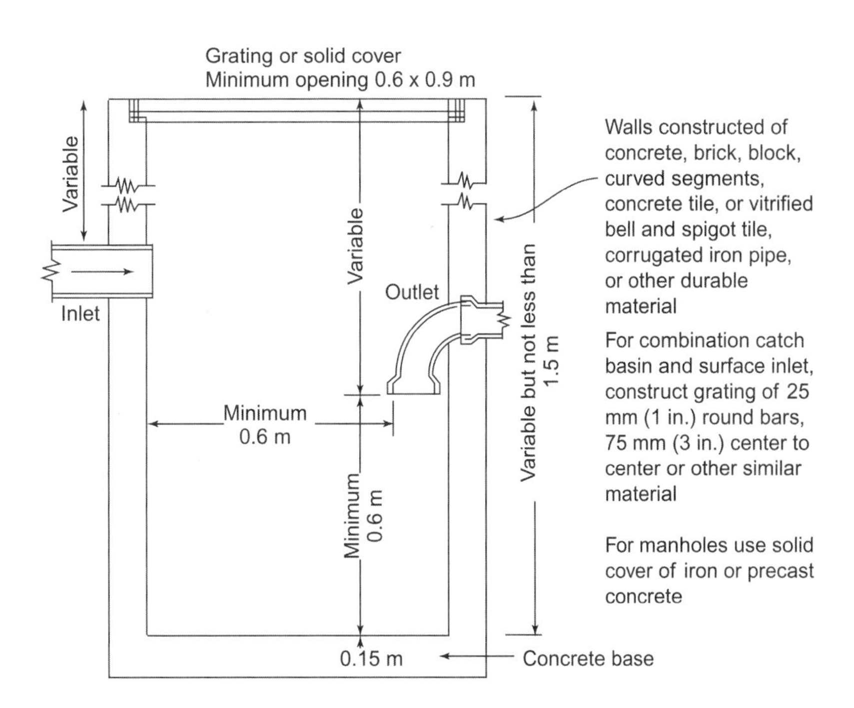

Figure 14–13
Sediment trap for pipe drains.

for retarding the outflow when the basin becomes filled with sediment and to reduce the floating trash entering the drain. Where structures do not have this feature, the tendency is to neglect the cleaning of the basin.

Junction boxes may be installed where several lines join at different elevations. Except for the catch basin, they are similar to the structure shown in Figure 14-13. To facilitate surface operations, the top of the manhole should be placed 0.3 m or more below the surface.

14.18 Control Structures

Control structures are placed in the WTM system to maintain the water table at a specified level. The functioning of such structures is similar to that for control dams in open ditches described in Chapter 13. Controlled drainage is essential in the management of organic soils because of its effectiveness in controlling subsidence. Commercial water level control structures and valves for pipe drains are available for subirrigation or controlled drainage.

Installation

14.19 Machinery

Laterals are generally installed with a laser-controlled drain plow or trencher (Broughton & Fouss, 1999). The drain plow became practical with the development of corrugated plastic pipe. Drain plows are normally designed to install drains less than 150 mm in diameter. Larger-diameter drains are typically installed with a chain trencher, wheel trencher, or backhoe.

The installation starts at the outlet to allow all excess water to flow to a free outlet, so a trencher or backhoe is initially used to place the mains in the field. The main is the largest pipe used in the system, which is typically larger than most drain plows can accept. The drain plow or trencher is then used to install the laterals. Table 14-5 gives an indication of the relative speed of installation by the different drainage machines. The handling of the flexible pipe is important, as stretch above about 5 percent is detrimental to the strength of the pipe. Pulling force as well as bending of the pipe may introduce stretch. Pipe placed with feeders are often used to reduce the stretch.

TABLE 14-5 Rates of Installation of Subsurface Drainpipe

Machine Type	*Pipe-Laying Rate (m/h)*
Backhoe	12–24
Wheel trencher	60–150
Chain trencher	100–200
Drain plow	400–1250

Source: Broughton and Fouss (1999).

14.20 Grade Control

Drains should be laid on true grade; however, this is rarely possible. Drainage machines must be capable of excavating a trench or laying drainage pipe to a grade that does not deviate from the grade established for the drainage work plan by more than half the pipe diameter and no greater than 50 mm. The allowable grade deviations should be checked at intervals of 10 m (van Zeijts & Zijlstra, 1999).

The bottom of the trench should be shaped with a supporting groove to provide good alignment and bottom support. Laser or manual methods may be used to establish grade. A laser system consists of a rotating laser beam that can give either a level or a sloping plane of reference. A receiver on the trenching machine or drain plow picks up the beam and automatically adjusts the machine to follow the laser beam. There are attachments to the laser system that can be used to produce a different grade than the rotating laser beam grade. For hand trenching, or for machines not equipped with an automatic grade-control system, a sight or string line is a simple grade-control method (Broughton & Fouss, 1999).

14.21 Junctions

A properly made junction is an essential part of the drain system. Manufactured fittings should be used to make all connections in the field (ASAE, 1998b). A backhoe is generally used to open the trench to allow personnel to access the junction site. The bottom of the trench is very important to the support of the flexible conduit. The bottom should be shaped to support the pipe or the bedding compacted along the pipe. Manufactured fittings should be used when changing pipe size. End caps or plugs are important for preventing soil from entering the end of the lateral.

14.22 Documentation

A suitable map should be made of the drainage system and filed with the deed to the property. Global Position Systems (GPS) and Geographic Information Systems (GIS) technologies are encouraged for the correct identification of the drainage system. Much time, effort, and money have been wasted because the location of old lines was not known. A record of the drainage system is also of great value to present and future owners for planning new lines, maintenance, and repair. The drainage system may be mapped by aerial photography, which shows greater detail than a sketch. Aerial photos should be taken shortly after installation. Aerial photos and satellite imagery can be used to locate existing, unmapped drainage systems because drier soil or differences in vegetation can be identified over the drains. Drier soil is evident in a drained field soon after a soaking rain.

Conduit Loads

Loads on underground conduits include those caused by the weight of the soil and by concentrated loads resulting from the passage of equipment or vehicles. At shallow depths, concentrated loads from field machinery largely determine the strength requirements of conduits; at greater depths the load due to the soil is the most significant. Loads for both conditions should be determined, particularly where the

TABLE 14–6 Recommended Maximum Trench Depths (m) for Plastic Pipe Buried in Loose, Fine-Textured Soil

Pipe Diameter (mm)	*Pipe Quality*	*Trench Width at Top of Pipe (m)*				
		0.2	*0.3*	*0.4*	*0.6*	*0.8 or greater*
100	Standard	[a]	3.9	2.1	1.7	1.6
	Heavy duty	[a]	[a]	3.0	2.1	1.9
150	Standard	[a]	3.1	2.1	1.7	1.6
	Heavy duty	[a]	[a]	2.9	2.0	1.9
200	Standard		[a]	2.2	1.7	1.6
	Heavy duty		[a]	3.0	2.1	1.9
250			[a]	2.8	2.0	1.9
300					2.0	1.9
380					2.1	1.9

[a] Any depth is permissible at this width or less. Minimum side clearance between the pipe and trench should be about 0.08 m.

Source: ASAE (1998a).

depth of the controlling load is not known. The supporting shape of the bottom trench is the controlling factor in the life of the pipe for both rigid and flexible pipe. The weight of the soil usually determines the load for drains. Maximum trench depths for the installation of corrugated plastic pipe buried in loose, fine-textured soil are presented in Table 14-6. Because the maximum depth is a function of soil type and density, corrugation design and pipe stiffness, pipe manufacturer, and installation techniques; depths approaching the recommended maximums in Table 14-6 should be checked with formulas presented in ASAE (1998a). For many installation and drainage conditions, the effects of surface loads on agricultural subsurface drains, particularly small-diameter lateral drains, can be neglected. Generally, additional loads caused by surface traffic are small where lateral drains have a 0.7 m or more of cover (Schwab et al., 1993).

Internet Resources

University of Minnesota Drainage site
http://d-outlet.coafes.umn.edu

USDA-ARS Soil Drainage Research Unit
http://www.ag.ohio-state.edu. Key search term: ARS Drainage Research

References

American Society of Agricultural Engineers (ASAE). (1998a). *Design of Subsurface Drains in Humid Areas.* EP480 MAR98. St. Joseph, MI: Author.

———.(1998b). *Construction of Subsurface Drains in Humid Areas.* EP481 MAR98. St. Joseph, MI: Author.

———.(1999a). *Design of Agricultural Drainage Pumping Plants.* EP369.1 DEC99. St. Joseph, MI: Author.

———.(1999b). *Design, Installation and Operation of Water Table Management Systems for Subirrigation/Controlled Drainage in Humid Regions.* EP479 DEC99. St. Joseph, MI: Author.

———.(2001). *Design, Construction, and Maintenance of Subsurface Drains in Arid and Semi-arid Areas.* EP463.1 DEC01. St. Joseph, MI: Author.

Broughton, R. S., & J. L. Fouss (1999). Subsurface drainage installation machinery and methods. In R. W. Skaggs, & J. van Schilfgaarde, eds., *Agricultural Drainage,* pp. 927–966. Agronomy Monograph 38. Madison, WI: ASA, CSSA, SSSA.

Ernst, L. F. (1975). Formula for groundwater flow in areas with subirrigation by means of open conduits with a raised water level. Misc. Reprint 178. Wageningen, The Netherlands: Institute of Land Water Management Research.

Evans, R. O., & N. R. Fausey. (1999). Effects of inadequate drainage on crop growth and yield. In R. W. Skaggs & J. van Schilfgaarde, eds., *Agricultural Drainage,* pp. 13–54. Agronomy Monograph 38. Madison, WI: ASA, CSSA, SSSA.

Gilliam, J. W., J. L. Baker, & K. R. Reddy. (1999). Water quality effects of drainage in humid regions. In R. W. Skaggs & J. van Schilfgaarde, eds., *Agricultural Drainage.* pp. 801–830. Agronomy Monograph 38. Madison, WI: ASA, CSSA, SSSA.

Madramootoo, C. A. (1999). Planning and design of drainage systems. In R. W. Skaggs & J. van Schilfgaarde, eds., *Agricultural Drainage,* pp. 871–892. Agronomy Monograph 38. Madison, WI: ASA, CSSA, SSSA.

Schwab, G.O., D.D. Fangmeier, W. J. Elliot, & R.K.Frevert (1993). *Soil and Water Conservation Engineering,* 4th ed. New York: Wiley.

Skaggs, R. W. (1999a). Drainage simulation models. In R. W. Skaggs & J. van Schilfgaarde, eds., *Agricultural Drainage,* pp. 469–500. Agronomy Monograph 38. Madison, WI: ASA, CSSA, SSSA.

———. (1999b). Water table management: Subirrigation and controlled drainage. In R. W. Skaggs & J. van Schilfgaarde, eds., *Agricultural Drainage,* pp. 695–718. Agronomy Monograph 38. Madison, WI: ASA, CSSA, SSSA.

Stuyt, L. C. P. M., & L. S. Willardson. (1999). Drain envelopes. In R. W. Skaggs & J. van Schilfgaarde, eds., *Agricultural Drainage,* pp. 927–1004. Agronomy Monograph 38. Madison, WI: ASA, CSSA, SSSA.

van der Ploeg, R. R., R. Horton, & D. Kirkham. (1999). Steady flow to drains and wells. In R. W. Skaggs & J. van Schilfgaarde, eds., *Agricultural Drainage,* pp. 213–264. Agronomy Monograph 38. Madison, WI: ASA, CSSA, SSSA.

van der Molen, W. H., & J. Wesseling. (1991). A solution in closed form and a series solution to replace the tables for thickness of the equivalent layer in Hooghoudt's drain spacing formula. *Agricultural Water Management, 19,* 1–16.

van Schilfgaarde, J. (1963). Tile drainage design procedure for falling water tables. *American Society for Civil Engineering Proceedings,* 89, (No. IR) 2:1–13 June and discussion December 1963, March and December 1964.

van Zeijts, T. E. J., & G. Zijlstra. (1999). Quality assurance and control. In R. W. Skaggs & J. van Schilfgaarde, eds., *Agricultural Drainage,* pp. 1005–1022. Agronomy Monograph 38. Madison, WI: ASA, CSSA, SSSA.

Problems

14.1 Calculate the discharge of a 150-mm diameter corrugated plastic pipe flowing full if the slope is 0.4 percent.

14.2 Calculate the total area that could be drained with a 254-mm pipe with a slope of 0.3 percent if the pipe is installed in mineral soil in your area and field crops are to be grown.

14.3 What size of pipe is required to remove surface water with a surface inlet if the runoff accumulates from 8.0 ha and the slope in the drain is 0.4 percent? Design for field crops in your area.

14.4 What slope is required to provide a velocity of 0.5 m/s at full flow in a 100-mm corrugated plastic pipe? In a 300-mm pipe?

14.5 A drainage system flows full for 3 days following a period of heavy rainfall over a 20-ha field. Compute the discharge volume during this period if the system was designed for a drainage coefficient of 13 mm/d.

14.6 A drainage design calls for 300-mm pipe on a grade of 0.4 percent to carry 30 L/s of flow. The landowner has some extra 150-mm diameter pipe and does not want to purchase the more expensive 300-mm diameter pipe. Determine the number of 150-mm pipes that would be required to carry the flow.

14.7 Compute the drain spacing for a clay loam soil having a drainable porosity of 0.06 m^3/m^3, a hydraulic conductivity of 0.6 m/d, and an impermeable layer 2.4 m below the drains. Drainage requirements for the crop are such that the initial water table depth below the surface is 0.15 m, with a rate of drop of 0.21 m for the first day. Because of outlet depth, the maximum depth of the drain is limited to 1.2 m measured to the bottom of the 125-mm drains.

14.8 Determine the optimum drain spacing for a subirrigation system using the data given in Problem 14.7. Maintain a water table between 0.3 and 0.45 m from the surface with an evaporation demand of 7 mm/d.

14.9 Determine the pipe size at the outlet of a 20-ha field if the grade is 0.1 percent, the soil drainable porosity is 0.03 m^3/m^3, and the optimum rate of drop of the water table is 0.15 m in the first 12 h following a heavy rainfall.

CHAPTER

15 Irrigation Principles

Human dependence on irrigation can be traced to prehistoric times. Irrigation in very early times was practiced in Egypt, Asia, and the Americas. For the most part, water supplies were available to these people only during periods of heavy runoff. Modern practices of irrigation have been made possible by deep well pumps and by the storage of large quantities of water in reservoirs. Thus, by using either underground or surface reservoirs, it is now possible to bridge over the years and provide consistent water supplies.

Increasing demands for water, limited availability, and concerns about water quality, make effective use of water essential. Because irrigation is a major water user, it is essential that irrigation systems be planned, designed, and operated efficiently. This requires a thorough understanding of the relationships among crops, soils, water supply, and system capabilities.

Plant Water Needs

The water requirement and time of maximum demand vary with different crops. Although growing crops are continuously using water, the rate of evapotranspiration depends on the kind of crop, the degree of maturity, soil characteristics, and atmospheric conditions, such as radiation, temperature, wind, and humidity (see Chapter 4). Where sufficient water is available, the soil water content should be maintained for optimum growth and yield. The rate of growth with different soil water contents varies with different soils and crops. Some crops are able to withstand drought or high water contents better than others. During the early stages of growth the water needs are generally low but increase rapidly during the maximum growing period to the fruiting stage. During the later stages of maturity, water use decreases and irrigation is usually discontinued while the crops are ripening. Water stress during late growing and early flowering stages are most likely to decrease yields.

The water needs of landscape plants are typically lower than for agricultural crops because maximum yield is not required. Landscape plants are irrigated to maintain health, appearance, and reasonable growth (Chapter 4).

15.1 Crop Water Requirements

To make maximum use of available water, the irrigator must have a knowledge of the seasonal water requirements of crops at all growth stages. The seasonal requirement is necessary to select the crop to meet the water supply. Table 15-1 illustrates seasonal evapotranspiration and water requirements for crops grown near Deming, New Mexico. It should be noted that effective rainfall is considered in determining the field irrigation requirement. The duration and length of periods of inadequate precipitation during the growing season in humid and subhumid regions largely determine the economic feasibility of irrigation. In the northern hemisphere, water deficiency during the months of June, July, and August is more serious than in earlier or later months.

Estimates of evapotranspiration must be known when planning and managing an irrigation system. The evaporation pan and Penman-Monteith methods have been discussed in Chapter 4 (see ASCE, 2005, Allen et al., 1998, Hoffman et al., 1990, or Jensen et al., 1990, for these and other methods). Evapotranspiration may be determined by field measurements as shown in Figure 15-1 for three crops grown in the Salt River Valley of Arizona. Evaporation in mm/day is shown by the curves in Figure 15-1, and the monthly water use in mm is shown along the horizontal axis. The bar chart in the upper left gives the percentage of the seasonal water use that is removed from various depths of the root zone. Wheat, being a fairly short season crop in this region, has the lowest seasonal use of 655 mm, with the high water requirement occurring during March and April. Alfalfa, a long-season crop, has a seasonal use of 1888 mm and in this area grows during the entire year with the exception of December and January. Cotton, a tropical crop, has its highest seasonal use during the hottest portion of the summer and a seasonal use of 1046 mm. Similar data are available for other crops and regions.

TABLE 15–1 Seasonal Evapotranspiration and Irrigation Requirements for Crops Near Deming, New Mexico[a]

Crop	*Length of Growing Season (days)*	*ET*[b] *(mm)*	P_e[c] *(mm)*	$ET–P_e$ *(mm)*	E_a[d] *(%)*	*IR*[e] *(mm)*
Alfalfa	197	915	152	763	70	1090
Beans (dry)	92	335	102	233	65	358
Corn	137	587	135	452	65	695
Cotton	197	668	152	516	65	794
Grain (spring)	112	396	33	363	65	558
Sorghum	137	549	135	414	65	637

[a]Average frost-free period is April 15 to October 29. Irrigation prior to the frost-free period may be necessary for some crops.

[b]Evapotranspiration.

[c]Effective rainfall.

[d]Water application efficiency.

[e]Irrigation requirement.

Source: Jensen (1973).

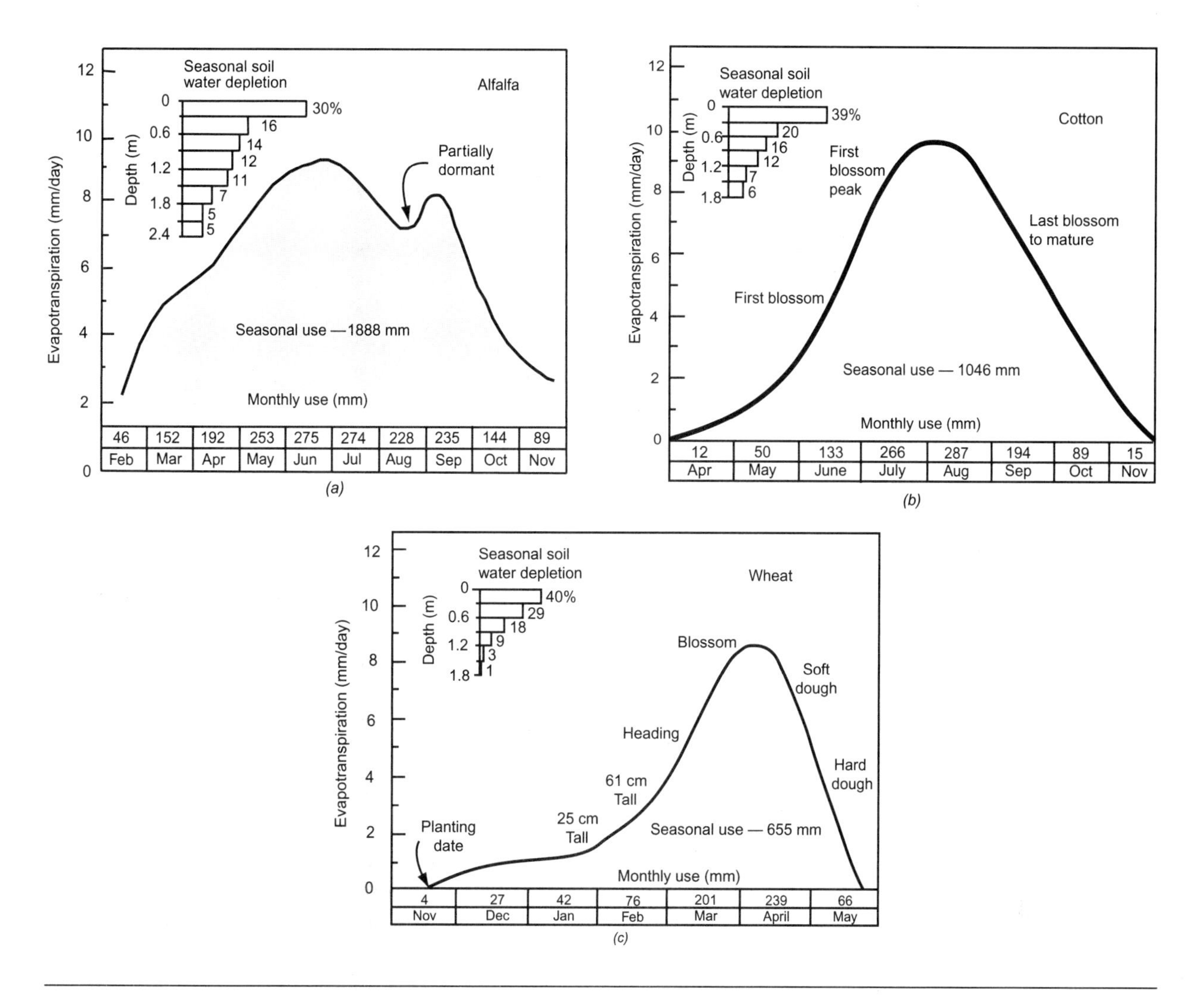

Figure 15–1 Average evapotranspiration and seasonal water depletion with depth for (a) alfalfa, (b) cotton, and (c) wheat at Mesa and Tempe, Arizona. (Redrawn and adapted from Erie et al., 1982)

15.2 Effective Rainfall

Rainfall must be considered when determining the crop water needs that must be supplied by irrigation. Not all rainfall is effective, only the portion that contributes to evapotranspiration. Rainfall on a wet soil is ineffective in meeting evapotranspiration but may contribute to the leaching requirement. Rainfall that produces runoff has reduced effectiveness.

A commonly used method for estimating effective rainfall was developed by the SCS (1970). This method bases effective rainfall on monthly evapotranspiration, monthly rainfall, and the soil water deficit, which may also be the net irrigation depth. The following equations describe the relationships:

$$P_e = f(D)[1.25P_m^{0.824} - 2.93][10^{0.000955\,ET}] \quad \textbf{15.1}$$

$$f(D) = 0.53 + 0.0116D - 8.94 \times 10^{-5}D^2 + 2.32 \times 10^{-7}D^3 \quad \textbf{15.2}$$

where P_e = estimated effective rainfall for a 75-mm soil water deficit (mm),
$f(D)$ = adjustment factor for soil water deficits or net irrigation depths (equals 1.0 for D = 75 mm),
P_m = mean monthly rainfall (mm),
ET = average monthly evapotranspiration (mm),
D = soil water deficit or net irrigation depth (mm).

Note that P_e is taken as the lowest of P_m, ET, or P_e from Equations 15.1 and 15.2 because effective precipitation cannot exceed the monthly rainfall or evapotranspiration.

Example 15.1

Determine the effective rainfall for the monthly ET and rainfall values below. The average depth of irrigation is 50 mm.

Solution. The effective rainfall from Equations 15.1 and 15.2 for May is

$$f(50) = 0.53 + 0.0116(50) - 8.94 \times 10^{-5}(50)^2 + 2.32 \times 10^{-7}(50^3) = 0.92$$

$$P_e = 0.92\,[1.25(100)^{0.824} - 2.93][10^{0.000955(75)}] = 57 \text{ mm}$$

Month	ET (mm)	Rainfall (mm)	Calculated P_e (mm)	Actual P_e (mm)
May	40	90	57	40
June	150	80	67	67
July	225	70	68	68
August	210	50	40	40
September	75	20	31	20
Total	700	310	263	235

Note that the actual P_e is taken as the minimum of ET, monthly rainfall, and the P_e calculated from Equations 15.1 and 15.2, and totals 235 mm for the season.

Soils And Salinity

Knowledge of soil characteristics is very important for crop production. The soil is a reservoir for water and chemicals including plant nutrients, plus provides a medium to support the plants. For irrigation, the water-holding capacity and salt content of the soil must be considered.

15.3 The Soil Water Reservoir

In planning and managing irrigation, it is helpful to think of the soil's capacity to store available water as the soil water reservoir. The reservoir is filled periodically by applications of water, which may contain salts and other chemicals. Water is slowly depleted by evapotranspiration. Water application in excess of the reservoir capacity is wasted unless it is used for leaching (Section 15.5). Irrigation is usually scheduled to prevent the soil water reservoir from becoming so low as to cause plant stress.

Irrigation can raise the soil water content to the field capacity. With either sprinkler or surface irrigation, the infiltration capacity and the permeability of the soil will determine how fast water can be applied. In sprinkler irrigation, the water should be applied at a rate lower than the infiltration capacity. In surface irrigation, the soil surface is flooded to allow water to enter the soil. In some cases where early spring runoff is available beyond that which may be stored in surface reservoirs, fields are sometimes flooded with surface runoff. The water is channeled through the irrigation ditches and onto the fields to fill the soil water reservoir and to conserve other water supplies for use later in the season.

For irrigation design and management, the water-holding capacity of the soil reservoir must be known. Representative values of soil physical properties including their water-holding abilities are listed in Table 15-2 for selected soil textures (Saxton, 2005; based on over 2000 samples used in the soil water characteristics portion of the SPAW model, http://hydrolab.arsusda.gov/SPAW/Index.htm). The values in Table 15-2 may vary with organic matter content, compaction, and soil salinity.

TABLE 15–2 Representative Physical Properties of Soils for Selected Soil Textures

Soil Texture	*Total Pore Space (% by vol)*	*Apparent Specific Gravity, A_s*	*Field Capacity, FC_v (% by vol)*	*Permanent Wilting, PWP_v (% by vol)*	*Available Water (mm/m)*
Sandy loam	39	1.58	16	7	80
	(37–40)	(1.56–1.59)	(11–22)	(3–12)	(50–110)
Sandy clay loam	41	1.57	26	16	100
	(38–42)	(1.53–1.60)	(20–32)	(13–19)	(70–120)
Loam	42	1.55	25	12	130
	(40–43)	(1.50–1.58)	(18–31)	(7–16)	(110–150)
Silt loam	43	1.52	29	11	180
	(40–46)	(1.44–1.59)	(16–36)	(3–16)	(130–230)
Silt	40	1.58	29	6	230
	(39–42)	(1.55–1.61)	(25–32)	(4–8)	(210–250)
Silty clay loam	47	1.40	37	20	180
	(45–50)	(1.33–1.47)	(34–40)	(17–22)	(160–200)
Clay loam	44	1.47	34	20	140
	(42–47)	(1.41–1.53)	(30–37)	(17–22)	(130–160)
Clay	49	1.35	42	28	140
	(44–56)	(1.19–1.32)	(36–47)	(23–33)	(130–150)

Note: Numbers are rounded, and normal ranges are shown in parentheses.

Source: Saxton (2005).

Total pore space (porosity) is the percentage of the bulk soil that is pores; however, the pores may be filled with water. Apparent specific gravity is the dry bulk density of the soil divided by the density of water. (*Note:* Apparent specific gravity and dry bulk density in g/cm^3 have identical numerical values.) Field capacity, *FC*, is the water content after a soil is well wetted and allowed to drain 1 to 2 days and represents the upper limit of water available to plants. Permanent wilting point, *PWP*, represents the lower limit of water available to plants. Neither *FC* nor *PWP* can be precisely defined, because both vary with soils and plants. The difference between *FC* and *PWP* is known as available water, *AW*, and can be estimated using

$$AW = (FC_v - PWP_v)D_r \tag{15.3}$$

where AW = depth of water available to plants (L),
FC_v = volumetric field capacity (L^3/L^3),
PWP_v = volumetric permanent wilting point (L^3/L^3),
D_r = depth of the root zone or depth of a layer of soil within the root zone (L).

Ranges of rooting depths for various crops are given in Table 15-3. Actual rooting depths will vary with soil conditions such as compacted layers, changes in soil texture, and water content. Because dry weight (gravimetric) water contents are easier to obtain, it is useful to relate volumetric water content θ_v to dry weight content θ_d in percent by $\theta_v = (\theta_d)(A_s)$ where A_s is the soil apparent specific gravity. Available water can be estimated from Table 15-2.

Plants can remove only a portion of the available water before growth and yield are affected. This portion is known as readily available water, *RAW*, and varies with the crop and evapotranspiration rate. Because the irrigation manager often decides how much water can be depleted, the portion depleted is called management-allowed depletion, *MAD*. Typical values for *MAD* are given in Table 15-4 for 9 rates of maximum evapotranspiration and 4 crop groups.

Readily available water is defined by

$$RAW = MAD \times AW \tag{15.4}$$

Example 15.2

Determine the readily available water for corn having a 1.5-m rooting depth in a fine sandy loam soil and a maximum *ET* of 7 mm/day. From physical sampling of the soil, the apparent specific gravity is 1.45, *FC* and *PWP* are 0.18 and 0.10 dry weight basis, respectively.

Solution. From Equation 15.3, determine the available water.

$$AW = (0.18 \times 1.45 - 0.10 \times 1.45)1.5 \text{ m} = 0.174 \text{ m or } 174 \text{ mm}$$

From Table 15-4 read MAD = 0.5, and determine RAW from Equation 15.4.

$$RAW = 0.5 \times 174 \text{ mm} = 87 \text{ mm}$$

TABLE 15–3 Ranges of Effective Rooting Depths and Salt Tolerance Values as a Function of the Electrical Conductivity of the Soil Saturation Extract for Selected Mature Crops

Crop	*Ranges of Effective Rooting Depths,* (m)	*Salt Tolerance Threshold*[a] c_t (dS/m)	*Percent Yield Decline*[b] *D* [%/(dS/m)]	*Qualitative Salt Tolerance Rating*[c]
Alfalfa	1.0–2.0	2.0	7.3	MS
Almond	~1.5	1.5	19	S
Apple	~1.5			S
Apricot	~1.5	1.6	24	S
Barley (grain)	1.0–1.3	8.0	5.0	T
Bean	0.4–0.8	1.0	19	S
Beet, garden	0.6–1.0	4.0	9.0	MT
Broccoli	0.6	2.8	9.2	MS
Cabbage	0.6–1.0	1.8	9.7	MS
Carrot	0.5–1.0	1.0	14	S
Clover, *Trifolium* spp.	0.6–0.9	1.5	12	MS
Corn (grain)	1.0–1.7	1.7	12	MS
Cotton	1.0–2.0	7.7	5.2	T
Cucumber	0.7–1.2	2.5	13	MS
Date palm	1.5–2.5	4.0	3.6	T
Grape	1.0–2.0	1.5	9.6	MS
Grapefruit	1.2–1.5	1.8	16	S
Lemon	1.2–1.5			S
Lettuce	0.03–0.5	1.3	13	MS
Melons	1.0–1.5			MS
Olive	~1.5			MT
Onion	0.8–2.0	1.2	16	S
Orange	1.2–1.5	1.7	16	S
Peach	~1.5	1.7	21	S
Peanut	0.5–1.0	3.2	29	MS
Pepper	0.5–1.0	1.5	14	MS
Plum	~1.5	1.5	18	S
Potato	0.4–0.8	1.7	12	MS
Rice, paddy	0.3	3.0	12	MS
Sorghum	1.0–2.0	6.8	16	MT
Soybean	0.8–1.5	5.0	20	MT
Spinach	0.3–0.5	2.0	7.6	MS
Squash, zucchini	0.6–0.9	4.7	9.4	MT
Strawberry	0.2–0.3	1.0	33	S
Sugarbeet	0.8–2.0	7.0	5.9	T
Sugarcane	1.2–2.0	1.7	5.9	MS
Sweet potato	1.0–1.5	1.5	11	MS

(Continued)

Crop	Ranges of Effective Rooting Depths,	Salt Tolerance Threshold[a] c_t	Percent Yield Decline[b] D	Qualitative Salt Tolerance Rating[c]
	(m)	(dS/m)	[%/(dS/m)]	
Tomato	0.7–1.5	2.5	9.9	MS
Wheat (grain)	1.0–1.5	6.0	7.1	MT

[a] Salt tolerance threshold is the mean soil salinity at initial yield decline.
[b] Percent yield decline is the rate of yield reduction per unit increase in salinity beyond the threshold.
[c] Qualitative salt tolerance ratings are sensitive (S), moderately sensitive (MS), moderately tolerant (MT), and tolerant (T).

Sources: Mass and Hoffman (1977); Doorenbos and Pruitt (1977); Tanji (1990).

Figure 15-2 illustrates the changes in soil water from field capacity as water is removed by *ET* during a growing season. It is assumed that the field is irrigated soon after planting so the season begins with the soil at field capacity. The arrows above the horizontal axis indicate dates of irrigation and rainfall. The readily available water is 80 mm after the roots have reached their nominal rooting depth. The two early-season irrigations reflect that the roots have not reached their nominal depth so light irrigations are needed. During the peak growing period the irrigations are more frequent because of higher *ET* rates. As the crop matures, *ET* and irrigation frequency both decrease. The figure illustrates an arid region as only one rainfall event occurs on Day 70. In regions where rainfall is expected, the soil would not be filled to field capacity in order to have storage capacity for the rainfall.

15.4 Salinity

The presence of soluble salts in the root zone can be a serious problem, especially in arid regions. In subhumid regions, where irrigation is provided on a supplemental

TABLE 15–4 Typical Management Allowed Depletion Values, *MAD*, for Maintaining Maximum Evapotranspiration Rates of Crops Grouped According to Stress Sensitivity

Crop Group	Maximum Evapotranspiration Rates, mm/day								
	2	3	4	5	6	7	8	9	10
	MAD								
1	0.50	0.43	0.35	0.30	0.25	0.23	0.20	0.20	0.18
2	0.68	0.58	0.48	0.40	0.35	0.33	0.28	0.25	0.23
3	0.80	0.70	0.60	0.50	0.45	0.43	0.38	0.35	0.30
4	0.88	0.80	0.70	0.60	0.55	0.50	0.45	0.43	0.40

Crop Group	Crops
1	Onion, pepper, potato
2	Banana, cabbage, pea, tomato
3	Alfalfa, bean, citrus, groundnut, pineapple, sunflower, watermelon, wheat
4	Cotton, sorghum, olive, grape, safflower, maize, soybean, sugar beet, tobacco

Source: Doorenbos and Kassam (1979).

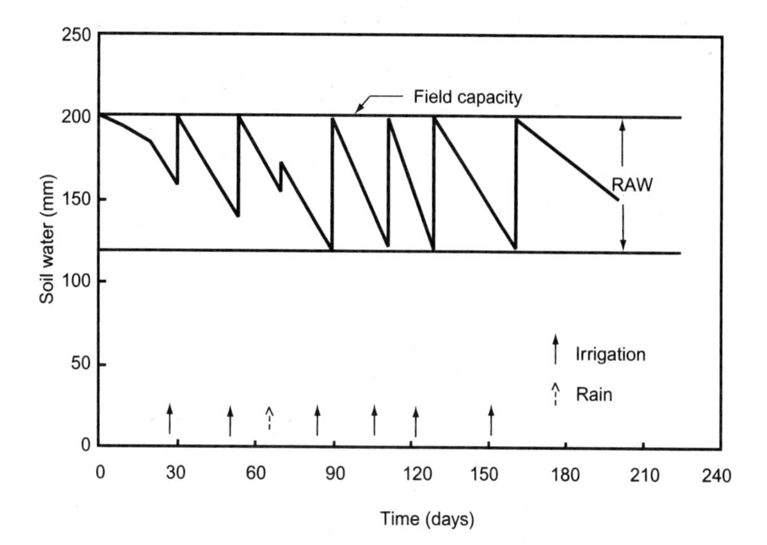

Figure 15–2
Illustration of the changes in soil water during a growing season.

basis, salinity is usually of little concern, because rainfall is sufficient to leach out any accumulated salts; however, all water from surface streams and underground sources contain dissolved salts. The salts applied to the soil with irrigation water remain in the soil unless flushed out in drainage water or removed in the harvested crop. Usually the quantity of salt removed by crops is so small that it will not make a significant contribution to salt removal or enter into determinations of leaching requirements.

The salt content of soils is determined by measuring its electrical conductivity, *EC*. Electrical conductivity is measured on a saturated extract of the soil in deciSiemens per meter (dS/m) at 25°C. One deciSiemen per meter represents about 670 ppm of dissolved salts. Salt-affected soils may be classified as saline, sodic, or saline-sodic soils.

Saline soils. These soils contain sufficient soluble salt to affect the growth of most plants. Sodium salts are in relatively low concentration in comparison with calcium and magnesium salts. Saline soils often are recognized by the presence of white crusts on the soil, by spotty stands, and by stunted and irregular plant growth. Saline soils generally are flocculated, and the permeability is comparable to that of similar nonsaline soils.

The principal effect of salinity is to reduce the availability of water to the plant. In cases of extremely high salinity, there may be curling and yellowing of the leaves, firing in the margins of the leaves, or actual death of the plant. Long before such effects are observed, the general nutrition and growth physiology of the plant will have been altered.

Sodic Soils. These soils are relatively low in soluble salts, but contain sufficient exchangeable (adsorbed) sodium to interfere with the growth of most plants. Exchangeable sodium is adsorbed on the surfaces of the fine soil particles. It is not readily leached until displaced by other cations, such as calcium and magnesium.

As the proportion of exchangeable sodium increases, soils tend to become dispersed, less permeable to water, and of poorer tilth. High-sodium soils usually are plastic and sticky when wet, and are prone to form clods and crusts on drying.

These conditions result in reduced plant growth, poor germination, and poor root development because of inadequate water penetration.

Saline-Sodic Soils. These soils contain sufficient quantities of both total soluble salts and adsorbed sodium to reduce the yields of most plants. As long as excess soluble salts are present, the physical properties of these soils are similar to those of saline soils. If the excess soluble salts are removed, these soils may assume the properties of sodic soils.

Both sodic and saline-sodic soils may be improved by the replacement of the excess adsorbed sodium by calcium or magnesium. This usually is done by applying soluble amendments that supply these cations. Acid-forming amendments, such as sulfur or sulfuric acid, may be used on calcareous soils, because they react with limestone (calcium carbonate) to form gypsum, a more soluble calcium salt.

Leaching is the only practical way to remove the salts added to the soil by the irrigation water. Sufficient water must be applied to dissolve the excess salts and carry them away by subsurface drainage.

With the necessity of using additional water beyond the needs of the plant to provide sufficient leaching, it is imperative under irrigation that there be adequate drainage of water passing through the root zone. Natural drainage through the underlying soil may be adequate. In cases where subsurface drainage is inadequate, open or pipe drains must be provided.

Water will rise 0.6 to 1.5 m or more in the soil above the water table by capillarity. The height to which water will rise above a free-water surface depends on soil texture, structure, and other factors. If the capillary rise only occurs in the bottom of the root zone, it may supply some water for plant growth; however, if the water contains dissolved salts, the salts will remain in the root zone. If the capillary rise reaches the soil surface, water evaporates leaving a salt deposit typical of saline soils. When this occurs, plant growth is probably reduced.

Some crops can tolerate large amounts of salt. Others are more easily injured. The mean soil salinity at which crop yields begin to decline is known as the salt tolerance threshold. The salt tolerance threshold of a number of mature crops is shown in Table 15-3. Similar data for a number of additional crops are available from Ayers and Westcot (1985) or Tanji (1990). The tolerance of crops listed may vary somewhat, depending on the particular variety grown, the cultural practices used, and climatic factors. Young plants are often more sensitive to salinity than mature plants. Table 15-3 also contains a qualitative salt tolerance rating and percent yield decline values or the rate of yield reduction per unit increase in salinity after the threshold is reached. The relative yield can be defined by

$$Y_r = \begin{cases} 100 & 0 \le c \le c_t \\ 100 - D(c - c_t) & c_t \le c \le c_0 \\ 0 & c \ge c_0 \end{cases} \tag{15.5}$$

where Y_r = relative yield,

D = yield decline from Table 15-3, percent/(dS/m),

c = average root zone salinity (dS/m),

c_t = salt tolerance threshold from Table 15-3 (dS/m),

c_o = soil salinity above which the yield is zero (dS/m).

The value of c_o is obtained by setting $Y_r = 0$ into the middle Equation 15.5 with D and c_t values for the crop from Table 15-3 and solving for c.

Example 15.3

What is the relative yield of corn if the average soil salinity is 2.7 dS/m. At what soil salinity will the yield be zero?

Solution. From Table 15-3 read c_t is 1.7 dS/m and D is 12 percent/(dS/m). Substitute into Equation 15.5 and solve for relative yield

$$Y_r = 100 - 12(2.7 - 1.7) = 88\%$$

The soil salinity for zero yield is obtained from Equation 15.5 with $Y_r = 0$

$$c_o = \frac{100 + 12 \times 1.7}{12} = 10.0 \text{ dS/m}$$

Thus if the average soil salinity is 10 dS/m or greater, there will be no yield.

15.5 Leaching

The traditional concept of leaching involves the application of water to achieve more or less uniform salt removal from the entire root zone. However, salt accumulations can take place in the lower root zone without adverse effects if the upper root zone is maintained at a high–water content (Hoffman et al., 1990; Ayers & Westcot, 1985). As can be noted in the bar chart (upper left) in Figure 15-1, most of the water transpired by the plant is normally taken from the upper portion of the root zone. This area will be leached to a considerable degree by normal applications of irrigation water and by rainfall that may come at any time of the year. The same amount of water, when applied with more frequent irrigations, is more effective in removing salts from this critical upper portion of the root zone than from the lower root zone. Thus, high-frequency sprinkler, microirrigation, or surface irrigation should be effective for salinity control. Other concepts that may be helpful in controlling salinity are the use of soil or water amendments, deep tillage, and irrigation before planting.

Hoffman et al. (1990) present a graphical solution for determining the leaching requirement L_r (Figure 15-3) as a function of the salinity of the water and the crop salt tolerance threshold, as given in Table 15-3. Figure 15-3 is based on a model by Hoffman & van Genuchten (1983) that determines the linearly averaged mean root zone salinity by solving the continuity equation for one-dimensional vertical flow of water in the soil with an exponential soil water removal function. The results were related to field trials by compensating for different field leaching criteria.

The leaching requirement L_r represents that portion of the infiltrated water that must pass through the soil and percolate below the root zone. Thus, the depth of water infiltrated must equal $(1 + L_r)$ times the soil water deficit to maintain a salt balance in the soil appropriate to the crop and the accepted potential yield reduction.

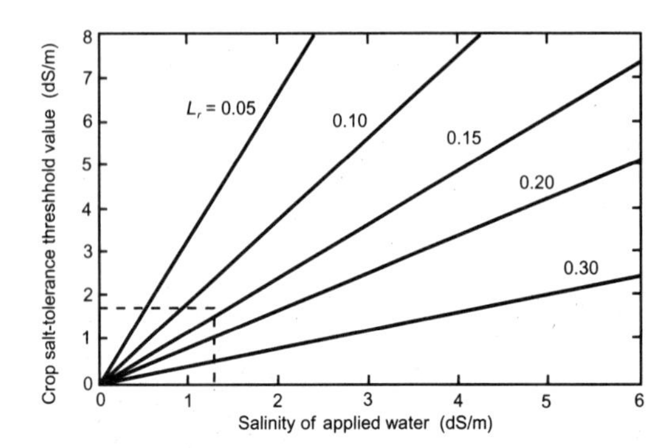

Figure 15–3
Leaching requirement L_r as a function of the salinity of the applied water and the salt-tolerant threshold value for a crop. The dashed lines are from Example 15.8. (Hoffman & van Genuchten, 1983)

With most irrigation applications, sufficient water moves through the root zone to leach at least some of the salts, and less additional water for leaching is needed. It is good practice to monitor the actual salt content of the soil through chemical analyses of soil samples to assure that neither inadequate nor excessive quantities of water are being applied.

Irrigation Management

Irrigation system designers must consider the operation and management requirements for their designs. The performance or efficiency of the system, water delivery requirements, and irrigation scheduling must be understood. A comprehensive review of irrigation management considerations is given by Hoffman et al. (1990).

15.6 Irrigation Efficiency and Uniformity

Efficiency is an output divided by an input and is usually expressed as a percentage. An efficiency figure is meaningful only when the output and input are clearly defined. There are three basic irrigation efficiency concepts:

1. Water conveyance efficiency, E_c

$$E_c = 100\,W_d/W_i \tag{15.6}$$

where W_d = water delivered by a distribution system,
W_i = water introduced into the distribution system.

The water conveyance efficiency definition can be applied along any reach of a distribution system. For example, a water conveyance efficiency could be calculated from a pump discharge to a given field or from a major diversion work to a farm turnout.

2. Water application efficiency E_a

$$E_a = 100W_s/W_d \qquad \textbf{15.7}$$

where W_s = water stored in the soil root zone by irrigation,
W_d = water delivered to the area being irrigated.

This efficiency may be calculated for an individual furrow or border strip, for an entire field, or an entire farm or project. When applied to areas larger than a field, it overlaps the definition of conveyance efficiency.

3. Water use efficiency E_u

$$E_u = 100W_u/W_d \qquad \textbf{15.8}$$

where W_u = water beneficially used,
W_d = water delivered to the area being irrigated.

The concept of beneficial use differs from that of water stored in the root zone in that the leaching water would be considered beneficially used, although it moved through the soil water reservoir. Sometimes water use efficiency is based on the dry plant weight produced by a unit volume of water.

Example 15.4

If 42 m^3/s is pumped into a distribution system and 38 m^3/s is delivered to a turnout 3 km from the pumps, what is the conveyance efficiency of the portion of the distribution system used for conveying this water?

Solution. Substituting into Equation 15.6,

$$E_c = 100 \times 38/42 = 90\%$$

Example 15.5

Delivery of 0.5 m^3/s to a 30-ha field is continued for 40 h. Soil water measurements before and after the irrigation indicate that 0.16 m of water was stored in the root zone. Compute the application efficiency.

Solution. Apply Equation 15.7:

$$W_d = (0.5\ m^3/s)\ (3600\ s/h)\ (40\ h) = 72\ 000\ m^3$$

$$W_s = (0.16\ m)\ (30\ ha)\ (10\ 000\ m^2/ha) = 48\ 000\ m^3$$

$$E_a = 100\ (48\ 000/72\ 000) = 67\%$$

The difference between 72 000 – 48 000 = 24 000 m^3 is deep percolation or runoff.

Another useful measurement of the effectiveness of irrigation is the uniformity of water distribution. The uniformity coefficient, *UC*, is usually applied to sprinkler irrigation and is given by

$$UC = 1 - \frac{\left(\sum_{i=1}^{n} |y_i - d|\right)/n}{d} \qquad \textbf{15.9}$$

where y_i = measured depth of water caught or infiltrated (L),
d = average depth of water caught or infiltrated (L),
n = number of samples collected.

This coefficient indicates the degree to which water has been applied to a uniform depth throughout the field. (*Note:* Each value should represent an equal area.) Values of *UC* above 0.8 are acceptable.

Example 15.6

A solid-set sprinkler system with 15- by 12-m spacing was operated for 4 h. Twenty catch cans were placed under the system on a 3- by 3-m spacing, and the following depths of water (in mm) were measured in the cans immediately after irrigation stopped. Assuming evaporation was equal from all cans, determine the uniformity coefficient.

Depth	Deviation	Depth	Deviation	Depth	Deviation	Depth	Deviation
50	9	44	3	40	1	42	1
44	3	42	1	39	2	40	1
43	2	37	4	30	11	32	9
48	7	42	1	36	5	40	1
47	6	44	3	35	6	45	4

Solution. Apply Equation 15.9.

$$\text{Average depth} = (50 + 44 + \ldots + 45)/20 = 820/20 = 41$$

$$\left(\sum_{i=1}^{n} |y_i - d|\right)/n = (|50 - 41| + |44 - 41| + \ldots + |40 - 41| + |45 - 41|)/20$$

$$= (9 + 3 + \ldots + 1 + 4)/20 = 80/20 = 4$$

$$UC = 1 - 4/41 = 0.90$$

A second measure of uniformity is the distribution uniformity low quarter *(DULQ)*, which is usually applied to surface or sprinkler irrigation.

$$DULQ = \frac{\text{average low-quarter depth}}{\text{average depth}} \tag{15.10}$$

Average low-quarter depth is the average of the lowest one fourth of all values caught or infiltrated, where each value represents an equal area. Average depth is the average of all values caught or infiltrated. Values of *DULQ* above 0.7 are considered acceptable.

Example 15.7 A uniformity check was taken by probing for water penetration depth at 16 equally spaced stations down one border strip. The depths of penetration (m) were recorded as follows:

0.95	0.98	0.98	0.92
0.89	0.89	0.83	0.79
0.74	0.68	0.77	0.83
0.78	0.72	0.84	0.88

Solution. Apply Equation 15.10 using the four lowest depths.

$$\text{Average low-quarter depth} = (0.74 + 0.68 + 0.77 + 0.72)/4 = 0.73$$

$$\text{Average depth} = (0.95 + 0.98 + \ldots + 84 + 0.88)/16 = 0.84$$

$$DULQ = 0.73/0.84 = 0.87$$

(*Note:* Depths of penetration are not as reliable as measured depths of infiltrated water but are often easier to obtain.)

15.7 Irrigation Requirement

The irrigation requirement, *IR*, is the total amount of water that must be supplied over a growing season to a crop that is not limited by water, fertilizer, salinity, or diseases.

$$IR = \frac{(ET - P_e)(1 + L_r)}{E_a} \tag{15.11}$$

where IR = seasonal irrigation requirement (L),
ET = seasonal evapotranspiration (L),
P_e = effective rainfall from Equations 15.1 and 15.2 (L),
L_r = leaching requirement as defined by Equation 15.5 and Figure 15-3,
E_a = application efficiency (decimal).

Equation 15.11 assumes that the soil water contents at the beginning and end of the season are similar.

Example 15.8 Corn is sprinkler irrigated with water having an electrical conductivity of 1.2 dS/m; the depth of irrigation is 50 mm, and the application efficiency is 70 percent. Use the rainfall, *ET*, and effective rainfall data from Example 15.1. Compute the seasonal irrigation requirement.

Solution. Assume no yield reduction is preferred, and from Table 15-3 note the threshold salinity value is 1.7 dS/m; then from Figure 15-3 read $L_r = 0.13$. Now solve Equation 15.11 for the irrigation requirement.

$$IR = \frac{(700 - 235)(1 + 0.13)}{0.70} = 750 \text{ mm}$$

15.8 Irrigation Scheduling

Irrigations must be scheduled according to water availability and crop need. If adequate water supplies are available, irrigations are usually provided to obtain optimum or maximum yield. However, overirrigation should be avoided, as this can decrease yields by reducing soil aeration and leaching fertilizers while increasing water and energy costs. In addition, overirrigation can contribute to high water tables and water pollution. If water supplies are limited and/or expensive, the irrigation scheduling strategy becomes one of maximizing economic return. In practice many irrigations are applied on a routine schedule based on the experience of the farm manager and water availability.

Irrigation scheduling requires knowing when to irrigate and how much water to apply. When to irrigate can be determined from plant or soil indicators or water balance techniques. How much water to apply can be based on soil water measurements or water balance techniques.

Since the objective of irrigation is to provide water for plant growth, plant indicators can directly show the need for water. Growth and appearance are visual indicators of water status. Slow growth of leaves and stems may indicate water stress. Irregular growth patterns may suggest excessive wet and dry cycles. Plant wilting and dark color are also indicators of water stress. Leaf or canopy temperatures, which can be easily measured with infrared thermometers, are good indicators. Increases in leaf temperature relative to air temperatures indicate transpiration is slowing, hence leaves are hotter than those of well-watered plants. Leaf water potentials become lower (more negative) when plants are water stressed; however, these measurements are difficult to perform. Care must be taken when interpreting these signs, because diseases or nutrient deficiencies may cause similar reactions.

Soil indicators include feel and appearance, tensiometers, porous blocks, gravimetric sampling, neutron probes, and time domain reflectometry (TDR). Feel and appearance requires obtaining a soil sample with a probe and judging the water content by color and consistency. With experience this method can indicate both the need for irrigation and the amount to apply. Tensiometers measure soil water tension in the wet range (0 to 80 kPa) and can serve to indicate the need for irrigation. Gravimetric sampling or laboratory measurements are required to convert tension readings to depths to be applied. Gravimetric sampling is a direct method for determining the water content of soil samples and the soil water deficit, but is labor intensive. Porous blocks change their water content according to that of the surrounding soil. Changes in electrical resistance or thermal conductivity are sensed to indicate the soil water status. These devices must be calibrated for reliable measurements of water content, but are simple to read. Neutron probes indicate water content or water depth, but should be calibrated. In addition, training and/or certification in radiation safety are required. Time domain reflectometry measures the propagation of an electrical signal on metal electrodes placed in the soil. The signal is dependent on the dielectric constant of the soil, which is affected by the soil water content and the electrical conductivity. Thus, this method can potentially measure both water content and salinity of the soil.

Water balance techniques can be used to obtain a record of the estimated water in the root zone. For example, irrigation is scheduled when the soil water depletion equals the readily available water, *RAW*, as defined by Equation 15.4. Equation 15.4 includes *MAD* (management-allowed depletion), which represents the irrigation manager's decision on what portion of the available water can be removed from the soil before growth or yield are affected. Crop and soil characteristics, water availability and costs, irrigation system capabilities, and other factors influence the value selected for *MAD*. The soil water balance may be determined from soil water measurements. If estimates of evapotranspiration are available, the soil water balance can be calculated from

$$D_i = D_{i-1} + (ET - P_e)_i \quad \mathbf{15.12}$$

where D_i, D_{i-1} = soil water deficit at the end of day i and i–1 (L),
ET = evapotranspiration calculated from climatic data for day i (L),
P_e = effective rainfall for day i (L).

Historic data can be used to estimate ET and P_e and to develop a projected irrigation schedule. The soil water balance can be updated from current climatic data or soil water measurements. Irrigation applications are scheduled when D_i reaches the threshold value defined by *RAW*. Soil water measurements can be used to verify or adjust the calculated data.

In some cases irrigation districts, government agencies, or irrigation consultants provide irrigation water management services to farm operators. These services can provide the irrigation manager with recommendations for effective management of irrigation activities. In several states, weather station networks have been established. From these data the potential evapotranspiration rates and/or crop *ET* values are provided for the regions surrounding each weather station. Computers may be utilized by irrigators to directly acquire these data from websites or other computers. Programs are available that collect current climatic data or use historic data to calculate evapotranspiration, compute the soil water balance, and project when to irrigate.

Example 15.9

Determine the date and amount of water to apply to a field having a soil water deficit of 50 mm at the end of the day on June 14. The soil is a silt loam, effective root zone depth is 1.0 m, and *MAD* is 50 percent of the available water.

Solution. For the *ET* and P_e values given, apply Equation 15.12 to calculate D_i. For example,

$$D_{15} = 50 + 7 - 0 = 57, \; D_{16} = 57 + 6 - 5 = 58, \text{ etc.}$$

Date (June)	ET (mm)	P_e (mm)	D_i (mm)
15	7	0	57
16	6	5	58
17	8	0	66
18	7	0	73
19	8	6	75
20	9	0	84
21	8	0	92
22	10	0	102

From Table 15-2, the available water is 180 mm/m or 180 mm for the 1-meter root zone. $RAW = 180 \times 0.5 = 90$ mm. Irrigation should begin July 20 with a net application of 84 mm. For a large field, irrigation may begin earlier to assure that the crop in the last portion irrigated is not stressed.

15.9 Flow Rate Required

The flow rate required for an irrigation system depends on the depth required, area irrigated, and time allowed for irrigation. These are related to plant water use, readily available water, and irrigation system characteristics. The flow rate can be determined from

$$Q = \frac{D \times A}{t \times E_a} \tag{15.13}$$

where Q = flow rate (L^3/T),

D = net depth to be applied (L),

A = area to be irrigated (L^2),

t = time available for irrigation (T),

E_a = system application efficiency (decimal).

Example 15.10 Determine the flow rate required for Example 15.9 if the field area is 20 ha, the application depth is 84 mm, water is available for 48 hours, and the system has an efficiency of 70 percent.

Solution. Substitute into Equation 15.13.

$$Q = \frac{84\,\text{mm}/1000\,\text{mm/m}\ 20\,\text{ha} \times 10\,000\ \text{m}^2/\text{ha} \times 1000\ \text{L}/m^3}{48\,\text{h} \times 3600\ \text{s/h} \times 70/100} = 139\,\text{L/s}$$

This can be rounded to 140 L/s for the system flow rate.

Irrigation Methods

The methods of applying water may be classified as subirrigation, surface irrigation, sprinkler irrigation, and microirrigation.

15.10 Subirrigation

In special situations water may be applied below the soil surface by developing or maintaining a water table that allows water to move up through the root zone by capillary action. This is essentially the same practice as controlled drainage or water table management discussed in Chapter 14. Controlled drainage becomes subirrigation if water must be supplied to maintain the desired water level. Water may be introduced into the soil profile through open ditches or pipe drains. In some river valleys and near lakes, subirrigation is a natural process. Water table maintenance is suitable where the soil in the plant root zone is highly permeable and there is either a continuous impermeable layer or a natural water table below the root zone. Because subirrigation allows no opportunity for leaching and establishes an upward movement of water, salt accumulation is a hazard; hence the salt content of the water should be low.

15.11 Surface Irrigation

Surface irrigation is used worldwide, especially in arid regions where large dams and water distribution systems have been built. With this method water is applied by flooding the surface. Common surface irrigation methods include level basins, border strips, furrows, and corrugations. With basins and border strips the flow is contained by soil levees or dikes, which may be straight or on the contour. For highest efficiency and uniformity, surface-irrigated fields should be graded to a smooth and uniform slope and the rate of water application should be measured and controlled (Chapter 16).

15.12 Sprinkler Irrigation

Lightweight portable pipes with slip joint connections were common for water distribution; however, because labor costs are high for moving these systems, they are becoming limited to high-value crops. Mechanical-move systems are now common. These may be either intermittent- or continuous-move. Solid-set and permanent systems are suitable for intensively cultivated areas growing a high-income crop, such as flowers, fruits, and vegetables. Sprinkler irrigation systems provide a reasonably uniform application of water. On coarse-textured soils, water application efficiency may be twice as high as with surface irrigation.

Sprinkler irrigation can be used for temperature control. Irrigation water, especially if supplied from wells, is often considerably warmer than the soil and air near the surface under frost conditions. The heat of fusion released by water freezing on plant parts keeps the temperature from falling below 0°C. Conversely, irrigation may be used for cooling, particularly when germination occurs under high temperatures, or to delay premature blossoming of fruit trees when warm weather occurs before the frost danger has past. The cooling effect of evaporation lowers the temperature of the plant parts (Chapter 17).

15.13 Microirrigation

Increasing use is being made of microirrigation (trickle, drip, or subsurface) systems that apply water at very low rates, often to individual plants. Such rates are achieved through the use of specially designed emitters or special tubes. A typical emitter might apply water at 2 to 10 L/h and is usually installed on or just below the soil surface. Tubes with in-line emitters or porous tubes supplying 1 to 5 L/min per 100 m of tube and are often installed 0.1 to 0.3 m below the soil surface (subsurface drip irrigation). These systems provide an opportunity for efficient use of water because of minimum evaporation losses and because the irrigation is limited to the soil in the active root zone under and near the plants. Microirrigation systems were limited to high-value crops because of high initial costs; however, decreasing costs have widened their usage. Since the distribution pipes are usually at or near the surface, operation of field equipment may be difficult. Both sprinkler and microirrigation systems are well adapted to application of agricultural chemicals, such as fertilizers or pesticides, with the irrigation water (Chapters 17 and 18).

15.14 Comparison of Irrigation Methods

Table 15-5 compares different types of irrigation systems in relation to site and situation factors.

Efficient surface irrigation requires grading the land surface to control the flow of water. The extent of grading required depends on the topography. In some soil and topographic situations, the presence of unproductive subsoils may make grading for surface irrigation unfeasible. The utilization of level basins where large streams of water are available generally provides high irrigation efficiencies. If the water is delivered by gravity flow from a reservoir, surface irrigation has no energy costs for pumping.

Sprinkler irrigation is particularly adaptable to sloping or undulating land where grading for surface irrigation is not feasible. It is appropriate for most circumstances where the infiltration rate exceeds the rate of water application. With sprinkler irrigation, the rate of water application can be easily controlled. Simple sprinkler irrigation systems have relatively low initial costs while mechanical-move systems have higher initial costs. With mechanical-move systems labor can be substantially reduced. In some cases disease problems have resulted from moistened foliage. Evaporation losses with sprinkler irrigation are not excessively high, even in arid regions. A well-designed and managed sprinkler system can provide a high efficiency of water application.

A well-designed microirrigation system can provide a high efficiency of water application. It is especially well suited to tree fruit and high-value crops. Water must be clean and uncontaminated, usually achieved by a filtration system. Microirrigation lends itself well to automation and has a low labor requirement, if it is well designed and maintained. Since microirrigation systems usually operate at low pressure, energy requirements are generally lower than with sprinkler systems. Some low-pressure sprinklers operate at pressures comparable to those of microirrigation systems.

TABLE 15–5 Comparison of Irrigation Systems in Relation to Site and Situation Factors

	Improved Surface Irrigation Systems		***Sprinkler Irrigation Systems***			***Microirrigation Systems***
Site and Situation Factors	***Redesigned Surface Systems***	***Level Basins***	***Intermittent Mechanical-Move***	***Continuous Mechanical-Move***	***Solid-Set and Permanent***	***Emitters and Porous Tubes***
Infiltration rate	Moderate to low	Moderate	All	Medium to high	All	All
Topography	Moderate slopes	Small slopes	Level to rolling	Level to rolling	Level to rolling	All
Crops	All	All	Generally shorter crops	All but trees and vineyards	All	High value required
Water supply	Large streams	Very large streams	Small streams nearly continuous	Small streams nearly continuous	Small streams	Small streams, continuous and clean
Water quality	All but very high salts	All	Salty water may harm plants	Salty water may harm plants	Salty water may harm plants	All
Efficiency	Average 60–70%	Average 80%	Average 70–80%	Average 80%	Average 70–80%	Average 80–90%
Labor requirement	High training required	Low, some training	Moderate, some training	Low, some training	Low to seasonal high, little training	Low to high, training required
Capital requirement	Low to moderate	Moderate	Moderate	Moderate	High	High
Energy requirement	Low	Low	Moderate to high	Moderate to high	Moderate	Low to moderate
Management skill	Moderate	Moderate	Moderate	Moderate to high	Moderate	High
Machinery operations	Medium to long fields	Short fields	Medium field length, small interference	Some interference circular fields	Some interference	May have considerable interference
Duration of use	Short to long	Long	Short to medium	Short to medium	Long term	Variable
Weather	All	All	Poor in windy conditions	Better in windy conditions than other sprinklers	Windy conditions reduce performance, good for cooling	All
Chemical application	Fair	Good	Good	Good	Good	Very good

Source: Fangmeier and Biggs (1986).

Internet Resources

Natural Resources Conservation Service Irrigation Page
http://www.wcc.nrcs.usda.gov/nrcsirrig/

Selected University Irrigation Sites
http://www.ianr.unl.edu (University of Nebraska). Key search term: irrigation
http://www.clemson.edu (Clemson University). Key search term: irrigation
http://www.engineering.usu.edu (Utah State University). Key search term: biological and irrigation engineering
http://www.itrc.org (California State Polytechnic University at San Luis Obispo)
http://irrigation.tamu.edu (Texas A & M University)

Source for SPAW model
http://hydrolab.arsusda.gov/SPAW/Index.htm

References

Allen, R. G., L. S. Pereira, D. Raes, & M. Smith. (1998). *Crop Evapotranspiration—Guidelines for Computing Crop Water Requirements.* FAO Irrigation and Drainage Paper No. 56. Rome: Food and Agriculture Organization (FAO).

ASCE Standardization of Reference Evapotranspiration Task Committee (ASCE). (2005). *The ASCE Standardized Reference Evapotranspiration Equation.* Report of Task Committee.

Ayers, R. S., & D. W. Westcot. (1985). *Water Quality for Agriculture.* FAO Irrigation and Drainage Paper No. 29, Rev. 1. Rome: Food and Agriculture Organization (FAO).

Doorenbos, J., & A. H. Kassam. (1979). *Yield Response to Water.* FAO Irrigation and Drainage Paper No. 33. Rome: Food and Agriculture Organization (FAO).

Doorenbos, J., & W. O. Pruitt. (1977). *Guidelines for Predicting Crop Water Requirements.* FAO Irrigation and Drainage Paper No. 24. Rome: FAO.

Erie, L. J., O. F. French, D. A. Bucks, & K. Harris. (1982). *Consumptive Use of Water by Major Crops in the Southwestern United States.* USDA-ARS Cons. Res. Rep. No. 29. Washington, DC: Author.

Fangmeier, D. D., & E. N. Biggs. (1986). *Alternative Irrigation Systems.* Rep. 8555. Tucson: Cooperative Extension Service. University of Arizona.

Hoffman, G. J., J. D. Rhoades, J. Letey, & F. Sheng. (1990). Salinity management. In G. J. Hoffman, T. A. Howell, & K. H. Solomon, eds., *Management of Farm Irrigation Systems,* Monograph. St. Joseph, MI: ASAE.

Hoffman, G. J., & M. Th. van Genuchten. (1983). Soil properties and efficient water use: Water management for salinity control. In H. M. Taylor, W. Jordan, & T. Sinclair, eds., *Limitations to Efficient Water Use in Crop Production,* pp. 73–85. Madison, WI: American Society of Agronomy.

Jensen, M. E. (1973). *Consumptive Use of Water and Irrigation Water Requirements.* New York: ASCE.

Jensen, M. E., R. D. Burman, & R. G. Allen, eds. (1990). *Evapotranspiration and Irrigation Water Requirements.* New York: ASCE.

Mass, E. V., & G. J. Hoffman. (1977). Crop salt tolerance—Current assessment. *Journal of Irrigation and Drainage Division,* ASCE, 103, 115–134.

Saxton, K. (2005). Personal communication. USDA-ARS. Pullman, WA.

Tanji, K. K., ed. (1990). *Agricultural Salinity Assessment and Management.* New York: ASCE.

U.S. Soil Conservation Service (SCS). (1970). *Irrigation Water Requirements.* Tech. Release No. 21. Washington, DC: Author.

Problems

15.1 Estimate the readily available water for wheat having a 0.8-m root zone depth in a clay loam soil. Maximum ET is 6 mm/day. What is the time between irrigations? Assuming the water application efficiency is 70 percent, how much water should be delivered?

15.2 A 75-mm application of water measured at the pump increased the average water content of the top 1.0 m of soil from 18 to 23 percent (dry-weight basis). If the average dry bulk density of the soil is 1400 kg/m^3, what is the water application efficiency?

15.3 A flow of 5 m^3/s is diverted from a river into a canal. Of this amount 4 m^3/s is delivered to farmland. The surface runoff from the irrigated area averages 0.7 m^3/s and the contribution to ground water is 0.4 m^3/s. What is the water conveyance efficiency? What is the water application efficiency?

15.4 Determine the water application efficiency, distribution uniformity low quarter, and uniformity coefficient if a stream of 85 L/s is delivered to the field for 2 hours, runoff averaged 42 L/s for 1 hour, and the depth of penetration of the water varied linearly from 1.6 m at the upper end to 1.2 m at the lower end of the field. The root zone depth is 1.6 m.

15.5 Determine the leaching requirement and water application to an alfalfa field if the salinity of the irrigation water is 1.2 dS/m, 15 days have elapsed since the last irrigation, and the average evapotranspiration rate of alfalfa is 9 mm/day. Assume no rainfall, application efficiency is 80 percent, and the target yield is 100 percent.

15.6 A field to be developed for irrigation has an average soil salinity of 5 dS/m. Which crops from Table 15-3 could be grown without a yield reduction? Assume water is not available for leaching.

15.7 Soil samples from an irrigated soybean field show the average soil salinity is 6 dS/m. What yield decline is expected?

15.8 Determine the irrigation requirement for furrow-irrigated cotton using the evapotranspiration data from Figure 15-1b. Assume the only rainfall is 50 mm during the month of August. The irrigation water has an EC_i = 2.3 dS/m, and 100 mm is the net application depth. Assume no yield reduction due to salinity and the application efficiency is 70 percent.

15.9 Determine the date and amount of the next irrigation for cotton if the last irrigation was July 31, evapotranspiration is 7.2 mm/day, effective rainfall was 20 mm on August 5 and 15 mm on August 12, and the soil is a clay loam 1-m deep.

15.10 Redraw the graph in Figure 15-2 to reflect a rainfall of 60 mm on Day 70.

15.11 In Figure 15-2, why was irrigation water applied on Day 28?

CHAPTER

16 Surface Irrigation

Surface irrigation is an important method of irrigation in the United States and in most other countries with large irrigated areas. A 2000 U.S. irrigation survey ("2000 Irrigation Survey," 2001) reported 44.9 percent of the irrigation was accomplished with surface methods. In the western states, where this percentage is higher, the major water supply for irrigation is surface runoff, usually stored in reservoirs. Because this water must be conveyed for considerable distances, conveyance canals and control structures are key parts of most irrigation systems in arid regions. The hydraulic principles involved in the design of control structures are presented in Chapter 9 and the design of canals in Chapter 6. Ground water also provides an important source of water for surface irrigation (Chapter 11). Walker & Skogerboe (1987) present a comprehensive review of surface irrigation.

Application of Water

The various surface methods of applying water to field crops are illustrated in Figure 16-1.

16.1 Flooding

Ordinary flooding is the application of irrigation water from field ditches that may be nearly on the contour or up and down slope. After water leaves the ditches, no attempt is made to control the flow by means of levees or land shaping. For this reason ordinary flooding is frequently referred to as "wild flooding." Ordinary flooding is not a recommended practice, because of its poor efficiency and the high demand for limited water supplies.

16.2 Graded Border Strips

The graded border strip method of irrigation consists of dividing the field into a series of smooth uniformly graded strips separated by low soil ridges or levees and flooding the area between the levees. Normally, the direction of the strip is in the

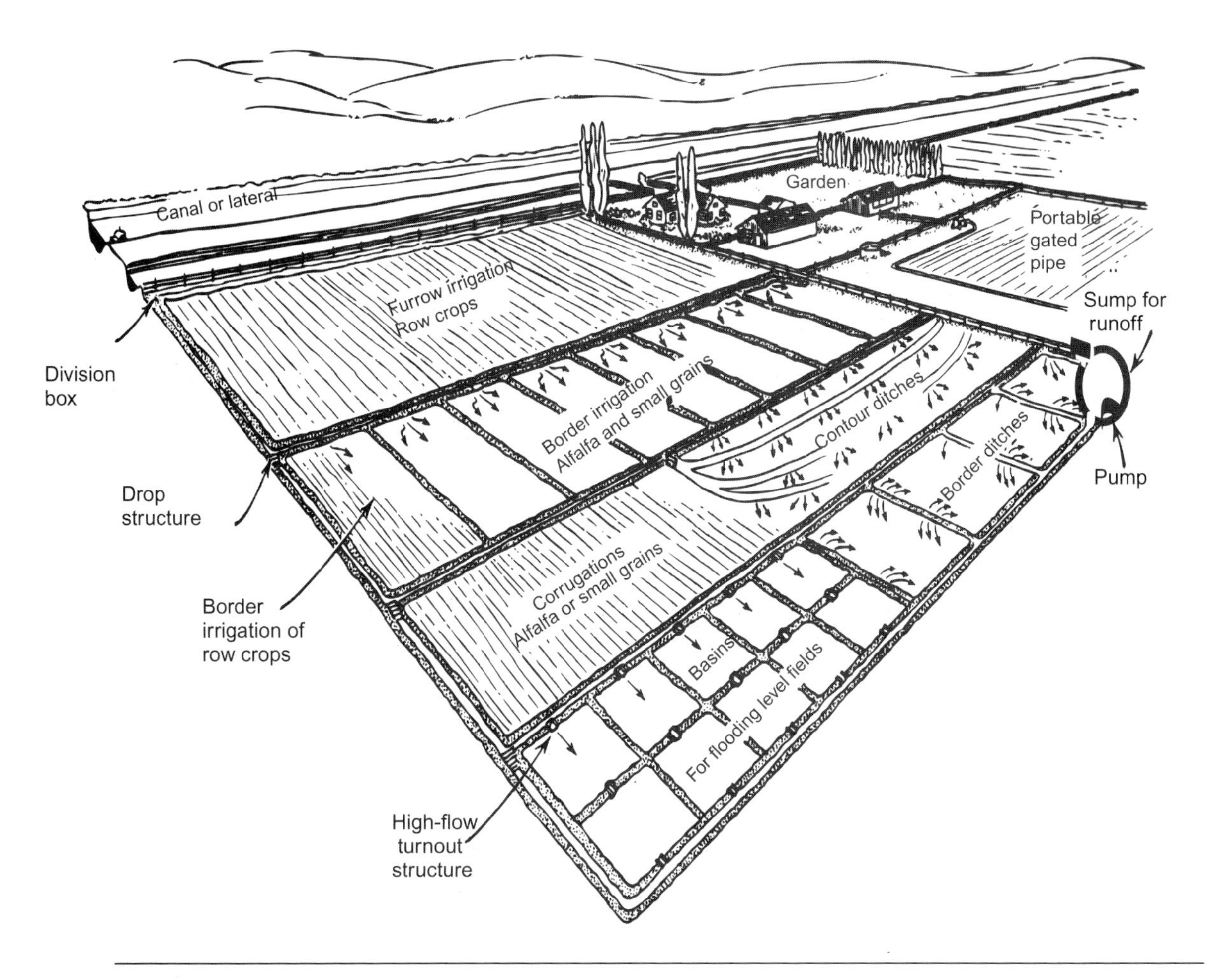

Figure 16–1 Surface methods of applying water to field crops. (Redrawn and modified from SCS, 1947)

direction of greatest slope, but in some cases, the strips are placed nearly on the contour. Slopes less than 0.5 percent are best suited for border strip irrigation, however, slopes up to 4 percent may be used if erosion can be controlled. The strips usually vary from 10 to 30 m in width and 100 to 400 m in length. Soil ridges or levees between border strips should be sufficiently high to prevent overtopping during irrigation. The SCS (1974) recommended that the border ridges should be high enough that the flow depth at the head of the border is 25 percent less than the ridge height. To prevent water from concentrating on either side of the strip, the land should be level or nearly level (less than a 0.05-m difference in elevation) perpendicular to the flow down the strip. Where row crops are grown in the border strip, furrows confine the flow and eliminate problems with cross slope.

16.3 Level Border Strips and Level Basins

The layout of level border strips is similar to that described for graded border strips, except that the surface is leveled within the area to be irrigated. These areas may be long and narrow, or they may be nearly square (basins). Laser leveling techniques are required to prepare the smooth, level surfaces necessary for this method of irrigation.

Where relatively large rates of flow are available, the field can be quickly covered resulting in high application efficiencies. Level border strips and level basins also lend themselves well to preplant irrigation using stream flow diverted during periods of high runoff. In orchard irrigation, small level basins may contain only one tree.

16.4 Furrows

Irrigation by furrows submerges only from one fifth to one half the surface, resulting in less evaporation and improved soil structure, and permitting cultivation sooner after irrigation than flooding methods. Furrows vary in size and are up and down slope or on the contour. Furrows 80 to 300 mm deep are especially suited to row crops because the furrow can be constructed with normal tillage. Contour furrow irrigation may be practiced on slopes up to 12 percent, depending on the crop, the erodibility of the soil, and size of the irrigation stream. Small, shallow furrows, called corrugations, are particularly suitable for relatively irregular topography and close-growing crops, such as meadow and small grains. Corrugations are often used in poorly leveled border strips.

Furrows may have significant quantities of runoff if a constant inflow rate is maintained throughout the application interval. Application efficiencies tend to be highest when runoff and deep percolation values are similar. To reduce runoff, the design inflow stream can be *cutback* (reduced) when water reaches the end of the field, which greatly improves the application efficiency. This procedure increases the labor requirement because cutback must be made during the irrigation. Unless the supply flow can be decreased, additional labor is required to utilize the flow remaining from the cutback stream. An alternative to cutback is to install a runoff reuse system.

In cases where water advance is too slow because the infiltration rate is high, *surge irrigation* may be advantageous. Here intermittent rather than continuous streams are delivered to furrows. Between surges most or all of the water infiltrates. The next surge advances faster across the wetted portion because the infiltration rate is lower and roughness may be reduced. Fast advance improves uniformity, and proper design and management can reduce runoff and deep percolation, which increases application efficiencies. Special surge valves are available to facilitate surge irrigation. These valves alternate the flow between two sets of furrows to form the surges. The water is delivered to the furrows through gated pipe connected to the surge valves.

Surface Irrigation and the Environment

Runoff and deep percolation frequently occur with surface irrigation and may affect water quality and the environment. Runoff from the end of an irrigated field usually contains low concentrations of dissolved chemicals unless the source water has chemicals or the runoff contains detached soil particles. Chemicals such as fertilizers or pesticides may be carried by the detached soil particles. Environmental effects can be minimized by reusing the runoff on downstream fields, using small stream sizes to minimize erosion and sediment transport by the runoff water, and timely application of fertilizers in amounts to just meet crop needs. If runoff water is reused or controlled, more runoff is probably preferable to deep percolation.

Deep-percolation waters move through the root zone and carry water-soluble fertilizers, pesticides, and/or salts that were leached to maintain a salt balance in the soil. These chemicals may be transported to the ground water supply or to a drainage system. The environmental effects of deep percolation can be minimized by timely application of fertilizers in amounts that just meet crop needs, by minimizing deep percolation through good irrigation design and management, and by applying pesticides several days before irrigation. In general, good water management is important for reducing the environmental impacts of surface irrigation.

Design and Evaluation

Since the hydraulics of surface irrigation are complex, empirical procedures have been employed in designing surface irrigation systems. More rigorous approaches are currently available such as volume balance models and dynamic simulations of surface flow. In designing a system, it is recommended that local practices be explored and locally available data be considered. For example, extension services, experiment stations, and the NRCS have prepared "irrigation guides" suggesting design and management procedures for many states. Selection of design values, such as border strip width, depends on judgment and experience in managing water under specific soil, slope, and crop conditions. For example, border strip width is often based on some multiple of the width of machinery required for crop tillage, planting, or harvesting.

16.5 Surface Irrigation Variables

Recognition and understanding of the variables involved in the hydraulics of surface irrigation are essential to effective design. The pertinent variables (Figure 16-2) are (1) inflow rate and time, (2) infiltration rate, (3) soil and vegetative roughness, (4) channel shape (furrows or borders), (5) field slope, (6) field length, (7) applications depth, (8) flow depth, (9) advance rate, and (10) erosion potential.

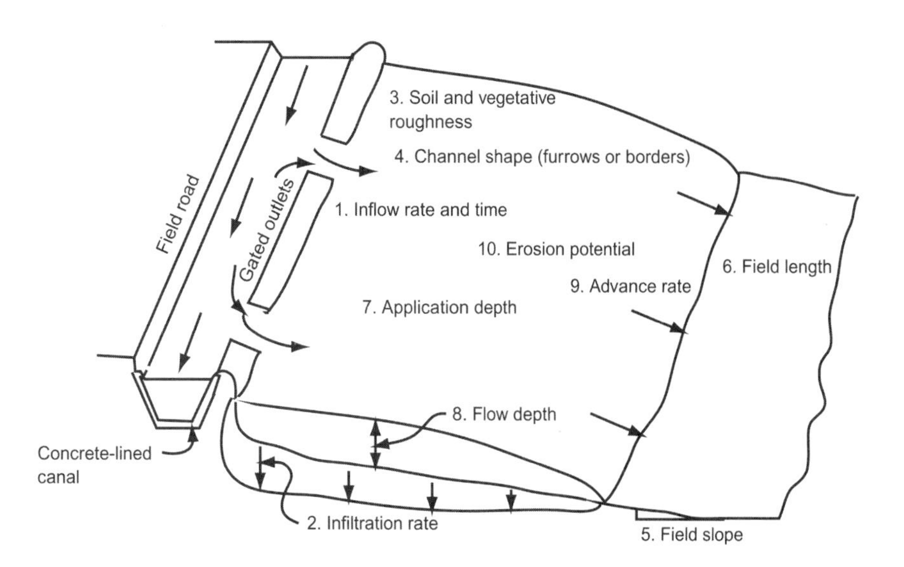

Figure 16–2

Schematic view of flow in surface irrigation indicating the variables involved.

The available stream size may be defined by the delivery system; however, the inflow rate and inflow time to a border strip or a furrow are design variables. The field length and slope are generally defined by property boundaries and the field conditions. The application depth is determined by the crop rooting and soil depths, crop water use, crop response to water stress, water storage capacity of the soil, water availability, and irrigation system (see Chapter 15). The field efficiency will depend on management practices of the irrigator, preparation of the field, and quality of the design.

The effect of roughness is included in the Manning equation

$$Q = \frac{C_u A \ R^{2/3} \ S_f^{1/2}}{n} \tag{16.1}$$

See Chapter 6 for definitions of the variables. See Appendix B for recommended values of the Manning roughness coefficient n.

Infiltration is one of the most important variables in surface irrigation design and performance, and one of the most difficult variables to determine. Infiltration varies throughout the season and is affected by soil texture and structure, compaction, tillage, water content, and channel shape. General soil survey data may provide estimates of infiltration. Some of the variability in infiltration can be overcome by designing separate fields or areas having nearly uniform soil types. Field measurements of infiltration with ponded water tend to underestimate the actual infiltration during irrigation; however, infiltration measurements with flowing water are expensive or can only be made after the irrigation system is installed. Although many equations are used to describe infiltration (Chapter 5), irrigation engineers tend to use some form of the modified Kostiakov equation (Clemmens et al., 2000)

$$z = k\tau^a + b\tau + c \tag{16.2}$$

where z = infiltrated volume per unit area, or depth (L^3/L^2),

k, a, b, c = empirical constants derived for the soil to be irrigated [k (L/T^a), a (dimensionless), b (L/T), c (L)],

τ = infiltration opportunity time (T).

In Equation 16.2, the constant c is large for cracking soils where most of the infiltration occurs in the cracks. The constant b approximates the steady-state infiltration rate after a long period of wetting. The constants k and a define the curvature of the infiltration, primarily after initial wetting. The constant a varies from 0.3 to 0.8, with the small values of a predicting higher infiltration rates for short times of wetting.

16.6 Surface Irrigation Hydraulics

The flow of water across an infiltrating surface is unsteady and nonuniform. The mathematical equations describing the flow of water over the soil surface are based on the principles of conservation of mass and Newton's second law. The resulting nonlinear, parabolic, partial differential equations, can be solved numerically with a computer; however, they are not often solved because it is time consuming and simplified versions yield sufficiently accurate solutions (Strelkoff & Katopodes, 1977). It was found that the inertia (acceleration) terms could be neglected. When computer speed was limiting, even simpler versions, such as kinematic wave and volume balance, were found to be adequate under certain conditions. Zero-inertia models can now be solved on a personal computer and are most frequently used to simulate surface irrigation applications. These models do not directly produce a design, but through multiple simulations a design can be developed. Strelkoff (1990) developed a dimensionless model SRFR for simulating the irrigation of basin, border strips, or furrows with a personal computer. Sets of data covering wide ranges of field and irrigation parameters were obtained from hundreds of simulations with SRFR or similar computer models. These data became the basis for two additional computer programs, BASIN (Clemmens et al., 1995) and BORDER (Strelkoff et al., 1996), that query the data and provide information on irrigation performance for a design. The programs SRFR, BASIN, and BORDER can be downloaded from http://www.uswcl.ars.ag.gov.

16.7 General Design Approach

Before a surface irrigation system can be properly designed, soil, crop, field, and water supply conditions must be known or determined regardless of the design method selected. Soil properties required include depth, water storage characteristics, and infiltration characteristics. Crop properties required include planting dates and characteristics, tillage and harvesting operations, rooting depth, plant density, and water requirement. Size, shape, and slope are important field properties. The location, flow rate, times available, cost, and quality of the water supply must be known. Several of these parameters require field measurements, whereas others may be calculated or obtained from NRCS, universities, consultants, water districts, or other sources.

The major objective of irrigation system design is to provide adequate water to the crop with high efficiency and uniformity. High uniformity of surface irrigation is achieved if the advance curve (time of the advancing water front versus advance distance) and recession curve (time when water disappears from the soil surface versus distance) of the irrigation water are parallel (Figure 16-3). Note that the difference between the advance and recession curves is the infiltration opportunity time. High efficiency is achieved if the depth of application beyond the water

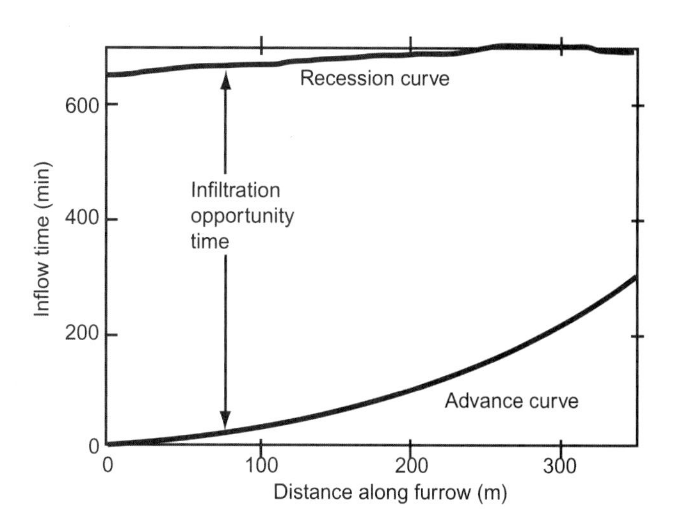

Figure 16–3
Typical advance and recession curves for a furrow irrigated field. (Simulated with SRFR)

storage capacity of soil in the root zone and runoff are minimized. An example of the relationship among the design parameters is shown in the set of performance curves in Figure 16-4 from the BORDER program. This figure illustrates the effect of inflow and border length on potential application efficiency and gives a large range of design solutions in a single graph for one set of field irrigation parameters. A ridge of high efficiency occurs and widens from the lower left to the upper right as the inflow rate and border strip length are both increased. From this type of graph the designer can select the combination of variables, in this case length and inflow rate, that produces the best design for the field situation. This figure demonstrates only one example, because the required depth or slope might be varied and the length fixed by land ownership or terrain. Another useful feature of the BORDER program is the ability to select an inflow–length coordinate on the graph; the resulting infiltration depth profile will appear as shown in Figure 16-5.

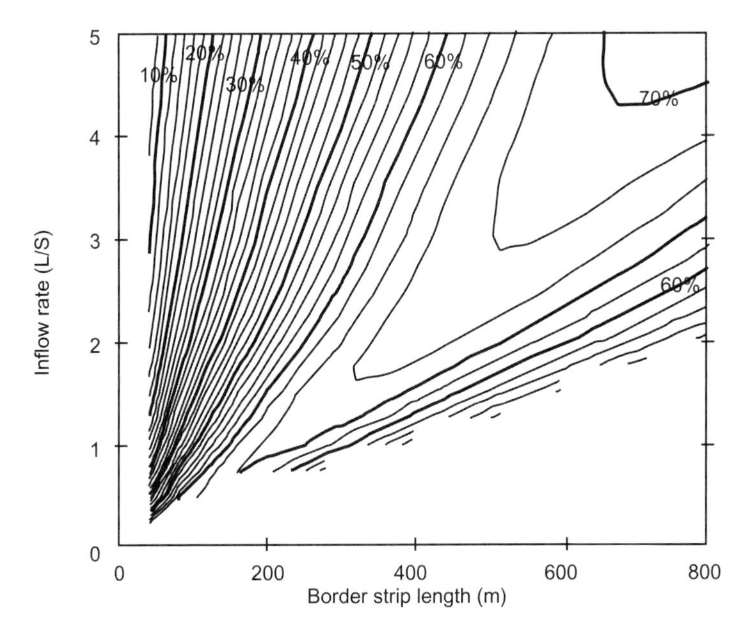

Figure 16–4

Application efficiency contours for border strips with k = 35 mm/h^a, a = 0.55, n = 0.15, slope = 0.002 m/m, and required depth = 75 mm. (From the BORDER program)

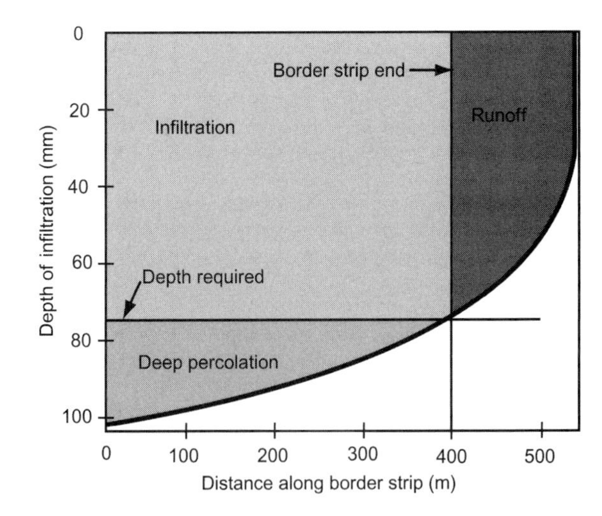

Figure 16–5

Distribution of infiltrated depths for irrigation parameters in Figure 16-4 with length = 400 m and inflow = 2.2 L/s per m of border strip width. (From the BORDER program)

Figure 16-5 shows infiltration depth versus distance down a border strip. Because of the assumed initial conditions, the minimum depth infiltrated equals the required depth of infiltration. The portion of the infiltration curve below the required depth represents the volume of deep percolation. The portion of the infiltration curve beyond the strip length represents the runoff volume. Thus, Figure 16-5 shows a graphical representation of the distribution of water after an irrigation, based on the assumption that the input parameters are actual field conditions.

16.8 NRCS Design Methods

Commonly used references in the United States for the design of border irrigation systems have been Chapter 4 "Border Irrigation," of the *National Engineering Handbook Section 15 Irrigation* published by the SCS (1974) and Chapter 5, "Furrow Irrigation" (SCS, 1984) for graded furrow irrigation systems. The SCS design procedures can also be found in Hart et al. (1983) and Schwab et al. (1993). These publications present empirical equations relating the infiltration, roughness, irrigation efficiency, length of run, time of application, and stream size.

On gently sloping, well-leveled, uniformly graded border strips, application efficiencies of 60 to 75 percent are usually feasible. For design purposes, Manning n values of 0.04 are used for smooth, bare soil surfaces; 0.10 for small grains and similar crops if the rows run lengthwise with the border strip. An n value of 0.15 is suggested for alfalfa and broadcast small grains, while small-grain crops drilled across the border strip and dense sod have an n value of 0.25.

16.9 Assumed Volume Balance Method

In a volume balance method, the sum of the surface volume and the infiltrated or subsurface volume during advance must equal the inflow volume and can be expressed by

$$V_{in}(t) = V_y(t) + V_z(t) \tag{16.3}$$

where $V_{in}(t)$ = inflow volume at time t (L^3),
$V_y(t)$ = volume in surface storage at time t (L^3),
$V_z(t)$ = infiltrated volume at time t (L^3).

Clemmens et al. (2000) proposed using an assumed surface volume approach for surface irrigation system design. Here certain assumptions are made regarding the surface and subsurface volumes in order to determine the advance distance (distance of the advancing front of water on the soil surface). For this case the inflow volume is

$$V_{in}(t) = Q_{in} \times t_x \tag{16.4}$$

where Q_{in} = the inflow rate (L^3/T),
t_x = the time for water to advance to a distance x (T).

The volume in surface storage is

$$V_y(t) = \sigma_y \times A_0(t_x) \times x \tag{16.5}$$

where σ_y = surface storage shape factor defined as the ratio of the average cross-sectional area of flow along the field to the cross-sectional area of flow at the upstream end of the field and generally varies between 0.7 and 0.8, with 0.75 the typical value,

$A_o(t_x)$ = cross-sectional flow area at the inlet at time t_x, typically determined from the Manning equation (Equation 16.1) (L^2).

When infiltration is defined by Equation 16.2, the infiltrated volume during advance is given by

$$V_z(t_x) = W\left[\sigma_z \times k \times t_x^a + \left(\frac{h}{1+h}\right)b(t_x) + c\right]x \qquad \textbf{16.6}$$

where σ_z = subsurface shape factor,

W = the furrow spacing (L).

The subsurface shape factor σ_z is the field average infiltrated depth divided by the infiltrated depth at the head of the field. Advance is often described by

$$t_x = s \times x^h \qquad \textbf{16.7}$$

where t_x is the time for water to advance to a distance x, and s and h are constants for a specific irrigation event. The subsurface shape factor is obtained by integrating the depth infiltrated by the kt^a term in Equation 16.2 over the advance distance, with advance defined by Equation 16.7, then dividing by the length to obtain the average depth infiltrated. The average infiltrated depth is divided by the depth infiltrated at the upstream end of the field to obtain σ_z, which becomes

$$\sigma_z = \frac{h + a(h-1) + 1}{(1+a)(1+h)} \qquad \textbf{16.8}$$

The design procedure requires an iterative approach because of the interrelationships among the field parameters. In general, advance is first computed from Equations 16.1, 16.2, and 16.5 through 16.8. It is frequently assumed that the flow depth at the inlet is at normal depth for the entire time of inflow and the friction slope is assumed equal to the bottom slope, $S_o = S_f$. These assumptions are generally valid for field slopes greater than 0.05 percent. Since the minimum depth of infiltration usually occurs at the downstream end of the field, the required infiltration opportunity time there must equal the time to infiltrate the required depth. This means the recession time at the downstream end equals the advance time plus the required infiltration opportunity time. The time of cutoff of inflow must be sufficient to meet the downstream required recession time and must consider the effects of water in surface storage at cutoff and available for infiltration. Once the cutoff time is known, the recession times are determined, which allows calculation of the infiltration opportunity times and infiltrated depths or volumes over the entire field. Finally, runoff, deep percolation, and efficiency can be estimated. If the efficiency is not satisfactory, the inflow rate is typically changed and the process repeated. The efficiency may also be improved by changing border strip or furrow length or slope, or the required depth of application.

16.10 Assumed Volume Balance Furrow Irrigation Design

Application efficiencies for graded furrows are usually in the 50 to 60 percent range unless a cutback, a surge flow, or a reuse system is used. Suggested furrow lengths for various slopes and soil textures are given in Table 16-1 and vary from 60 to 800 m. Furrow slopes of 2 percent or less are recommended to avoid excessive erosion. In humid regions slopes must be smaller, and maximum slopes can be estimated from SCS (1984) and Hart et al. (1983)

$$S_{max} < 67/\,(P_{30})^{1.3} \qquad \textbf{16.9}$$

where P_{30} is the 30-min, 2-year rainfall (mm) (Chapter 3) and S_{max} is the maximum allowable furrow slope in percent. The maximum allowable stream size to avoid excessive erosion in furrows can be obtained from

$$Q_{max} < 0.60/S \qquad \textbf{16.10}$$

where Q_{max} is the maximum furrow inflow in L/s and S is the furrow slope in percent. Inflows into furrows with small slopes are often limited by capacity, and inflow rates are lower than those computed from Equation 16.10.

Infiltration in furrows is more complicated than on flat surfaces such as in borders and basins. The infiltration from furrows is two-dimensional with downward flow in the furrow bottom and lateral flow in the sides of the furrow, depending on the furrow shape. In coarse-textured sandy soils, gravitational forces predominate and flow is mostly downward and proportional to the wetted width. In fine-textured soils, capillary forces predominate and the lateral flows from adjacent furrows will meet and infiltration becomes proportional to the furrow spacing. In many cases both gravitational and capillary forces affect the flow; however, it is still possible to define infiltration based on furrow spacing with

TABLE 16–1 Suggested Maximum Lengths of Cultivated Furrows

Furrow	Average Depth of Water Applied (mm)											
	75	150	225	300	50	100	150	200	50	75	100	125
Slope (%)	Clays				Loams				Sands			
	Length (m)											
0.05	300	400	400	400	120	270	400	400	60	90	150	190
0.1	340	440	470	500	180	340	440	470	90	120	190	220
0.2	370	470	530	620	220	370	470	530	120	190	250	300
0.3	400	500	620	800	280	400	500,	600	150	220	280	400
0.5	400	500	560	750	280	370	470	530	120	190	250	300
1.0	280	400	500	600	250	300	370	470	90	150	220	250
1.5	250	340	430	500	220	280	340	400	80	120	190	220
2.0	220	270	340	400	180	250	300	340	60	90	150	190

Source: Booher (1974).

adequate measurements. Because the flow depth decreases along a furrow, the infiltration also tends to decrease. To define the wetted width, an average flow depth is recommended.

The furrow shape is required in order to calculate the flow area. A trapezoidal shape is assumed, while parabolic shapes are possible. With the furrow shape, Q, S_f, and n known, the flow area at the head of the furrow can be determined by trial and error from Equation 16.1. It is assumed that the infiltrated volume per unit length does not change with flow depth or wetted width. This assumption is not valid for coarse-textured soils, however, the effects are small if the runoff is not minimized.

The design is essentially trial and error for a given set of field length, slope, inflow rate, infiltration, roughness, and required infiltration depth. The procedure starts with determination of advance time with calculation of h based on estimated time for the advancing stream to reach half the field length and the end of the field from

$$h = \log_{10}(t_{L_F}/t_{L/2})/\log_{10}(x_{L_F}/x_{L/2}) \quad \textbf{16.11}$$

which can be derived from Equation 16.7.

The minimum depth of infiltration typically occurs at the downstream end of the field. To assure that the desired depth of infiltration occurs at the downstream end of the field, the cutoff time of inflow must include the advance time and the time to infiltrate the desired depth. The cutoff time is calculated from

$$t_{co} = t_A(L_F) + \tau_{req} \quad \textbf{16.12}$$

where t_{co} = time of cutoff of the inflow (T),
$t_A(L_F)$ = time of advance to the end of the field (T),
τ_{req} = time required to infiltrate the design depth (T).

In Equation 16.12 the water in surface storage in the furrow at cutoff is neglected. This is a valid assumption for sloping furrows with free runoff or no ponding at the downstream end of the field; however, if the slope is small, this volume cannot be neglected. The corrected cutoff time can be obtained by including a term for the surface storage, then Equation 16.12 becomes

$$t_{co} = t_A(L_F) + \tau_{req} - V_y(L_F)/Q_{in} = t_A(L_F) + \tau_{req} - \sigma_y A_0 L_F/Q_{in} \quad \textbf{16.13}$$

where $V_y(L_F)$ is the surface volume at time of cutoff. The recession curve is assumed to vary linearly from the value obtained from Equation 16.13 at the upper end to the value obtained from Equation 16.12 at the lower end of the field.

For design procedures for furrow irrigation systems with cutback or for reuse systems, see Clemmens et al. (2000) or other references.

Example 16.1

Design a sloping-furrow irrigation system for a field in an arid region, length is 350 m, slope is 0.003 m/m, n = 0.04, field-measured infiltration based on the furrow spacing is z(mm) = 20(mm/h$^{0.45}$) t(h)$^{0.45}$ + 5 (mm/h) t (h), furrow spacing is 0.9 m, and the desired application depth is 80 mm. A stream of 120 L/s is available. Assume the furrow shape is a trapezoid, with bottom width of 100 mm and sideslopes of 1.5:1 (H:V), and $\sigma_y = 0.75$. Use a spreadsheet to perform the calculations.

Solution.

(1) Enter input data.

(2) Iterate Equation 16.2 and find $\tau_{req} = 398$ min for $z_{req} = 80$ mm.

$$z = 20\left(\frac{398}{60}\right)^{0.45} + 5\left(\frac{398}{60}\right) + 0 = 46.8 + 33.2 + 0 = 80 \text{ mm}$$

(3) Calculate Q_{max} from Equation 16.10.

$$Q_{max} = \frac{0.60}{0.003 \times 100} = 2.0 \text{ L/s}$$

(4) Assume $Q_{in} = 1.2$ L/s. Now iterate the flow depth in Equation 16.1 until it gives the assumed Q_{in}. The results are

Flow depth $y_n = 0.0503$ m.

$A_o = 0.0088 \text{ m}^2$, wetted perimeter $= 0.2814$ m

$$Q_{in} = 0.0088\left(\frac{0.0088}{0.2814}\right)^{2/3} \times 0.003^{1/2}/0.04 = 0.0012 \text{ m}^3/\text{s} = 1.20 \text{ L/s}$$

(5) Choose initial advance times with time to advance to the end of the field (L) two to three times the time chosen to advance one half the field ($L_F/2$). Thus 50 min was chosen for advance to 175 m ($L_F/2$) and 150 min for advance to 350 m (L_F) for starting values to calculate an initial h from Equation 16.11 and initial σ_z from Equation 16.8.

$$h = \log_{10}(150/50) \,/\, \log_{10}(350/175) = 1.585$$

$$\sigma_z = \frac{1.585 + 0.45(1.585 - 1) + 1}{(1 + 0.45)(1 + 1.585)} = 0.760$$

Perform volume balance calculations for chosen initial advance times to $L_F/2$ and L_F.

For 50 minutes, $$V_{in}(t) = \frac{1.2(\text{L/s})\left[60 \text{ (s/min)}\right]\left[50 \text{ (min)}\right]}{1000 \text{ (L/m}^3)} = 3.6 \text{ m}^3$$

From Equation 16.5, with $x = 175$, $V_y(t) = 0.75\,[0.00881\,(\text{m}^3)][175\,(\text{m}^3)]$
$= 1.156 \text{ m}^3$

From Equation 16.6,

$$V_z(t) = 0.9(\text{m}) \times \left[\frac{\left(0.760 \times 20 \times \left(\frac{50}{60}\right)^{0.45} + \left(\frac{1.585}{1.0 + 1.585}\right) \times 5 \times \left(\frac{50}{60}\right) + 0\right)(\text{mm})}{1000(\text{mm/m})} \right] \times 175\ (\text{m}) = 2.607\ \text{m}^3$$

$$\text{Volume error} = [V_{in}(t) - V_y(t) - V_z(t)]/V_{in}(t)$$

$$= 100[3.60 - 1.156 - 2.607]/3.60 = 4.5 \text{ percent.}$$

The results for x = 175 and x = 350 m are shown in the following table.

x(m) Input	*t_A (min) First choice*	*Vol. error (%)*	*$V_{in}(t)$ inflow volume (m³) Equation 16.4*	*$V_y(t)$ surface storage (m³) Equation 16.5*	*$V_z(t)$ infiltrated volume (m³) Equation 16.6*
0	0				
175	50	−4.5%	3.600	1.156	2.607
350	150	−10.7%	10.800	2.312	9.645

(6) To obtain a better volume balance, divide the field into 8 segments. Substitute the values from the table in Step 5. Calculate a new volume balance for x =175 and 350 m by iterating t_A until the volume error is less than 0.1 percent. This results in the advance times of 55.4 and 192 min to 175 and 350 m, respectively. Recalculate h from Equation 16.11 and σ_z from Equation 16.8 for each iteration. For the final iteration, the values of h and σ_z are

$$h = \log_{10}(192/55.4) \,/\, \log_{10}(350/175) = 1.794$$

$$\sigma_z = \frac{1.793 + 0.45(1.793 - 1) + 1}{(1 + 0.45)(1 + 1.793)} = 0.7778$$

The following volume balance table is obtained by iterating t_A for the remaining segments to the nearest 0.1 minute. Note that h and σ_z do not need to be recalculated because they depend only on the values for x = 175 and 350 m. One aid to the iteration process can be obtained by estimating t_A with Equation 16.7. For x = 43.75 the calculation is

$$t_x = t_L\left(\frac{x}{L_F}\right)^h \text{ or } t_{43.75} = 192.0\left(\frac{43.75}{350}\right)^{1.794} = 4.61 \text{ min}$$

All calculated values of t_A are shown in the seventh column of the table below and are used as a first estimate in the second column, which are then changed manually until the volume error is acceptable. A second aid is to use a solver analysis. The volume error is squared in the last column. The volume error is summed, which becomes the target cell for solver. Solver is programmed to change t_A until the target cell reaches an acceptable value with the restriction that the advance time must be greater than zero.

x (m)	t_A (min)	Vol. error (percent)	$V_{in}(t)$ (m^3)	$V_y(t)$ (m^3)	$V_z(t)$ (m^3)	Calc. t_A (min)	$(Vol\text{-}err)^2$
0	0						
43.75	7.6	0.09	0.547	0.290	0.258	4.6	2.11E-04
87.5	19.4	−0.04	1.398	0.579	0.819	16.0	5.49E-03
131.25	35.2	−0.04	2.537	0.869	1.669	33.1	6.76E-03
175.0	55.4	0.04	3.986	1.158	2.830	55.4	6.55E-05
218.75	80.4	0.00	5.789	1.448	4.341	82.7	2.38E-04
262.5	110.9	0.00	7.985	1.737	6.248	114.6	1.84E-04
306.25	147.7	−0.02	10.634	2.027	8.610	151.1	1.62E-05
350.0	192.0	0.00	13.824	2.317	11.507	191.9	3.60E-05
						Sum	1.30E-02

(7) Calculate cutoff time t_{co} from Equation 16.12 and volume applied from Equation 16.4, with $t_x = t_{co}$

$$t_{co} = 192 + 398 = 590 \text{ min} = 9.83 \text{ h}$$

$$V_{in} = (1.2)(60)(590)/1000 = 42.480 \text{ m}^3 \text{or an average depth of } 134.9 \text{ mm}$$

(8) Compute infiltrated depths and volumes assuming recession occurs at 590 minutes everywhere along the furrow.

x (m)	t_A (min)	t_{rec} (min)	τ (min)	z (mm)	V_z (m3)
0.0	0.0	590	590	105.1	
43.75	7.6	590	582	104.2	4.120
87.5	19.4	590	571	102.7	4.072
131.25	35.2	590	555	100.6	4.002
175.0	55.4	590	535	98.1	3.912
218.75	80.4	590	510	94.8	3.798
262.5	110.9	590	479	90.9	3.656
306.25	147.7	590	442	86.0	3.482
350.0	192.0	590	398	80.0	3.269
				Total	30.311

Note that z is calculated from Equation 16.2 with $\tau = t_{rec} - t_A$ and V_z from the depths z.

(9) Calculate irrigation performance. (Note that furrow spacing is 0.9 m.)

Total infiltration $= 30.31\ m^3$ or average depth infiltrated $= 96.2$ mm,

Runoff $= 42.48 - 30.31 = 12.17\ m^3$ or 38.6 mm,

Deep percolation $= 96.2 - 80 = 16.2$ mm $= 5.1\ m^3$,

Application efficiency $= 80/(96.2 + 38.6) = 59.3$ percent.

(10) Solve Equation 16.13 to determine if t_{co} should be adjusted for surface storage.

$$t_{co} = 192 + 398 - 0.75 \times 0.0088 \times 350 \times 1000/(1.2 \times 60) = 558 \text{ min}$$

This is 32 minutes less than from Equation 16.12, and the recession curve should be adjusted.

(11) Calculate the new recession curve and infiltrated depths and volumes.

x (m)	t_A *(min)*	t_{rec} *(min)*	τ *(min)*	*z (mm)*	V_z (m^3)
0.0	0.0	558	558	101.0	
43.75	7.6	562	554	100.6	3.969
87.5	19.4	566	546	99.6	3.941
131.25	35.2	570	535	98.1	3.891
175.0	55.4	574	519	96.0	3.821
218.75	80.3	578	498	93.3	3.726
262.5	110.9	582	471	89.8	3.604
306.25	147.7	586	438	85.5	3.451
350.0	192.0	590	398	80.0	3.258
				Total	29.662

(12) Recalculate irrigation performance.

Total infiltration $= 29.66\ m^3$ or equivalent to an average depth of 94.2mm,

Runoff $= [(1.2)(558)(60)/1000] - 29.66 = 10.5\ m^3$ or 33.3 mm,

Deep percolation $= 4.46\ m^3$ or 14.2 mm,

Application efficiency $= 80/(94.2 + 33.4) = 62.7$ percent.

Distribution uniformity low quarter,

$$DULQ = \left[\frac{\frac{(89.8 + 85.5)}{2} + \frac{(85.5 + 80.0)}{2}}{2} \right] \frac{1}{94.2} = 0.90$$

The application efficiency improved from 59.3 to 62.7 or 3.4 percent. This difference is small and may not be significant compared to potential infiltration inaccuracies. Typically the runoff and deep percolation volumes should be similar for highest efficiency. Here runoff is higher and could potentially be reduced by decreasing the inflow.

(13) Reduce the inflow and repeat the calculations. An inflow of 1 L/s is the accepted design inflow rate with the following results.

$$t_{co} = 646 \text{ min}$$

Total infiltration $= 32.18$ m^3 or an average infiltrated depth of 102.1 mm,

Runoff $= 6.6$ m^3 or 20.9 mm,

Deep percolation $= 6.98$ m^3 or 22.1 mm,

Application efficiency $= 80/(102.1 + 20.9) = 65$ percent.

$DULQ = 0.87$

(14) With a total flow of 120 L/s available, 120 furrows will be irrigated in each set. To account for infiltration and other variations between furrows and during the season, an experienced irrigator will adjust the furrow inflow rates by changing the flow to individual furrows by adjusting the elevation of the outlet of the siphon tube, or changing the gate opening on gated pipe, or changing the number of furrows irrigated in a set to obtain the highest efficiency for each irrigation application.

16.11 Furrow Irrigation with Surge Flow

With surge flow irrigation of furrows, water is typically cycled between two sets of furrows for intermittent advance and wetting of the soil (Figure 16-6) (Walker & Skogerboe, 1987).

Intermittent wetting tends to reduce the infiltration rate of the soil, particularly for medium to coarse textured soils (Stringham, 1988; Clemmens et al., 2000). The reduced infiltration on high infiltration rate soils potentially reduces deep percolation and improves irrigation uniformity. Surge irrigation in furrows has gained acceptance with the development of special valves for flow control. Commercial surge flow valves control the cycles and have the additional advantage of providing a cutback flow by cycling more frequently (typically 10- to 20-minute cycles) after advance is completed. The cutback flow potentially reduces runoff and increases efficiency. Application efficiencies of 80 percent are typical with surge irrigation.

The time between surges should be sufficient to allow essentially all the water on the surface to infiltrate. How this reduces infiltration is not entirely understood, but is believed to be a combination of consolidation of soil particles at the surface, a reduction in the water potential gradient, and other factors. Two approaches have been proposed for modeling infiltration during surge flow. Izuno & Podmore (1985) proposed using the Kostiakov equation during the first surge and a constant infiltration rate during all subsequent wetting cycles. Blair & Smerdon (1987)

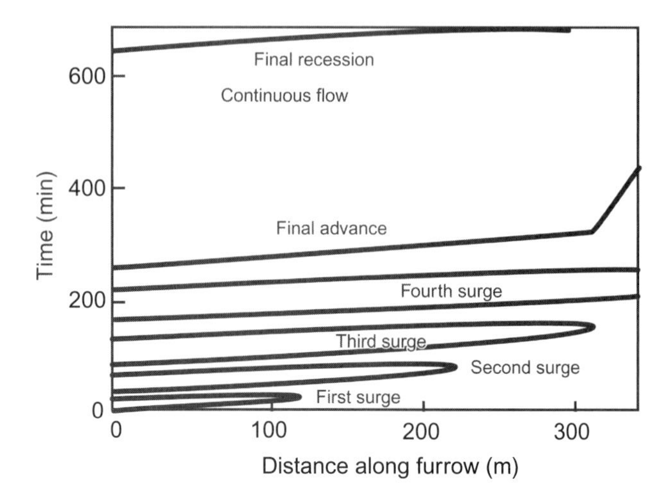

Figure 16–6
Example of the advance and recession curves for surge flow in furrows. (Simulated with SRFR)

proposed that the infiltration curve applies throughout the irrigation, but water only accumulates in the soil when water is on the surface; i.e., time continues from first wetting, but infiltration only occurs during wetting. The first model is easier to use, but the second seems to have better agreement with field observations. Both models are complicated to use with the assumed volume balance approach.

Design procedures have been difficult to develop. Because most surge systems are improvements to existing systems, it has proven satisfactory to use the first set of an irrigation to adjust the surge valve to the actual field conditions. The inflow rates with surge irrigation are larger than with normal furrow irrigation to assure fast advance and an adequate stream during the continuing phase, but rates should not exceed the maximum nonerosive stream. The surge inflow times in the surge valves are adjusted to give four to six surges (Figure 16-6) before water reaches the end of the field. Efficiency is improved if the next-to-last advance stops near the field end so that the wetting time of the last advance over the dry portion is maximized. The cycle times should be adjusted so that the surges during advance do not overlap but do overlap during the continuing phase. After the last surge, the valve is adjusted to cycle about every 10 to 20 minutes so that the inflow to each furrow is essentially equivalent to half the initial surge flow rate. Thus, a cutback flow rate is achieved. These 10- to 20-minute cycles are continued until the desired amount of water has been applied.

Reducing the infiltration of fine-textured soils may not be useful; however, the mechanization of cutback with surge valves may save labor and reduce runoff. Instead of repeated surges, only one surge is applied to each set. Thus water is applied until it reaches the end of the furrows for the first set, then diverted to the furrows in the second set. After water reaches the end of the furrows in the second set, the valve is adjusted to cycle about every 10 to 20 minutes to provide half the flow rate to all furrows in both sets.

16.12 Border Strip Irrigation Design

Border strip lengths and slopes are generally defined by the inflow stream, infiltration rate, field size, and land slope. The application efficiency will depend on the design, the management practices of the irrigator, and the preparation of the field for irrigation. On gently sloping, well-leveled, uniformly graded fields, an efficiency of 60 to 75 percent is usually feasible. SCS (1974) gives suggested design efficiencies.

Some design limits are necessary to prevent erosion and ensure minimal depths of flow. The maximum nonerosive stream for crops that are not sod-forming, such as small grains and alfalfa, is

$$q_{max} = 0.00018/S^{0.75} \tag{16.14}$$

where q_{max} is the maximum inflow rate in $m^3\ s^{-1}\ m^{-1}$ of border strip width or m^2/s (border strips are designed using flow per unit width q), S is the border strip slope in m/m. For crops that form a dense sod, Equation 16.14 can be increased by a factor of 2. Minimum flow depths are needed to adequately spread water over the entire soil surface. Thus a minimum inflow is required, which can be computed from

$$q_{min} = 6.0 \times 10^{-6} L \times S^{0.5}/n \tag{16.15}$$

where q_{min} is the minimum inflow rate in m^2/s, L is the border strip length in m, S is the border strip slope in m/m, and n is the Manning roughness coefficient.

16.13 Assumed Volume Balance Border Strip Irrigation Design

The assumed volume-balance design process for sloping border strips with free outflow is similar to that used for sloping furrows (Clemmens et al., 2000) and is based on Equations 16.3 through 16.8. One simplification is to use a unit width so that the Manning equation for normal depth can be written as

$$y_n = \left(\frac{q_{in}\, n}{S_o^{1/2}}\right)^{3/5} \tag{16.16}$$

where y_n is the normal depth in m, q_{in} is the inflow per unit width of strip in $m^3\ s^{-1}\ m^{-1}$ or m^2/s, and S_o is the field slope in m/m.

In border strip irrigation the recession curve varies much more with distance than for furrow irrigation. The recession time at the downstream end is set to the time to infiltrate the required depth of irrigation there. This is a reasonable assumption, except on short strips with steep slopes where the minimum depth may be at the upstream end. Even under this condition, the lowest quarter of infiltrated depths is split between the upstream and downstream ends and the area of the field that might be underirrigated is small. For details of this method, see Clemmens et al. (2000).

16.14 Border Strip Irrigation Design Based on Computer Simulations

Numerous simulations of border strip irrigations with various field conditions were made by Shatanawi (1980) using a zero-inertia model producing infiltrated depth profiles. These results were utilized in the border strip irrigation design program BORDER (Strelkoff et al., 1996). With this program the user enters the field conditions and evaluates the influence of various design parameters on performance.

Example 16.2 Design a border strip irrigation system for a field that is 400 m long, slope is 0.002 m/m, alfalfa is being grown so $n = 0.15$, infiltration is given by $z(\text{mm}) = 35(\text{mm/h}^{0.55})\ t(\text{h}^{0.55})$, the desired depth of application is 75 mm, and a stream of 120 L/s is available. Assume typical machinery width is 6.5 m. Use the BORDER program to evaluate and select a design.

Solution.

(1) Determine q_{max} and q_{min} from Equations 16.4 and 16.5.

$$q_{max} = \frac{0.00018}{0.002^{0.75}} = 0.019\ \text{m}^3\text{s}^{-1}\ \text{m}^{-1} = 19\ \text{Ls}^{-1}\ \text{m}^{-1}$$

$$q_{min} = \frac{0.000006 \times 400 \times 0.002^{0.5}}{0.15} = 0.0007\ \text{m}^3\text{s}^{-1}\ \text{m}^{-1} = 0.7\ \text{Ls}^{-1}\ \text{m}^{-1}$$

(2) Open the BORDER program.

(3) Choose Options.

Units: Select Metric and Hours for time,

Precision: Select Fine for Performance Grid Resolution, Contour Interval, and Iteration Tolerance,

Grid selection: Select Performance grid.

(4) Choose Field Characteristics.

Enter: Infiltration variables $k = 35\ \text{mm/h}^a$, $a = 0.55$,

Enter: Manning roughness coefficient $n = 0.15$,

Enter: Slope $S = 0.002$ m/m,

These are the same data used in Figure 16.4.

(5) Select Application, then Operation Evaluation.

Enter: Target depth = 75 mm,

Enter: Length = 800 m,

Enter: Border width = 1.0 m,

Enter: Inflow 3 L/s,

Enter cutoff time = 4 hours,

Choose the Minimum Depth option.

(6) Select Application, then Physical Design.

Enter: Border Width = 1.0 m.

(7) Select Execute, then Options.

Select Performance Parameter = Potential Application Efficiency,

Enter Display Ranges Minimum Inflow = 0.0, Maximum Inflow = 5.0 L/s,

Enter: Minimum Length = 0.0, Maximum Length = 800 m.

A flow rate less than q_{max} and a length longer than the actual field length are entered to obtain a better view of the performance curves.

(8) Select Execute, then GO. A graph of performance contours appears and is shown in Figure 16-4. This graph presents a range of solutions for the field parameters entered. The advantage of these curves is that the efficiency is displayed for a wide range of inflows and border strip lengths, so that results of other design choices can be easily observed.

(9) Place the cursor on the 400-m-length grid line and inflow of 2.2 L/s, which is about the peak application efficiency. When the left mouse button is clicked, a figure similar to the infiltration profile of Figure 16-5 appears. Data to the right of the figure (not shown) indicate an application efficiency of 66.9 percent. And a time of application of 340 minutes. The same efficiency is obtained for inflows of 2.1 and 2.3 L/s. These flow rates are greater than q_{min} and less than q_{max}, and the design is acceptable.

(10) Check the normal depth from Equation 16.16

$$y_n = \left(\frac{2.3/1000 \times 0.15}{0.002^{1/2}}\right)^{3/5} = 0.054 \text{ m}$$

The border levee, including freeboard, must be 0.054/0.75 = 0.072 m in height.

(11) A different set of curves is obtained if the Application-Management option is selected. Enter Border Length = 400 m, Border Width = 1 m, and select the Cutoff Time option. Under Execute-Options, set Performance Parameters = Potential Application Efficiency, select Minimum Depth, enter Minimum Inflow = 0.7 L/s, Maximum Inflow = 19.0 L/s, Minimum Cutoff Time = 0.0, and Maximum Cutoff Time = 7.0 hours. Run the program to display Efficiency contours for inflow rate versus cutoff time. Efficiencies above 70 percent are possible. For example, if the intersection for an inflow of 8.0 L s^{-1} m^{-1} and cutoff time of 53.6 min is clicked, an infiltrated depth profile will be displayed showing an application efficiency of 75.7 percent. The normal depth y_n for 8.0 L s^{-1} m^{-1} is 0.114 m, requiring a minimum levee height of 0.152 m, which is usually feasible. The minimum depth of application is 48.6 mm, which occurs at the upper end of the field. This illustrates that for this field situation the application efficiency can be increased if the depth of application is decreased; however, more frequent applications are required, which increases labor costs. In addition, flows this large for short times are more difficult for the irrigator to manage.

(12) The border strip width is usually two or four times the machinery width—in this case 13 or 26 m. The final decision may depend on the quality of field

preparation. If there is no cross slope, the 26-m width is preferable because fewer border levees are required. With 120 L/s available, an inflow rate of 60 L/s per border strip 26 m wide or 2.31 L s^{-1} m^{-1} border width is selected for the design and application efficiency = 66.8 percent. This is within the range calculated above and acceptable. Thus the field would be divided into border strips 26 m wide and 2 strips irrigated simultaneously for 324 min with the flow divided equally.

This example illustrates how the BORDER program can aid the design of border strip irrigation systems. The program will also evaluate existing systems by entering measured field data for an irrigation, and viewing alternative management or re-design options, and comparing them with the field measurements.

16.15 Basin Irrigation Design

The efficiencies of basin irrigation systems can exceed 90 percent with proper design and management. The volume balance approach can be used for basin irrigation design or the program BASIN (Clemmens et al., 1995) can be used. A few differences between the design of level basins and sloping furrows or border strips should be noted. First, the basins have no slope and are surrounded by levees, so there is no runoff and recession will occur nearly simultaneously over the basin. Second, the Manning equation does not apply to zero slopes, thus instead of the field slope, the water surface slope is used. The water surface slope is typically based on upstream depth and advance distance during the advance phase and on upstream and downstream depths if inflow continues after water reaches the end of the basin.

16.16 Evaluation of Existing Systems

The design of new surface irrigation systems constitutes an important engineering activity. Since few new surface systems are being installed, even greater opportunities are often available in the evaluation and improvement of existing systems. Many of these older systems were designed and installed before the current design tools were available and when water supplies were not as limited as they are now. Fields or portions of fields that do not receive sufficient water have reduced production. Excessive irrigation not only wastes water, but may leach water-soluble chemicals and nutrients into the ground water or a drainage system, or may carry chemicals and other nutrients in the surface runoff.

Evaluation of existing systems can be approached in several ways. Observations of variable crop growth and high (dry) and low (wet) areas of the field indicate the need for releveling. The simplest measurement, but very valuable, is to probe the soil the day after irrigation with a rod or soil probe. The purpose is to identify the depth of the wetting front to determine if the field has been adequately irrigated and to find the uniformity of application over the field. The rod will penetrate easily to the depth of the wetted soil but will be stopped by dry soil. Thus, it will easily show areas of underirrigation. If the rod penetrates easily everywhere, it probably indicates overirrigation. This method cannot detect if the soil was wet from previous irrigation, and it cannot detect differences in soil type. Also, the rod may be stopped by rocks. While the irrigator is walking the field probing the soil, additional observations of plant growth, diseases, and insects will be beneficial to overall crop management.

Flow measurement devices should be installed permanently to measure the inflow so the irrigator can determine the average depth applied and water costs. An additional measuring device should be installed to measure runoff. The difference between inflow and runoff is the water infiltrated, which should be compared to the soil water deficit prior to irrigation. From these measurements, the irrigator can estimate the application efficiency.

The soil water deficit can be estimated from ET calculations (Chapter 4) or more accurately obtained by soil measurements. Soil water measurements before irrigation show the effects of previous irrigations, show differences in soil types, and may explain crop growth patterns. The difference between measurements taken before irrigation and one to two days after irrigation will be the depth of water stored in the soil by irrigation. These depths can estimate the uniformity of application and, when combined with flow measurements described above, permit calculation of application efficiency.

A volume balance may be performed on an actual field irrigation for evaluation purposes using inflow, runoff, and soil water deficit data. In addition, advance and recession times at equally spaced locations along the field are required. The differences between the advance and recession times are the infiltration opportunity times. If the advance and recession curves are plotted, the irrigator has a graphic representation of the uniformity of the irrigation. Parallel curves occur for a uniform irrigation, assuming uniform soils and infiltration rates. From the average depth infiltrated and the average infiltration opportunity times, one point on the infiltration curve is obtained. An infiltration curve for the field can be obtained from a detailed volume balance; however, flow depths also must be obtained during the irrigation. The volume balance method of evaluating an irrigation is the most accurate but also the most time consuming. For details on various diagnostic procedures, see Merriam & Keller (1978). Results from these evaluations are important sources of information for the redesign of a surface system or the development of improved management criteria.

16.17 Installation of Surface Irrigation Systems

The installation of surface irrigation systems begins with the topographic survey of the fields, which gives the slopes required for designing a system. The completed design will include the new field elevations according to the design slope. Leveling the field to the design elevations is the first step in installation of a surface irrigation system. If concrete-lined ditches are to be installed, some of the soil from the field will be used to form soil levees (ditch pads) for the ditches. These levees are installed at the design slope and elevation of the ditch. The elevation of the ditch should be sufficient to provide head for water delivery from the ditch to the field surface.

Distribution of Water

The water supply for surface irrigation is normally delivered either from surface storage by conveyance ditches or from irrigation wells. Sometimes surface storage and underground supplies are combined to provide an adequate water supply.

16.18 Surface Ditches

A system of open ditches often distributes the water from the source to the field as shown in Figure 16-1. These ditch systems should also be carefully designed, as discussed in Chapter 6, to provide adequate head (elevation) and capacity to supply water for all areas to be irrigated. The amount of land that can be irrigated is often limited by the quantity of water available as well as by the design and location of the ditch system. Where irrigation is used as an occasional supplement to rainfall, these ditches may be temporary. To minimize water losses, ditches may be lined with impermeable materials. This practice is particularly applicable in the more arid regions, where irrigation water supplies are limited and crop needs are largely dependent on irrigation water.

16.19 Devices to Control Water Flow

Control structures, such as illustrated in Figure 16-7, are essential in open ditch systems to (1) divide the flow into two or more ditches, (2) lower the water elevation without erosion, and (3) raise the water level in the ditch so that it will have adequate head for removal. Various devices are used to divert water from the

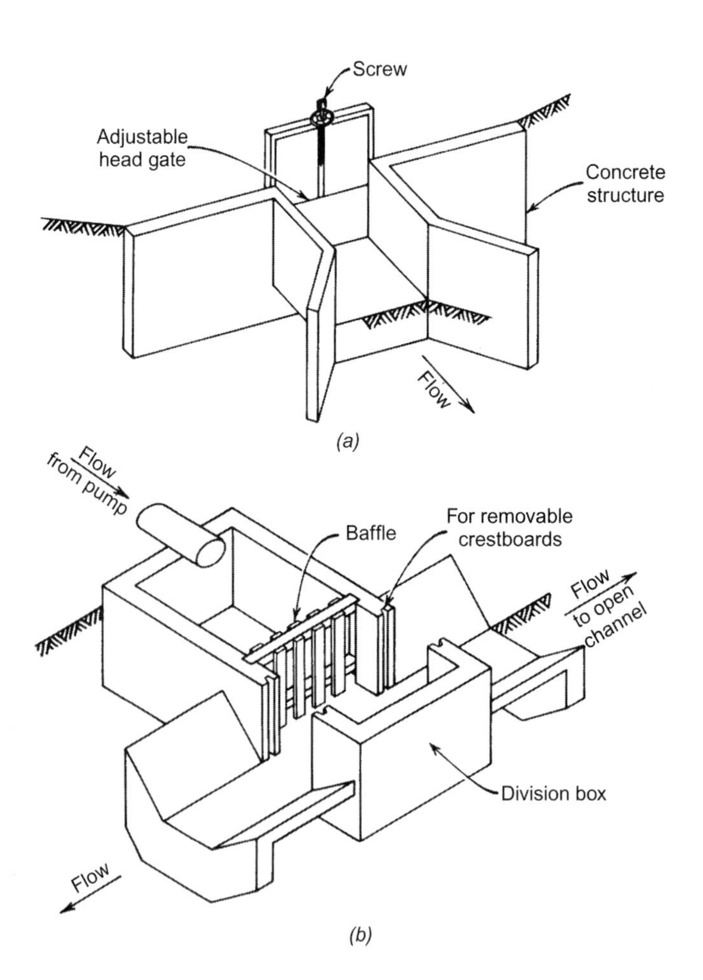

Figure 16–7

Concrete irrigation structure. (a) Canal flow control gate. (b) Pump outlet and division box. (Redrawn from SCS, 1979)

irrigation ditch and to control its flow to the appropriate basin, furrow, or border strip. Valves may be installed in the side or bottom of the ditch during construction. Other devices shown in Figure 16-8 are spiles, gate takeouts for border strip irrigation, and siphon tubes. These siphons, usually aluminum or plastic, carry the water from the ditch to the surface of the field and have the advantage of metering the quantity of water applied. Figure 16-9 gives the rate of flow that can be expected from siphons of various diameters and head differences between the inlet and outlet.

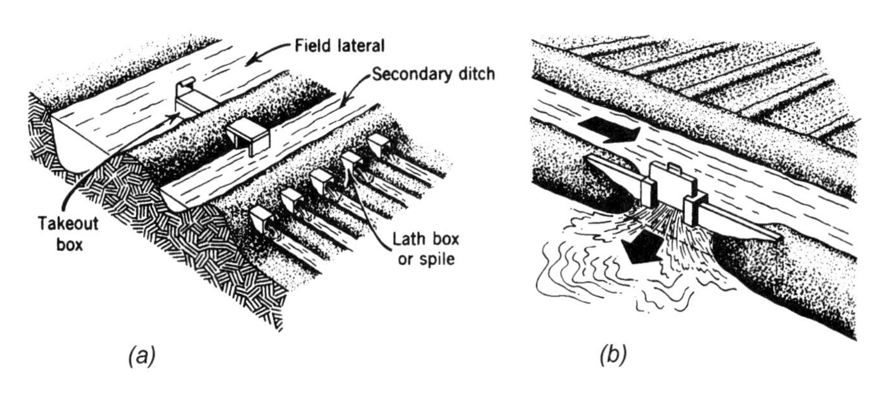

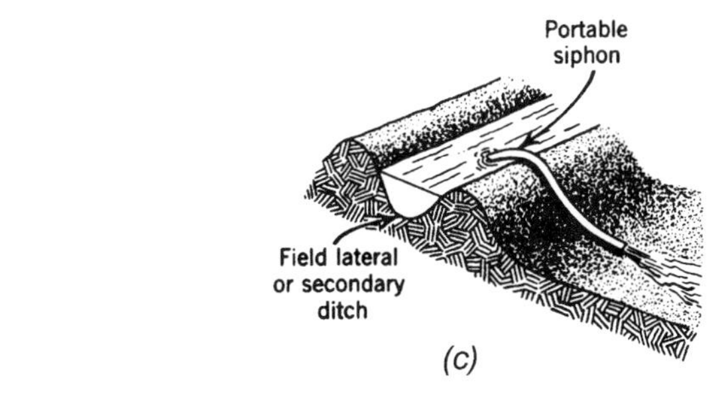

Figure 16–8

Devices for the distribution of water from irrigation ditches into fields. (a) Spile or lath box. (b) Border takeout. (c) Siphons. (Adapted from SCS & USBR, 1959)

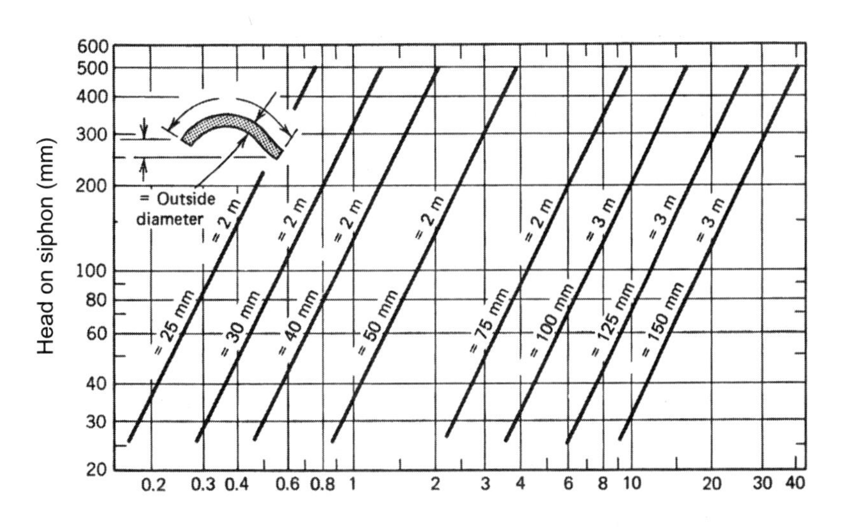

Figure 16–9

Discharge of aluminum siphon tubes as a function of head. (Adapted from SCS, 1984)

16.20 Underground Pipe

Because open-channel distribution systems present continuing problems of maintenance, constitute an obstruction to farming operations, and have a water surface subject to evaporation losses, underground pipe distribution systems shown in Figure 16-10a are becoming increasingly popular. In these systems, water flows from the distribution pipe upward through riser pipes and irrigation valves to appropriate basins, border strips, or furrows. In Figure 16-10b a multiple-outlet riser controls the distribution of water to several furrows. Alfalfa valves (Figure 16-10a) regulate the flow to a basin or border strip or to a header ditch from which water can be distributed to the field.

16.21 Portable Pipe

Surface irrigation with portable plastic or aluminum pipe or large-diameter plastic tubing may be advantageous, particularly in circumstances where water is applied infrequently. Lightweight gated pipe (Figure 16-10c) provides a convenient and portable method of applying water in furrow irrigation. Portable flumes are sometimes used, particularly with high-value crops, where they may be removed from the

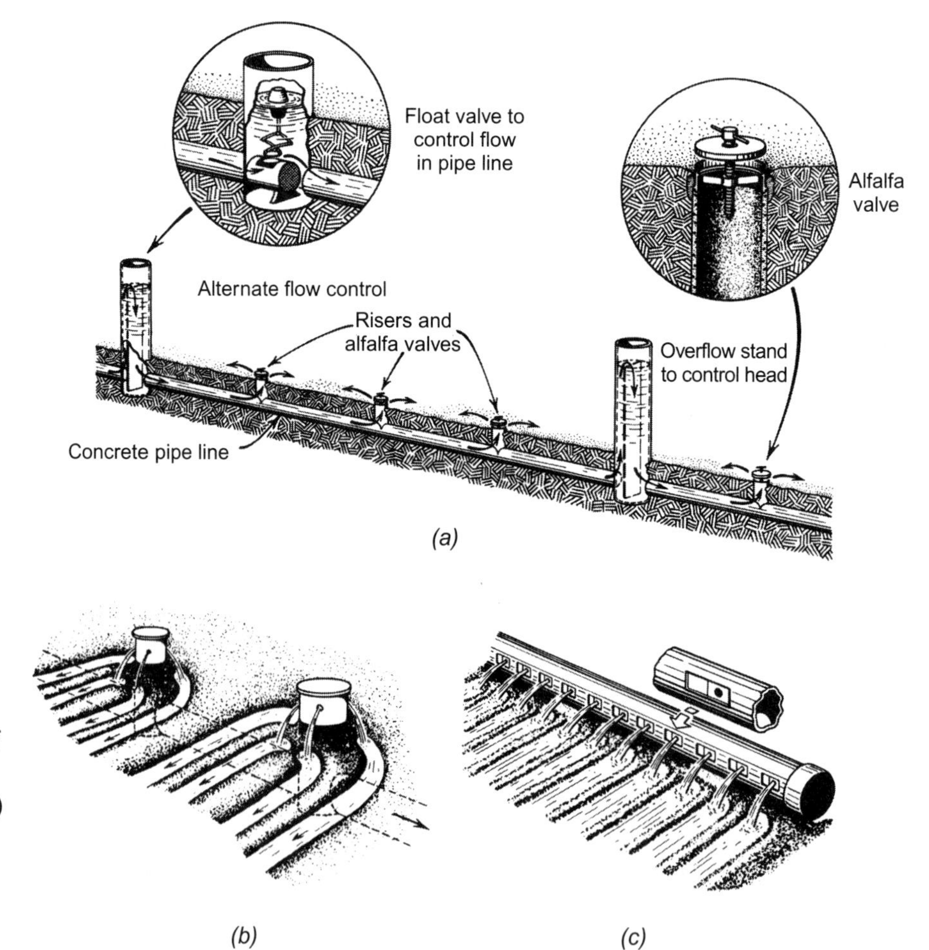

Figure 16–10
Methods of distribution of water from (a) low-pressure underground pipe, (b) multiple-outlet risers, and (c) portable gated pipe. (Adapted from SCS & USBR, 1959)

fields during certain field operations. Portable pipe lines are typically designed with one or at most two pipe sizes for convenience of the irrigators. Although gated pipe can be decreased in diameter as the flow decreases, it is inconvenient for the irrigators to keep multiple pipe sizes in the proper order for loading and unloading. Also, the gated pipe may be the delivery system and carry the design flow nearly its entire length.

Internet Resources

Natural Resources Conservation Service Irrigation page
http://www.wcc.nrcs.usda.gov/nrcsirrig/

Selected university irrigation sites
http://www.ianr.unl.edu (University of Nebraska). Key search term: irrigation publications
http://www.engineering.usu.edu (Utah State University). Key search term: irrigation
http://www.itrc.org (California State Polytechnic University at San Luis Obispo)
http://irrigation.tamu.edu (Texas A & M University)

Source for BASIN, BORDER, and SRFR software
http://www.uswcl.ars.ag.gov

Irrigation equipment and dealers
http://www.irrigation.org

References

2000 Irrigation Survey. (2001). *Irrigation Journal, 51*(1), 12–30, 40, 41.

Blair, A. W., & E. T. Smerdon. (1987). Modeling surge irrigation infiltration. *Journal of Irrigation and Drainage Engineering, ASCE, 113*(4), 497–515.

Booher, L. J. (1974). *Surface Irrigation.* FAO Pub. No. 95. Rome: Food and Agriculture Organization (FAO).

Clemmens, A. J., A. R. Dedrick, & R. J. Strand. (1995). *BASIN: A Computer Program for the Design of Level-Basin Irrigation Systems.* Version 2.0. WCL Report 19. Phoenix, AZ: USDA, Agricultural Research Service, U.S. Water Conservation Laboratory.

Clemmens, A. J., W. R. Walker, D. D. Fangmeier, & L. A. Hardy. (2000). Design of surface irrigation systems. Draft chapter submitted for ASAE monograph *Design and Operation of Farm Irrigation Systems,* 2nd ed. G. J. Hoffman, D. L. Martin, & R. Elliot, eds. ASAE Monograph. St. Joseph, MI: ASAE.

Hansen, V. E., O. W. Israelsen, & G. E. Stringham. (1980). *Irrigation Principles and Practices,* 4th ed. New York: Wiley.

Hart, W. E., H. G. Collins, G. Woodard, & A. S. Humphreys. (1983). Design and Operation of Gravity or Surface Systems. In M. E. Jensen, ed., *Design and Operation of Farm Irrigation Systems,* pp. 501–580. ASAE Monograph 3. St. Joseph, MI: ASAE.

Izuno, F. T., & T. H. Podmore. (1985). Kinematic wave model for surge irrigation research in furrows. *Transactions of the ASAE, 28*(4), 1145–1150.

Merriam, J. L., & J. Keller. (1978). *Farm Irrigation System Evaluation: A Guide for Management.* Agricultural and Irrigation Engineering Department, Utah State University.

Schwab, G. O., D. D. Fangmeier, W. J. Elliot, & R. K. Frevert. (1993). *Soil and Water Conservation Engineering,* 4th ed. New York: Wiley.

Shatanawi, M. R. (1980). *Analysis and Design of Irrigation in Sloping Borders.* Doctoral dissertation, Davis: University of California at Davis.

Strelkoff, T. (1990). *SRFR: A Computer Program for Simulating Flow in Surface Irrigation—Furrows-Basins-Borders*. WCL Report No. 17. Phoenix, AZ: U.S. Water Conservation Laboratory.

Strelkoff, T. S., A. J. Clemmens, B. V. Schmidt, & E. J. Slosky. (1996). *BORDER, a Design and Management Aid for Sloping Border Irrigation Systems*. Version 1.0. WCL Report No. 21. Phoenix, AZ: U.S. Water Conservation Laboratory, USDA, ARS.

Strelkoff, T., & N. D. Katopodes. (1977). Border irrigation hydraulics with zero inertia. *Journal of the Irrigation and Drainage Division, ASCE, 103*(IR-3), 325–342.

Stringham, G. E. (ed.). (1988). *Surge Flow Irrigation*. Research Bulletin 515. Utah Agricultural Experiment Station. Logan: Utah State University.

U.S. Soil Conservation Service (SCS). (1947). *First Aid for the Irrigator*. USDA Miscellaneous Publication 624.

———. (1974). Border irrigation. In *National Engineering Handbook*, Sec. 15, *Irrigation*, Chap. 4. Washington, DC: U.S. Government Printing Office.

———. (1984). Furrow irrigation, 2nd ed. In *National Engineering Handbook*, Sec. 15, *Irrigation*, Chap. 5. Washington, DC: U.S. Government Printing Office.

———. (1979). *Engineering Field Manual for Conservation Practices*. (Litho.). Washington, DC: USDA–Soil Conservation Service.

U.S. Soil Conservation Service (SCS) & U.S. Bureau of Reclamation (USBR). (1959). *Irrigation on Western Farms*. Agricultural Information Bulletin 199. Washington, DC: USDA–Soil Conservation Service and USDI–Bureau of Reclamation.

Walker, W. R., & G. V. Skogerboe. (1987). *Surface Irrigation: Theory and Practice*. Englewood Cliffs, NJ: Prentice Hall.

Problems

16.1 Repeat Example 16.1 with a field length of 600 m.

16.2 Design a concrete-lined irrigation canal including freeboard for 120 L/s. Assume the channel slope is 0.0004 m/m, 1:1 sideslopes, and 300-, 450-, or 600-mm bottom widths are available.

16.3 Determine the inflow rate and time, deep percolation, runoff, and maximum application efficiency for furrows 400 m long, slope is 0.006 m/m, infiltration is $z = 25(\text{mm/h}^{0.6})\ t(\text{h})^{0.6}$, row width is 1 m, desired depth of application is 100 mm, roughness coefficient is 0.04, and furrow shape is a trapezoid with 1:1 sideslopes and 200-mm bottom width.

16.4 Design a furrow irrigation system for a field 600 m by 900 m with slopes of 0.0003 m/m and 0.002 m/m, respectively, and the water supply is a well at the high corner with a flow rate of 50 L/s. The infiltration is $z = 36(\text{mm/h}^{0.4})\ t(\text{h})^{0.4}$, row width is 0.9 m, desired depth of application is 80 mm, roughness coefficient is 0.04, and furrow shape is a trapezoid with 1.5:1 sideslopes and 100-mm bottom width. Determine the inflow rate and time, deep percolation, runoff, and application efficiency. Gated pipe will be used for irrigating the furrows. Include a design of the pipe delivery system.

16.5 The water surface elevation in a canal is 0.8 m, the elevation of the inlet to a 50-mm dia. 2-m long siphon tube is 0.6 m, the elevation of the outlet of the siphon tube is 0.4 m, and the elevation of the water surface elevation in the furrow at the outlet is 0.3 m. What is the discharge rate of the siphon? What is the discharge rate if the water surface elevation in the furrow is 0.5 m?

16.6 What diameter of siphon tube would you recommend that the irrigator purchase for the final flow rate of 1 L/s in Example 16.1? Assume the maximum head available is 150 mm.

16.7 Determine the discharge of 25-mm and a 75-mm-diameter siphon tube 2 m long for a head of 200 mm from the graph in Figure 16-9. Compare the relative flows of both Figure 16-9 and the theoretical flow rates, assuming the same friction loss in both tubes and the Manning equation applies. *Note:* Do the analysis first neglecting entrance losses and, second, considering entrance losses.

16.8 Derive Equation 16.16. Assume a border strip is a wide, very shallow, open channel.

16.9 Repeat Example 16.2 with a roughness $n = 0.25$.

16.10 Determine the time of application time and efficiency, and stream size for a 10-m wide border strip, length is 450 m with a slope of 0.3 percent, the depth of application is 100 mm, the infiltration is $z = 80(\text{mm/h}^{0.6})\ t(\text{h})^{0.6}$, and the roughness coefficient is 0.04.

16.11 Design a border strip irrigation system for a field 600 m by 1000 m with slopes of 0.0003 m/m and 0.001 m/m, respectively, and the water supply is at the high corner with $Q = 1.1\ \text{m}^3/\text{s}$. The infiltration is $z = 60(\text{mm/h}^{0.5})\ t(\text{h})^{0.5}$, desired depth of application is 100 mm, and roughness coefficient is 0.15.

16.12 If the flow into the border strip described in Example 16.2 is discharged by 100-mm diameter, 3-m long siphon tubes under a head of 80 mm, how many siphon tubes are required to irrigate each border strip?

16.13 Determine the basin size and flow rate for the field in Problem 16.11 if the field has no slope and the design efficiency is to be 85 percent.

CHAPTER

17 Sprinkler Irrigation

Sprinkler irrigation is a versatile means of applying water to any crop, soil, and topographic condition. It is popular because (1) surface ditches are not necessary, (2) prior land preparation is minimal, and (3) pipes are easily transported and provide no obstructions to farm operations when irrigation is not needed. According to the "2000 Irrigation Survey" in *Irrigation Journal* (2001), about 12 500 000 ha or 49.9 percent of the irrigated land in the United States is irrigated with sprinkler systems.

Sprinkling is suitable for sandy soils or other soil and topographic conditions where surface irrigation may be inefficient or expensive, or where erosion may be particularly hazardous. Low rates and amounts of water may be applied, such as are required for seed germination, frost protection, delay of fruit budding, and cooling of crops in hot weather. Fertilizers and soil amendments may be dissolved in the water and applied through the irrigation system. The major concerns of sprinkler systems are investment costs and labor requirements (Chapter 15).

17.1 Sprinkler Systems

The major components of sprinkler systems are pumps, main lines, lateral lines, riser pipes or drop tubes, and sprinkler heads. These are illustrated for hand-move portable sprinkler systems in Figure 17-1. Additional important components include valves, flow meters, chemical injectors, and pipe fittings.

Basic examples of sprinkler heads and pipe couplers are shown in Figure 17-2. Details of these components vary considerably, depending on the manufacturer. For rotating sprinklers, the sprinkler heads usually have two nozzles, one to apply water at a considerable distance from the sprinkler (range nozzle) and the other to cover the area near the sprinkler (spreader nozzle) (Figure 17-2b). Of the devices to rotate the sprinkler, the most common taps the sprinkler head with a small hammer activated by the force of the water on a small vane connected to it. Most rotating sprinklers cover a full circle, but some may be set to operate for any portion of a circle (part-circle). A number of sprinkler heads are available for special purposes. Some provide a low-angle jet for use in orchards. Some work at pressures as low as 35 kPa.

Giant sprinkler units discharge 20 to 30 L/s at pressures of 550 to 700 kPa and spray the water up to 50 m. With the large wetted diameters, distances between

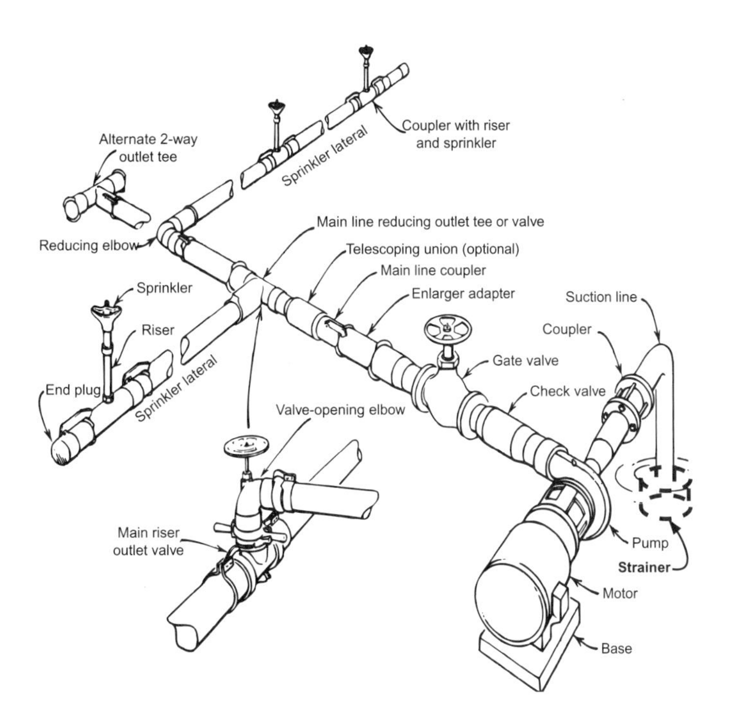

Figure 17–1
Components of a hand-move portable sprinkler irrigation system.

sprinkler setups are much greater than for standard sprinklers. Because giant sprinklers produce large droplets, their impact may damage plants or the structure of bare soils.

Low-pressure sprays (Figure 17-2c) operate from 70 to 210 kPa. They were developed primarily for use on center-pivot and linear-move systems to reduce the pressure requirement. Droplet sizes are small, but application rates tend to be high unless small trusses or spray booms are attached to the lateral. Spray heads are mounted on the booms both ahead and behind the lateral to increase the width of the wetted area and decrease the application rate.

Riser pipes attach the sprinkler to the lateral and raise heads above the crop so the crop does not interfere with the spray, which distorts the spray pattern. Drop tubes are used on center-pivot or linear-move systems to lower the sprinkler heads below the lateral, but above the crop. This is often done to reduce wind drift.

Laterals are usually 6.1-, 9.1-, or 12.2-m lengths of aluminum or other lightweight pipe connected with couplers for hand-move systems. The laterals are usually galvanized steel pipe for center-pivot and linear-move systems.

The main line may be either movable or permanent. Movable mains generally have a lower first cost and can be more easily adapted to a variety of conditions. Permanent buried mains offer savings in labor and reduce obstruction to field operations. Water is taken from the main, either through a valve placed at each point of junction with a lateral, or in some cases, either with an L- or a T-fitting that has been supplied in place of one of the couplings along the main.

A pump is required to overcome the elevation difference between the water source and the sprinkler nozzle, to counteract friction losses, and to provide adequate pressure at the nozzle for good water distribution. As discussed in Chapter 19, the type of pump will vary with the discharge, pressure, and the vertical distance to the

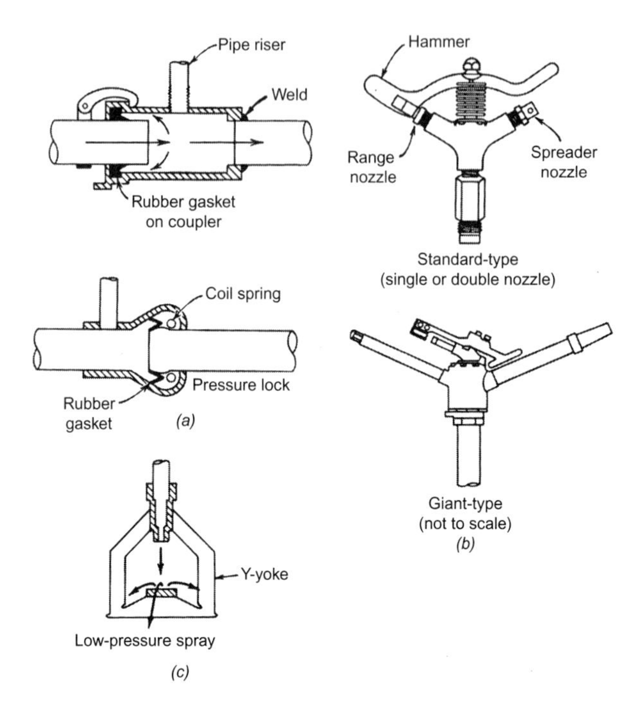

Figure 17–2

Examples of (a) quick-connect couplers, (b) sprinkler heads, and (c) low-pressure spray head.

source of water. The pump should have adequate capacity to meet future needs and to allow for wear. For most farm irrigation systems, capacity generally varies from 6 to 60 L/s.

In general, systems are described according to the method of moving the lateral lines, on which various types of sprinklers are attached. Common systems are shown in Figure 17-3. During normal operation, laterals may be moved by hand or mechanically. The sprinkler system may cover only a small part of the field at a time or be a solid-set system in which laterals and sprinklers are placed over the entire field. With the solid-set system, all or some of the sprinklers may be operated at the same time.

Hand-move laterals have the lowest investment cost but the highest labor requirement. With giant sprinklers, the spacing can be increased, thereby reducing labor, but higher pressures are required, which increases the pumping cost. These sprinklers are pulled or transported from one location to another or moved continuously. Hand-move laterals with standard sprinklers are most suitable for low-growing crops, but are impractical in tall crops because of adverse conditions for moving the pipe. Mechanically moved systems have higher initial costs but lower labor requirements.

17.2 Evaporation and Wind Drift

Evaporation and wind drift from sprinklers has proven difficult to measure or predict. Early researchers found that evaporation from their catchment devices was larger than from the spray droplets. Frost & Schwalen (1955) conducted extensive field tests that eliminated or accounted for catchment evaporation losses. Evaporation and wind drift are highest with high pressures, small nozzle diameters, high wind speeds, and high vapor pressure deficits (Frost, 1963). Thus sprinklers should be operated at the low end of their recommended operating range under windy conditions. It is unclear what the effects of drift may be on downwind portions of the field.

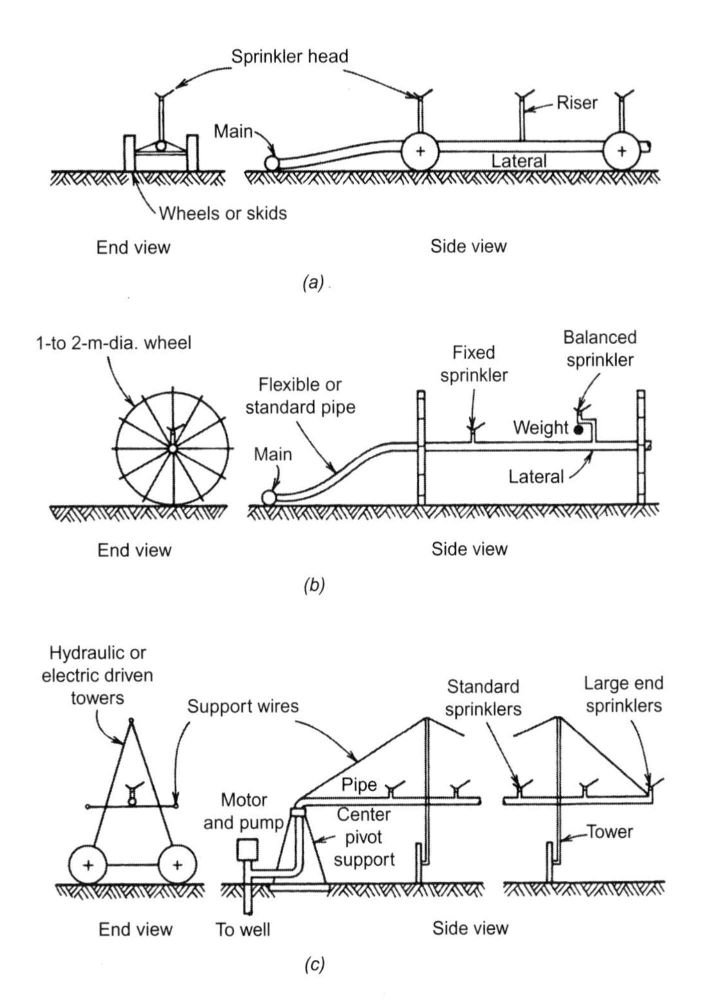

Figure 17–3

Mechanical-move sprinkler systems. (a) End-pull lateral. (b) Side-roll lateral. (c) Center-pivot lateral.

Generally evaporation and wind drift are 3 to 8 percent of the sprinkler discharge and range from near zero to over 20 percent. With a plant canopy, the net loss may be lower because the energy reaching the plant is reduced by the energy absorbed by evaporation from the spray and hence evapotranspiration is lowered. A bare soil tends to increase evaporative losses. Evaporation and wind drift are often accounted for by slightly lowering the design application efficiency.

17.3 Distribution Pattern of Sprinklers

The factors that influence the distribution pattern of sprinklers are nozzle pressure, wind velocity, speed of rotation, and the sprinkler design. Inadequate pressure will result in a "ring-shaped" distribution and a reduction in the area covered (Figure 17-4); high pressures produce smaller drops with high application rates near the sprinkler and a smaller diameter of coverage.

Wind will cause a variable diameter of coverage and also somewhat higher rates near the sprinkler. A high speed of rotation of the sprinkler greatly reduces the area covered and causes excessive wear of the sprinkler. The variation in speed of rotation is not due to wind but to changes in frictional resistance attributed to lack of

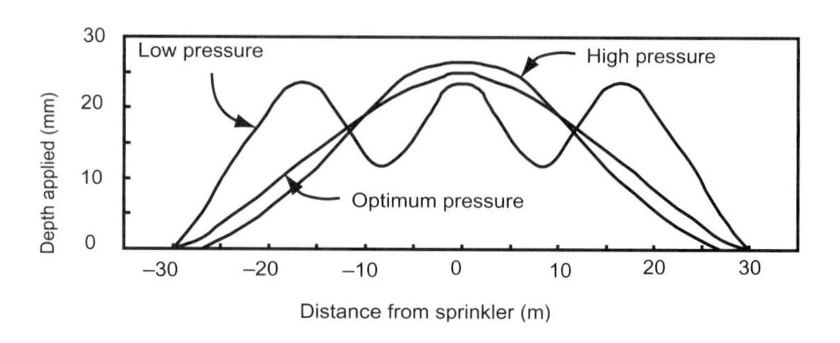

Figure 17–4
Water distribution patterns from a typical sprinkler operated at low, optimal, and high pressures.

precision in manufacture and to wear. Sprinkler design determines the nozzle type, spray pattern, number of nozzles, nozzle angle, and speed of rotation—all influencing the water distribution pattern.

The effect of wind for a single sprinkler is illustrated in Figure 17-5. The effect of the orientation of the sprinkler lateral relative to wind direction is also shown in Figure 17-5 and demonstrates why laterals should be installed perpendicular to the prevailing wind direction. Because one sprinkler does not apply water uniformly over the area, sprinkler patterns are overlapped to provide more uniform coverage. The distribution pattern shown in Figure 17-6 illustrates how overlapping the optimal pattern (Figure 17-4) of several sprinklers gives a relatively uniform distribution between sprinklers. Although Figure 17-6 shows relatively uniform distribution over the area, wind will skew the pattern which typically leads to a less uniform distribution.

Uniformity of application of a sprinkler system can be expressed by the uniformity coefficient given in Chapter 15. In applying the coefficient to evaluate a sprinkler system, the depths can be the depth of water applied or the depth of water penetration in the soil. Depth of water applied evaluates the sprinkler system alone. Depth of penetration combines the effects of the sprinkler system, the topography, and the soil conditions. To measure the uniformity of application, observations at many points are made in the field with cans uniformly spaced in the sprinkler area bounded by four adjacent regularly spaced sprinklers. An estimate of the uniformity coefficient can also be determined graphically if the distribution pattern of a single sprinkler is known and then overlapping, as shown in Figure 17-6. An absolutely uniform application would give a uniformity coefficient of 1.0. Uniformity coefficients of 0.8 or more are acceptable. Distribution in wind may be improved by placing the laterals midway between their locations during the previous irrigation, by irrigating at night when wind is low, or by using only the range nozzle on two-nozzle sprinklers.

17.4 System Requirements

Sprinkler systems should be properly designed hydraulically and economical in cost. Selection of the system should also consider the availability and training of labor for moving and maintaining the system. The design of a sprinkler irrigation system involves the maximum rate of application, the irrigation period, and the depth of application. Application rates in excess of the soil infiltration capacity result in runoff with accompanying poor distribution of water, loss of water, and soil erosion. Systems producing runoff have lower application efficiencies. Maximum water application rates for various soil conditions are given in Table 17-1. These values may serve as a guide where reliable local recommendations are not available.

Applications at rates well below the maximum have been found beneficial. Rates of one half the infiltration rate of the soil combined with nozzle pressures that provide a fine spray have resulted in improved maintenance of soil structure and minimization of soil compaction.

The depth of application and the irrigation period are closely related. Irrigation period is the time required to irrigate an area with one application of water. The depth of application will depend on the available water-holding capacity of the soil and crop needs.

Under humid conditions, rains may bring the entire field up to a given water level. As the plants use this water, the soil water content for the entire field decreases, thus irrigation must start soon enough to complete an application before plants in the last portion to be irrigated suffer from water deficiency. In this situation the irrigation period is set so that the entire irrigated area will be covered before the last portion to be irrigated reaches a water level 10 percent below the readily available water.

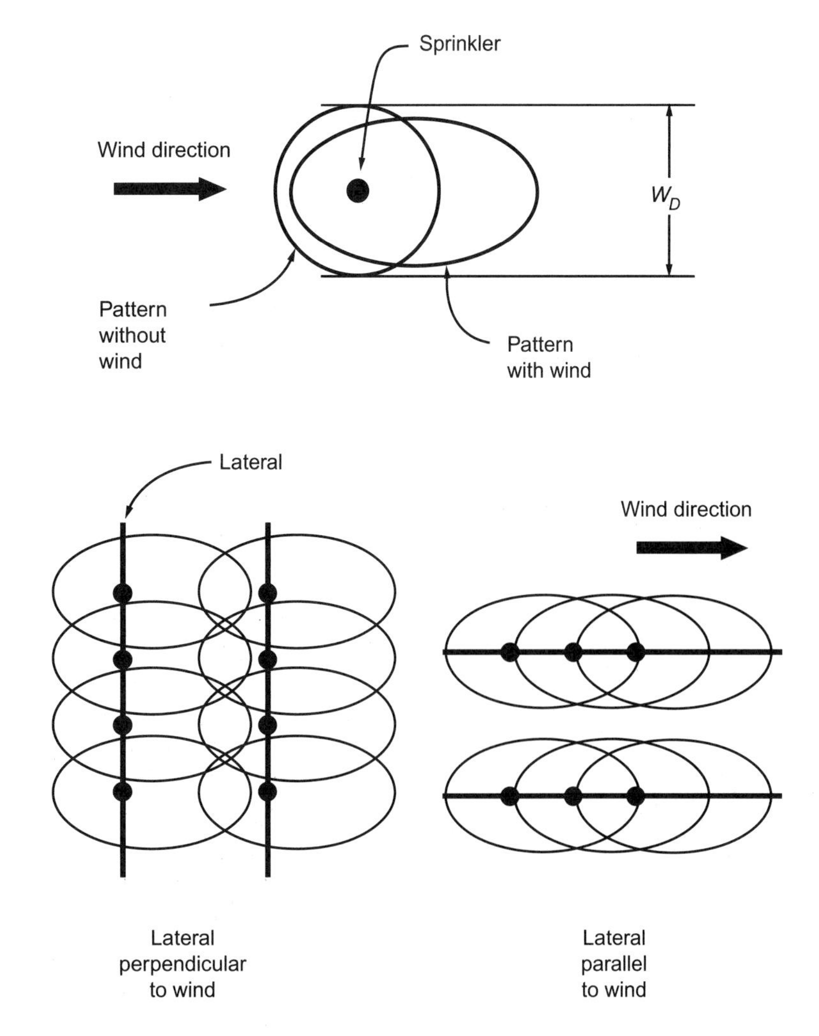

Figure 17–5
Illustration of the effect of wind on water distribution with a single sprinkler and with laterals perpendicular and parallel to the wind. (Martin et al., 2000)

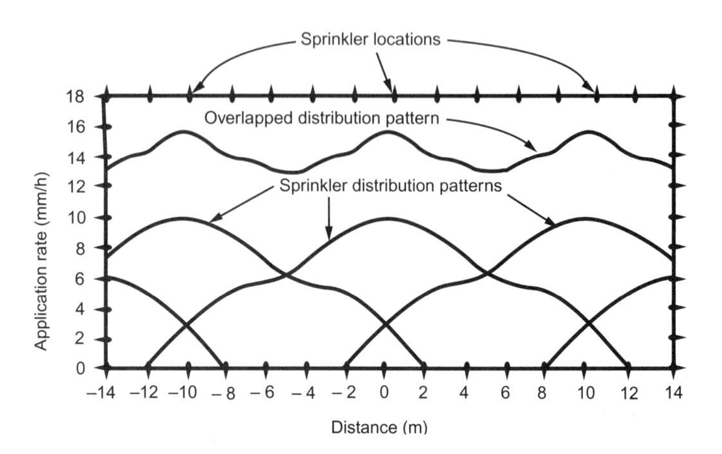

Figure 17–6
Distribution pattern from a sprinkler showing overlapping to give a relatively uniform combined distribution for no wind.

TABLE 17–1 Suggested Maximum Water Application Rates for Sprinklers for Average Soil, Slope, and Cultural Conditions

Soil Texture and Profile Conditions	*Maximum Water Application Rate for Slope and Cultural Conditions (mm/h)*			
	0% Slope		*10% Slope*	
	With Cover	*Bare*	*With Cover*	*Bare*
Light sandy loams uniform in texture to 2 m	32	20	19	11
Light sandy loams over more compact subsoils	25	15	13	8
Silt loams uniform in texture to 2 m	16	10	10	6
Silt loams over more compact subsoil	10	6	5	3
Heavy-textured clays or clay loams	5	3	3	2

Source: Adapted from SCS (1983).

Typical water-holding capacities are given in Table 15-2. Average root depths of crops are given in Table 15-3, and recommended values of Management Allowed Depletion, *MAD,* are given in Table 15-4.

Example 17.1

A sprinkler irrigation system is required to irrigate 16 ha of tomatoes having a maximum ET rate of 5 mm/day on a field with a silt loam soil to a depth of 0.9 m and coarse sand below. The measured available water-holding capacity of the silt loam soil is 120 mm/m. Determine the limiting rate of application, the irrigation period, the net depth of water for application, the depth of water pumped per application, and the required system capacity in hectares per day.

Solution. From Table 17-1, the limiting rate of application is 16 mm/h and the depth of the root zone from Table 15-3 is about 1.1 m; however, since the soil is only 0.9-m deep, 0.9 m is taken as the root depth. The total available water-holding capacity is (0.9 m × 120 mm/m) 108 mm. From Table 15-4, read *MAD* as 0.4. The net depth of application is 108 mm × 0.4 = 43.2 mm. Assuming a water application efficiency of 70 percent, the depth of water pumped per application is

43.2 mm/0.70 = 62 mm. The irrigation period is 43.2 mm/5 mm/day = 8.6 days or round to 8 days. To cover the field in 8 days, the system must be able to pump and discharge (5 mm/day × 8 days)/0.7 = 57.1 mm and irrigate 16 ha/8 days = 2.0 ha/day. This information is then used as a guide in the selection of equipment.

Intermittent- or Set-Move Systems

17.5 General Rules for Sprinkler System Design and Layout

In the absence of experience, the following rules may be helpful for design and layout of hand-move or mechanical-move systems other than center-pivot and linear-move. Equipment companies, consultants, and government agency personnel may also be of great assistance.

In general, sprinkler spacing along the lateral and lateral spacing along the main should be as wide as possible to reduce equipment and labor costs. Since wider spacings require higher pressures and thus higher pumping costs, these wide spacings are more easily justified where power costs are low. Wider spacings also result in higher application rates in which cases the infiltration rate may become a limiting factor. The following general rules should be kept in mind: (1) Mains should be laid uphill and downhill. (2) Laterals should be laid across slope or nearly on the contour. (3) In windy areas, laterals should be laid perpendicular to prevailing winds during the irrigation season. (4) For multiple lateral operation, lateral pipe sizes should be limited to not more than two diameters. (5) If possible, a water supply nearest to the center of the area should be chosen. (6) For balanced design, lateral operation as shown in Figure 17-7 should be provided. (7) Layout should facilitate and minimize lateral movement during the season. (8) Differences in the number of sprinklers operating for the various setups should be minimized. (9) Booster pumps should be considered where small portions of the field require a high pressure at the pump. (10) Layout should be modified to apply different rates and amounts of water where soils differ greatly in the design area.

17.6 Layouts for Hand-Move, Side-Roll, and End-Pull Systems

The number of possible arrangements for mains, laterals, and sprinklers is practically unlimited. The arrangement selected should allow a minimal investment in irrigation pipe, have a low labor requirement, and provide for an application of water over the total area in the required period of time. The most suitable layout can be determined only after a careful study of the conditions to be encountered. The choice will depend to a large extent on the types and capacities of the sprinklers and the pressure required. For typical medium-pressure systems, the laterals are moved about 18 m at each setting and the sprinklers are spaced about 12 m along each lateral. The design layout may also depend on property boundaries, or the owner's desire for a field road along the main line or a road along the field boundary, or if chemicals are being applied through the system.

Typical layouts for sprinkler irrigation systems are shown in Figure 17-7. The layout in Figure 17-7a is suitable where the water supply can be obtained from a stream or canal alongside the field to be irrigated. This arrangement either eliminates the mainline or requires a relatively short main, depending on the number of moves for

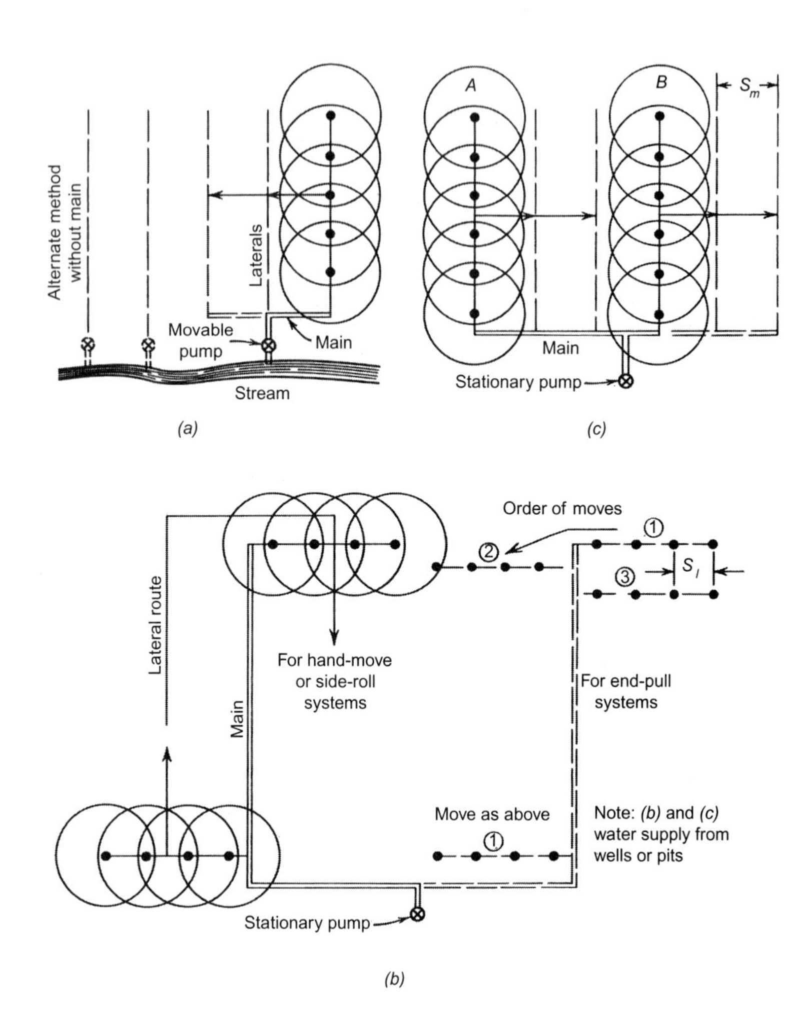

Figure 17–7

Field layouts of main and laterals for hand-move, side-roll, and end-pull sprinkler systems. (a) Fully portable. (b) Portable or permanent (buried) main and portable laterals. (c) Portable main and laterals.

the pump. Less pipe is required for this method than for any of the others. The layouts illustrated in Figures 17-7b and 17-7c are suitable where the water supply is from a well or pond. In Figure 17-7b the two laterals are started at opposite ends of the field and are moved in opposite directions. Since the far half of the main supplies a maximum of one lateral at a time, the diameter of that section can be reduced. The system shown in Figure 17-7c can be designed for continuous operation. While line A is in operation, the operator moves line B. When the required amount of water has been applied, line B is turned on and then line A is moved. With this procedure the capacity of the pump needs to be adequate to supply only one lateral.

17.7 Layout for Solid-Set Systems

Layouts of solid-set systems are similar to those for hand-move systems. Sufficient pipes and sprinklers are installed so that the entire field or any portion can be irrigated without moving laterals. This greatly reduces labor for moving the system, but increases the cost for equipment and initial installation; also, higher flow rates are required. Solid-set systems are frequently used for environmental control where large areas must be irrigated simultaneously. Above-ground solid-set systems typically use aluminum pipe, whereas permanent systems use buried plastic pipe.

17.8 Capacity of a Sprinkler System

The minimum capacity of a sprinkler system depends on the area to be irrigated, the depth of water application at each irrigation, the efficiency of application, and the application time. The actual capacity of the system is the sum of the discharges of the maximum number of sprinklers operating simultaneously.

17.9 Application Rate and Sprinkler Discharge

When the sprinkler discharge and the sprinkler spacing have been selected, the average application rate can be computed with the equation

$$r = \frac{q}{S_l \times S_m} \tag{17.1}$$

where r = average application rate (L/T),
q = discharge of each sprinkler (L^3/T),
S_l = sprinkler spacing along the lateral (L),
S_m = sprinkler spacing between laterals or along the main (L).

For example, a sprinkler discharging 0.6 L/s at a spacing of 12.2 × 16.3 m produces an application rate of 10.9 mm/h.

The theoretical discharge of a nozzle may be computed from the orifice flow equation (Equation 6.18)

$$q = aC\sqrt{2gh} \tag{17.2}$$

For simplification of calculations, this equation becomes

$$q = 0.00111C \times d_n^2 \times P^{1/2} \tag{17.3}$$

where q = nozzle discharge (L/s),
C = coefficient of discharge,
d_n = diameter of the nozzle orifice (mm),
P = pressure at the nozzle (kPa) (1 m of water head = 9.81 kPa).

The coefficient of discharge for well-designed, small nozzles varies from about 0.95 to 0.98. Some nozzles have coefficients as low as 0.80. Normally larger nozzles have smaller coefficients. When the sprinkler has two nozzles, the total discharge is the combined discharge of both nozzles.

17.10 Sprinkler Selection and Spacing

The actual selection of the sprinkler is based largely on design information furnished by the manufacturers of the equipment. The choice depends primarily on the diameter of coverage required, pressure available, and capacity of the sprinkler. The data given in Table 17-2 may assist in selecting the pressure and spacing.

Sprinkler discharges and diameters of coverage for one sprinkler head are given in Table 17-3. Each manufacturer will recommend a combination of nozzle sizes and pressures to give the best breakup of the stream and distribution pattern for uniform

TABLE 17–2 Classification of Sprinklers and Their Adaptability

Type of Sprinkler	*Very Low Pressure (34–105 kPa)*	*Low Pressure (100–210 kPa)*	*Medium Pressure (210–415 kPa)*	*High Pressure (350–690 kPa)*	*Very High Pressure (550–830 kPa)*	*Undertree Low Angle (70–350 kPa)*
General characteristics	Special thrust springs or reaction-type arms	Usually single-nozzle oscillating or long-arm dual nozzle design	Either single- or dual-nozzle design	Either single- or dual-nozzle design	One large nozzle with smaller supplemental nozzles to fill in pattern gaps; small nozzle rotates the sprinkler	Designed to keep stream trajectories below fruit and foliage by lowering the nozzle angle
Range of wetted diameters	6–15 m	18–24 m	23–37 m	34–70 m	60–120 m	12–27 m
Recommended application rate	10 mm/h	5 mm/h	5 mm/h	13 mm/h	16 mm/h	8 mm/h
Jet characteristics (assuming proper pressure–nozzle size relations)	Water drops are large because of low pressure	Water drops are fairly well broken	Water drops are well broken over entire wetted diameter	Water drops are well broken over entire wetted diameter	Water drops are extremely well broken	Water drops are extremely well broken
Water distribution pattern (assuming proper spacing and pressure–nozzle size relation)	Fair	Fair to good at upper limits of pressure range	Very good	Good *except* where wind velocities exceed 6 km/h	Acceptable in calm air, severely distorted by wind	Fairly good; diamond pattern recommended where laterals are spaced more than one tree interspace
Adaptations and limitations	Small acreages confined to soils with intake rates exceeding 13 mm/h and to good ground cover on medium- to coarse-textured soils	Primarily for undertree sprinkling in orchards; can be used for field crops and vegetables	For all field crops and most irrigable soils; well adapted to overtree sprinkling in orchards and groves and to tobacco shades	Same as for medium pressure sprinklers except where wind is excessive	Adaptable to close-growing crops that provide a good ground cover; for rapid coverage and for odd-shaped areas; limited to soils with high intake rates	For all orchards or citrus groves; in orchards where wind will distort overtree sprinkler patterns; in orchards where available pressure is not sufficient for operation of overtree sprinklers

Source: SCS (1983).

TABLE 17–3 Example Manufacturer's Sprinkler Characteristics for a Head with Two Nozzles

Nozzle Pressure kPa	Nozzle Diameters, mm					
	3.97 x 3.18		4.76 x 3.97		6.35 x 3.97	
	Dia.[a]	L/s	Dia.[a]	L/s	Dia.[a]	L/s
207	25	0.37	26	0.52	28	0.76
276	27	0.43	28	0.61	31	0.90
345	28	0.47	30	0.68	34	1.00
414	30	0.52	31	0.74	36	1.10

Note: Pressures to left and below dashed line recommended for best breakup of stream.

[a]Diameter of coverage in m.

application. Single-nozzle sprinklers are generally recommended only for small nozzles with limited diameter of coverage or for part-circle sprinklers. Because the distribution and coverage depends on the angle of the stream from the horizontal and the rate of rotation, sprinklers should be selected from manufacturers' tables.

The maximum sprinkler spacings for satisfactory uniformity depend on sprinkler design, operating pressure, and wind speed. Because winds distort the distribution, wind speed must be considered when selecting and spacing the heads. Operating pressures at the low end of the operating range are often recommended because this produces larger droplets, which are less affected by wind. Also, nozzles with lower trajectory angles are available and the resulting stream is less susceptible to wind drift. If possible, a sprinkler should be tested on site to determine the distribution pattern. These patterns can be overlapped at various spacings to determine if the sprinkler is appropriate and if the resulting uniformity coefficient is 0.8 or greater. If distribution data are not available, the values in Table 17-4 can be used as a guide for sprinkler

TABLE 17–4 Suggested Maximum Sprinkler Spacings for Various Wind Conditions with Lateral Perpendicular to the Prevailing Winds

Wind Velocity (km/h)	Sprinkler Spacings in Percent of the Diameter of No-Wind Sprinkler Pattern	
	S_l, Along the Lateral	S_m, Along the Main
0–2	50	65
2–6	40	60
	45	55
6–16	35	55
	45	50
16–24	30	50
	40	45
24–32	30	45

Sources: Martin et al. (2000) and Kincaid (2000).

spacing. It should be noted that the values in Table 17-4 are based on averages of many sprinklers and thus may not apply to a particular sprinkler head. These values may be applied in design when uniformity coefficients are not available and where high uniformity is not crucial. In cases where high-value crops are grown and/or where chemicals are applied, on-site testing is required or highly recommended. When low-angle nozzles and straightening vanes in the nozzles are used, the values in Table 17-4 for the next lower wind speed may be selected (Keller & Bliesner, 1990).

17.11 Sizing of Laterals and Mains

The diameters and lengths of aluminum and plastic pipe are limited to standard dimensions. The standard diameters of plastic and aluminum pipe are given in Appendix D. For side-roll systems the wall thickness is greater because the pipe is the axle of the system and must withstand the torque for movement.

Laterals and mains should provide the required rate of flow with a reasonable head loss. For laterals, the sections at the distal end of the line have less water to carry and may therefore be smaller; however, many authorities advise against "tapering" of pipe diameters in laterals in portable systems, as it then becomes necessary to keep the various pipe sizes in the same relative position. The system may also be less adaptable to other fields and situations.

The total pressure variation in the laterals should not be more than ± 10 percent of the design pressure or not more than a 20 percent variation between the maximum and minimum pressures. If the lateral runs up or downhill, allowance for the elevation difference should be made in determining the variation in head. If the lateral runs uphill, less pressure will be available at the nozzle; if it runs downhill, there will be a tendency to balance the head loss resulting from friction.

The friction or head loss in pipe can be estimated by the Hazen-Williams equation

$$H_f = \frac{1.21 \times 10^{10} L\left(\dfrac{Q}{C}\right)^{1.852}}{D^{4.87}} \qquad \textbf{17.4}$$

where H_f = friction loss in pipe with the same flow throughout (m),
L = pipe length (m),
Q = flow (L/s),
C = pipe roughness coefficient,
D = actual inside pipe diameter (mm).

Tables of friction loss per 100 units of length for various pipe materials and sizes are often available in irrigation and technical handbooks.

Recommended values of C are 145 to 150 for plastic pipe and 120 for aluminum pipe with couplers and new or coated steel pipe. In addition to the type of material, values of C increase with pipe diameter, decrease as the number of couplers increases, and increase as the inside walls become smoother.

Although Equation 17.4 is for a mainline with the same flow the entire length, it may be adapted to lateral pipes with uniformly spaced sprinkler outlets by multiplying the friction loss H_f by a factor F to obtain the friction loss in the lateral and given the symbol h_f. Values of F are given in Table 17-5 for the Hazen-Williams equation with exponent = 1.852 for the first sprinkler a full interval from the main and for the first sprinkler at half the sprinkler interval from the main. If the head loss for

TABLE 17–5 Correction Factor F for Friction Losses in Pipes with Multiple Outlets for Use with Equation 17.4 and Exponent of 1.852.

Number of Sprinklers	Correction Factor, F: First Sprinkler One Sprinkler Interval from Main	Correction Factor, F: First Sprinkler One-Half Sprinkler Interval from Main	Number of Sprinklers
2	0.64	0.52	2
3	0.53	0.43	3
4	0.49	0.40	4
5	0.46	0.39	5
6	0.44	0.38	6
7	0.43	0.37	7, 8
8	0.42	0.36	9–30
9	0.41	0.35	> 30
10, 11	0.40		
12–15	0.39		
16–21	0.38		
22–36	0.37		
37–129	0.36		
>130	0.35		

For center-pivots:

F = 0.56 with an end-gun operating

F = 0.54 without an end-gun

Source: Adapted from Martin et al. (2000).

a main line is 70 kPa, the loss for the same-length lateral with eight sprinklers is only 29 kPa (0.42 × 70) when the first sprinkler is one interval from the main.

The diameter of the main should be adequate to supply the laterals in each of their positions. The rate of flow required for each lateral may be determined by the total capacity of the sprinklers on the lateral. The position of the laterals that gives the highest friction loss in the main should be used for design purposes. The friction loss in the main may be computed by Equation 17.4. Allowable friction loss in the main varies with the cost of power and the price differential between different diameters of pipe. For small systems with few irrigations per season, an approximate maximum friction loss is 4 kPa per 10-m length of pipe. The most economical size should be determined by balancing the increase in pumping costs against the amortized cost difference of the pipe. Procedures for economic analyses are given by Keller & Bliesner (1990).

The design capacity for sprinklers on a lateral with uniform spacing should be based on the average operating pressure. The average pressure for design in a sprinkler line (Figure 17-8) can be obtained by integrating the pressure distribution in the line over the length and dividing by the length (Martin et al., 2000). The result is

$$H_a = H_d + \frac{F\,H_f}{3.852} + \frac{S_e\,L_L}{2} = H_d + 0.26\,h_f + \frac{S_e\,L_L}{2} \quad \textbf{17.5}$$

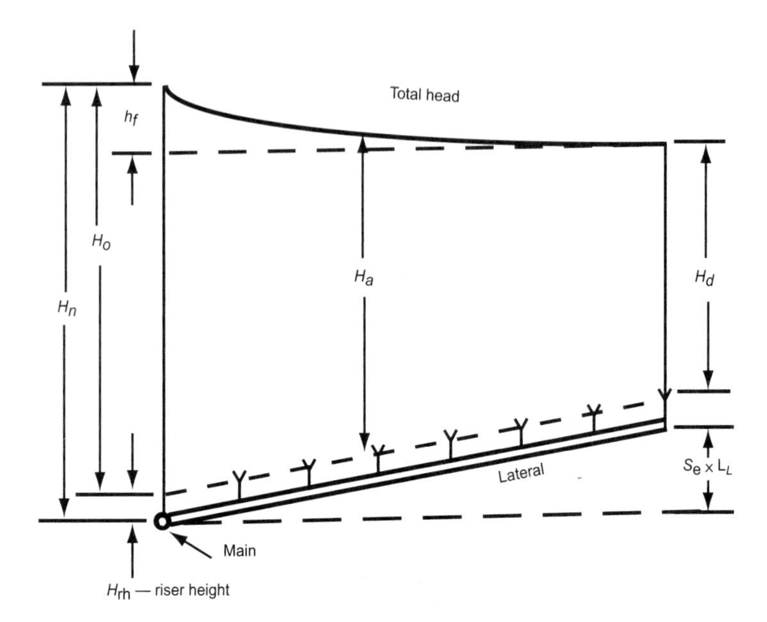

Figure 17–8

Pressure profile of a sprinkler lateral with flow uphill.

where H_a = average sprinkler pressure of a lateral (L),
H_d = sprinkler pressure at the distal end of the lateral (L),
S_e = uniform slope of the lateral from the inlet—positive slope is uphill (L/L),
L_L = lateral length (L).

The location of the average sprinkler pressure occurs at about 0.4 × lateral length from the inlet for a horizontal lateral. The location moves further from the inlet if the lateral slope increases uphill. The location moves closer to the inlet if the lateral slope increases downhill until $-S_e \times L_L = h_f/2$; then the average sprinkler pressure occurs at two locations along the lateral.

The sprinkler pressure at the inlet to the lateral H_o can be obtained from (Martin et al., 2000)

$$H_o = H_a + 0.74\,h_f + S_e\frac{L_L}{2} \tag{17.6}$$

The sprinkler pressure at the distal end of the lateral H_d is obtained from Equation 17.5

$$H_d = H_a - 0.26\,h_f - S_e\frac{L_L}{2} \tag{17.7}$$

Equations 17.6 and 17.7 can be used to determine the maximum and minimum pressures in the lateral to assure that the pressure variation is less than ± 10 percent of the average or design pressure. The maximum pressure occurs at the inlet if the lateral runs uphill or if the downslope is less than the average friction slope ($-S_e \times L_L < h_f$). The maximum pressure occurs at the distal end of the lateral if the downslope is greater than the average friction slope ($-S_e \times L_L > h_f$).

The pressure or head required at the junction of the lateral and the main H_n is the sprinkler pressure at the inlet to the lateral plus the riser height or

$$H_n = H_o + H_{rh} \tag{17.8}$$

where H_{rh} is the riser height (L).

17.12 Pump and Power Units

In selecting a suitable pump (Chapter 19), it is necessary to determine the maximum total head against which the pump is working. This head may be determined by

$$H_t = H_n + H_m + H_j + H_s \tag{17.9}$$

where H_t = total design head against which the pump is working (L),
H_m = maximum friction loss in the main and the suction line (L),
H_j = elevation difference between the pump and the junction of the lateral and the main (L),
H_s = elevation difference between the pump and the water supply after drawdown (L).

The amount of water that will be required is the sum of the capacity of all sprinklers to be operated at the same time. When the total head and rate of pumping are known, the pump may be selected from rating curves or tables furnished by the manufacturer as discussed in Chapter 19.

The following example illustrates the design of a simple sprinkler system.

Example 17.2

Design a side-roll irrigation system to irrigate the level, square 16-ha field in Example 17.1 for a maximum rate of application of 16 mm/h, for applying 57.1 mm of water in 8 days or 2.0 ha per day. Assume a wind velocity of 5 km/h, H_j = 1.0 m, H_s = 3.0 m, riser extends 0.2 m above the lateral, and well location is at the center of the field.

Solution. Begin the design process by selecting a sprinkler and spacing, then determining if they are acceptable for the field conditions.

(1) From Table 17-3 select the sprinkler with 6.35 × 3.97 mm nozzles at 276 kPa with a flow of 0.9 L/s and 31 m wetted diameter. From Table 17-4 for the wind speed of 5 km/h, read 40 percent and 60 percent of wetted diameter for sprinkler spacings along the lateral and main, respectively. The maximum sprinkler spacing along the lateral is 0.40 × 31 = 12.4 m; however, common pipe lengths are 6.1, 9.1, 12.2, and 18.3 m. Therefore a 12.2-m sprinkler spacing is assumed. The maximum lateral spacing is 0.6 × 31 = 18.6 m, and a spacing of 18.3 m is selected.

(2) Calculate the application rate from Equation 17.1.

$$r = \frac{\dfrac{0.9(\text{L/s}) \times 3600(\text{s/h}) \times 1000(\text{mm/m})}{1000(\text{L/m}^3)}}{[12.2(\text{m}) \times 18.3(\text{m})]} = 14.5 \text{ mm/h}$$

(3) This is acceptable, because it is less than 16 mm/h.

(4) The set-time or the operating time per lateral to apply 57.1 mm is 57.1/14.5 = 3.93 h/set.

(5) Assume the main runs along the center of the field as in Figure 17-9. The number of sprinklers per lateral is 200/12.2 = 16.4 or round to 16 sprinklers. If the first sprinkler is 12.2 m from the main, the last sprinkler will be 16 × 12.2 = 195 m from the main or 5 m from the boundary. The last sprinkler will spray 10.5 m ($\frac{31}{2} - 5$) beyond the field boundary. For simplicity this is accepted, however, the first sprinkler could be moved to one half the spacing (6.1 m) from the main to reduce the overspray to 4.4 m, or a part-circle head could be placed at the end of the lateral, or one sprinkler could be removed from the lateral.

(6) The number of lateral locations is given by 400/18.3 = 21.86, which is rounded to 21 sets to avoid spray beyond the field boundary. (*Note:* With 21 sets there are 20 spaces between set locations or 20 × 18.3 = 366 m plus the sprinklers will spray one half the wetted diameter on each side of the field, so the total coverage is 366 + 31 = 397 m.)

(7) Determine the number of laterals needed. From Step 4 with 3.93 hours per set, 42 sets, and assuming it takes 1 hour to move one lateral, it will require 42 sets × 3.93 h/set + 1 h/move × 42 moves = 207 h or 8.6 days. Since this is greater than the 8 days allowed and does not include time for repairs or maintenance, two laterals are required. An alternative is to use one long lateral with the main along one side.

(8) With two laterals the operating time can be calculated as follows: 3.93 h/set × 3 sets/day + 3 moves/day × 2 h/move = 17.8 h/day. With 3 sets/day and 21 sets, it will require 7 days to irrigate the field, which is less than the 8 days allowed. The eighth day can be used for repairs, moving the laterals back to their starting positions, or applying more water if the ET rate is higher than normal.

(9) Because the 276 kPa pressure selected is the minimum recommended for that head, it is assumed that this will be the pressure at the distal end of the lateral. The average pressure in the lateral needs to be calculated; however, this depends on the lateral size. Assume the lateral is aluminum pipe with a 101.6-mm outside diameter with a wall thickness of 1.83 mm (this is the minimum diameter recommended for side-roll systems in Appendix D). Begin the iterative process for calculating the average pressure and lateral flow rate by using 0.9 L/s for 16 sprinklers or a flow of 14.4 L/s. Calculate the head loss in the lateral with Equation 17.4 and an F value of 0.38 from Table 17-5.

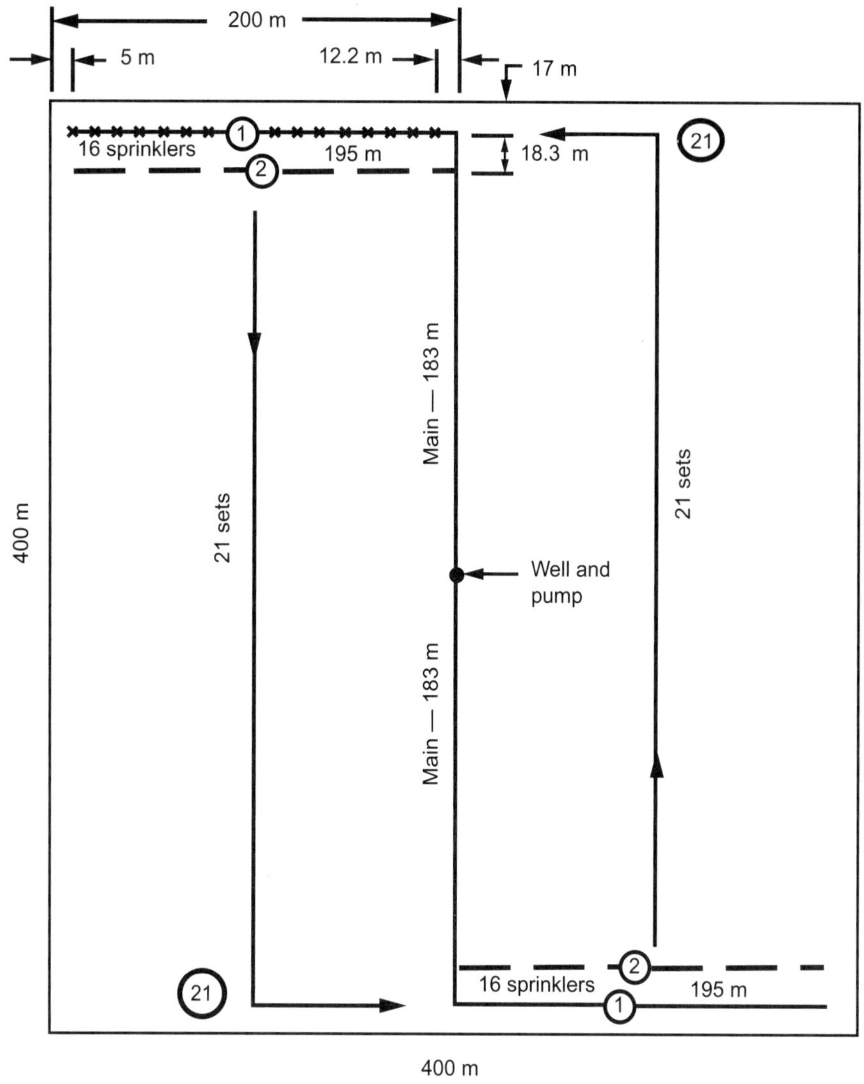

Figure 17–9
Side-roll sprinkler layout for the 16-ha field described in Example 17.2.

$$h_f = \frac{0.38 \times 1.21 \times 10^{10} \times 195 \times \left(\frac{14.4}{120}\right)^{1.852}}{(101.6 - 2 \times 1.83)^{4.87}} = 3.56 \text{ m}$$

Calculate the average pressure head using Equation 17.5 (H_o in m = 276 kPa/9.81 m/s²).

$$H_a = \frac{276}{9.81} + 0.26 \times 3.56 + 0 \times \frac{195}{2} = 29.1 \text{ m}$$

Recalculate the sprinkler flow for the average pressure.

$$q = 0.9(9.81 \times 29.1/276)^{0.5} = 0.915 \text{ L/s}$$

Recalculate the application rate = 14.8 mm/h, which is acceptable. Recalculating the friction loss in the lateral with the new average pressure gives 3.67 m. Recalculating the average pressure gives 29.08 m or 29.1 m, and another iteration is not required.

(10) Calculate the pressure at the inlet to the lateral from Equation 17.6.

$$H_o = 29.1 + 0.74 \times 3.67 + 0 \times 195/2 = 31.8\text{m}$$

(11) Determine if the pressure variation in the lateral is less than 20 percent of the average pressure. The pressure variation is 3.7 m which is within 0.2×29.1 = 5.8 m. Therefore, a 101.6-mm lateral is acceptable. If $h_f > 0.2\ H_a$, a larger pipe size would need to be selected.

(12) Determine the mainline size using Equation 17.4 for 101.6- and 127.0-mm OD aluminum pipe with 1.3 mm wall thickness. From Step 5 and the layout in Figure 17-9, the main is 183 m long on each side of the well and pump.

$$H_f = \frac{1.21 \times 10^{10} \times 183\left(\dfrac{14.64}{120}\right)^{1.852}}{(101.6 - 2 \times 1.3)^{4.87}} = 8.6 \text{ m}$$

$$H_f = \frac{1.21 \times 10^{10} \times 183\left(\dfrac{14.64}{120}\right)^{1.852}}{(127.0 - 2 \times 1.3)^{4.87}} = 2.8 \text{ m}$$

For lower pumping costs and better uniformity, the 127-mm pipe is selected. An economic analysis could be performed to verify this selection.

(13) Determine the wheel diameter for the system. Common wheel diameters are 1.17, 1.47, 1.63, and 1.93 m. Four revolutions of the 1.47-m diameter wheel will give a travel distance of $4 \times \pi \times 1.47$ = 18.5 m or 3 revolutions of the 1.93-m wheel will give 18.2 m. A 1.47-m wheel is selected, as it will provide sufficient clearance between the lateral and the tomatoes and be less susceptible to wind damage or movement.

(14) Determine the head required at the junction of the lateral and the main by substituting into Equation 17.8. The riser height is one half the wheel diameter (1.47/2) plus the length of the riser above the lateral (0.2 m).

$$H_n = 31.8 + 1.47/2 + 0.2 = 32.7 \text{ m}$$

(15) Determine the total head and capacity required at the pump with the laterals at the most remote location and at the highest elevation by substituting into Equation 17.9.

$$H_t = 32.7 + 2.8 + 1.0 + 3.0 = 39.5 \text{ m}$$

Select a pump from manufacturer's characteristic curves to deliver 29.3 L/s at a total head of 39.5 m with an efficiency as high as possible. Allow some factor of safety for possible operation of the pump under other conditions and for loss of efficiency with wear and age.

(16) Determine the size of the power unit. Assume a pump efficiency of 70 percent and substitute into Equation 19.4.

$$BP = 9.81\left(\frac{29.3 \text{ L/s}}{1000 \text{ L/m}^3}\right)\left(\frac{39.5 \text{ m}}{0.7}\right) = 16.2 \text{ kW}$$

Select a power unit capable of continuously delivering 16.2 kW. Assuming a water-cooled internal combustion engine will continuously deliver 70 percent of its rated output, a 23.1 (16.2/0.7) kW engine or larger is required.

Continuous-Move Systems

17.13 Center-Pivot Systems

The development of center-pivot systems has dramatically changed sprinkler irrigation practices (Martin et al., 2000). These systems consist of a lateral pipe supported by a tower and truss system that rotates around a pivot point (Figure 17-10). The towers are moved by electric or hydraulic motors. The speed of rotation is determined by the speed and timer setting of the outer tower. The other towers are controlled by a system of switches that automatically activate to maintain alignment of the lateral. For most electric systems, a one-minute timer cycle controls the speed of rotation. If the timer of the outer tower is set at 50 percent, the tower will move at a constant velocity for 30 seconds, then stop for 30 seconds. Most hydraulic and some electric systems provide continuous movement of the outer tower at variable speeds. For a given set of sprinklers, the speed of rotation determines the depth of application.

Sprinkler spacings along the lateral commonly have varying intervals, unless furrow drops are used, then the spacing will be uniform. Each sprinkler covers an area given by

$$A_r = 2\pi R S_l \qquad \textbf{17.10}$$

where A_r = wetted area for the sprinkler at distance R from the pivot,
S_l = sum of half the distances from R to the two adjacent sprinklers as shown in Figures 17.3c and 17.10.

A large part-circle sprinkler (end-gun) is often placed at the end of the lateral and automatically turned on to irrigate more of the corners of the field. Some manufacturers provide a fold-back extension arm that attaches to the end of the lateral to cover most of the corner areas of a square field. An end-gun is then located at the end of the extension arm. These systems will also cover odd-shaped fields. Some center pivot laterals can be towed to another field by rotating the tower wheels 90 degrees, but during normal sprinkling the lateral rotates about the center-pivot point as before.

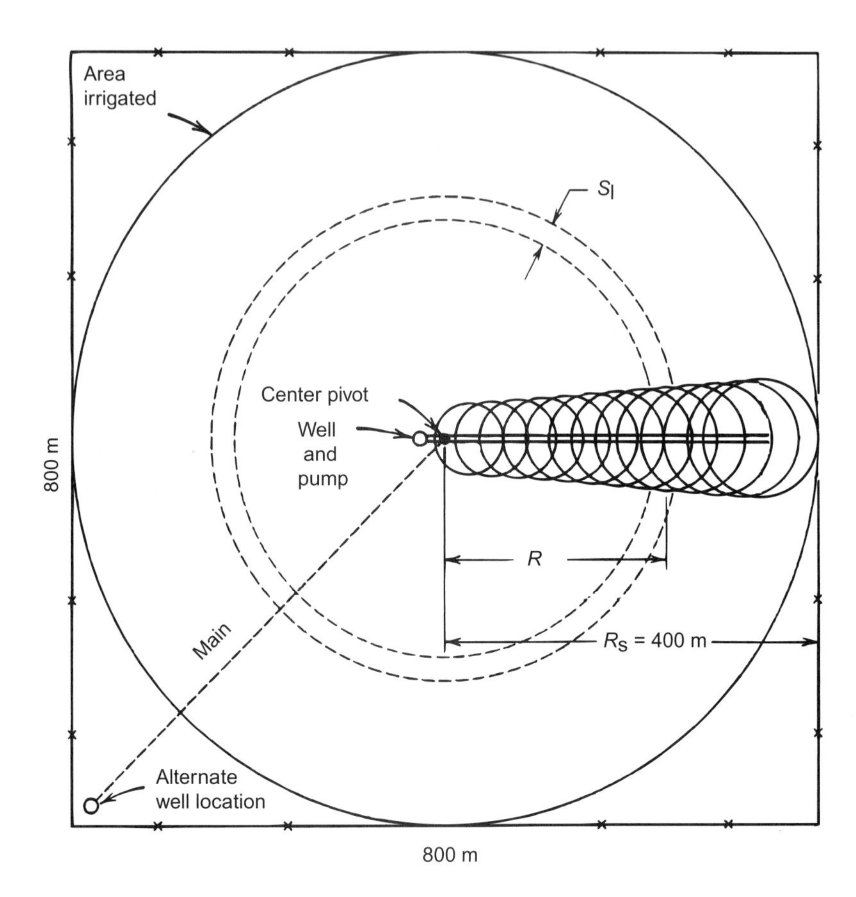

Figure 17–10

Center-pivot sprinkler system for a square 64-ha field. Alternate well location is for convenience of supplying water for up to four such fields.

17.14 Capacity of a Center-Pivot System

The capacity of a center-pivot system is given by

$$Q_s = \frac{\pi R_s^2 \times d_a}{t \times E_a} \tag{17.11}$$

where Q_s = capacity of the system (L^3/T),
R_s = radius of the area irrigated without the corners(L),
d_a = net depth of application per revolution (L),
t = time of operation per revolution (T),
E_a = application efficiency (decimal).

The numerator on the right side of Equation 17.11 is the net volume of water applied. Dividing the numerator by the application efficiency and time per revolution gives the gross flow rate or system capacity. The discharge required for a sprinkler on the lateral is given by

$$q_R = \frac{2Q_s \, R \, S_l}{R_s^2} \tag{17.12}$$

where q_R (L^3/T)is the discharge of a sprinkler located at the radial distance R (L) from the pivot. Discharge from Equation 17.12 is theoretically correct for an infinite number of sprinklers applying the average depth of water uniformly along the lateral. In practice the discharge determined from Equation 17.12 cannot be met because of the limited number of sprinklers and nozzle diameters available; however, Chu & Moe (1972) found that observed data were sufficiently close to a uniform distribution to be acceptable.

17.15 Size of Center-Pivot Laterals

Chu & Moe (1972) found $F = 0.54$ for center-pivot systems without an end gun, whereas Pair et al. (1983) give $F = 0.56$ when an end gun is operating. These F values are higher than those in Table 17-5 because sprinkler discharge along the lateral is not uniform with less flow removed near the pivot. The size of the lateral is determined by pipe cost, energy cost, structural strength of the pipe, and limits on allowable friction loss adjusted for elevation differences along the line. The selected pipe diameter should be the closest nominal size commercially available.

In order to select the nozzle sizes along the lateral, the pressure in the lateral must be known. Chu & Moe (1972) derived the following equation for pressure distribution along a lateral for a level field.

$$H_R = H_d + h_f[1 - 1.875(x - 2x^3/3 + x^5/5)] \qquad \textbf{17.13}$$

where H_R = pressure at a distance R from the pivot (L),
H_d = pressure at the end of the lateral (L),
h_f = friction loss in the center-pivot lateral (L),
x = R/R_s, distance from the pivot divided by the radius of the irrigated field.

From Equation 17.13, nozzle pressure can be computed for a selected sprinkler location along the lateral. With knowledge of the pressure, the nozzle size can be selected and the area covered determined. Most center-pivot manufacturers have developed computer programs to aid in designing the sprinkler package.

17.16 Center-Pivot Application Rates

The peak application rate along the lateral for a specific sprinkler package with an elliptical sprinkler pattern can be obtained from (Martin et al., 2000)

$$r_p = 4\left(\frac{Q_s}{\pi R_s^2}\right)\frac{R}{W_R} \qquad \textbf{17.14}$$

where r_p = peak application rate at radius R (L/T),
R = distance from pivot (L),
W_R = wetted radius of the sprinkler at radius R (L).

The term $Q_s/\pi R_s^2$ is the gross application rate of the system. Equation 17.14 shows that the peak application rate increases linearly with both the system capacity and distance from the pivot and inversely with the wetted radius of the sprinkler package. The peak application rate for the system occurs at the end of the lateral where $R = R_s$ because the area covered is largest. As the system passes over a particular point the application rate is low and increases until the lateral is overhead and then decreases as shown in Figure 17-11. The shaded area in Figure 17-11 denotes the portion of the applied water that exceeds the infiltration rate of the soil and becomes runoff from the field or to another portion of the field.

The time to reach the peak application rate t_p (T) from first wetting is determined from

$$t_p = \frac{W_R \quad d_a}{2\pi R\left(\dfrac{Q_s}{\pi R_s^2}\right)E_a} \tag{17.15}$$

This equation shows that the time to reach the peak application rate is (1) linearly related to both the wetted radius of the sprinkler package and the gross application depth and (2) inversely related to both the distance from the pivot and the gross application rate of the system. The total time water is applied to a point is twice t_p. The peak application rate and the time to peak application rate should be compared to the infiltration characteristics of the soil and surface storage as in Figure 17-11 to determine if excessive runoff will occur and if the system is suitable for that field. For more detailed procedures see Martin et al. (2000). If the application rate is too high, tillage systems that increase surface storage may be utilized or booms (small trusses) are sometimes suspended under the pivot lateral at about a 45-degree angle to the lateral. These booms have three to seven low-pressure heads and extend the area of application while often decreasing the pressure and energy requirement.

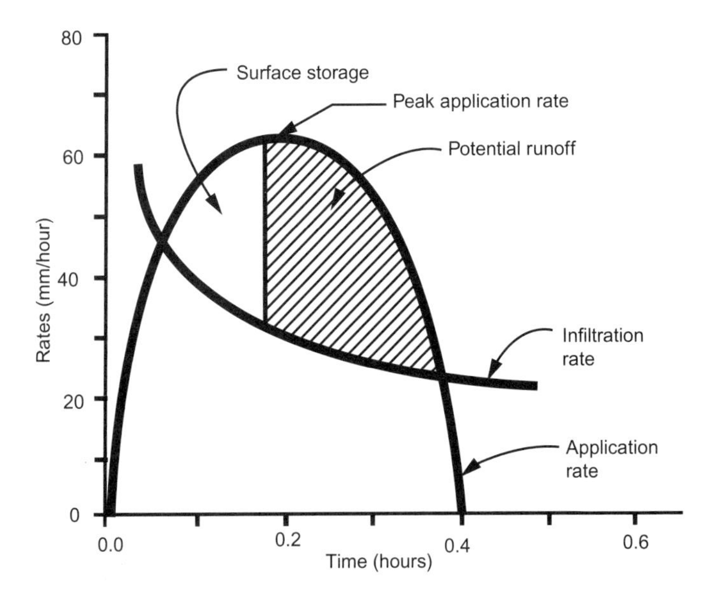

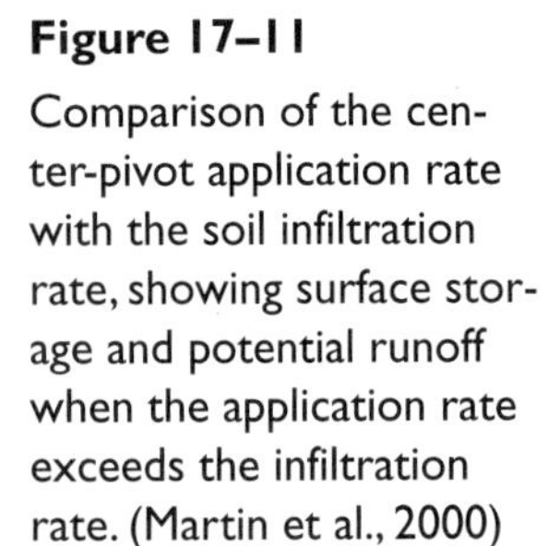
Figure 17–11

Comparison of the center-pivot application rate with the soil infiltration rate, showing surface storage and potential runoff when the application rate exceeds the infiltration rate. (Martin et al., 2000)

Example 17.3 Design a center-pivot irrigation system to irrigate the circular area of a square 800-m field (Figure 17-10) at a maximum rate of application of 25 mm/h with a net depth of application of 18 mm. The corners are not irrigated. The field is to be irrigated in 66 hours. Assume an application efficiency of 75 percent. The highest point of the field is 1 m above the pivot point, and no spray is allowed beyond the field boundary. Galvanized steel pipe is used for the lateral.

Solution.

(1) Determine the theoretical capacity of the system from Equation 17.11

$$Q_s = \frac{\pi R_s^2 (\text{m}^2) d_a\,(\text{mm})\ 1000\ (\text{L/m}^3)}{t(\text{h})3600(\text{s/h})\ 1000\ (\text{mm/m})E_a} = \frac{3.14 \times 400^2 \times 18}{3600 \times 66 \times 0.75} = 50.7\ \text{L/s}$$

The actual discharge of the system will depend on the sum of the discharges of the actual sprinklers selected.

(2) Assume a sprinkler spacing of 2.6 m at the end of the lateral. The discharge needed for the last sprinkler is obtained from Equation 17.12.

$$q_R = \frac{2 \times 50.7(\text{L/s}) \times 400(\text{m}) \times 2.6(\text{m})}{400^2(\text{m}^2)} = 0.66\ \text{L/s}$$

(3) Select a sprinkler from a manufacturer's catalog to deliver about 0.66 L/s. Assume a sprinkler is found that requires a head of 20 m and has a 13-m wetted radius. Because no spray is allowed beyond the field boundary the actual lateral length is 400 – 13 = 387 m. The capacity of the system is now recalculated.

$$Q_s = \frac{3.14 \times 387^2 \times 18}{3600 \times 66 \times 0.75} = 47.5\ \text{L/s}$$

The new discharge for the last sprinkler is

$$q_R = \frac{2 \times 47.5 \times 387 \times 2.6}{387^2} = 0.64\ \text{L/s}$$

(4) Determine the peak application rate at the end of the system from Equation 17.14.

$$r_p = 4\left(\frac{47.5\ (\text{L/s}) \times 3600(\text{s/h}) \times 1000(\text{mm/m})}{\pi\ 387^2(\text{m}^2) \times 1000(\text{L/m}^3)}\right)\frac{387(\text{m})}{13(\text{m})} = 43.3\ \text{mm/h}$$

Since this exceeds 25 mm/h, the sprinkler system must be modified or surface storage evaluated to estimate the potential for runoff.

(5) From Equation 17.15, the time to peak application is

$$t_p = \frac{W_r(\text{m})}{2\,\pi R(\text{m})\left(\dfrac{Q_s(\text{L/s}) \times 3600(\text{s/h})}{\pi R_s^2(\text{m}^2) \times 1000(\text{L/m}^3)}\right)} \quad \frac{d_a(\text{mm})}{E_a \times 1000(\text{mm/m})}$$

$$= \frac{W_r}{7200\pi R\left(\dfrac{Q_s}{\pi R_s^2}\right)}\frac{d_a}{E_a} = \frac{13}{7200 \times \left(\dfrac{47.5}{387}\right)}\frac{18}{0.75} = 0.35 \text{ h}$$

The wetting time at the outer end of the system is twice the time to peak or 0.7 h.

(6) Determine the friction loss in the 177.8-mm OD galvanized steel lateral (2.79 mm wall thickness) with $F = 0.54$ and $C = 120$ from Equation 17.4

$$h_f = \frac{0.54 \times 1.21 \times 10^{10} \times 387(47.5/120)^{1.852}}{(177.8 - 2 \times 2.79)^{4.87}} = 5.9 \text{ m}$$

If a head loss of 5.9 m is too large, the next larger pipe size might be used for the initial portion or the entire lateral.

(7) Assume the lateral is 4 m above the soil surface and the sprinkler heads are attached below the lateral on 2-m-long drop tubes. The total head at the pivot including 4.0 m lateral height minus 2.0 m drop and $H_e = 1.0$ m is

$$H_t = 20 + 5.9 + (4.0 - 2.0) + 1.0 = 28.9 \text{ m}$$

(*Note:* For convenience and simplicity it was assumed that the lateral was a constant diameter. In actual systems there is usually about a 12-m overhang from the last tower that is a 101- or 114-mm-diameter pipe.)

(8) The nozzle size may be determined at any point along the lateral from the pressure head available and the required sprinkler capacity. For example, at a distance of 200 m from the pivot for $x = 0.517$ (200/387) the nozzle pressure from Equation 17.13 is

$$H_R = 20 + 5.9[1 - 1.875(0.517 - 2 \times 0.517^3/3 + 0.517^5/5)] = 21.1 \text{ m}$$

The required capacity for 2.6-m spacing of the sprinkler at $R = 200$ from Equation 17.12 is

$$q_r = \frac{2 \times 47.5 \times 200 \times 2.6}{387^2} = 0.31 \text{ L/s}$$

The calculated nozzle diameter can be determined from Equation 17.3 with $C = 0.97$

$$d_n = \left[\frac{0.31}{0.00111(0.97)(21.1 \times 9.81)^{0.5}}\right]^{0.5} = 4.46 \text{ mm}$$

The actual nozzle and sprinkler selected will be the closest diameter available; however, the diameter of adjacent nozzles must also be considered to account for under- or over-sizing.

Selecting sprinklers and nozzle diameters for a system is frequently performed with a special computer program that starts at the distal end of the lateral and calculates the pressure, discharge, and required nozzle at each sprinkler location. Center-pivot systems are usually sold as a complete system from each manufacturer, thus the example above is intended to show only some of the important design parameters.

17.17 Linear-Move Systems

Linear-move systems use much the same hardware as center-pivot systems except they move linearly across the field. Linear-move systems are more expensive than center-pivot systems because they must include a water supply along the entire distance of move. The most common supplies are from (1) a concrete-lined ditch where a special moving dam, suction pipe, pump, electrical generator, and engine are attached to a special tower unit or (2) a pressurized pipeline with several hydrants and a long hose that delivers water to the lateral. Linear-move systems are somewhat simpler to design than center-pivots because each section of the lateral irrigates the same area. This feature also makes linear-move systems potentially more suitable for delivering crop nutrients and other chemicals.

17.18 Precision Application Systems

Modified center-pivot or linear-move systems have low-pressure water application devices attached to drop tubes from the laterals. These devices are designed to apply small, frequent applications near the soil surface often to individual furrows. The advantages of these systems are that they have a low energy requirement, reduce the potential for wind losses or drift, and reduce evaporation from soil and plant surfaces. Since the application is to a small area, soil surface management is required to increase infiltration and surface storage. Pits or basins are installed in the furrows and/or crop residue is managed to increase retention and infiltration of both irrigation and rainfall. For design details, see Martin et al. (2000) or Lyle & Bordovsky (1981). Modifications also are under development to add special pipes and nozzles to apply pesticides with these systems. Sensors also are being added that detect crop and soil conditions as the system moves and automatically adjust water and fertilizer applications.

Sprinkler Irrigation and the Environment

17.19 Sprinkler Systems for Environmental Control

Sprinkling has been successful for protecting small plants from wind damage, soil from blowing, and plants from frosts or freezing, and for reducing high air and soil temperatures. Since the entire area usually needs protection at the same time, solid-set systems are required. The rate of application should be as low as possible or just sufficient to achieve the desired control. Small pipes and low-volume sprinklers may be desired to reduce costs, but normal sprinkler systems may be modified for dual use. Because water is applied without regard to irrigation requirements, natural drainage should be adequate or a good drainage system should first be installed.

Especially in organic or sandy soils where onions, carrots, lettuce, and other small seeded crops are grown, the soil dries quickly and the seed may be blown away or covered too deeply for germination. When such plants are small, they are easily damaged by wind-blown particles. Protection for such conditions can be provided with a sprinkler system that will apply low rates up to 2.5 mm/h. Operation at night, when winds are usually at a minimum, will provide more uniform coverage.

Low-growing plants can be protected from freezing injury, which is likely in either the early spring or the late fall. Sprinkling has been most successful against radiation frosts. Water must be applied continuously at about 2.5 mm/h until the plant is free of ice. Sprinkling should be started before the temperature reaches 0°C at the plant level. Strawberries have been protected from temperatures as low as −6°C. Tomatoes, apples, cranberries, cherries, and citrus have been successfully protected. Tall plants, such as trees, may suffer limb breakage when ice accumulates, but in some areas low-level under-tree sprinkling has provided some control. Rates of application may be reduced by increasing the normal sprinkler spacing. A slightly higher pressure may be desirable to increase the diameter of coverage and to give a better breakup of the water droplets.

Sprinkling during the day to reduce plant heat stress has been successful with many plants, such as lettuce, potatoes, green beans, small fruits, tomatoes, cucumbers, and cantaloupe. This practice is sometimes called "misting" or "air-conditioning" irrigation. Maximum stress in the plant usually occurs at high temperatures, at low humidity, with rapid air movement, on bright cloudless days, and/or with rapidly growing crops on dry soils. Under these conditions crops at a critical stage of growth as during emergence, flowering, or fruit enlargement may benefit greatly from low applications of water during the midday. At 27°C, water loss was reduced 80 percent with an increase in humidity from 50 to 90 percent. Measured temperature reductions in the plant canopy of about 11°C were attained by misting in an atmosphere of 38 percent relative humidity. Green bean yields were increased 52 percent by midday misting during the bloom and pod development period. Potatoes and corn respond to sprinkling, especially when temperatures exceed 30°C. The tasseling period is a critical time for corn. Small quantities of water applied frequently to strawberries increased the quality and the yield by as much as 55 percent. For low-growing crops the same sprinkler system can be adapted for frost protection. Misting in a greenhouse or under a lath house to reduce transpiration of nursery plants for propagation increases plant growth and root development.

During periods of high incoming radiation, soil temperature may be 20°C greater than the ambient air temperature. Seedlings emerging through soil with temperatures as high as 50°C frequently die as a result of high transpiration. Small applications of water at this critical period often ensure emergence and good stands. Another benefit from sprinkling at this stage is enhancement of the effectiveness of herbicides applied to control weeds.

17.20 Sprinkler Systems for Fertilizer or Chemical Applications

Fertilizers, soil amendments, and pesticides may be injected into the sprinkler line as a convenient means of applying these materials to the soil or crop. This method primarily reduces labor costs and in some cases may improve the effectiveness and timeliness of application. Equipment should be resistant to corrosion from chemicals that may be present in the water and a backflow protection device is required to protect against water-source contamination.

Liquid and dry fertilizers have been successfully applied with sprinklers. Dry material must first be dissolved in a supply tank. The liquid may be injected on the suction side of the pump, forced under pressure into the discharge line, or injected into the discharge line by a differential pressure device such as a Venturi section. The material is applied for a short time during the irrigation set. Sprinkling should be continued for at least 30 minutes after the material has been applied to rinse the supply tank, pipes, and crop. For center-pivot or other moving systems, the material must be applied continuously or until the field has been completely covered.

17.21 Waste Application

Liquid manure, sewage wastes, and other wastes are applied with sprinklers for disposing of unwanted material. A large amount of specialized commercial equipment is available. Wastes are usually sprinkled on wooded areas or on land in permanent grass. Application rates should be lower than the infiltration rate so that the water does not run off. Excessive application depths should be avoided to prevent ground water contamination. Federal and state guidelines must be consulted regarding the application of chemicals and waste to crops and soils.

For waste disposal with sprinkler systems, the solids should be well mixed and small enough so as not to plug the nozzles, which necessitates an effective nonplugging screen on the suction side of the pump. The pump selected must be designed to handle solids. Nozzles must not contain straightening vanes and must be checked periodically for plugging by debris or salt precipitates. Liquid must be stored in a lagoon or other holding pond. Storage must be sufficient to contain wastes produced during nonoperational periods such as subfreezing temperatures or when rainfall has filled the soil profile.

17.22 Effects of Sprinkler Irrigation on Soils and Plants

Soil compaction may be caused by the impact of large droplets that damage the soil structure. This reduces the infiltration rate and may lead to runoff and erosion, and less water stored in the root zone. This effect is most likely to occur on a bare soil. Leaving mulch on the soil surface will reduce the potential damage. Vegetation may absorb some of the energy; however, seedlings and other tender vegetation may be damaged. Increasing the operating pressure will reduce the droplet size, but droplet size may still be too large with large sprinklers.

Chemical elements such as sodium and chloride are absorbed by plant leaves. If sufficient amounts are absorbed, plant growth and yield are reduced. Some plants are more sensitive than others. Trace elements, including arsenic, boron, cadmium, chromium, lead, molybdenum, and selenium, are also toxic to some plants. These elements mainly originate in the geologic materials at the site instead of the irrigation water. Some of the effects of these elements can be reduced by sprinkling at night when evaporation losses are low. Thus it is important during the planning stages for an irrigation system to carefully check the chemical content of the water supply and the soil and to determine the chemical sensitivity of any crops to be grown. Frequent applications increase foliar absorption, and the moist conditions may promote diseases.

Sprinkler Systems for Turf

Sprinkler systems are generally used to irrigate turf, such as on golf courses, parks, athletic fields, and lawns. These systems are usually permanent installations with buried plastic pipe lines. Many sources of information are available on turf irrigation systems, including Rochester (1995), Watkins (1987), Pair et al. (1983), and manufacturers' booklets. The design and layout criteria are similar to those for agricultural crops; however, there are several important differences:

1. Triangular sprinkler layouts may be used. Triangular layouts tend to give a higher uniformity, but are not convenient for rectangular areas.
2. Sprinkler heads are installed flush with the soil surface so they do not interfere with the use and maintenance of the turf. So that the spray is not affected by the grass, sprinklers usually have a "pop-up" feature where the nozzles rise up to 100 mm above the base when operating (Figure 17-12). Heads are available with a smooth cap so the head is protected from damage and turf users are not exposed to projecting surfaces.
3. Small and/or irregular-shaped areas make designs more difficult. Many landscaped areas are around buildings and include walkways or special areas added to increase the attractiveness of the site. These situations require more effort in the design and layout of the sprinklers. Fortunately, sprinkler manufacturers have produced special heads to aid in irrigating oddly shaped turf areas.
4. Spray is not allowed beyond the turf boundaries onto roads, walkways, buildings, or adjacent boundaries. Fines or liability problems may occur if spray or runoff affects buildings, streets, vehicles, or pedestrians. Wind must also be considered to prevent wind drift from affecting these areas. During design and layout, a critical side—preferably straight—is usually selected and serves as the starting line for the design.
5. The water supply for many turf systems is from a municipal water system, so water is available on demand. Because of daytime use, some turf can only be irrigated at night, and the system must be designed to accommodate this restriction.
6. Most turf systems are automated to provide better water control and nighttime irrigation. Many electronic controllers are available that will control one to many valves. Wires must be sized and then installed from the controller to the electric solenoid valves for each section. The valve station for each section typically contains a valve and may contain a filter and a pressure regulator if required. A series of valves may be installed near the controller and pipe installed from each valve to the area it controls. Alternatively, wires may be run to valves at various locations. The selection is determined by economics, location of water and electrical sources, and aesthetics.
7. Since ornamentals or plants besides turf are often planted in the same area, it is highly desirable for plants with similar water requirements to be irrigated together. This is usually accomplished by having a separate valve and pipe for different plantings. If only a few plants are involved, it is probably more economical to use one system, but carefully select heads that will deliver the appropriate amount of water to each plant type. Microirrigation may be more suitable for ornamentals and other widely spaced plants.

(a)

(b)

Figure 17–12 Examples of "pop-up" sprinkler heads for turf applications: (a) spray head and (b) rotary head. (Photographs published with the permission of RainBird Inc.)

8. System design and installation must meet standards and codes for safety and to prevent contamination of the water supply. For example, backflow prevention devices are required so that water cannot flow back into the supply system. In small systems, simple vacuum-breaker backflow devices are frequently installed with each valve. In large systems, one large, spring-loaded device is installed at the water connection.
9. To conserve freshwater supplies, treated waste effluent waters may be applied to turf. Special precautions are required, such as installing separate pipe systems. Specially colored heads may be required to indicate that nonpotable water is being used. Standards and codes defining the use of treated waste effluent must be followed.

Internet Resources

Websites for equipment and information
- http://www.clemson.edu Key search term: irrigation
- http://www.irrigation-mart.com
- http://www.irrigation.org
- http://www.wcc.nrcs.usda.gov/nrcsirrig/

Websites for sizing irrigation pipelines
- http://eesc.orst.edu Key search terms: sizing pipes, PNW290

Websites for friction loss tables
- http://www.commerce.state.wi.us Key search term: pipe friction loss
- http://www.homepower.com Key search term: pipe friction loss

References

Chu, S. T., & D. L. Moe. (1972). Hydraulics of a center pivot system. *ASAE Transactions, 15,* 894–896.

Frost, K. R. (1963). Factors affecting evapotranspiration losses during sprinkling. *ASAE Transactions, 6,* 282–287.

Frost, K. R., & H. C. Schwalen. (1955). Sprinkler evaporation losses. *Agricultural Engineering, 36*(8), 526–528.

2000 Irrigation Survey. (2001). *Irrigation Journal, 51*(1), 12–29, 40, 41.

Keller, J., & R. D. Bliesner. (1990). *Sprinkle and Trickle Irrigation.* New York: Van Nostrand Reinhold.

Kincaid, D. C. (2000). Personal communication. USDA-ARS. Kimberly, ID.

Lyle, W. M., & J. P. Bordovsky. (1981). Low Energy Precision Application (LEPA) Systems. *ASAE Transactions, 24*(5), 1241–1245.

Martin, D. L., D. C. Kincaid, & W. M. Lyle. (2000). Design and operation of sprinkler systems. Draft chapter for ASAE Monograph. *Design and Operation of Farm Irrigation Systems,* 2nd ed. G. J. Hoffman, D. L. Martin, & R. L. Elliott, eds. ASAE Monograph. St. Joseph, MI: ASAE.

Pair, C. H., W. H. Hinz, K. R. Frost, R. E. Sneed, & T. J. Schlitz, eds. (1983). *Irrigation,* 5th ed. Silver Springs, MD: Sprinkler Irrigation Association.

Rochester, E. W. (1995). *Landscape Irrigation Design.* ASAE Publication No. 8. St. Joseph, MI: ASAE.

U.S. Soil Conservation Service (SCS). (1983). Sprinkler irrigation. In *National Engineering Handbook,* Sect. 15, *Irrigation,* Chap. 11. Washington, DC: U. S. Government Printing Office.

Watkins, J. A. (1987). *Turf Irrigation Manual.* Dallas, TX: Telcso Industries.

Problems

17.1 Determine the required discharge of a sprinkler system to apply water at a rate of 15 mm/h. Two 186-m sprinkler laterals with 16 sprinklers, each at 12.2 m spacing on the line and 18.3 m spacing between laterals, are required.

17.2 Allowing 1 h for moving each 186-m sprinkler lateral described in Problem 17.1, how many hours would be required to apply a 60-mm application of water to a square 16-ha field? How many 10-h days are required?

17.3 Determine the discharge rate for one sprinkler operating at 270 kPa and having two nozzles, 4.0 mm (range), and 2.4 mm (spreader) in diameter, with discharge coefficients of 0.96.

17.4 Compute the depth rate of application for a 0.6-L/s sprinkler head if the sprinkler spacing is 9.1×12.2 m.

17.5 Compute the total friction loss for a sprinkler system having a 101.6-mm diameter main 250 m long and one 76.2-mm lateral 122 m long with 10 sprinkler heads. The pump delivers 8.0 L/s. Sixteen-gauge aluminum pipe (1.3 mm wall thickness) and standard couplers are used.

17.6 Design a sprinkler irrigation system for a square 16-ha field to irrigate the entire field within a 14-day period. Assume no wind and the first sprinkler is one half interval from the main, maximum elevation of main above pump is 2 m, laterals are level, riser height is 1 m, friction loss in the suction line is 1 m, and pump efficiency is 65 percent. Only 16.5 h per day are available for moving pipe and sprinkling. Depth of application is to be 60 mm at each setting at a rate not to exceed

9.0 mm/h. A well 25 m deep located at the center of the field has the following drawdown-discharge characteristics obtained from a well test: 10 m, 12 L/s; 15 m, 15 L/s; 20 m, 19 L/s.

17.7 Determine the total pumping head for a sprinkler irrigation system on level land. The average operating pressure at the nozzle is 276 kPa; the friction loss is 5 m in the main and 3.7 m in the lateral, the drawdown in the well is 4.3 m at the required discharge of 32 L/s; the riser height is 1.5 m; and friction loss in all valves and suction line is 3.0 m. Determine the power requirement for the pump if it operates at 70 percent efficiency.

17.8 Derive Equation 17.2 from conservation of mass and energy, and then derive the equation for sprinkler nozzle discharge $q = 0.00111\, C\, d_n^2\, P^{1/2}$.

17.9 Assuming a triangular depth-distribution pattern for a rectangular placement of sprinklers and a diameter of coverage of 30 m, compute the uniformity coefficient for 12 × 18 m sprinkler spacing using a 3-m grid for depth measurements. Assume that the depth of application at the sprinkler is 30 mm and is zero at 15 m.

17.10 A 36-ha field is to be irrigated at a maximum rate of 10 mm/h with a sprinkler system. The root zone is 1 m deep, and the available water capacity of the soil is 200 mm/m of depth. The water application efficiency is 70 percent, and the soil is to be irrigated when 45 percent of the available water capacity is depleted. The peak rate of water use is 5.0 mm/day. Determine the net depth of application, depth of water to be pumped, days to cover the field, and area to be irrigated per day.

17.11 If the pressures at opposite ends of a sprinkler lateral are 290 and 260 kPa, what would be the discharge of the distal sprinkler if the sprinkler at the 290-kPa end discharged 0.8 L/s?

17.12 Twenty sample cans are uniformly spaced in the area covered by four sprinklers. The following depths were caught in the cans: 27, 24, 21, 27, 24, 22, 22, 26, 21, 20, 19, 23, 20, 18, 16, 20, 26, 22, 15, and 17 mm. Determine the uniformity coefficient and the distribution uniformity low quarter.

17.13 A center-pivot system is to irrigate the circular area of a square 64-ha level field. Peak water use is 6 mm/day, application efficiency is 75 percent, and maximum operating time is 22 h/day. Assume 161.6-mm inside diameter lateral (steel tube), 4-m lateral height, 3-m sprinkler spacing, 16-m wetted diameter with 30-m pressure at the distal end of the lateral. Determine total discharge, pressure at the pivot, discharge of the distal sprinkler, and the peak application rate. Also find the pressure in the lateral at 350-m radius.

CHAPTER

18 Microirrigation

Microirrigation is a method for delivering slow, frequent applications of water to the soil using a low-pressure distribution system and special flow-control outlets. Microirrigation is also referred to as drip, subsurface, bubbler, low-flow, low-pressure, or trickle irrigation, and all have similar design and management criteria.

These systems deliver water to individual plants or rows of plants. The outlets are generally spaced at short intervals along small tubing, and unlike surface or sprinkler irrigation, only the soil near the plant is watered. The outlets include emitters, orifices, bubblers, and sprays or microsprinklers with flows ranging from 2 to over 200 L/h.

According to Karmeli & Keller (1975), microirrigation research began in Germany about 1860. In the 1940s it was introduced in England especially for watering and fertilizing plants in greenhouses. With the increased availability of plastic pipe and the development of emitters in Israel in the 1950s, it has since become an important method of irrigation in Australia, Europe, Israel, Japan, Mexico, South Africa, and the United States. According to the "2000 Irrigation Survey" in *Irrigation Journal* (2001) California had 675 000 ha, Florida had 270 000 ha, and the U.S. total was over 3 000 000 ha of microirrigation.

Microirrigation has been accepted mostly in the more arid regions for watering high-value crops, such as fruit and nut trees, grapes and other vine crops, sugar cane, pineapples, strawberries, flowers, melons, vegetables, and landscape plants. Microirrigation has also been successfully used on row crops such as corn, cotton, sorghum, and tomatoes. In Arizona, subsurface systems have been used for cotton, melons, vegetables, and wheat in a crop rotation system for over 10 years (Wuertz, 2001). Special equipment was developed for tillage and field operations without removing or damaging the tubing.

18.1 Advantages and Disadvantages of Microirrigation

With microirrigation only the root zone of the plant is supplied with water, and with proper system management, deep-percolation losses are minimal. Soil evaporation may be lower because only a portion of the surface area is wet. Like solid-set sprinkler systems, labor requirements are lower and the system can be readily automated.

Reduced percolation and evaporation losses result in high efficiencies of water use. Weeds are more easily controlled, especially for the soil area that is not irrigated. Bacteria, fungi, and other pests and diseases that depend on a moist environment are reduced as the above-ground plant parts normally are completely dry; however, harmful soil organisms may be enhanced in the frequently wetted soil. Because soil is kept at a high water level and the water does not contact the plant leaves, use of more saline water may be possible with less stress and damage to the plant, such as leaf burn. Field edge losses and spray evaporation, associated with sprinklers, are reduced with these systems. Low rates of water application at lower pressures are possible, eliminating runoff. With some crops, yields and quality are increased, probably as a result of maintenance of a high temporal soil water level adequate to meet transpiration demands. Experiments have shown crop yield differences varying from little to more than a 50 percent increase compared with other methods of irrigation. Some fertilizers and pesticides may be injected into the system and applied in small quantities, as needed, with the water. With good system design and management, this practice can minimize chemical applications and reduce chemical movement to the ground water supply. Except for acids and chlorine, the system should be operated for sufficient time after injection to assure that chemicals have been flushed from the lines.

The major disadvantages of microirrigation are initial cost and potential clogging of system components, especially emitters, by particulate, biological, and chemical matter. A good filter system is required to remove particulate matter from the water. Chlorine is frequently injected into the water for biological control. Acids are injected to control pH and chemical precipitation, including salt accumulations at the emitter outlets. Special precautions are required for the handling and injection of chemicals (ASAE, 2001c).

Emitters may not be well suited to certain crops and special problems may be caused by salinity. Salt tends to accumulate along the fringes of the wetted surface strip (Figure 18-1). Because these systems normally wet only part of the potential soil-root volume, plant roots may be restricted to the soil volume near each emitter as shown in Figure 18-1. The dry soil area between emitter lateral lines may lead to

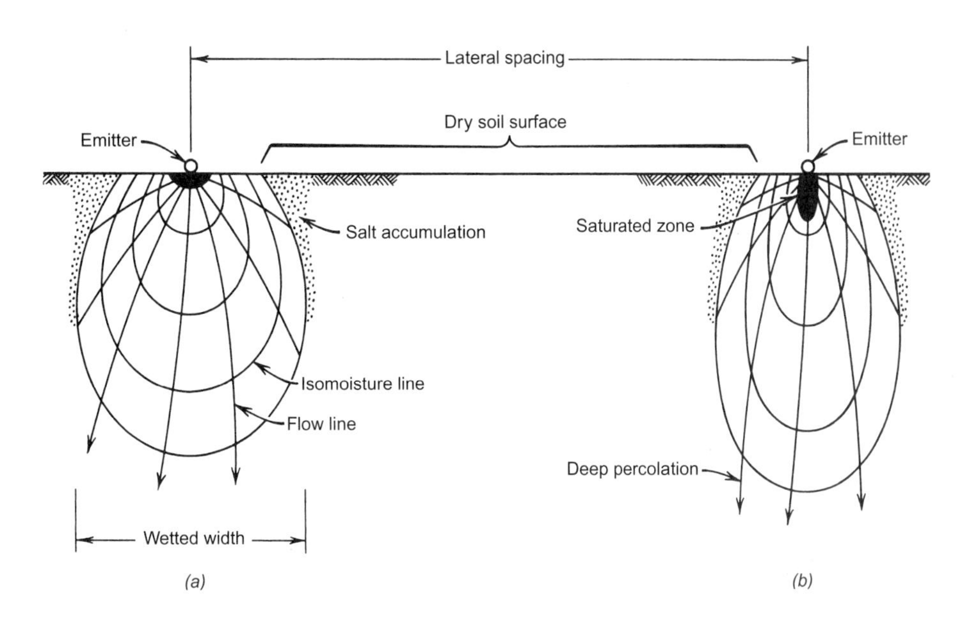

Figure 18–1
Soil-wetting pattern with microirrigation. (a) Medium and heavy soils. (b) Sandy soils. (Adapted from Karmeli & Keller, 1975)

dust formation from tillage operations and subsequent wind erosion. Compared with surface and sprinkler irrigation systems, more highly skilled labor is required to operate and maintain the filtration equipment and other specialized components. Rodents sometimes damage tubing or other plastic components.

18.2 Layout and Components of Microirrigation Systems

System layouts are similar to sprinkler systems (Chapter 17). As with sprinkler systems many arrangements are possible. The layout given in Figure 18-2 shows split-line operation for the upper left quadrant of a 16-ha orchard. The well is located at the center of the larger field. The layout would be similar for the other quadrants. Tree rows are parallel to the laterals. Subunits 1, 2, 3, and 4 of the 4-ha quadrant in Figure 18-2 could be operated independently of each other or in any combination, because each subunit has its own control valve.

As shown in Figures 18-2 and 18-3, the primary components of a microirrigation system are the control head, the main and submain, a manifold, and lateral lines to which the emitters are attached. The control head may consist of the pump, filters, injectors, backflow preventers, pressure gauges and regulators, water meter, flushing valves, air relief valves, and programmable control devices. The manifold is a line to which the laterals are connected. The manifold, submains, and main may be on the surface or buried underground. The manifold is usually flexible pipe if laid on the surface or rigid pipe if buried. The main lines may be any type of pipe, such as polyethylene (PE), polyvinyl chloride (PVC), galvanized steel, or

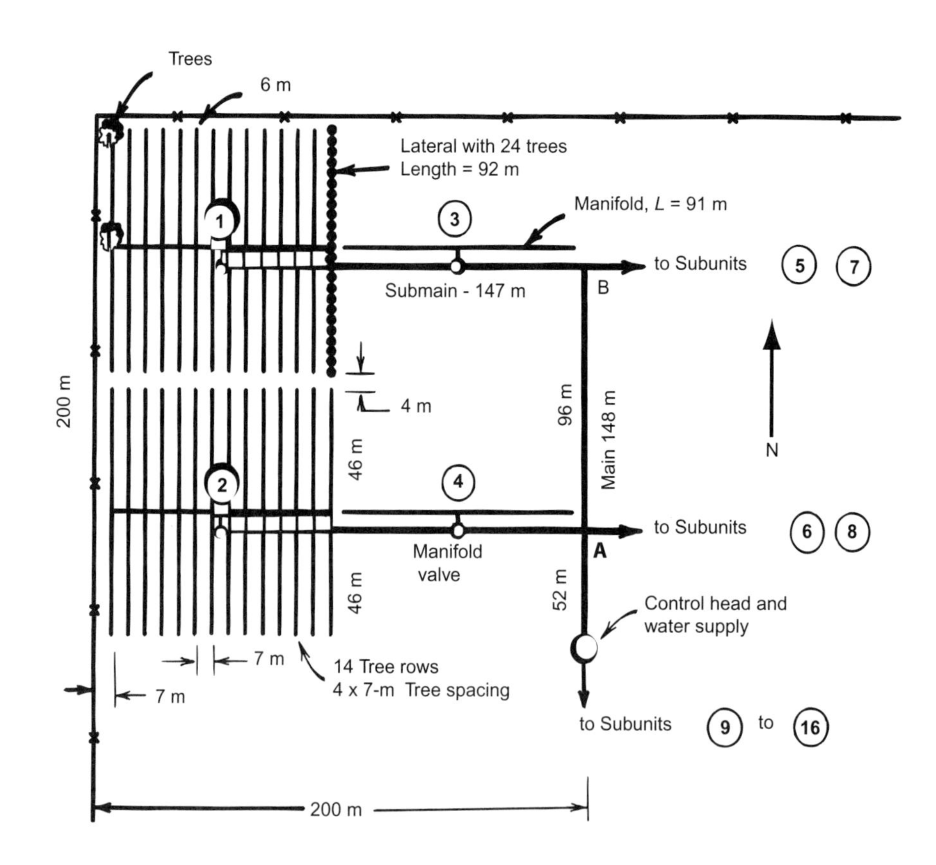

Figure 18–2
Microirrigation system layout for a 4-ha quadrant of a 16-ha orchard with the control head and water supply in the center of a square field.

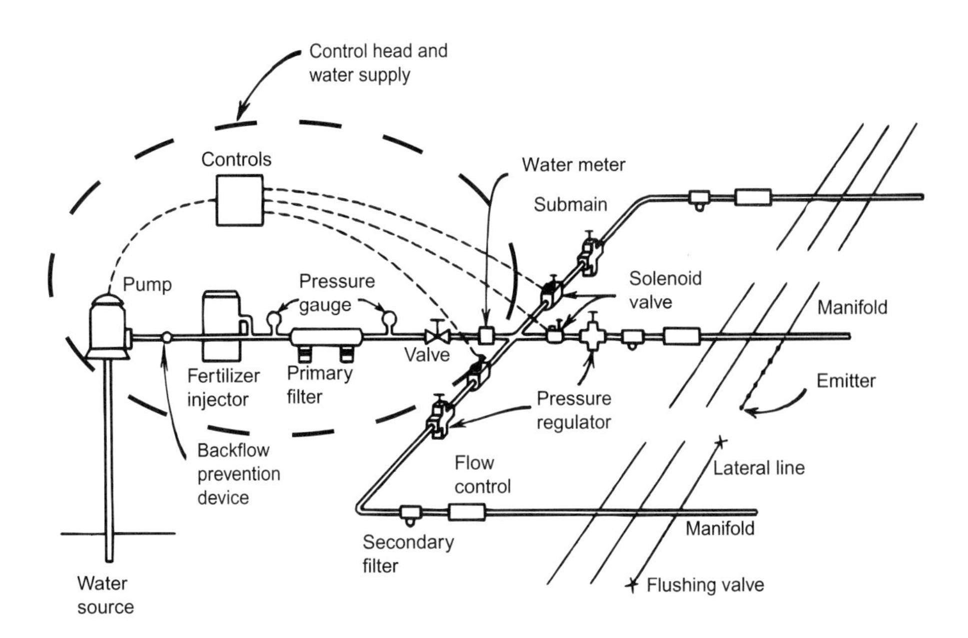

Figure 18–3

Components of a microirrigation system.

aluminum. See Appendix D and ASAE (1998, 2001b) for information on plastic pipe. The lateral lines that have emitters are usually PE or flexible PVC tubing. They generally range from 10 to 32 mm in diameter and have emitters spaced at short intervals appropriate for the crop to be grown. A flushing device at the end of each lateral or a manifold, with an outlet valve, connected to the ends of several laterals is necessary to flush sediment and debris from the laterals. Sometimes secondary filters and pressure regulators are installed at the inlet to each manifold. Valves also may be installed for air and/or vacuum relief.

An efficient filter to prevent emitter clogging is the most important component of the microirrigation system. Particulate removal is more important than for drinking water. Microirrigation systems generally require screen, gravel, or graded sand filters (Evans et al., 2000). Recommendations of the emitter manufacturer should be followed in selecting the filtration system. In the absence of such recommendations, the net opening diameter of the filter should be smaller than one tenth to one fourth of the emitter opening diameter (Evans et al., 2000). For clean ground water, an 80- to 200-mesh filter may be adequate. This filter will remove soil, sand, and debris, but should not be used with high-algae water. For water with high silt and algae contents, a sand filter followed by a screen filter is recommended. A sand separator ahead of the filter may be necessary if the water contains considerable sand. In-line strainers with replaceable screens and cleanout plugs may be adequate with small amounts of sand. These are recommended as a safety precaution should accidents during cleaning or filter damage allow particles or unfiltered water to enter into the system. Filters must be cleaned and serviced regularly. Pressure loss through the filter should be monitored to determine the need for maintenance. (See ASAE, 2000, for details on the testing and performance of media filters.)

Lateral lines may be located along the tree row with several emitters required for each tree as shown in Figure 18-4. Many laterals have multiple emitters, such as the "spaghetti" tubing or "pigtail" lines shown in Figure 18-4c. One or more laterals per row (Figure 18-4a or b) may be provided, depending on the size of the trees. A single line is adequate for small trees. For row crops, a lateral is usually installed for each row.

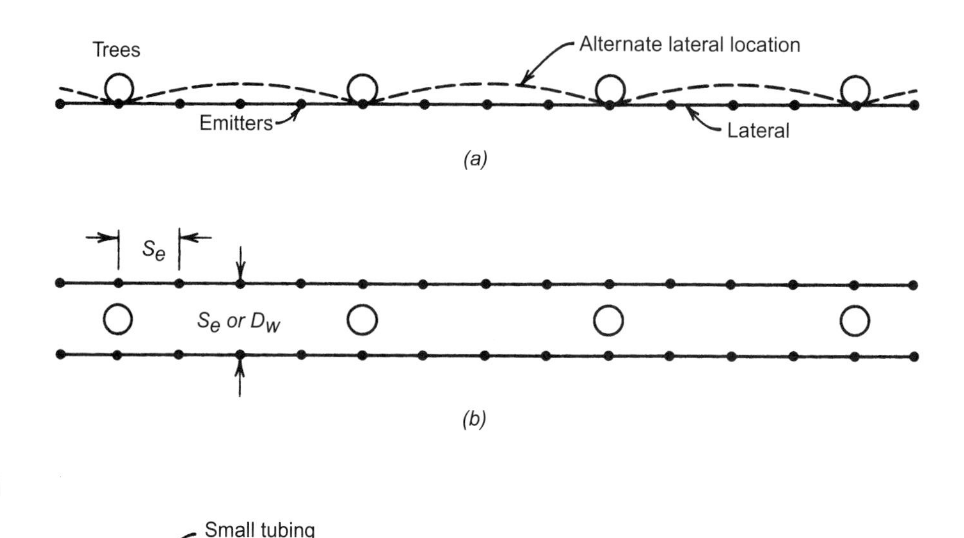

Figure 18–4
Lateral and emitter locations for an orchard or vineyard. (a) Single lateral for each row of plants. (b) Two laterals for each plant row. (c) Multiple outlet layouts.

Many types and designs of emitters are commercially available, some of which are shown in Figure 18-5. The emitter controls the flow from the lateral and reduces the pressure with small openings, long passageways, tortuous passageways, vortex chambers, manual adjustment, or other mechanical devices. Some emitters (Figure 18-5f) may be pressure-compensating by changing the length or by mechanically changing the cross section of passageways or size of orifice, and give a nearly constant discharge over a wide range of pressures. Some are self-cleaning and flush automatically. Porous pipe and tubing may have many small openings as shown in Figure 18-5h to j. The actual size of the opening is much smaller than indicated in the drawing. Some holes are barely visible to the naked eye. The double-tube lateral shown in Figure 18-5j has more openings in the outer channel than in the main flow channel. Such tubes have thin walls and are inexpensive. Tubes may be discarded after the crop is harvested and replaced with new lines. Emitters may be placed on the soil surface, or they may be buried at shallow depths for protection (subsurface drip irrigation, SDI).

18.3 Emitter Discharge

In an orifice-type emitter the flow is fully turbulent and the discharge can be determined from the sprinkler nozzle equation (Equation 17.3). The discharge of any emitter may be expressed by the power-curve equation (Karmeli & Keller, 1975)

$$q = Kh^x \qquad \textbf{18.1}$$

where q = emitter discharge (L^3/T),
K = constant for each emitter,

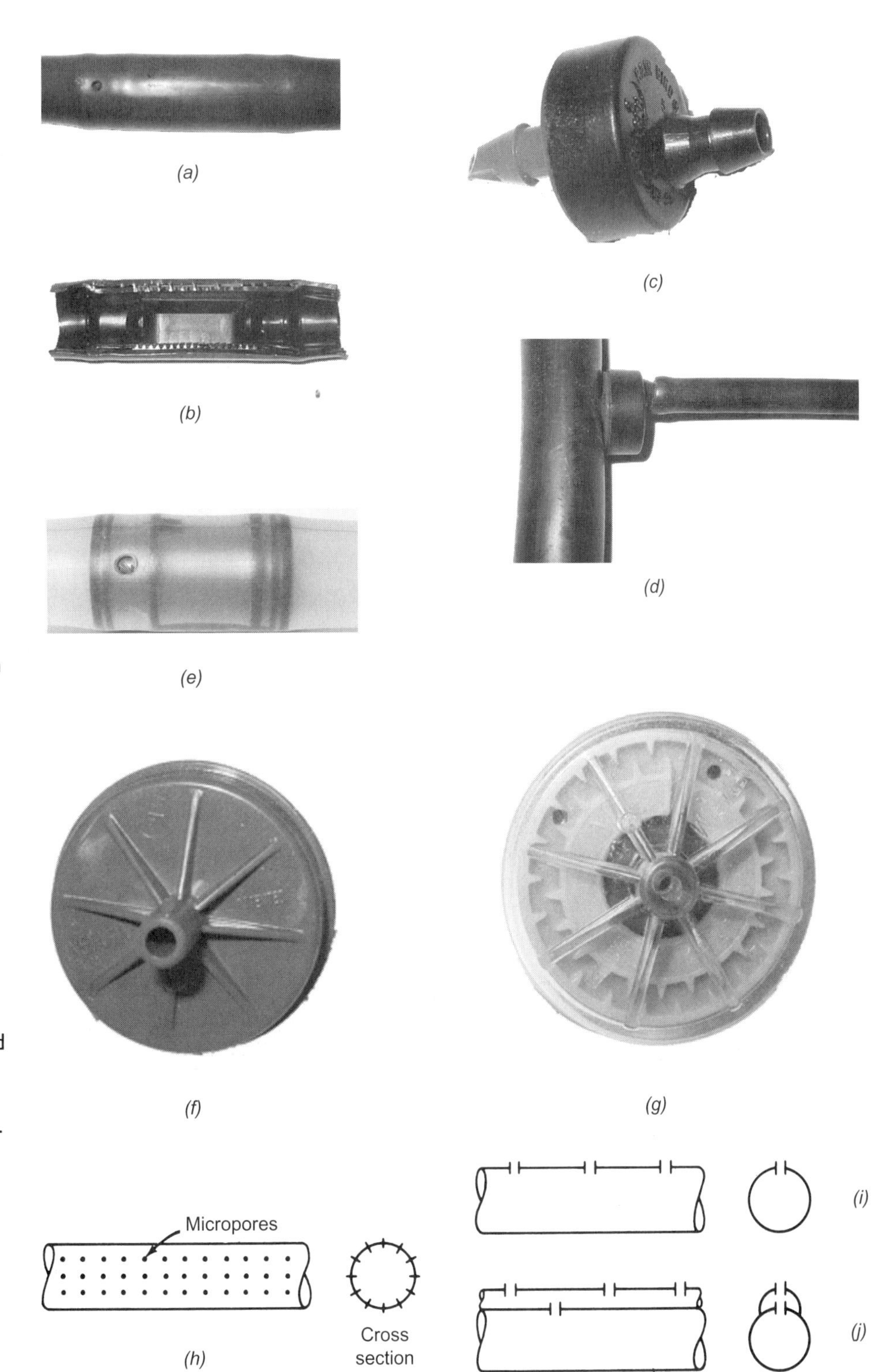

Figure 18–5

Types of microirrigation laterals and emitters. (a) In-line emitter formed in the lateral during production. (b) Cross section of emitter (a). (c) Emitter with barb for attachment. (d) Emitter (c) attached to a lateral and with a small tube attached to deliver water to the desired location. (e) Pre-molded in-line emitter is a transparent tube. (f) Pressure compensating emitter. (g) Emitter (f) with a transparent cap to show the pressure reducing path. (h) Porous tubing. (i) Single-tube lateral. (j) Double-tube lateral.

h = pressure head (L),

x = emitter discharge exponent.

The exponent x and constant K can be determined from a log–log plot of head versus discharge. With x known, K can be determined from Equation 18-1. With fully turbulent flow $x = 0.5$, and in a laminar flow regime $x = 1.0$. In a fully pressure-compensating emitter, K is a constant for a wide range of pressures and $x \approx 0$. Because of the large number of emitters available, it may be more convenient to determine discharge directly from manufacturers' curves. Some examples are shown in Figure 18-6, where K and x are for use in Equation 18.1. Double-tube laterals and porous tubing are typically rated as discharge per length, for example, 3 L/min per 100 m at 70 kPa pressure. Manufacturers' data should be obtained for the actual discharge-versus-pressure rating. Emitter discharge usually varies from about 1 to 30 L/h and pressures range from 15 to 280 kPa. Average diameters of openings for emitters range from 0.0025 to 0.25 mm. Emitters made from thermoplastic materials may vary in discharge depending on the temperature. Thus discharge curves should be corrected for temperature.

18.4 Water Distribution from Emitters

Microirrigation was developed to provide more efficient application of water and to increase yields. An ideal system should provide a uniform discharge from each emitter. Application efficiency depends on the variation of emitter discharge, pressure variation along the lateral, and seepage below the root zone or other losses, such as soil evaporation. Emitter discharge variability is greater than that for sprinkler nozzles because of smaller openings and lower design pressures. Such variability may result from the design of the emitter, materials, and care of manufacture. Solomon (1979) found that the statistical coefficient of variation may range from 0.02 to 0.4. The coefficient of variation (C_v = standard deviation/mean) should be available for emitters and provided by the manufacturer. ASAE (2001a) guidelines for classification of emitter uniformity are shown in Table 18-1.

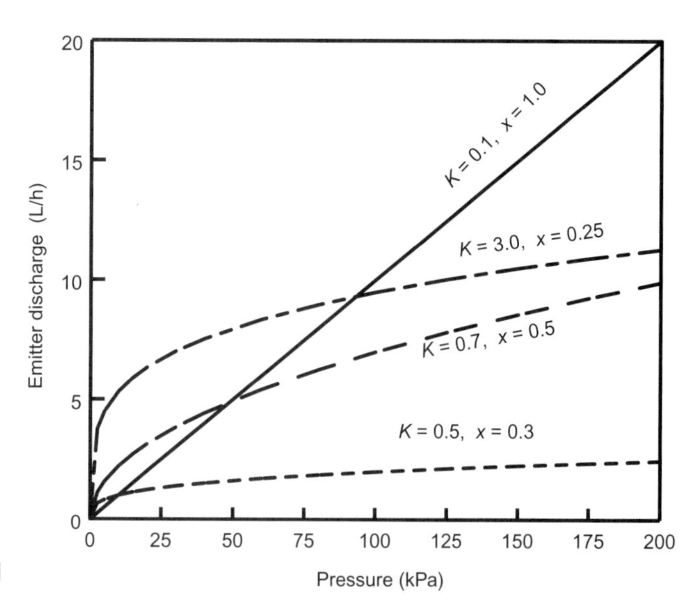

Figure 18–6

Examples of emitter discharges for various values of constant K and exponent x.

TABLE 18–1 Recommended Classification of Manufacturer's Coefficient of Variation, C_v

Emitter Type	*C_v Range*	*Classification*
Point source	<0.05	Excellent
	0.05–0.07	Average
	0.07–0.11	Marginal
	0.11–0.15	Poor
	>0.15	Unacceptable
Line source	<0.10	Good
	0.10–0.20	Average
	>0.20	Marginal to unacceptable

Source: ASAE (2001a).

Microirrigation systems must deliver the water required to each plant with minimum losses to obtain high efficiencies. This can be achieved by having a high uniformity of water delivery by each subunit of the system that has a separate control valve. Thus Subunits 1, 2, 3, and 4 in Figure 18-2 should each be designed for a high uniformity. The uniformity varies with pressure, emitter variation, and number of emitters per plant. This is defined by the emission uniformity (ASAE, 2001a).

$$EU = 100\left[1 - 1.27\frac{C_v}{n^{0.5}}\right]\frac{q_{min}}{q_{avg}} \tag{18.2}$$

where EU = emission uniformity (percent),

C_v = manufacturer's coefficient of variation,

n = number of emitters per plant for trees and shrubs or 1 for line sources,

q_{min} = minimum emitter discharge rate for the minimum pressure in the subunit (L^3/T),

q_{avg} = average or design emitter discharge rate for the subunit (L^3/T).

Recommended design values for EU are shown in Table 18-2.

TABLE 18–2 Recommended Ranges of Design Emission Uniformity, *EU*

Emitter Type	*Spacing (m)*	*Topography*	*Slope (%)*	*EU Range (%)*
Point source on perennial crops	>4	Uniform	<2	90–95
		Steep or undulating	>2	85–90
Point source on perennial or semipermanent crops	<4	Uniform	<2	85–90
		Steep or undulating	>2	80–90
Line source on annual or perennial crops	All	Uniform	<2	80–90
		Steep or undulating	>2	70–85

Source: ASAE (2001a).

Example 18.1 Determine the emission uniformity of a system subunit that uses an emitter with $K = 0.3$, $x = 0.57$, $C_v = 0.06$, and two emitters per plant with average pressure of 100 kPa and minimum pressure of 90 kPa.

Solution. Substitute Equation 18.1 into Equation 18.2

$$EU = 100\left[1 - \frac{1.27(0.06)}{2^{0.5}}\right]\frac{0.3(90)^{0.57}}{0.3(100)^{0.57}} = 89 \text{ percent}$$

18.5 Microirrigation System Design

A major difference between microirrigation and other systems is that not all the soil will be irrigated, especially for widely spaced plants. A minimum of 30 percent of the area should be irrigated. For mature trees 75 percent of the area may need to be irrigated, whereas nearly 100 percent of the area is irrigated for closely spaced plants in arid regions.

Emitter spacing and numbers required depends on the wetting pattern and plant spacing. To obtain data on horizontal and vertical water movement, field tests are preferred and should be conducted at several representative locations. If field measurements are not available, estimates may be obtained from Table 18-3 for the maximum horizontal wetted diameter, D_w, from a single emitter. For a line source, the emitter spacing S_e should be less than or equal to 0.8 D_w to overlap the wetting patterns of adjacent emitters along a lateral. For double laterals, a spacing of D_w between laterals will adequately wet the area; however, if emitters are individually spaced, the spacing can be D_w in both directions. If the water is saline, the spacing between laterals should be reduced to S_e. These closer spacings reduce dry areas between emitters where salts might accumulate. The number of emitters required per plant, n, can be obtained from

$$n = \frac{p_w \times \text{area/plant}}{\text{effective area wetted by one emitter}} \qquad \textbf{18.3}$$

where p_w is the percentage of the total area to be irrigated, "area/plant" is based on the plant spacing, and "effective area wetted by one emitter" depends on the wetted diameter, emitter layout, and water quality.

Example 18.2 For the layout in Figure 18-2 with mature orange trees and one lateral per row, determine the number of the emission devices needed per tree if 40 percent of the area is to be irrigated, salt content of the irrigation water is low, soil is layered, medium density, with coarse texture to a 2-m depth, and emitters will be installed on a "pigtail" as in Figure 18-4c.

Solution. From Table 15-3 determine that the effective rooting depth is 1.5 m; then, from Table 18-3, D_w = 1.8 m. The effective wetted area is assumed to be 0.8(1.8 m) $\times$ 1.8 m, which is substituted into Equation 18.3.

$$n = \frac{0.40 \times 4 \times 7}{0.8(1.8) \times 1.8} = 4.3, \text{ or round to 5 emitters per tree}$$

Note that if the irrigation water is saline, the effective wetted area is reduced to 0.8(1.8) × 0.8(1.8), which provides overlap of the wetting patterns and reduces salt accumulations within the wetted area.

The evapotranspiration of crops under microirrigation is not well defined because the area is not entirely shaded and it is generally somewhat less than under conventional systems that irrigate the entire area. Local or regional data for measured water use with microirrigation may be available. If local water use data are not available for microirrigation, Keller & Bliesner (1990) suggested the following for average peak transpiration rate for microirrigation design:

$$ET_t = ET_p \times 0.1 \times p^{0.5} \qquad \textbf{18.4}$$

where ET_t = average peak transpiration of crops under microirrigation (L/T),
ET_p = average peak conventional ET rate for the crop (L/T),
p = percentage total area shaded by the crop.

For example, if a mature orchard shades 70 percent of the area and the average peak conventional *ET* is 7 mm/day, the net microirrigation design rate is 5.9 mm/day ($7 \times 0.1 \times 70^{0.5}$).

The diameter of the laterals and manifolds should be selected to satisfy Equation 18.2 and the appropriate *EU* from Table 18-2. In addition, the velocity should be less than 1.5 m/s to limit friction loss and to prevent pipe damage from water hammer and surge pressures. Water hammer and surge pressures can be reduced by installing slow-acting valves and by controlling the flow when the supply pump is started.

TABLE 18–3 Estimated Maximum Diameter of the Wetted Circle Formed by a Single Emission Outlet Discharging 4 L/h on Various Soils

		Varying Layers	
Soil or Root Depth and Soil Texture	***Homogeneous Soil (m)***	***Generally Low Density (m)***	***Generally Medium Density (m)***
Depth 0.75 m			
Coarse	0.45	0.75	1.05
Medium	0.90	1.2	1.5
Fine	1.05	1.5	1.8
Depth 1.5 m			
Coarse	0.75	1.4	1.8
Medium	1.2	2.1	2.7
Fine	1.5	2.0	2.4

Source: Adapted from SCS (1984).

The maximum difference in pressure in a subunit usually occurs between the control point at the inlet and the emitter farthest from the inlet. In Figure 18-2 with no slope, the maximum difference in pressure to the farthest emitter is that for one half the lateral length plus one half the manifold length. Where the manifold is connected to the end of each lateral and submain is connected to the end of the manifold, the head loss would be computed for their entire length.

For minimum cost, Karmeli & Keller (1975) recommended that on a level area, 55 percent of the allowable head loss should be allocated to the lateral and 45 percent to the manifold. As in sprinkler laterals, allowable head loss should be adjusted for elevation differences along the lateral and along the manifold, unless pressure-compensating devices are used.

The friction loss for mains and submains can be computed from the Hazen-Williams or the Darcy-Weisbach equations. The Hazen-Williams equation as used for sprinkler design (Equation 17.4) is

$$H_f = 1.21 \times 10^{10} \frac{L}{D^{4.87}} \left(\frac{Q}{C}\right)^{1.852} \qquad \textbf{18.5}$$

where H_f = head loss in the pipe (m),
L = pipe length (m),
D = actual i.d. (inside diameter) of the pipe (mm),
Q = flow rate in the pipe (L/s),
C = Hazen-Williams roughness coefficient.

Equation 18.5 applies for continuous lengths of pipe. For smooth plastic pipe, C = 150. The same F factor for pipes with multiple outlets in Table 17-5 is valid for microirrigation laterals. Also, Equations 17.5, 17.6, and 17.7 can be applied to microirrigation laterals and are repeated below with $h_f = F \times H_f$.

$$H_a = H_d + 0.26 h_f + \frac{S_o L_L}{2} \qquad \textbf{18.6}$$

$$H_o = H_a + 0.74 h_f + \frac{S_o L_L}{2} \qquad \textbf{18.7}$$

Solving Equation 18.6 for H_d yields

$$H_d = H_a - 0.26 h_f - \frac{S_o L_L}{2} \qquad \textbf{18.8}$$

where H_a = average pressure in the lateral (L),
H_d = pressure at the distal end of the lateral (L),
S_o = slope of the lateral (positive uphill),

L_L = lateral length (L),
H_o = inlet pressure of the lateral (L).

For in-line emitters (Figures 18-5a and 18-5b), on-line emitters (Figure 18-5d), and other connectors, the head loss should be increased. Such losses may be expressed as equivalent length of pipe. This increase in length L_e can be estimated as follows (Karmeli & Keller, 1975),

(1) L_e = 1.0 to 3.0 m for each in-line emitter (Figure 18-5a),

(2) L_e = 0.1 to 0.6 m for each on-line emitter (emitter attached by insert through pipe wall, Figure 18-5d),

(3) L_e = 0.3 to 1.0 m for a solvent-welded T-connector.

The design process for a subunit requires determination of the peak *ET*, water required per plant, emitter selection, design operating pressure, minimum emitter discharge in the subunit, average emitter discharge in the subunit, and the emission uniformity, *EU*, of the subunit. Determining the average and minimum emitter discharges in a subunit requires calculating the average pressure in each lateral in the subunit. The process is usually started at the farthest lateral of the subunit having the lowest pressure because of distance from the water source and highest elevation. The design pressure and emitter discharge for the farthest lateral is based on the average emitter flow needed to meet ET. The pressure at the inlet to the next lateral upstream is determined from the pressure at the inlet to the first lateral and the friction loss in the manifold to the next lateral. A spreadsheet can be set to iterate for the average emitter pressure and emitter discharge in the lateral. The procedure continues until the average pressure and emitter discharge are calculated for all the laterals in the subunit.

If *EU* is too low, the options are to (1) choose another emitter with a smaller discharge exponent, manufacturing coefficient of variation, or both; (2) increase the number of emitters per plant; (3) redesign the system with a higher h_{avg}, which may require selecting a different emitter; or (4) reduce the design *EU* (SCS, 1984). If *EU* is high, the system may be overdesigned and more costly than necessary. Pipe sizes may also be changed to change *EU*.

Example 18.3

Design a microirrigation system for the orchard layout shown in Figure 18-2, assuming mature orange trees. The field is level with trees spaced 4-m apart in rows on a 7-m spacing and a manifold in the center of 92-m long laterals. Assume the maximum time for irrigation is 22 hours/day, 40 percent of the area is to be irrigated, and average peak transpiration rate corrected for shaded area is 5.9 mm/day. Subunits 1 and 2 are irrigated together every 2 days; Subunits 3 and 4 are irrigated on the alternate days. In addition, each quarter of the field has similar subunits irrigated every other day. Determine the emitter discharge, the required discharge for each subunit, and the total pumping head and flow rate to irrigate the field.

Solution.

(1) Use a spreadsheet to perform the calculations. From Table 18-2 the minimum *EU* is 90 percent. The volume of water required per tree per day is

$$\text{Volume} = \frac{5.9\,(\text{mm/day}) \times 4\,(\text{m}) \times 7\,(\text{m}) \times 1000\,(\text{L/m}^3)}{1000\,(\text{mm/m}) \times 90/100} = 184\ \text{L/day}$$

(2) Since each tree is irrigated every other day, 2 × 184 = 368 L must be delivered at each irrigation to each tree to meet the average peak transpiration. (*Note:* If the irrigation water contained sufficient salts, extra water would be required for leaching (Chapter 15; Keller & Bliesner, 1990; Clark et al., 2005)). From Example 18.2, five emitters are used per tree and each emitter must deliver 368/5 = 73.6 L in 22 h or 73.6/22 = 3.35 L/h is the required emitter flow rate.

(3) Determine if the soil has sufficient capacity to store the water applied. From Example 18.2, the rooting depth of oranges is 1.5 m and the expected wetted diameter is 1.8 m. From Table 15.4, *MAD* for Group 3 with 6 mm/day is 0.45. From Table 15-2 for a sandy loam soil, the available water is 12 percent on a volumetric basis. The potential wetted soil volume from five emitters is $5 \times \pi \times (1.8/2)^2 \times 1.5 = 19.1$ m^3. The maximum change in water content from an irrigation is soil volume × available water × *MAD* or 19.1 × 0.12 × 0.45 = 1.0 m^3 or 1000 L, which is very adequate compared to the 368 L to be applied. Although it appears that fewer emitters might be used, this is not recommended because sufficient soil must be wetted for the trees to develop an adequate root system for stability and for water and nutrient uptake.

(4) An emitter is selected with $K = 0.151$, $x = 0.63$ for q in L/hour, h in kPa, and a coefficient of variation of 0.05. *Note:* This information should be available from the manufacturer; if not, it can be obtained by testing several emitters.

(5) Substituting into Equation 18.1, $h = (3.35/0.151)^{1/0.63} = 137.0$ kPa or 13.96 m. This is the average operating pressure of the lateral with the lowest pressure, usually the farthest lateral.

(6) For 12 trees per half-lateral, $F = 0.36$ from Table 17-5. Calculate h_f from Equation 18.5. The discharge in the first lateral is 3.35 × 5 × 12 = 201 L/h or 0.056 L/s. A T-connection at each tree connects the pigtail to the lateral and has an equivalent length of 2 m of pipe. Assume an initial lateral inside diameter of 15.8 mm. The equivalent lateral length is 46 + 12 × 2 = 70 m.

$$h_f = 0.36 \times 1.21 \times 10^{10} \times \frac{70}{15.8^{4.87}} \left(\frac{0.056}{150}\right)^{1.852} = 0.20 \text{ m}$$

(7) Calculate the pressure at the distal end of the first lateral from Equation 18.8.

$$H_d = 13.96 - 0.26 \times 0.20 - 0 \times 46/2 = 13.91 \text{ m}$$

(8) The discharge of the last emitter for this case is q_{min}.

$$q_{min} = 0.151(13.91 \times 9.81)^{0.63} = 3.34 \text{ L/h}$$

(9) Calculate the inlet pressure for the first lateral using Equation 18.7.

$$H_o = 13.96 + 0.74 \times 0.20 + 0 \times 46/2 = 14.11 \text{ m}$$

(10) Calculate the head loss in the section of manifold between laterals 1 and 2 using Equation 18.5. The manifold is 26.6 mm in diameter, the equivalent

length for the fitting is 0.5 m so the manifold length is taken as 7.5 m, and the discharge in the manifold is $0.056 \times 2 = 0.11$ L/s because of the flow to the other half lateral.

$$H_f = 1.21 \times 10^{10} \times \frac{7.5}{26.6^{4.87}}\left(\frac{0.11}{150}\right)^{1.852} = 0.02 \text{ m}$$

(11) The pressure at the inlet to lateral 2 is $0.02 + 14.11 = 14.13$ m.

(12) Calculate the average pressure in lateral 2 from Equation 18.7. Because the discharge in lateral 2 is not yet known, the value for h_f for lateral 1 is assumed. Iterate until the values for h_f and emitter discharge converge. (The spreadsheet can perform the iterations.) In this case the head loss in the manifold is very small and one iteration is sufficient.

$$H_a = 14.13 - 0.74 \times 0.20 - 0 \times 46/2 = 13.98 \text{ m}$$

$$q = 0.151 \times (13.98 \times 9.81)^{0.63} = 3.35 \text{ L/s}$$

$$h_f = 0.36 \times 1.21 \times 10^{10} \times \frac{70}{15.8^{4.87}}\left(\frac{3.35 \times 12 \times 5/3600}{150}\right)^{1.852} = 0.20 \text{ m}$$

$$H_a = 14.13 - 0.74 \times 0.20 - 0 \times 46/2 = 13.98 \text{ m}$$

The average lateral pressures match and the lateral discharge is 3.35 L/h.

(13) Calculate the total discharge in lateral 2:

$$Q = 3.35 \times 12 \times 5 = 201 \text{ L/h or } 0.056 \text{ L/s}$$

(14) The discharge in the manifold between laterals 2 and 3 is

$$Q = 0.11 + 0.056 \times 2 = 0.22 \text{ L/s}$$

(15) Repeating the process from Step 8 for the laterals produces the following table. The solution is obtained by iterating in the spreadsheet until the values for average emitter flow and h_f in each row are within prescribed tolerances.

Lateral No.	*Lateral Inlet Head (m)*	*Average Lateral Head, H_a (m)*	*Average Emitter Flow (L/h)*	*Total Flow in Lateral (L/s)*	*Head Loss in Lateral, h_f (m)*	*Flow in Manifold (L/s)*	*Manifold Head Loss, H_f (m)*	*Manifold Velocity (m/s)*
1	14.11	13.96	3.35	0.056	0.20	0.11	0.02	0.20
2	14.13	13.98	3.35	0.056	0.20	0.22	0.06	0.40
3	14.19	14.04	3.36	0.056	0.20	0.34	0.13	0.60
4	14.31	14.17	3.38	0.056	0.20	0.45	0.22	0.81
5	14.53	14.38	3.41	0.057	0.20	0.56	0.33	1.01
6	14.87	14.71	3.46	0.058	0.21	0.68	0.47	1.22
7	15.34	15.18	3.53	0.059	0.22	0.80	0.64	1.43

Note: Manifold velocity does not exceed 1.5 m/s; however, adding another lateral would increase the velocity beyond the limit and the manifold diameter would need to be increased.

(16) Determine the average emitter discharge for the seven laterals, then calculate *EU* for Subunit 1 from Equation 18.2. *Note*: If the field was not level, the average emitter discharge would need to be calculated for both uphill and downhill half-laterals to calculate *EU* for Subunit 1.

$$q_{avg} = (3.35 + \ldots + 3.53)/7 = 3.41$$

$$EU = 100\left[1 - \frac{1.27(0.05)}{5^{0.5}}\right]\frac{3.34}{3.41} = 95.2 \text{ percent}$$

This is greater than 90 percent determined as acceptable in Step 1, so, the subunit design is acceptable.

(17) Compute the friction loss in each main and submain from Equation 18.5 by determining the diameter that keeps the velocity below 1.5 m/s and the pressure loss within reasonable limits. Assume PVC for the manifold and mains.

Line	*Q (L/s)*	*Pipe i.d.*[a] *(mm)*	*Velocity (m/s)*	*Pipe Length (m)*	H_f *(m)*
Submain, Subunit 1 to B	1.59	52.5	0.73	147	1.64
Main, B to A	3.18	62.7	1.03	96	1.63
Main, A to control head	6.36	77.9	1.34	52	1.11

[a] Inside diameters of Schedule 40 PVC pipe.

(18) The total head required at the outlet from the control head is the pressure required at the inlet to the manifold from Step 15 plus the friction loss in the submain and main lines from Step 17. Assume the head loss through the valve for Subunit 1, obtained from a manufacturer's table, is 1.2 m.

$$15.3 + 1.2 + 1.6 + 1.6 + 1.1 = 20.8 \text{ or use } 21\,\text{m}$$

To irrigate half of the field, the pump must deliver 12.7 L/s (2 × 6.36). The total head requirement for the pump is 21 m plus allowances for pumping lift, pump wear, pump losses, and losses through filters, pressure regulators, valves, and other devices.

18.6 Subsurface Drip Irrigation

Subsurface drip irrigation (SDI) laterals and pipelines in the cropped area are buried. The laterals are placed to deliver water within the plant root zone. Both tubing and tapes are used for SDI laterals. Tubing has thicker walls similar to polyethylene pipe and can be buried deeper. Tapes have thin walls and are less expensive than tubing, but are not as durable and may collapse from soil compaction and traffic. Thicker walled tapes are available and have been successfully used for several years. Thin-walled tapes are used when laterals are replaced after one or two seasons. Pressure-compensating emitters are available with both tubing and tapes.

Evans et al. (2000) indicate the following advantages for properly designed and managed SDI systems: (1) yield may increase; (2) field access is not limited by the irrigation system; (3) limited wetting of the soil surface reduces weed growth and better weed control with minimal chemical application and reduced incidences of diseases; (4) system efficiently and effectively applies labeled chemicals for disease and pest control; (5) reduced handling and exposure of workers to chemicals; (6) permanent beds, minimum tillage, and multiple cropping systems are possible with special equipment; and (7) water and emitter temperature extremes are buffered by the soil, which minimizes the associated flow rate changes.

The disadvantages of SDI systems include (1) initial cost may be high; (2) emitters may be plugged by roots or chemical precipitates; (3) potential rodent or insect damage may occur; (4) salts may accumulate above the laterals, which may interfere with root development; (5) upward movement of water is low in coarse-textured soils; (6) vehicular traffic or roots may pinch the tubing; and (7) management is crucial and somewhat less is known about SDI than other irrigation systems.

The design requirements for SDI systems are similar to those for other microirrigation systems. Special attention must be given to water filtration and treatment, proper location of check and vacuum relief valves, and flushing of the laterals. Because the laterals are buried, it is more difficult to see plugged outlets and to repair plugged emitters or damaged tubing. Check and vacuum relief valves are necessary to prevent negative pressures from developing in the laterals. Negative pressures may cause soil particles to be pulled into the laterals and increase plugging. Emitter outlets should face up so that debris along the bottom of the lateral is less likely to flow into the emitter. Because the ends of the laterals are buried, it is desirable to install a manifold at the downstream end of the laterals for flushing, if laterals are closely spaced. This extra manifold also tends to balance the flow in the laterals and supplies water to both ends of plugged or broken laterals.

The depth of SDI laterals is a balance between avoiding damage by traffic or tillage equipment and the water distribution to the active root zone. It is generally recommended that laterals be installed as shallow as possible with minimal surface wetting because most of the biological activity, air movement, and rooting occurs in the shallow soil layers. With annual and shallow rooted crops, the lateral depth may need to be shallow to wet the soil surface for germination and crop growth. Soil surface wetting (surfacing) is minimized if the laterals are placed deeper, which minimizes evaporation and potential weed growth. Laterals are generally installed at shallower depths in coarse-textured soils and deeper in fine-textured soils. SDI laterals are typically installed at 0.025 to 0.075 m for shallow-rooted crops, at 0.3 to 0.5 m for crops such as cotton, potatoes, sugar beets, and maize, and at 0.15- to 0.2-m depths for many vegetable crops. Emitters may be spaced 0.2 to 1.5 m apart along the lateral depending on the crop and soil characteristics. If a line source is needed, the emitters should be spaced sufficiently close to overlap wetting patterns. Field testing of soil water movement is recommended to determine emitter spacing.

Salinity requires special attention in semiarid and arid regions, because salts tend to reach high concentrations above the lateral and are leached if rainfall occurs. The leached salts will move into the active root zone and may severely affect plant growth. One management solution is to turn on the irrigation system, which will continue to move the salts downward through the soil past the root zone.

Example 18.4

Design a subsurface drip irrigation system for a subunit that is one eighth (100 rows) of the 400-m square field in Figure 18-2. Assume cotton is grown in Arizona with 1-m row spacings and the soil is fine with varying layers of generally low density to 1 m. The field slope is 0.002 down from north to south with no cross slope and laterals run on the slope. Assume the laterals are 15.8 mm in diameter with emitters molded onto the exterior of the pipe with Hazen-Williams $C = 140$ and will be buried 0.3 m below the soil surface. The laterals are to be supplied from manifolds at both ends to allow occasional flushing of the laterals. Assume minimum *EU* and application efficiency are 90 percent.

Solution. The procedure is similar to Example 18.3.

(1) Use a spreadsheet to perform the calculations. From Figure 15-1, read that the peak *ET* is 10 mm/day and the rooting depth is 1.8 m; however, in this case the rooting depth is limited by the soil to 1 m.

(2) Assume the initial design will use an emitter discharging 2 L/h emitter with $K = 0.5$ and $x = 0.3$ for q in L/h and h in kPa. Field measurements show the average wetted diameter for this emitter is 0.75 m. Select a 0.5-m emitter spacing to ensure the wetted diameters overlap.

(3) Assume a 1-day irrigation frequency. The actual frequency can be changed with a corresponding change in duration. Select a lateral length of 190 m to allow roadways at each end of the subunit. Because water is being supplied from a manifold at both ends, the design will be based on a 95-m half-lateral length with water flowing uphill. The volume of water per half-lateral is

$$d = \frac{10 \text{ mm/day} \times 95 \text{ m} \times 1 \text{ m} \times 1000 \text{ L/m}^3}{1000 \text{ mm/m} \times 0.90} = 1056 \text{ L/day}$$

(4) Substituting into Equation 18.1, $h = (2.0/0.5)^{1/0.3} = 101.6$ kPa or 10.36 m. This is the average operating pressure in the lateral.

(5) With 380 emitters per lateral discharging 2 L/h, the irrigation time is 1056 L/day/(2 L/h $\times$ 190) = 2.8 h/day. If the field is divided into 8 subunits, the total irrigation time is $2.8 \times 8 = 22.4$ h/day. If the irrigator wants more time for maintenance or repairs, a different layout would be needed.

(6) Calculate the head loss in the half-lateral from Equation 18.5. From Table 17-5, the *F* value for 190 emitters per lateral is 0.35. The discharge is $2 \times 190 = 380$ L/h or 0.106 L/s.

$$h_f = 0.35 \times 1.21 \times 10^{10} \frac{95}{15.8^{4.87}} \left(\frac{0.106}{140} \right)^{1.852} = 0.97 \text{ m}$$

(7) Calculate the pressure at the distal end of the farthest lateral from Equation 18.8

$$H_d = 10.36 - 0.26 \times 0.97 - 0.002 \times 95/2 = 10.01 \text{ m}$$

(8) The discharge for the last emitter for this lateral is q_{min}.

$$q_{min} = 0.5(10.01 \times 9.81)^{0.3} = 1.98 \text{ L/h}$$

(9) Calculate the inlet pressure for the first lateral using Equation 18.7

$$H_o = 10.36 + 0.74 \times 0.97 + 0.002 \times 95/2 = 11.17 \text{ m}$$

(10) Calculate the head loss in the section of the manifold between laterals 1 and 2 using Equation 18.5. The manifold is 62.71 mm (60 mm nominal) diameter, the connection has no head loss, and the flow in the manifold is 0.211 L/s.

$$H_f = 1.21 \times 10^{10} \times \frac{1}{62.71^{4.87}}\left(\frac{0.106}{150}\right)^{1.852} = 0.00003 \text{ m}$$

(11) The pressure at the inlet to lateral 2 is 11.17 + 0.00003 = 11.17 m.

(12) Calculate the average pressure in lateral 2 from Equation 18.7. Because the discharge in lateral 2 is not known, the value for h_f for lateral 1 is assumed. Iterate until the values for h_f and emitter discharge converge. (The spreadsheet can perform the iterations.) Here the head loss in the manifold is very small and one iteration is sufficient.

$$H_a = 11.17 - 0.74 \times 0.97 - 0.002 \times 95/2 = 10.36 \text{ m}$$

$$q = 0.5(10.36 \times 9.81)^{0.3} = 2.0 \text{ L/h}$$

$$Q_{lateral} = 2.0 \times 190/3600 = 0.106 \text{ L/s}$$

$$h_f = 0.35 \times 1.21 \times 10^{10} \times \frac{95}{15.8^{4.87}}\left(\frac{0.106}{140}\right)^{1.852} = 0.97 \text{ m}$$

$$H_a = 11.17 - 0.74 \times 0.97 - 0.002 \times 95/2 = 10.36 \text{ m}$$

The average pressures match and the average emitter discharge is 2.00 L/h.

(13) Calculate the total discharge in lateral 2 = 2.00 × 190 = 380 L/h or 0.106 L/s.

(14) The discharge in the manifold between laterals 2 and 3 is 0.106 + 0.106 = 0.212 L/s.

(15) Repeating the process from Step 8 produces a table in the spreadsheet of these values for each lateral. Assuming 100 laterals, the velocity nears the limit in the 60-mm manifold so the manifold is increased to a 100-mm nominal diameter at lateral 40. The following table shows the calculations for a few selected laterals.

Lateral No.	Lateral Inlet Head (m)	Average Lateral Head, H_a (m)	Average Emitter Flow (L/h)	Total Flow in Lateral (L/s)	Head Loss in Lateral, h_f (m)	Flow in Manifold (L/s)	Manifold Head Loss, H_f (m)	Manifold Velocity (m/s)
1	11.17	10.36	2.000	0.106	0.97	0.106	0.00003	0.03
2	11.17	10.36	2.000	0.106	0.97	0.212	0.00011	0.07
39	11.52	10.69	2.019	0.107	0.99	4.14	0.0277	1.34
40	11.55	10.72	2.021	0.107	0.99	4.24	0.0027	0.52[a]
99	11.98	11.14	2.044	0.108	1.01	10.57	0.0145	1.29
100	11.99	11.15	2.045	0.108	1.01	10.68	0.0148	1.30

[a] Manifold diameter increased.

(16) The average emitter discharge in the 100 laterals is 2.0202 L/h. The emission uniformity is

$$EU = 100\left[1 - \frac{1.27 \times 0.05}{1^{0.5}}\right]\frac{1.98}{2.0202} = 91.8 \text{ percent}$$

This is greater than the desired 90 percent and the design is accepted.

If a design is not acceptable because *EU* is too low, redesign options include (1) increasing pressure, which increases the flow rate; (2) shortening the laterals; (3) selecting emitters that have more pressure compensation; (4) increasing lateral diameter; and (5) re-evaluating emitter wetting pattern to determine if emitter spacing can be increased.

18.7 Microirrigation of Landscape Plants

Landscape plantings, which are often arranged in small groups or as individual plants, are well suited to microirrigation. Water can be supplied to meet the needs of each plant. Simple systems may be connected to a valve on a house and consist of a backflow preventer, a small filter, a pressure regulator, and a lateral with the emitters. Complex systems must be designed for large areas such as parks and golf courses, around large structures, or along highways. In designing these systems, care must be taken to have plants with similar water requirements on the same control valve. Plants in greenhouses and nurseries also are easily irrigated with microirrigation systems since water and nutrients can be delivered to individual plants. Backflow prevention devices are required to prevent contamination of the water supply.

18.8 Microirrigation and the Environment

A properly designed and managed microirrigation system has several environmental advantages compared to surface and sprinkler systems. Because only the soil near plants is irrigated, smaller amounts of chemicals are applied and can be applied where they are most beneficial. Many plants are more tolerant of salinity when the soil water content is maintained at a high level with microirrigation, thus minimum leaching is required and less water and chemicals are carried below the root zone. Disposal of flush water from filters and the lateral lines can be an environmental problem, particularly if the water contains chemicals.

Internet Resources

General reference sites for equipment and information

- www.clemson.edu. Key search term: irrigation
- www.irrigation-mart.com
- www.irrigation.org
- www.wcc.nrcs.usda.gov/nrcsirrig/

References

American Society of Agricultural Engineers (ASAE). (1998). *Design, Installation and Performance of Underground, Thermoplastic Irrigation Pipelines.* S376.2. ASAE Standards. St. Joseph, MI: ASAE.

———. (2000). *Media Filters for Irrigation—Testing and Performance Reporting.* S539. ASAE Standards. St. Joseph, MI: ASAE.

———. (2001a). *Design and Installation Microirrigation Systems.* EP405.1. ASAE Standards. St. Joseph, MI: ASAE.

———. (2001b). *Polyethylene Pipe Used for Microirrigation Laterals.* S435. ASAE Standards. St. Joseph, MI: ASAE.

———. (2001c). *Safety Devices for Chemigation.* EP409.1. ASAE Standards. St. Joseph, MI: ASAE.

Clark, G. A., D. Z. Haman, J. F. Prochaska, & M. Yitayew. (2005). General system design principles. In F. Lamm, ed., *Microirrigation for Crop Production,* Chap. 5. Amsterdam: Elsevier.

Evans, R. G., I. P. Wu, & A. G. Smajstrala. (2000). Microirrigation systems. Draft chapter for ASAE Monograph. *Design and Operation of Farm Irrigation Systems,* 2nd ed. G. J. Hoffman, J., D. L. Martin, & R. L. Elliott, eds. ASAE Monograph. St. Joseph, MI: ASAE.

Karmeli, D., & J. Keller. (1975). *Trickle Irrigation Design.* Glendora, CA: Rain Bird Sprinkler Mfr. Corp.

Keller, J. D., & R. D. Bliesner. (1990). *Sprinkle and Trickle Irrigation.* New York: Van Nostrand Reinhold.

Solomon, K. (1979). *Manufacturing Variation of Trickle Emitters.* ASAE Trans, St. Joseph, MI: ASAE.

2000 Irrigation Survey. (2001). *Irrigation Journal, 51*(1), 12–29, 40, 41.

U.S. Soil Conservation Service (SCS). (1984). *Trickle irrigation.* In *National Engineering Handbook,* Sect. 15: *Irrigation,* Chap. 7. Washington, DC: U.S. Government Printing Office.

Wuertz, H. (2001) Personal communication. Sundance Farms, Coolidge, AZ.

PROBLEMS

18.1 Determine the emission uniformity for emitters with $K = 0.4$ and $x = 0.6$ for q in L/h and p in kPa if the manufacturer's coefficient of variation is 0.1. There are four emitters per plant, average operating pressure is 80 kPa, and the minimum pressure is 70 kPa. What is the emission uniformity if there are two emitters per plant? What is the emission uniformity with 4 emitters per plant, an average operating pressure is 110 kPa, and the minimum pressure is 100 kPa?

18.2 If the conventional peak *ET* of an orchard is 7.5 mm/day and 65 percent of the area is shaded by trees, determine the design *ET* rate, volume of water required per tree per day, and application rate in L/h per tree for a microirrigation system. Assume $EU = 92$ percent, a tree spacing of 3×6 m, 20 h/day operation, and irrigation interval of 2 days.

18.3 For the orchard in Problem 18.2, determine the number of emitters required per tree. Assume a coarse-textured soil of low density, a 1.5-m root zone, and 40 percent of the area is to be irrigated.

18.4 Determine the friction loss in the 15.8-mm-diameter lateral if 72 in-line emitters are uniformly spaced 1m apart along a line as in Figure 18-4a. Assume the average emitter discharge is 3.6 L/h and the equivalent pipe length for an emitter is 2 m.

18.5 Develop a spreadsheet for calculating *EU* for Subunit 1 in Example 18.3.

18.6 For the data in Example 18.3, determine *EU* for Subunit 2 by adding to the spreadsheet in Problem 18.5.

18.7 Determine *EU* for Subunit 1 of Example 18.3 if the field has a 1 percent slope uphill from the water supply and no cross slope.

18.8 From the data in Example 18.3, redesign the delivery system for laterals with 48 trees. Assume the manifolds remain 91 m long.

18.9 Design a microirrigation system for 10 rows of forty 20-L pots in a nursery. Assume the rows have a 1-m spacing and the pots a 0.5-m spacing in the row, and water is available at 15 m head. The pots are to receive 0.2 L of water twice a day.

18.10 Develop a spreadsheet to perform the calculations for Example 18.4.

18.11 Complete the design of Example 18.4 to determine the pipe diameter, head, and flow rate required at the outlet of the control head. Assume the flow rate in the manifold at the other end of the laterals is equal to the flow rate calculated for the manifold in the example.

18.12 Use the spreadsheet from Problem 18.10 to determine *EU* for the subunit in Example 18.4 if the emitter spacing is changed to 0.6 m. Can the field be irrigated in one day with this design? Explain your answer.

18.13 Use the spreadsheet from Problem 18.10 to determine *EU* for Example 18.4 if C_v is 0.07. What is the *EU* for Example 18.4 if the exponent x is changed to 0.4? What is *EU* in Example 18.4 if the lateral slope is 0.01? Explain why *EU* changes.

18.14 Use the spreadsheet from Problem 18. 10 to design one subunit of a SDI system for a 600-m square level field. Assume a row spacing of 0.8 m, peak *ET* is 7 mm/day, $K = 0.55$ and $x = 0.3$ in Equation 18.1 for q in L/h and h in kPa, and the operating time is a maximum of 22 hours/day.

CHAPTER

19 Pumps and Pumping

The purpose of a pump is to add energy to a fluid, i.e., the energy of the water leaving a pump (the sum of elevation potential, pressure, and kinetic energies) is higher than that of the water entering the pump. Engineers must be able to assess the performance requirements for a pump or pumping plant and determine the appropriate type, number, and sizes of pumps for an application, as well as the size and type of power units required to drive the pumps. In addition, they should be able to design the pumping plant installation, estimate the cost of operation, and supervise construction and operation of the plant.

Pumping plant installations are most often required in drainage and irrigation enterprises. Pumping plants for drainage may provide outlets for open ditches and pipe drains or lower the water table by pumping from shallow wells. In irrigation, pumps typically move water from wells and storage reservoirs into other reservoirs, irrigation canals, or pressurized pipe systems.

Types of Pumps

The most common pumps use rotating impellers or reciprocating pistons to transfer energy to the fluid. Reciprocating (piston) pumps are capable of developing high pressures but have relatively small capacities. They are not ordinarily suitable for drainage and irrigation, especially if sediments are present; however, small piston pumps may inject chemicals into pressurized irrigation systems. Other types of pumps include progressive cavity, diaphragm, rotary-lobe, gear, roller, and peristaltic, but none of those are commonly used for medium- to large-scale irrigation or drainage applications.

Flow through impeller pumps is classified as radial, axial, or mixed flow. In radial-flow pumps (Figure 19-1a), the fluid enters the pump at the center of the impeller and moves through the impeller perpendicular to the axis of rotation. In axial-flow pumps (Figure 19-1c), the fluid enters and leaves the impeller parallel to the axis of rotation of the impeller. In mixed-flow pumps (Figure 19-1b), the fluid leaves the impeller at an angle between radial and axial.

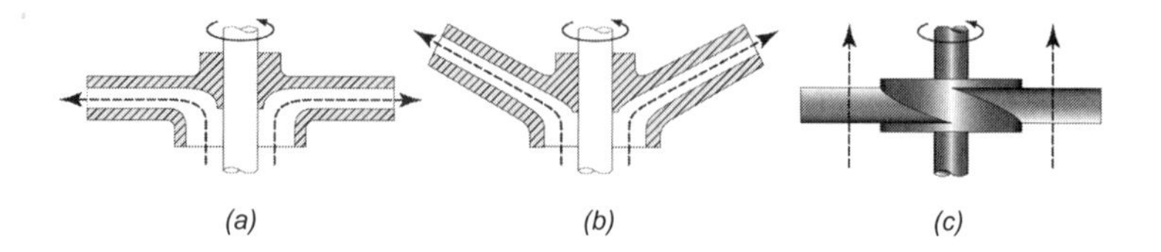

Figure 19–1

Flow directions through impellers with (a) radial flow, (b) mixed flow, and (c) axial flow.

Impeller pumps are commonly known as centrifugal, propeller, and turbine pumps. In centrifugal pumps, the flow from the impeller is radial; in propeller pumps, the flow from the impeller is axial. Most turbine pumps have mixed flow. Pumps may be selected from these three types for a wide range of discharge and head characteristics.

Centrifugal and Turbine Pumps

Centrifugal and turbine pumps are economical and simple in construction, yet they produce a smooth, steady discharge. They are small compared with their capacity, easy to operate, and capable of handling sediment and other foreign material. Since turbine pumps are frequently used in wells, they are also known as deep-well turbines.

19.1 Principles of Operation

Centrifugal and turbine pumps consist of two main parts: (1) the impeller or rotor that imparts energy to the water and (2) the casing that guides the water to and from the impeller. As shown in Figure 19-2, the water enters at the center or eye of the impeller and passes outward through the rotor to the discharge opening. For a particular pump, the discharge (Q), head (H), and power (P) vary with the pump speed (N) to the first, second, and third powers, respectively. These relations, called the Affinity Laws, are

$$\frac{Q_1}{Q_2} = \frac{N_1}{N_2} \qquad \frac{H_1}{H_2} = \left(\frac{N_1}{N_2}\right)^2 \qquad \frac{P_1}{P_2} = \left(\frac{N_1}{N_2}\right)^3 \qquad \textbf{19.1}$$

These relationships are valid over wide speed ranges, since pump efficiency changes little with speed.

Manufacturers typically offer families of pumps, i.e., a series of impellers of slightly different diameters or trims in a single casing. (Impeller trim is the amount of reduction of the diameter of an impeller as compared to the largest-diameter impeller available in the pump family. For example, a 240-mm impeller in a pump family where the largest impeller is 280 mm would have a 40-mm trim.) Knowing the discharge, head, and power requirement of one pump in the family allows one to calculate the operating characteristics of any other pump in the family. Since changing the diameter of the impeller (D) of a particular pump changes the peripheral velocity, it has essentially the same effect as changing the speed. This gives a second version of the Affinity Laws:

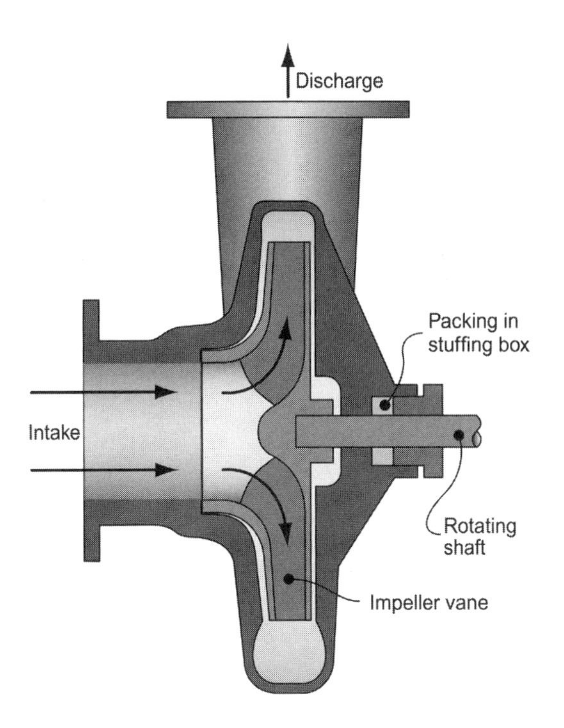

Figure 19–2

Cross section of a horizontal centrifugal pump (single-suction, radial flow, enclosed impeller).

$$\frac{Q_1}{Q_2} = \frac{D_1}{D_2} \qquad \frac{H_1}{H_2} = \left(\frac{D_1}{D_2}\right)^2 \qquad \frac{P_1}{P_2} = \left(\frac{D_1}{D_2}\right)^3 \qquad \textbf{19.2}$$

Equations 19.1 and 19.2 can be combined if both speed and impeller diameter are changed, though that is not often done.

As the diameter of the impeller decreases, the clearance between the impeller and the inside of the casing increases, lowering the efficiency of the pump. Therefore, the applicability of the second set of Affinity Laws is limited to trims of no more than 10 to 20 percent of the original diameter. It is best to consult the manufacturer's performance charts, since they will include variations in pump efficiency. Many manufacturers now have software available via the Internet that can aid in selecting the best pump(s) for a particular application.

Fluid mechanics texts (e.g., White, 1986) give Laws of Similarity for pumps and fans. The Laws of Similarity for speed are the same as Equation 19.1. The Laws of Similarity for diameter (actually, a characteristic linear dimension) are like Equation 19.2, but have exponents 3, 2, and 5 for Q, H, and P, respectively. Similarity requires that all linear dimensions change by the same factor. Impeller trim violates similarity since it alters only the diameter of the impeller, leaving all other dimensions unchanged.

19.2 Classification

With regard to the construction of the casing around the impeller, pumps are classified as volute or diffuser, as shown in Figure 19-3. The volute-type pump takes its name from the shape of the casing, i.e., a spiral with a cross-sectional area increasing toward the discharge opening. The radial pressure distribution in a simple

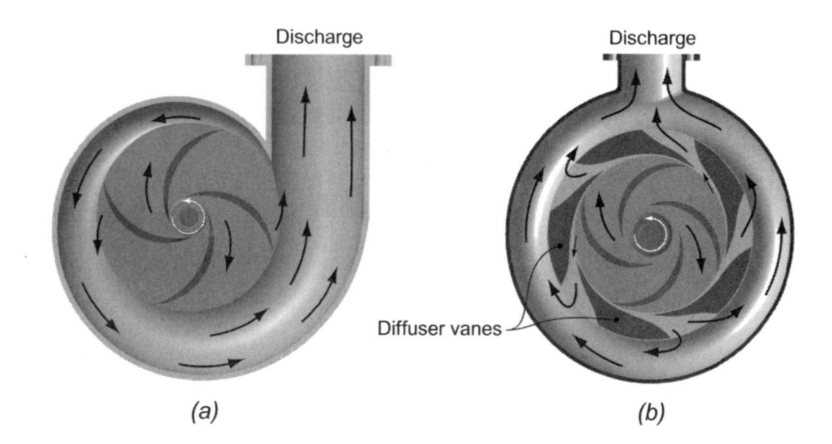

Figure 19–3
Cutaway views of (a) volute-type and (b) diffuser-type centrifugal pumps. Impeller rotation is counterclockwise.

volute is not balanced; causing a lateral load on the pump shaft that reduces bearing life. To achieve more uniform pressure distributions, some volutes are formed with two flow-conducting chambers, each receiving the flow from half the impeller. Diffuser-type pumps (which include many turbine-type pumps) have stationary guide vanes surrounding the impeller. As water leaving the periphery of the rotor flows through gradually enlarging passages between the vanes, the kinetic energy of the water is smoothly converted to pressure. As a result, diffuser-type pumps usually have relatively high efficiencies. They also have a radially uniform pressure distribution, so the lateral load on the pump shaft is minimal (Haman et al., 1994).

Centrifugal pumps are built with horizontal or vertical drive shafts and with different numbers of impellers and suction inlets. The suction inlet may be either single or double, depending on whether the water enters from only one side or from both sides of the impeller. Single-suction, horizontal centrifugal pumps are frequently used where the suction lift does not exceed 4 to 6 meters. Many pumps are designed with two or more impellers arranged in series. These are called multi-stage pumps. Both centrifugal and turbine pumps may have multiple impellers, but they are more common in the turbine type.

Many turbine and multistage centrifugal pumps are configured with a vertical axis of rotation. Some are designed to operate below the level of the source water (submersible), including in wells. Deep-well submersibles typically have a close-coupled electric motor in a slender configuration to fit inside the well casing. Other deep-well pumps are driven from the surface by a lineshaft, a long rod that transmits torque from the drive unit to the pump impellers. The well bore must be very straight to allow such a pump assembly to operate without excessive flexing of the shaft. Close-coupled submersibles are more adaptable in this sense, since they do not require a particularly straight well bore.

19.3 Centrifugal-Type Impellers

The design of the impellers greatly influences the efficiency and operating characteristics of the pump. Centrifugal-type impellers shown in Figure 19-4 are classified as (a) enclosed, (b) semienclosed, and (c) open. The open-type impeller has vanes that are open on all sides except where attached to the rotor. The semienclosed impeller has a shroud (plate) on one side. The enclosed impeller has shrouds

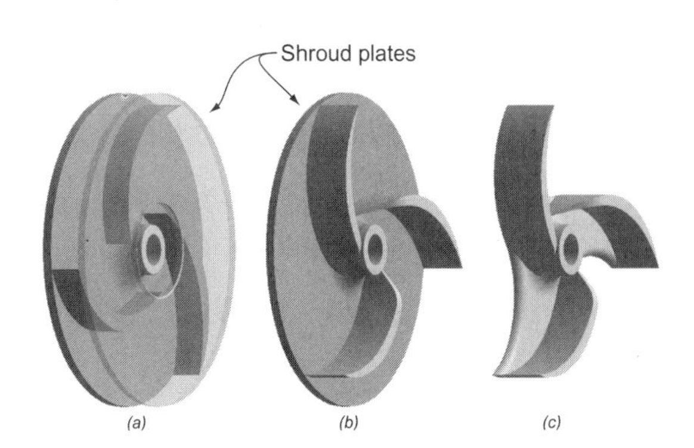

Figure 19–4

Centrifugal-type pump impellers: (a) enclosed, (b) semienclosed, and (c) open. The upper shroud plate of the enclosed impeller was rendered partially transparent to better show the vanes.

on both sides, thus enclosing the blades completely. The open and semienclosed impellers are most suitable for pumping suspended material or trashy water (e.g., containing fibrous materials like plant stems or animal wastes). Enclosed impellers are generally not suitable where suspended materials are carried in the water as they can greatly increase the wear on the impeller or clog it completely. Chopper pumps have either a cutter unit in front of the impeller or an integrated cutter-impeller to reduce the size of suspended materials in the flow so that they will pass through the rest of the system.

19.4 Performance Characteristics

Selection of a pump for a particular job requires knowledge of the head, discharge, and efficiency of the pump at different operating speeds. Curves that provide these data are called performance curves, as shown in Figure 19-5. The head–discharge curve shows the total head developed by the pump for a range of discharge rates. The head developed at zero flow is called the shutoff head. Pump efficiency accounts for energy losses internal to the pump, including fluid friction, shock losses due to sudden changes in velocity, leakage past the impeller, and mechanical friction. Every pump will have a best efficiency point (BEP), which defines its optimal operating conditions. The pump efficiency varies with discharge and can be found from the performance curve. A pump should be selected that will have a high efficiency for the range of discharges required by the job. For example, from the upper curves in Figure 19-5, efficiencies of 71 percent or greater can be obtained for discharges varying from 0.027 to 0.046 m^3/s at heads of 98 to 75 meters, respectively. In general, the pump should operate inside the range of 80–110 percent of the discharge at the BEP. There are sometimes reasons for selecting a pump having its BEP either slightly to the right or left of the design point, but they depend on the operating characteristics of both the pump and the system. Consult manufacturers, specialists, or advanced texts concerning specific applications.

Performance curves can be obtained from the pump manufacturer. Pump curves vary in shape and magnitude, depending on the size of the pump, type of impeller, and overall design. Performance characteristics are normally obtained by tests of a representative production line pump rather than by tests of each pump manufactured.

One of the biggest hazards to pumps is cavitation, which is the spontaneous formation of vapor bubbles where the pressure drops below the saturation vapor pressure of the water. The rapid collapse of those vapor bubbles where the pressure

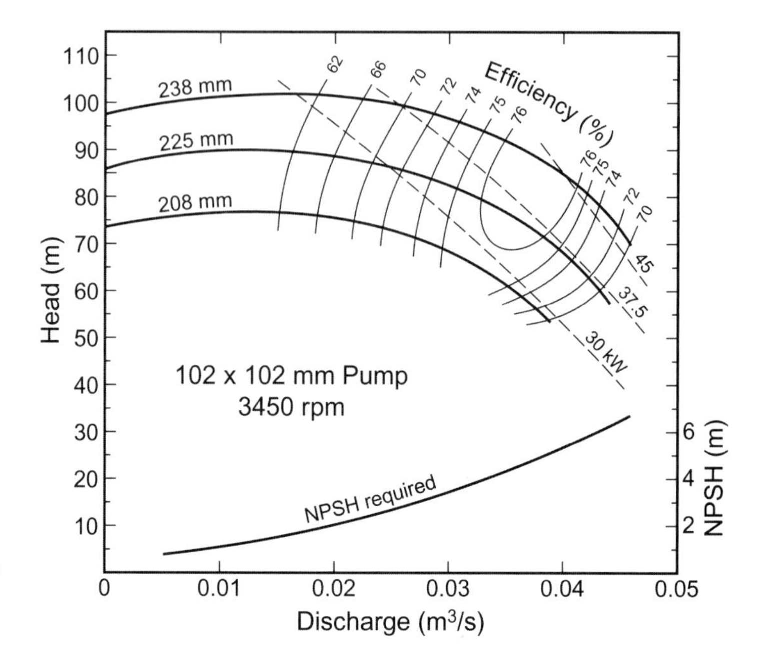

Figure 19–5

Performance characteristics of a centrifugal pump.

subsequently increases produces shock waves (in the immediate vicinity of the collapsed bubble) that create extremely corrosive conditions. Cavitation is most likely to occur at the pump inlet, near the eye of the impeller, and along the trailing sides of the impeller vanes.

To prevent damage to the pump, there must be sufficient absolute pressure at the inlet to push water into the pump without dropping the fluid pressure to the point where cavitation occurs. The inlet-side energy partitioning is depicted in Figure 19-6. The difference in elevation between the pump inlet and the free water surface is called the suction lift. The maximum practical suction lift can be computed as

$$H_s = H_t - H_f - e_s - NPSHr - F_s \qquad \textbf{19.3}$$

where H_s = maximum practical suction lift (L),

H_t = atmospheric pressure at the free water surface (L),

H_f = inlet-side friction losses (L),

e_s = saturated vapor pressure of the water (L),

$NPSHr$ = net positive suction head required (L),

F_s = safety factor, which may be taken as 0.6 m.

To correct H_t for altitude, subtract about 1.2 meters head per 1000 meters above mean sea level (H_t at sea level = 10.34 m). Friction losses and suction lift should be kept as low as possible. For this reason the suction line is usually larger than the discharge pipe, and the pump is placed close to the water supply. The head loss equivalent for the vapor pressure of the water must be considered to prevent cavitation, but it does not add to the total suction head when the pump is operating.

Net Positive Suction Head (NPSH) refers to the energy of the water at the inlet (suction side) of the pump. NPSHr (NPSH required) is a characteristic of the

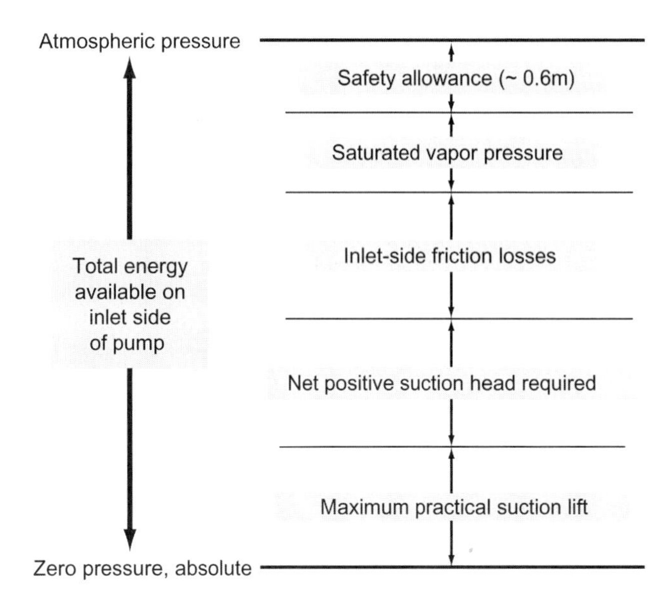

Figure 19–6
Inlet-side energy relationships.

pump. It is the head needed to push water into the eye of the impeller fast enough to prevent cavitation. Note that NPSHr increases as pump discharge increases. The manufacturer experimentally determines NPSHr and includes it with the performance curves, as shown in Figure 19-5. NPSHa (NPSH available) is the energy available at the inlet for a particular configuration. For safe operation, NPSHa must be greater than NPSHr. The following example illustrates the calculation of suction lift.

Example 19.1

An irrigator needs to pump 38 L/s of water from a stream into a distribution canal with a pump like those shown in Figure 19-5. The water temperature is 20°C and the elevation is 1460 m above sea level. The inlet-side friction losses at that flow rate are 1.25 m. What is the maximum height above the water surface at which the pump can be located?

Solution. The maximum practical suction lift is the largest elevation difference allowable between the pump inlet and the free surface of the water supply. The local atmospheric pressure is $H_t = 10.34 \text{ m} - (1.2 \text{ m}/1000 \text{ m}) \times 1460 \text{ m} = 8.59 \text{ m}$. The saturation vapor pressure (from Chapter 4) is 2.3 kPa = 0.23 m. In Figure 19-5, find the NPSHr for 38 L/s to be 4.8 m. Using the recommended safety factor of 0.6 m, Equation 19.3 gives

$$H_s = 8.59 \text{ m} - 1.25 \text{ m} - 0.23 \text{ m} - 4.8 \text{ m} - 0.6 \text{ m} = 1.7 \text{ m}$$

Fluctuations in supply water temperature and surface elevation must be considered when calculating the maximum suction lift for determining pump placement.

Pump operation at or near the maximum practical lift should be avoided. Accumulation of debris on the intake screen and increase of pipe friction with age could increase the inlet-side losses, leading to failure by cavitation. Vibration can also contribute to cavitation in pumping systems. Careful attention must be given to proper alignment of drive components to keep vibration and wear to a minimum.

Centrifugal pumps are generally not self-priming. To prime the pump, the suction line and pump should be nearly full of water. This can be accomplished by manually filling the pump with water or by removing the air with a suction pump or an engine exhaust primer. A gate valve on the discharge side of the pump and a check valve at the lower end of the suction line are essential for priming, except at very low suction heads. Submersible pumps are designed to operate below the free water surface, and so do not need priming. Self-priming pumps are available for low-capacity applications, but they tend to have relatively low efficiencies.

Pumps can be damaged if operated without water for more than a few seconds. Pumps should not be operated with the outlet closed (except very briefly for startup or similar purposes) because the water in the pump will rapidly be heated to boiling, causing vapor to displace the water. Without the water as a coolant, heat buildup in the bearings, packing, or wear rings will cause the pump to seize, severely damaging the pump and drive system (Driscoll, 1986).

Propeller Pumps

19.5 Principles of Operation

As distinguished from centrifugal pumps, the flow through the impeller of a propeller-type pump is parallel to the axis of the impeller rather than radial. The pumps are also referred to as axial-flow or screw-type pumps. The principle of operation is similar to that of a boat propeller except that the rotor is enclosed in a housing. Many propeller pumps have diffuser vanes mounted in the casing, similar to diffuser-type pumps.

19.6 Performance Characteristics

Propeller pumps are designed principally for low heads and large capacities. The discharge of these pumps varies from 0.04 to 4.4 cubic meters per second, with speeds ranging from 450 to 1750 rpm and heads usually not more than 9 meters. Although most propeller pumps are of the vertical type, the rotor may also be mounted on a horizontal shaft.

Example performance curves for a large propeller pump are shown in Figure 19-7. In comparison with centrifugal pumps, the efficiency curve is much flatter, heads are considerably lower, and the power curve is continually decreasing with greater discharge. With propeller pumps, the power unit may be overloaded by increasing the pumping head. These pumps are not suitable where the discharge must be throttled to reduce the rate of flow.

Pumping

Although pumping water for irrigation is similar in many respects to pumping for drainage, design requirements differ. For example, pumping heads in drainage

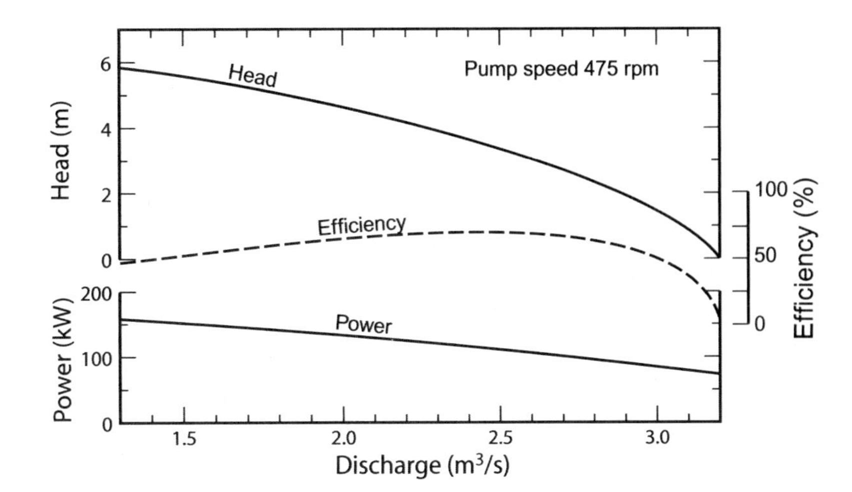

Figure 19–7
Performance curves for an axial-flow (propeller) pump.

applications are generally 6 meters or less, whereas heads up to 90 meters are common in irrigation applications.

19.7 Power Requirements and Efficiency

The power (energy per unit time) imparted to the water, i.e., water power, is the product of the discharge and change in energy status of the water between the inlet and outlet of the pump. Brake power is the power needed to drive the pump, i.e., the sum of the water power and all internal losses for the pump. The ratio of water power (power output) to brake power (power input) is the pump efficiency, E_p. Where the change in energy status is expressed in terms of head, the power relationship is given by

$$BP = \frac{WP}{E_p} = \frac{9.8QH}{E_p} \qquad \textbf{19.4}$$

where BP = brake power (kW),
WP = water power (kW),
E_p = pump efficiency,
Q = discharge (m^3/s),
H = total head (m).

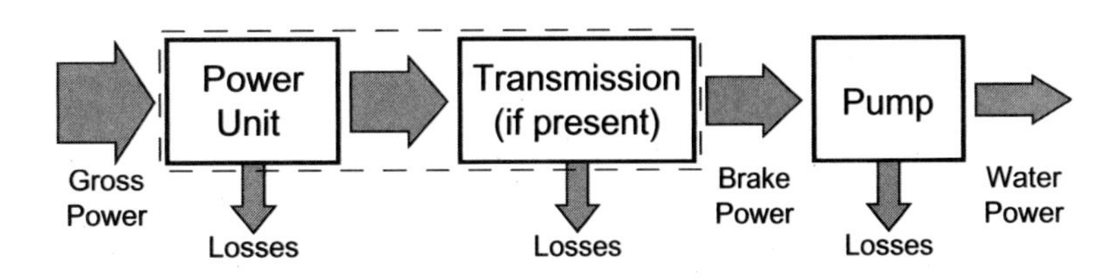

Figure 19–8
Energy flow through a pumping system.

In determining the total head for a system, the overall differences in elevation, pressure, and velocity, as well as all friction losses in the piping, valves, fittings, strainers, etc., must be considered. Energy flow through a pump system is shown in Figure 19-8.

Example 19.2 An irrigator must pump 40 L/s from a distribution canal to a sprinkler system. At the entrance of the sprinkler main, which is 12.0 m above the water level in the canal, the required pressure head is 64 m. The inlet-side friction losses at that flow rate are 1.1 m, and the friction losses between the pump and the sprinkler main are 3.7 m. The pump efficiency is 74 percent. Determine the water power output of the pump and the brake power requirement for the pump.

Solution. The total head is the sum of the changes in elevation, pressure, and velocity heads, plus the friction losses. The change in elevation head is 12 m. The change in pressure head is 64 m (taken relative to the free water surface, where the pressure head is zero). The velocity head in piping is usually negligible, assuming average flow velocities do not exceed 1.5 m/s.

$$\begin{aligned}\text{Total head} &= \Delta H_{elevation} + \Delta H_{pressure} + \Delta H_{velocity} + H_{friction} \\ &= 12\text{ m} + 64\text{ m} + (\approx 0\text{ m}) + (1.1\text{ m} + 3.7\text{ m}) \\ &= 80.8\text{ m}\end{aligned}$$

The water power and brake power are

$$WP = 9.8\,QH = 9.8(0.040)(80.0) = 31.4\text{ kW}$$

$$BP = \frac{WP}{E_p} = \frac{31.4\text{ kW}}{0.74} = 42.4\text{ kW}$$

A significant fraction of the input energy is lost to friction inside the pump. Pump efficiencies range from about 75 percent under favorable conditions to 20 percent or less under unfavorable conditions. A well-designed pump should have an efficiency of 70 percent or greater over a wide range of operating heads. It can be difficult to maintain high efficiencies in the field due to wear in the pump and other factors. It is good practice to periodically check the pump in the field. If efficiencies are low, the causes should be located and corrected. The overall efficiency of the installation, which includes the efficiency of the power plant, the pipe system, and the pump, should also be checked.

A simple evaluation of performance in the field requires measurements of the flow rate and the pressures at the pump inlet and discharge. Compare those values to the design operating point for the pump. If the head or discharge is not within the desired range, look for problems such as excessively worn impellers, clogged or corroded intake screens, air trapped in the piping, leaks in the suction line, valves not fully opened (or closed), incorrect pipe sizes, incorrect speed of the drive unit, etc. Pressure measurements at various points can help in locating problems in the distribution piping. Modifications to an existing system (e.g., adding more sprinkler heads) could alter the system enough to require a different pump.

19.8 Power Plants and Drives

Power plants for pumping should deliver sufficient power at the specified speed with maximum operating efficiency. Internal combustion engines and electric motors are by far the most common types of power units. The selection of the type of unit depends on (1) the amount of power required, (2) initial cost, (3) availability and cost of fuel or electricity, (4) annual use, and (5) duration and frequency of pumping (Table 19-1).

Internal combustion engines operate on a wide variety of fuels, such as gasoline, diesel oil, natural gas, and butane. Diesel engines are more expensive initially, but have lower fuel costs. Where the annual use is more than 800 to 1000 hours, the diesel engine may be justified. Otherwise, it may be more practical to use a gasoline engine. For continuous operation, water-cooled, gasoline engines may be expected to deliver 70 percent of their rated power, diesel engines 80 percent, and air-cooled gasoline engines 60 percent. Where a vertical centrifugal or a deep-well pump is to be driven with an internal combustion engine, right-angle gear drives are usually employed. In some instances, internal combustion engines are mounted with vertical crankshafts so they may be directly coupled to a vertical drive shaft. Belt drives may use either flat or V-belts suitable for driving vertical shafts or gear drives. In comparison to direct drives, which have a transmission efficiency of nearly 100 percent, the efficiency for gear drives ranges from 94 to 96 percent, for V-belts from 90 to 95 percent, and for flat belts from 80 to 95 percent. Table 19-1 gives typical service lives of the various system components.

Nebraska pumping plant performance criteria, developed at the University of Nebraska, provide standards for comparison of overall pumping plant performance. The criteria rate various power units (assuming 75 percent pump efficiency) according

TABLE 19–1 Service Life for Pumping Plant Components

Item	*Estimated Service Life*
Well and casing	20 yr
Plant housing	20 yr
Pump turbine	
Bowl assembly (about 50% of cost of pump unit)	16 000 h *or* 8 yr
Column, etc.	32 000 h *or* 16 yr
Pump, centrifugal	32 000 h *or* 16 yr
Power transmission	
Gear head	30 000 h *or* 15 yr
V-belt	6000 h *or* 3 yr
Flat-belt, rubber and fabric	10 000 h *or* 5 yr
Electric motor	50 000 h *or* 25 yr
Diesel engine	28 000 h *or* 14 yr
Gasoline engine	
Air-cooled	8000 h *or* 4 yr
Water-cooled	18 000 h *or* 9 yr
Propane engine	28 000 h *or* 14 yr

Source: SCS (1959).

TABLE 19–2 Nebraska Pumping Plant Performance Criteria

Energy Source	*b-kWh[a] per Unit of Energy*	*w-kWh[b] per Unit of Energy[c]*	*Energy Units*
Diesel	3.282	2.46	liter
Gasoline	2.273[d]	1.71	liter
Liquid propane	1.813[d]	1.36	liter
Natural gas	2.166[e]	1.62	m^3
Electricity	1.186[f]	0.885[g]	kWh

Note: Values have been converted to SI units.

[a]b-kWh (brake kilowatt-hours) is the work being accomplished by the power unit with only drive losses considered.

[b]w-kWh (water kilowatt-hours) is the work being accomplished by the pumping plant, power unit and pump.

[c]Based on 75% pump efficiency.

[d]Taken from Test D of Nebraska Tractor Test Reports. Drive losses are accounted for in the data. Assumes no cooling fan.

[e]Manufacturer's data corrected for 5% gear head drive loss with no cooling fan. Assumes natural gas energy content of 8230 kcal (9.57 kWh) per cubic meter. Unit energy content can vary from season to season.

[f]Assumes 88% electric motor efficiency.

[g]Direct connection, assumes no drive loss.

Source: NRCS (1997).

to energy delivered to the water per unit of fuel consumed. Comparing the actual performance of a pumping plant to the Nebraska criteria gives a performance rating (as a percentage). Since the Nebraska criteria represent a compromise between optimal and average performance, the performance rating may exceed 100 in some cases. Table 19-2 shows the pumping plant performance criteria. Assuming performance criteria of 100, Table 19-3 shows the overall efficiency of pumping plants. These values are useful for comparing alternative systems and checking performance of existing systems for excessive fuel consumption.

TABLE 19–3 Nebraska Performance Criteria versus Overall Efficiency[a]

Energy Type	*Unit of Energy*	*w-kWh per Unit of Energy*	*Performance Rating (%)*	*Overall Efficiency (%)*
Diesel	liter	2.46	100	23
Gasoline	liter	1.71	100	17
Propane	liter	1.36	100	18
Natural gas	m^3	1.62	100	17
Electric	kWh	0.885	100	66[b]

Note: Values have been converted to SI units.

[a]Efficiency given for electricity in *wire to water* efficiency, which is calculated at the pump site. Liquid or gas fuel is based on average energy content values.

[b]Overall efficiencies vary from 55% for 4 kW to 67% for 75 kW.

Source: NRCS (1997).

Example 19.3 A diesel engine driving a pump produces 74.6 water kW while consuming 33 liters of fuel per hour. Find the performance rating and the overall efficiency for the pumping plant.

Solution. The pumping plant performance is the water power produced per unit of fuel consumed. In this case,

$$\text{Perforamnce} = \frac{76.4 \text{ kW}}{33 \text{ L/h}} = 2.26 \frac{\text{kWh}}{\text{L}}$$

The performance rating is this value divided by the Nebraska performance criterion for diesel, 2.46 water kWh/L, i.e., 2.26/2.46 = 0.92, or 92 percent. The overall efficiency can then be estimated using Table 19-3. The overall efficiency for a diesel-powered pumping plant with a performance rating of 100 percent is 23 percent. For this pumping plant, the overall efficiency is $0.92 \times 23\% \approx 21\%$.

19.9 Pumps in Series and Pumps in Parallel

A single pump may not be sufficient or economically reasonable for many applications. Where the total head requirement is higher than a single pump can satisfy, multiple pumps may be arranged in series. Where the required capacity is high, or varies significantly, pumps may be arranged in parallel. In parallel arrangements, any number of pumps can be operated simultaneously to match the capacity needed. In series arrangements, all pumps should operate simultaneously.

For pumps in series, all water flows through each of the pumps in turn, so it is important that the capacities of the pumps be well matched. Many booster, turbine, and submersible pumps are arrangements of 2–15 identical pump units or stages (often called bowls) in close array on a single drive shaft. The head-discharge curve for a multistage pump can be approximated from that for a single stage by multiplying the head at a particular discharge times the number of stages. (Since efficiency tends to increase with multiple stages, this method will slightly underestimate performance.) For dissimilar pumps in series, the individual heads at a given discharge can be added to estimate the composite head (Figure 19-9a).

For pumps in parallel, the discharges are approximately additive. If the pumps are drawing from a common supply and discharging into a common outlet, they will have the same operating head. The composite discharge can be estimated by adding the individual discharges at a given head (Figure 19-9b). It is hard to obtain good performance from dissimilar pumps in parallel due to the difficulty of matching their performance characteristics. Using identical pumps in parallel arrangements eliminates that problem.

19.10 Pump Selection

Performance curves serve as a basis for selecting a pump to provide the required head and capacity for the range of expected operating conditions at or near maximum efficiency. The factors that should be considered include the head–capacity relationship of the well or sump from which the water is removed, space requirements of the pump, initial cost, type of power plant, and pump characteristics,

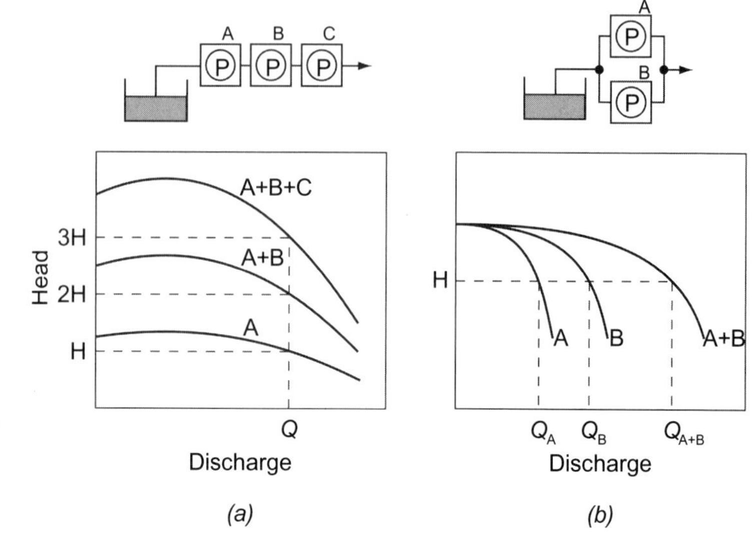

Figure 19–9
Composite pump curves for pumps in (a) series and (b) parallel.

as well as other possible uses for the pump. Storage capacity, rate of replenishment, and the well diameter may limit the pump size and type. For example, a large drainage ditch provides a nearly continuous source of water, whereas a well usually has a small storage capacity.

The initial cost that can be justified depends on the annual use and other economic considerations. The size of the power plant and type of drive should be adapted to the pump.

The centrifugal pump is suitable for low to high capacities at heads up to several hundred meters, the propeller pump for high capacities at low heads, and the mixed-flow pump for intermediate heads and capacities. Since the impellers of these pumps may be placed near or below the water level, the suction head developed may not be critical. Horizontal centrifugal pumps are best suited for pumping from surface water supplies, such as ponds and streams, provided the water surface does not fluctuate excessively. Where the water level varies considerably, a vertical centrifugal or deep-well turbine pump is more satisfactory.

19.11 Pumping Costs

The costs of pumping include fixed costs and operating costs. Fixed costs include interest on investment, depreciation, taxes, and insurance. The investment cost includes the construction and development of the well, pump, power plant, pump house, and water storage facilities. Construction costs for wells depend on the diameter and depth of the well, construction methods, nature of the material through which the well is drilled or dug, type of casing, length and type of well screen, and time required to develop and test the well (Chapter 11).

Table 19-1 gives the estimated service lives of various components of pumping plants for estimating depreciation. Taxes and insurance are approximately 1.5 percent of the total investment. Operating costs include fuel or electricity, lubricating oil, repairs, and labor for operating the installation. Fuel costs for internal combustion engines are generally proportional to the consumption, whereas the unit cost of electrical energy generally decreases with the amount consumed. A demand charge or minimum is made each month regardless of the amount of energy consumed. This demand charge may or may not include a given amount of energy. Longenbaugh & Duke (1983) provide more detail on economic analysis of pumping plants.

19.12 Design Capacity

The first consideration in selection of a pump and design of a pumping plant must be the determination of the required capacity. Irrigation pumping requirements depend on peak evapotranspiration rates, irrigation efficiencies, and the planned frequency and duration of pumping plant operation (Chapters 4 and 15–18). For example, a crop with peak *ET* of 6 mm/d with an irrigation system having overall efficiency of 69 percent requires a minimum pumping capacity (for continuous operation) of one liter per second per hectare. Maintenance and other downtime requirements will increase the required pump capacity.

Capacity of the pumping plant may also be dictated by characteristics of the water supply. For extended pumping from a well, the pump capacity should be matched to the well yield (Chapter 11). Where storage is part of the supply system (e.g., a pond or reservoir), the capacity of the pump may exceed the instantaneous rate of supply, but pumping will be intermittent.

Drainage coefficients for pumping plant design vary according to soil, topography, rainfall, and cropping conditions. Recommended coefficients for the upper Mississippi Valley range from 7 to 25 mm/day. Requirements in southern Texas and Louisiana are as high as 75 mm/day, including pumping discharge and reservoir storage. A drainage coefficient of 25 mm/day is suitable for field crops on organic soils in Florida. Fruit or vegetable crops may require and economically justify a drainage coefficient of as much as 75 mm/day. Additional recommendations for drainage coefficients may be obtained from Chapter 14, ASAE (1998), or state and local guides. Note that for ordinary drainage, this drainage coefficient is the same as that used for the design of ditch or pipe drainage systems.

19.13 Drainage Pumping Plants

In low-lying or very flat areas where gravity flow does not provide an adequate outlet, pumps may be used to lift drainage water into elevated drainage channels. Typical applications of pump drainage include land behind levees, outlets for pipe drains and open ditches, and flat land where the channel gradient is too small to discharge water at the required rate. Performance requirements for drainage pumping plants typically involve large capacities at low heads. (Similar performance requirements may be found with subirrigation systems using surface water supplies or irrigation systems transferring water between distribution canals.)

Figure 19-10 shows a pumping plant for pipe drains or open-ditch drainage of small areas. An inlet screen must be used with ditches to prevent weeds or other detritus from clogging the pump. The design of small-farm pumping plants may be simplified by using the design runoff rate recommended for open ditches plus 20 percent. Where electricity is available, electric motors with automatic controls are recommended for small installations. For drainage areas larger than 40 ha, storage capacity should be obtained by enlarging the open ditches or by excavating a storage basin. Constructed sumps as shown in Figure 19-10 are generally too expensive where the diameter is greater than 5 m. Where storage is available in open ditches or other reservoirs, sumps about 2 m in diameter are satisfactory. Most pumps operate only during the wet months, but some installations draining extremely low land may operate more or less continuously throughout the year. Usually, the plants operate intermittently and run only 10 to 20 percent of the time.

For large installations, it may be advantageous to use two pumps with different capacities, one for handling surface runoff during wet seasons and the other for seepage or flow from pipe drains. It is usually more economical to operate a small pump for long periods than to run a large pump for short periods.

Frequently, drainage installations lack storage capacity and the allowable variation of the water surface on the inlet side is small. Automatically controlled electric motors are especially suitable under these conditions. Internal combustion engines are usually started manually, but may be equipped with an automatic shutoff. The number of stopping and starting cycles should not exceed more than two per day for nonautomatic operation or 10 per hour for automatic operation (ASAE, 1999). Due to less frequent cycling, nonautomatic plants require more storage capacity than those with automatic controls.

Where a sump is used, such as for collecting water from a pipe drain system, it must have sufficient storage to avoid excessive starting and stopping. The cycle time is the sum of the running and stopping times, which are the storage capacity of the sump, S, divided by the net flow rate out of and into the sump, respectively. The cycle frequency, n, is the reciprocal of the cycle time and is usually expressed as cycles per hour.

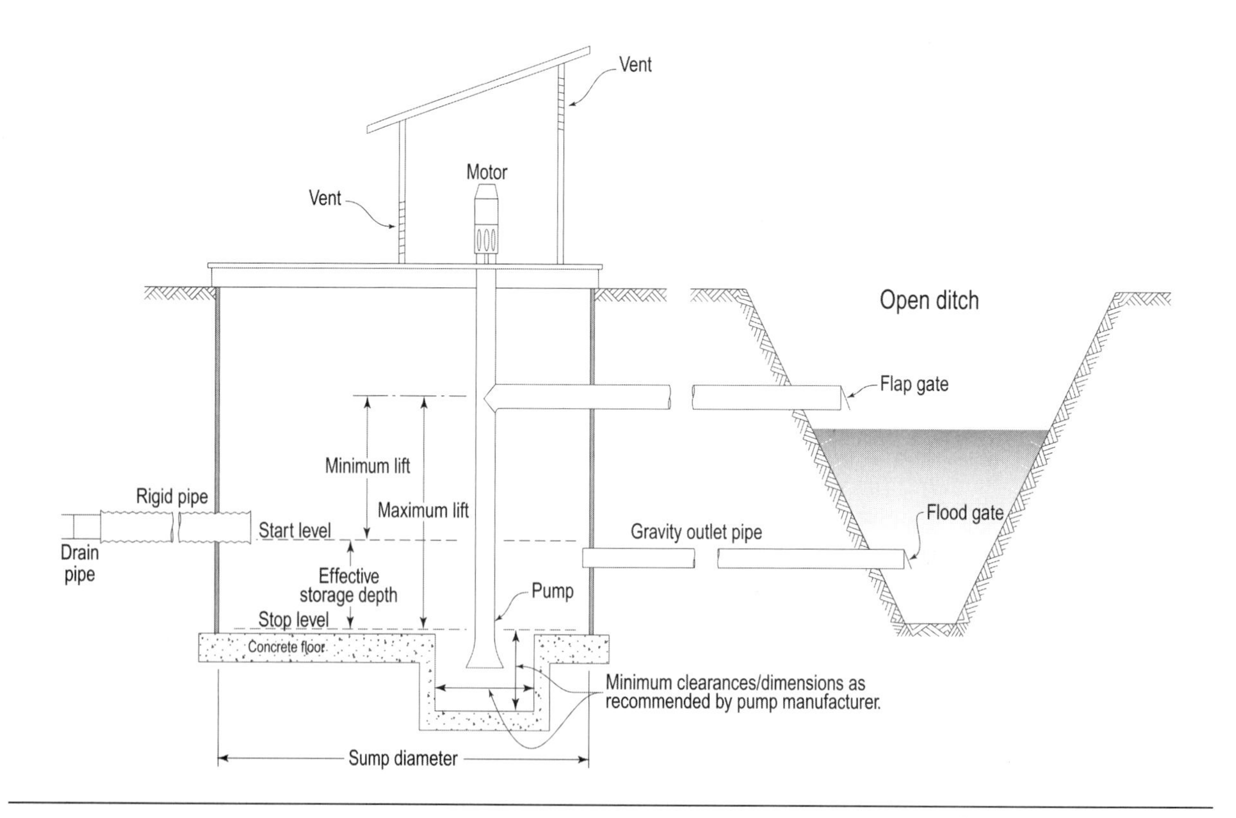

Figure 19–10 Small drainage pumping plant. Level sensors and motor controls are not shown.

$$\frac{1}{n} = \frac{S}{Q_p - Q_i} + \frac{S}{Q_i} \quad \textbf{19.5}$$

where n = cycle frequency (T^{-1}),
S = sump storage volume (L^3),
Q_i = inflow rate (L^3/T),
Q_p = average pumping rate (L^3/T).

For automatic operation, the minimum allowable sump storage volume can be found by maximizing the effective storage, $S \times n$. Solve Equation 19.5 for $S \times n$, assume constant Q_p, differentiate with respect to Q_i, and set the result equal to zero to find a maximum. This gives

$$Q_i = \frac{Q_p}{2} \quad \textbf{19.6}$$

Maximum effective storage occurs when the pumping rate, Q_p, is twice the inflow rate, Q_i. Substituting this expression for Q_i into Equation (19.5) gives

$$S = \frac{CQ_p}{n} \quad \textbf{19.7}$$

For S in cubic meters, Q_p in liters per second, and n in cycles per hour, the constant C is 0.9. For S in cubic feet, Q_p in gallons per minute, and n in cycles per hour, C is 2 (with less than 0.25 percent error).

The storage requirements depend on the discharge rate and the frequency of cycling. The pump will operate continuously if the inflow rate exceeds the pumping rate. If the inflow rate is less than the pumping rate, the pump will cycle.

Sumps should generally be shallow and as large as is reasonable. Recommended storage depths are 0.6 m for closed sumps and 0.3 m for open sumps. In earthen-walled storages, large and frequent fluctuations in the water level can contribute to bank sloughing and channel erosion (ASAE, 1999).

Example 19.4

Subsurface drainage water from 16 ha is collected in a closed sump. The drainage coefficient is 13 mm/d. Specify the capacity of the pump, the storage depth, and the diameter of the sump for this application.

Solution. The design inflow rate is the drainage area times the drainage coefficient.

$$Q_i = 16\ \text{ha}\frac{13\ \text{mm}}{d}\left(\frac{d}{86\ 400\text{s}}\right)\left(\frac{10\ 000\text{L}}{\text{ha-mm}}\right) = 24\ \text{L/s}$$

The pumping rate, Q_p, should be twice the inflow rate, so Q_p = 48 L/s. Using the recommended maximum fluctuation of 0.6 m for the storage depth and a maximum cycling rate of 10 per hour, calculate the storage volume using Equation 19.7.

$$S = \frac{0.9(48)}{10} = 4.3 \text{ m}^3$$

The area and diameter of the sump are then

$$A_{sump} = \frac{4.3 \text{ m}^3}{0.6 \text{ m}} = 7.2 \text{ m}^2$$

$$d_{sump} = \left(\frac{4A_{sump}}{\pi}\right)^{1/2} = 3.0 \text{ m}$$

This solution is not unique. Many combinations of sump depth and diameter are possible, provided the storage requirement is met and the other aspects of the design are physically and economically reasonable.

The pumping plant should be installed to provide minimum lift and located so that the pump house will not be flooded. A circular sump is recommended for such an installation since less reinforcement is required and it is easier to construct. The electric motor is usually operated automatically by means of electrode or float switches. Where open ditches provide storage, the sump can be reduced in size. The discharge pipe should outlet below the minimum water level in the outlet ditch, where practicable, to reduce the pumping head to a minimum. The gravity outlet pipe should be installed only when the water level in the ditch is lower than the pipe level for at least 10 percent of the flow period. The savings in pumping cost should justify the cost of the gravity outlet pipe and flood gate.

19.14 Irrigation Pumping Plants

Pumping requirements for irrigation depend on such factors as source of water, method of irrigation, and size of the irrigated area. Pumping from surface supplies is similar to pumped drainage except where pressure irrigation is practiced. Much higher heads are normally required where water is pumped from wells or pumped to pressurize irrigation systems. Wells located in the vicinity of the irrigated area have many advantages as compared with canals supplying water from distant sources. A typical deep-well turbine pump installation discharging into an underground pipeline is shown in Figure 19-11.

Water hammer is the term given to high-pressure waves that can propagate along a pipeline. The cause of water hammer is a sudden change in the velocity of the water, such as occurs when a valve is closed quickly. The kinetic energy of the water is converted to pressure, which reaches a peak at the closed end. The pressure builds, slightly compressing the water and slightly stretching the pipe. The compressed water then expands backward along the pipe. The momentum of this backward flow can cause a reduced pressure at the closed end. Either pressure extreme can damage the pipe. Joints are particularly vulnerable to damage because of the stress concentrations occurring there. Methods to compute the pressure rise and valve closing times to limit water hammer can be found in other texts (e.g., Cuenca, 1989; Brater & King, 1996).

Surge tanks can be installed in the discharge line, as illustrated in Figure 19-11, to attenuate the sharp pressure fluctuations that may develop from starting or stopping

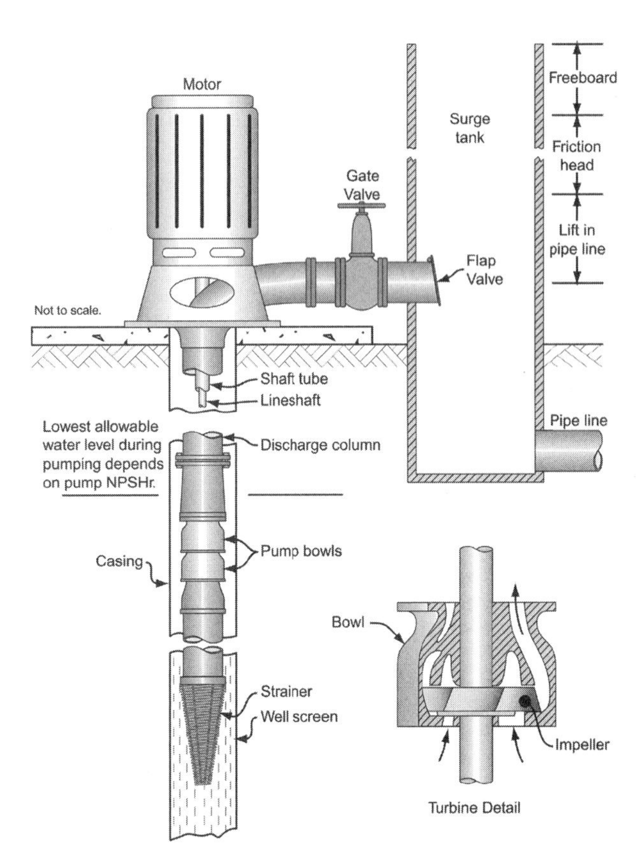

Figure 19–11

Deep-well turbine pump installation with a surge tank for low heads. (Modified from Rohwer, 1943, and Wood, 1950)

the pump when pumping into a pipeline. They are usually not required for portable sprinkler irrigation systems. Surge tanks that are open to atmospheric pressure are suitable for pumping into pipelines made of thin metal, concrete, or vitrified clay pipe. The storage tank must be high enough to allow for any increase in elevation of the pipeline, for friction head, and for reasonable freeboard. Where the total head exceeds 6 m, an enclosed metal tank having an air chamber is usually more practical than a surge tank. The maximum head at which irrigation pumping is practical varies with locality, value of the crop, energy costs, and other economic factors.

The capacity and head requirements of the irrigation system (Chapters 15–18) must be known before selecting a pump. The pump should be selected to match the head–capacity characteristics of the irrigation system and the water supply. The problem becomes more complicated where discharge requirements for the system are variable, which affects both the friction losses in the piping and the drawdown in the well. The pump should operate at or near its highest efficiency. A plot of the head–capacity curve for the pump with the system head–capacity curve can aid in pump selection. The system curve combines the head–discharge characteristics of the piping and distribution system (Chapters 17 and 18) with the drawdown–discharge characteristics of the well (Chapter 11). Such a set of curves for a typical field installation is shown in Figure 19-12.

The system curve will shift vertically with variations in the depth to the free water surface in the well, so high and low system curves will be needed to develop the best

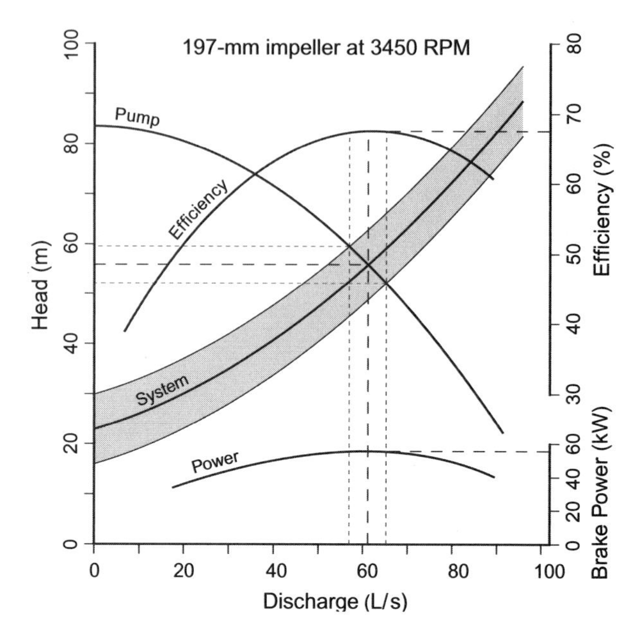

Figure 19–12

Fitting the pump to the system by combining pump performance curves with system head-capacity curves.

design. Figure 19-12 shows how a drop of 7 m in the free water surface would shift the system curve upward by 7 m of head (to the top of the shaded band) because an additional 7 m of head would be required to lift water out of the well. The intersection of the system curve and the pump curve would shift to the left, i.e., to a lower discharge. Conversely, a higher free water surface would shift the system curve downward and the intersection of the pump and system curves would occur at a higher discharge.

Example 19.5

An irrigation system is to be supplied by a well. The system head–capacity relationship is shown in Figure 19-12. The average discharge and head for the system are 60 L/s and 56 m. The head will vary from 52 to 60 m. Evaluate the pump shown in the figure for this application.

Solution. The head–capacity curve for a pump is overlain on the plot. The point where the two curves intersect is the operating point. The operating point will shift along the pump curve as the system head varies between 52 and 60 m head. The pump that is shown has its peak efficiency in the center of this range of heads and discharges. The power requirement is also very near its maximum. There is little danger of overloading the power unit since the power requirement drops off in either direction.

For the range of heads and discharges specified, this pump should give excellent service. It will operate very near its peak efficiency at all times. The only possible improvement would be to find a pump with a higher operating efficiency.

This procedure for selecting a pump emphasizes the importance of obtaining the head–capacity curve for the well before purchasing the pump.

In practice, it would be very rare to find a pump that exactly matched the desired outputs. However, the fit can often be improved by altering the pump speed or trimming the impeller. The pump curve shown in Figure 19-12 could be shifted down and to the left by either method. The advantage of speed changes over impeller trims is that speed changes can be reversed, whereas a trimmed impeller cannot be restored to its original size.

In recent years, variable-speed drives with load-sensing controls have become widely available. These allow the pump speed to be automatically adjusted in response to changes in the system. While variable-speed drives are more expensive than fixed-speed drives, their use can be justified where the savings from increased operating efficiency outweigh the increase in capital costs.

Internet Resource

The Hydraulic Institute
www.pumps.org

References

American Society of Agricultural Engineers (ASAE). (1998). *Design of Subsurface Drains in Humid Areas.* EP480. St. Joseph, MI: ASAE.

———. (1999). *Design of Agricultural Drainage Pumping Plants.* EP369.1. St. Joseph, MI: ASAE.

Brater, E. R., & H. W. King. (1996). *Handbook of Hydraulics,* 7th ed. New York: McGraw-Hill.

Cuenca, R. H. (1989). *Irrigation System Design: An Engineering Approach.* Englewood Cliffs, NJ: Prentice Hall.

Driscoll, F. G. (1986). *Groundwater and Wells,* 2nd ed. St. Paul, MN: Johnson Filtration Systems, Inc.

Haman, D. Z., F. T. Izuno, & A. G. Smajstrla. (1994). *Pumps for Florida Irrigation and Drainage Systems.* Circular 832. Gainesville: Florida Cooperative Extension Service, Institute of Food and Agricultural Sciences, University of Florida.

Longenbaugh, R. A., & H. R. Duke. (1983). Farm pumps. In M. E. Jensen, ed., *Design and Operation of Farm Irrigation Systems,* pp. 347–391. Monograph No. 3. St. Joseph, MI: ASAE.

Rohwer, C. (1943). *Design and Operation of Small Irrigation Pumping Plants.* USDA Cir. 678. Washington, DC: U.S. Department of Agriculture.

USDA Natural Resources Conservation Service (NRCS). (1997). Energy use and conservation. In *National Engineering Handbook,* Part 652, Irrigation Guide, Chap. 12. Washington, DC: Author.

U.S. Soil Conservation Service (SCS). (1959). Irrigation pumping plants. In *National Engineering Handbook* Section 15: "Irrigation" Chapter 8. Washington DC: U.S. Government Printing Office.

White, F. M. (1986). *Fluid Mechanics,* 2nd ed. New York: McGraw-Hill.

Wood, I. D. (1950). *Pumping for Irrigation.* SCS-TP-89 SCS. Washington, DC: USDA Soil Conservation Service.

Problems

19.1 A centrifugal pump operating at 1800 rpm with an efficiency of 70 percent discharges 30 L/s against a 25-m head and requires 10.5 kW. Assuming that the efficiency remains constant, estimate the discharge, head, and power if the speed is reduced to 1500 rpm.

19.2 From the performance curves for the centrifugal pump with a 238-mm impeller shown in Figure 19-5, what is the pump efficiency at a head of 90 m? What is the discharge? What are the efficiency and discharge if the head is reduced to 75 m?

19.3 What is the power requirement for pumping 70 L/s against a head of 50 m assuming a pump efficiency of 65 percent? What size electric motor is required assuming the efficiency of the motor is 90 percent?

19.4 A propeller pump in a drainage pumping plant has an average discharge of 63 L/s. What is the required depth between start and stop levels where the water is pumped from a sump 3.6 m in diameter? The maximum cycling rate is 10 per hour. At what inflow rate would maximum cycling occur?

19.5 Determine the maximum practical suction lift for a centrifugal pump if the discharge is 70 L/s, the water temperature is 25°C, friction losses in pipe and fittings are 2.6 m, the NPSHr of the pump is 3.4 m, and the altitude for operation is 900 meters above sea level.

19.6 For the centrifugal pump in Figure 19-5 with an impeller diameter of 208 mm, determine the head, efficiency, and power for a discharge of 30 L/s. Should this pump be recommended for these flow conditions?

19.7 A centrifugal pump in Figure 19-5 with a 238-mm impeller is operating against 90 m head. If the total head of the system is reduced to 80 m by reducing the pressure on the outlet, what happens to the pump operating conditions? Consider discharge, efficiency, power requirement, and NPSHr. How would this affect the power unit? If the system was originally configured with suction lift near the maximum, what are the consequences of the reduction in total head?

19.8 The axial-flow pump in Figure 19-7 is to be used in a large drainage pumping plant where the average lift is 3.0 m. What are the discharge and power? What would happen to the head and power if the outlet were restricted to reduce the discharge to 2.1 m^3/s? What would be the effect on the drive unit?

19.9 For the pump and system shown in Figure 19-12, determine the changes needed to obtain a discharge of 50 L/s by (a) trimming the impeller and (b) changing the pump speed. Also find the head and power for each of these alterations. Which method would you recommend? Why?

CHAPTER

20 Soil Erosion by Wind

In the arid and semiarid regions of the United States, large areas are affected by wind erosion. The Great Plains region, an area especially subject to soil movement by wind, represents about 20 percent of the total land area in the United States. Many humid regions are also damaged by wind erosion. The areas most subject to damage are the sandy soils along streams, lakes, and coastal plains and organic soils. Peats and mucks constitute about 10 million ha located in 34 states.

Wind erosion not only removes soil, but also damages crops, fences, buildings, and highways. Fine soil particles are lost along with nutrients, which can result in reduced crop yields. Eroded sediment particles are a nuisance for many people and can adversely affect the health of some individuals. There are also circumstances where eroded dust can obscure visibility. Such conditions can lead to fatal traffic accidents. For example, in 1991, 104 vehicles were involved in an accident on Interstate 5 in California, resulting in 15 deaths and 150 injuries.

Dust particles can travel far, even crossing oceans. It was the deposition of dust in Washington, D.C., from wind erosion in the Great Plains in the 1930s that resulted in the U.S. government establishing the Soil Conservation Service. Figure 20-1 shows the distribution of wind erosion hazard in the states west of the Mississippi River.

20.1 Air Quality

A major issue related to wind erosion is air quality (Saxton et al., 2000). The U.S. Environmental Protection Agency (EPA, 2003) has determined that particulate matter in the air can be detrimental to human health. Airborne particles are particularly dangerous for the elderly, individuals with cardiopulmonary diseases, and children. These particulates fall into two classes, coarse, less than 10 microns in diameter (PM-10), and fine, less than 2.5 microns (PM-2.5). The particles can come from a variety of sources. Fine particles are generally associated with automotive and other smoke emissions. Coarse particles are generated from unpaved roads, materials handling and grinding, and agriculture. Generally, agriculture is not a major contributor to poor air quality, but there are some areas, during some critical periods of the year, where this is not the case. Dust from wind erosion may further degrade air quality in

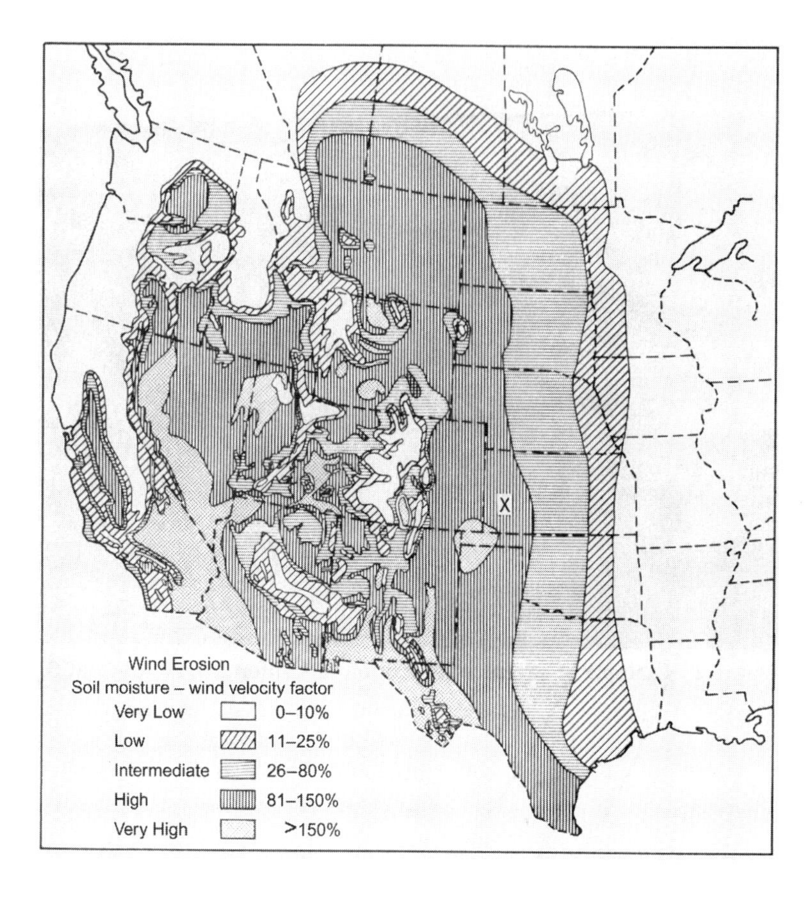

Figure 20–1
Relative potential soil loss by wind for the western United States and southern Canada as a percentage of that in the vicinity of Garden City, Kansas, marked by X. (Chepil et al., 1962)

some regions with air quality problems from other sources, such as the industrial areas around the Great Lakes. Readers are encouraged to determine the degree of local concern about airborne particles due to wind erosion.

The 1997 EPA standards for particles set the limit for PM-2.5 at 15 μg particulate matter per cubic meter of air annually, and 65 $\mu g\ m^{-3}$daily. The PM-10 values are set at 50 $\mu g\ m^{-3}$ annually and 150 $\mu g\ m^{-3}$ as a daily standard.

20.2 Wind and Water Erosion Processes

The physical processes causing particle detachment and transport in wind erosion are similar to the processes involved in rill and channel erosion by water. With wind, the fluid carrying the particles is a low-density gas, whereas in water it is a higher-density liquid. The density of the soil particles and the fluid, and the velocity and shear of the fluid on the particles affect the rate of particle detachment and subsequent transport. With both wind and water, the shear of the fluid varies greatly in time and location during an erosion event. Higher detachment rates result from pulses of high fluid shear, and deposition occurs when there is a drop in the shear further downstream or later in time. In both wind and water, small particles are more easily transported, particularly in suspension, whereas coarser, noncohesive particles tend to be more easily detached. Soil aggregates with lower densities are more likely to be transported than soil particles with higher densities. With wind erosion, moist soil has greater cohesion than dry soil and is less easily detached, whereas with streambank and other mass erosion, saturated soils have less cohesion than unsaturated soils and are more likely to fail.

20.3 Types of Soil Movement

Saltation, suspension, and surface creep are the three types of soil movement with both wind erosion and channel and rill erosion with water (Figure 20-2). Saltation is the process where fine particles (0.1 to 0.5 mm in diameter) are lifted from the surface and follow distinct trajectories under the influence of air resistance and gravity. When the particles return to the surface, they may rebound or become embedded when impacting the surface, but in either case they initiate movement of other particles to create an "avalanching" effect of additional soil movement. Most saltation occurs within 0.3 m of the surface. Saltation accounts for 55 to 72 percent of particle movement during wind erosion events. Suspended particles (0.02 to 0.1 mm in diameter) are dislodged by saltating particles and represent 3 to 10 percent of eroding particles. The smaller suspended particles may remain aloft for an extended period, traveling hundreds of kilometers. These suspended particles may become nuclei for raindrop formation. Sand-sized particles or aggregates (0.5 to 2 mm in diameter) are set in motion by the impact of saltating particles, and tend to roll or creep along the surface. Creep accounts for 7 to 25 percent of the soil movement.

20.4 Mechanics of Wind Erosion

For a precise understanding of the mechanics of wind erosion, an analysis must be made of the interactions among the climate, the soil, vegetation, and the length of exposed soil. Wind erosion may be divided into the three processes: (1) initiation of movement, (2) transportation, and (3) deposition.

Initiation of Movement. Soil movement is initiated from air turbulence and velocity. The fluid threshold velocity is defined as the minimum velocity required to produce soil movement by direct action of the wind, and the impact threshold velocity is the minimum velocity necessary to initiate movement from the impact of soil particles carried by saltation. The wind is always turbulent except near the

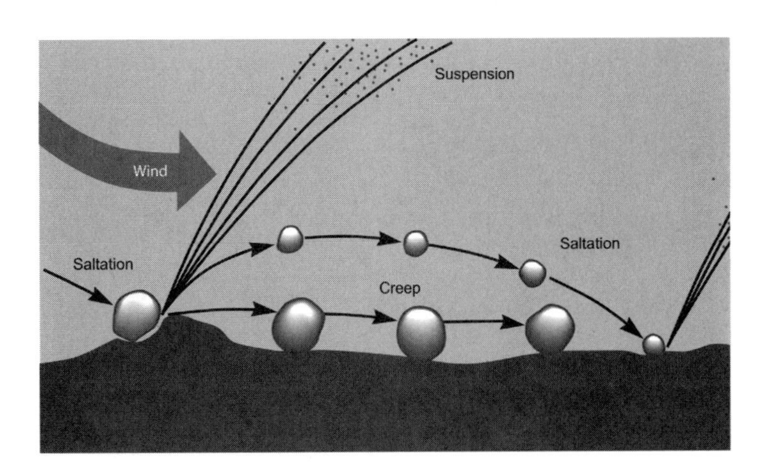

Figure 20–2
Main processes of soil movement by wind. (ARS, 2003)

surface and at low velocities (less than about 1 m/s). Wind speeds of 5 m/s or less at 0.3-m height are usually considered nonerosive for mineral soils.

Transportation. The quantity of soil moved is influenced by the aggregate size, texture, wind velocity, and distance across the eroding area. Winds, being variable in velocity and direction, produce gusts with eddies and cross currents that lift and transport soil. The quantity of soil moved varies as the cube of the excess wind velocity above the constant threshold velocity, the square root of the soil aggregate diameter, and the gradation of the soil.

The rate of soil movement increases with the distance from the windward edge of the field or eroded area. Fine particles drift and accumulate on the leeward side of the area or pile up in dunes. Increased rates of soil movement with distance from the windward edge of the area subject to erosion are the result of increasing amounts of erosive particles, thus causing greater abrasion and a gradual decrease in surface roughness.

The atmosphere has a tremendous capacity for transporting sediment in suspension, particularly those soil fractions less than 0.1 mm in diameter. It is estimated that the potential carrying capacity per cubic kilometer of the atmosphere is many tonnes of soil, depending on the wind velocity. For example, a dust storm originating in the Texas Panhandle deposited over 200 kg/ha in Iowa in 1937.

Deposition. Deposition of windborne sediment occurs when the gravitational force is greater than the forces holding the particles in the air. The process generally occurs when there is a decrease in wind velocity caused by vegetation or other physical barriers, such as ditches, vegetation, and snow fences. Raindrops may also remove dust from the air.

20.5 Estimating Wind Erosion

The most widely used method for estimating wind erosion is the Wind Erosion Equation method (WEQ) (NRCS, 2002). A computer model is now available, the Wind Erosion Prediction System (WEPS) model, that will eventually replace the WEQ method (ARS, 2003). A solution using the WEQ method is presented since it can be solved without the need for a computer. A spreadsheet solution to the WEQ, including a soil and climate database, is available (NRCS, 2003). Readers are encouraged to download the latest version of the WEPS computer model to compare to the predictions from examples and problems presented in this chapter. The WEQ method is presented as an interaction among five factors.

$$E = f(I,K,C,L,V) \tag{20.1}$$

where E = estimated average annual soil loss (Mg ha^{-1} y^{-1}),
I = soil erodibility index (Mg ha^{-1} y^{-1}),
K = ridge roughness factor,
C = climate factor,
L = unsheltered length of eroding field (m),
V = vegetative cover factor.

Erodibility Index, *I*. The above factors are not independent, but must be combined in a set of interacting equations to estimate wind erosion. The wind

TABLE 20–1 Typical Wind Erodibility Indices for Different Soil Textures

Predominant Soil Texture and Soil Erodibility Index	*Soil Erodibility Index I (Mg/ha-year)*
Loamy sands and sapric organic material[a]	360–700
Loamy sands	300
Sandy loams	200
Clays and clay loams	200
Calcareous loams	200
Noncalcareous loams, silt loam <20% clay, and hemic organic soils	125
Noncalcareous loams and silt loams >20% clay	100
Silt, noncalcareous silty clay loam and fibric organic soils	85
Wet or rocky soils not susceptible to erosion	—

[a] The *I* factors for Group I vary from 360 for coarse sands to 700 for very fine sands. Use 500 for an average.
Based on NRCS (2002).

erodibility, I, is a function of the soil aggregates greater than 0.84 mm in diameter. The following regression equation was developed from estimates of I given in Woodruff & Siddoway (1965).

$$I = 525\, e^{(-0.04F)} \tag{20.2}$$

where I is the wind erodibility, e is the natural logarithm base (2.718), and F is the percentage of dry soil fraction greater than 0.84 mm. The fraction of dry soil can vary during the season and can also be altered with changes in soil water content and organic matter. Table 20-1 summarizes typical soil wind erodibility values for different textures of soil.

Surface crusting caused by wetting and drying may reduce erosion on many soils, but is not considered in the WEQ method. Irrigation can also decrease erodibility, and suggested values for the I factor with irrigation are available (NRCS, 2002). The WEPS model accounts for crusting effects and the interaction between time of crusting and occurrence of wind erosion events (Hagen, 1991).

Roughness Factor, *K*. The roughness factor, K, is a measure of the effect of ridges made by tillage and planting implements on erosion rate. Ridges absorb and deflect wind energy and trap moving soil particles. Too much roughness, however, causes turbulence which may accelerate particle movement. Roughness can be due to natural undulations and the presence of knolls on the landscape where increased erosion has been observed, or to temporary ridges from tillage, which tend to decrease erosion rates. Table 20-2 provides adjustment factors that should be multiplied by the erodibility index I to account for the increased erosion on windward sides and tops of knolls. To estimate K, it is necessary to first estimate the ridge roughness from the equation

$$K_r = 4\frac{h^2}{d} \tag{20.3}$$

TABLE 20–2 Knoll Erodibility Adjustment Factors

Slope Change in Prevailing Wind Erosion Direction (%)	*Knoll Adjustment to I (factor)*	*Increase at Crest Area Where Erosion Is Most Severe (factor)*
3	1.3	1.5
4	1.6	1.9
5	1.9	2.5
6	2.3	3.2
8	3.0	4.8
10	3.6	6.8

Source: NRCS (2002).

where K_r = ridge roughness (mm),
h = ridge height (mm),
d = ridge spacing (mm).

From the ridge roughness K_r, the roughness factor K can be calculated by the regression relationship derived from Woodruff & Siddoway (1965).

$$K = 0.34 + \frac{12}{K_r + 18} + 6.2 \times 10^{-6} K_r^2 \qquad \textbf{20.4}$$

If the dominant wind direction is other than normal to the direction of the ridges, then the value of K is reduced, depending on the direction of the wind, and the value of I (NRCS, 2002).

Climate Factor, *C*. The climate factor is an index of climate erosivity, which includes the wind speed and the soil content at the surface. It is expressed as a fraction of the C factor for Garden City, Kansas. Figure 20-1 shows the distribution of C factors for the western United States. An interactive map to find C factors for the United States can be found online at http://nm6.ftw.nrcs.usda.gov/website/. Readers are encouraged to check with local agencies or state websites for local C factors. Methods to calculate the C factor considering wind speed, precipitation, and evapotranspiration are presented in NRCS (2002).

Unsheltered Distance, *L*. The L factor represents the unsheltered distance in meters along the prevailing wind erosion direction for the field or area to be evaluated. This distance is the length from a sheltered edge of a field, parallel to the direction of the prevailing wind, to the end of the unsheltered field.

Vegetative Cover Factor, *V*. The effect of vegetative cover in the wind erosion equation is expressed by relating the kind, amount, and orientation of vegetative material to its equivalent of small-grain residue. The small-grain equivalent (Skidmore, 1994) can be calculated from the relationship

$$SG = a\, R_w^b \qquad \textbf{20.5}$$

where SG = small-grain equivalent (kg/ha),
a, b = crop constants from Table 20-3,
R_w = quantity of residue to be converted to small-grain equivalent (kg/ha).

Where more than one crop is involved, such as growing a crop of sorghum with wheat residue, a weighted average of the coefficients is necessary so that

$$SG = a_1^{p_1} a_2^{p_2} (R_{w_1} + R_{w_2})^{b_1 p_1 + b_2 p_2} \quad \textbf{20.6}$$

Where p_1 and p_2 are the fractions of crop residue in each category (Skidmore, 1994). From the small-grain equivalent, the vegetative cover factor V (Mg/ha) can be calculated using

$$V = 2.533 \times 10^{-4} (SG)^{1.363} \quad \textbf{20.7}$$

Predicting Erosion. The prediction method presented here is based on Skidmore (1994) and is similar to that presented in the *NRCS National Agronomy Manual* (NRCS, 2002). To estimate annual erosion, the climate erosivity is estimated from Figure 20-1 or from a climate calculation. The soil erodibility index is determined from Equation 20.2 or Table 20-1. The effect of knolls may be included by multiplying I by the appropriate factor from Table 20-2. The ridge roughness factor is calculated from Equations 20.3 and 20.4. The estimated annual wind erosion can then be calculated by the following steps:

(1) The initial estimate of wind erosion E_1 is I, found from Equation 20.2 or Table 20-1 in Mg ha^{-1} y^{-1}

$$E_1 = I \quad \textbf{20.8}$$

(2) Calculate E_2 from Equation 20.8 and the soil and surface properties contained in Equation 20.4.

$$E_2 = I\,K \quad \textbf{20.9}$$

(3) Calculate E_3 by including the climate factor C presented in Figure 20-1 or from local information.

$$E_3 = I\,K\,C \quad \textbf{20.10}$$

(4) Calculate the maximum field length L_o for reducing the wind erosion estimate.

$$L_o\,(\mathrm{m}) = 1.56 \times 10^6 (E_2)^{-1.26} e^{(-0.00156\,E_2)} \quad \textbf{20.11}$$

(5) Calculate the field length factor WF

$$WF(\mathrm{Mg\ ha^{-1}\ y^{-1}}) = E_2 \times \left(1.0 - 1.2\left[\frac{L}{L_o}\right]^{-0.383}\right) e^{L/L_o} \quad \textbf{20.12}$$

where L = unsheltered distance (m).

(6) Calculate E_4 by combining the interaction of surface, soil, climate, and length effects.

$$E_4\,(\mathrm{Mg\ ha^{-1}\ y^{-1}}) = (WF^{0.348} + E_3^{\,0.348} - E_2^{\,0.348})^{2.87} \quad \textbf{20.13}$$

TABLE 20–3 Crop Residue Coefficients for Predicting Small-Grain Residue Equivalent (*SG*) (Kg/ha) from Crop Residues

Crop Residue	*Height (mm)*	*Length (mm)*	*Row Spacing (mm)*	*Orientation to Wind*	*Value a*	*Value b*
Surface Orientation: Standing						
Winter wheat	250	—	250	Normal	4.306	0.970
Rape	250	—	250	Normal	0.103	1.400
Cotton	340	—	750	Normal	0.188	1.145
Sunflowers	430	—	750	Normal	0.021	1.342
Forage sorghum	150	—	750	Normal	0.353	1.124
Silage corn	150	—	750	Normal	0.229	1.135
Surface orientation: Flat–random						
Winter wheat	—	250	—	—	7.279	0.782
Soybeans	—	250	—	—	0.167	1.173
Rape	—	250	—	—	0.064	1.294
Cotton	—	250	—	—	0.077	1.168
Sunflowers	—	430	—	—	0.011	1.368
Soybeans						
1/10 standing	60	—	750	Normal	0.016	1.553
9/10 flat–random	—	250	—	—	0.167	1.170
Ungrazed Rangeland						
Blue grama	300	—	—	—	0.60	1.39
Buffalograss	100	—	—	—	1.40	1.44
Properly Grazed Rangeland						
Big bluestem	150	—	—	—	0.22	1.34
Blue grama	50	—	—	—	1.60	1.08
Buffalograss	50	—	—	—	3.08	1.18
Little bluestem	100	—	—	—	0.19	1.37
Switchgrass	150	—	—	—	0.47	1.40
Western wheatgrass	100	—	—	—	1.54	1.17
Overgrazed Rangeland						
Big bluestem	25	—	—	—	4.12	0.92
Blue grama	25	—	—	—	3.06	1.14
Buffalograss	15	—	—	—	2.45	1.40
Little bluestem	30	—	—	—	0.52	1.26
Switchgrass	25	—	—	—	1.80	1.12
Western wheatgrass	25	—	—	—	3.93	1.07

Source: Skidmore (1994).

(7) Calculate the effects of vegetation through two factors based on the vegetative cover factor V.

$$a = e^{(-7.59V - 4.47\times10^{-1}V^2 + 2.95\times10^{-4}V^3)} \quad \textbf{20.14}$$

$$b = 1.0 + 8.93 \times 10^{-2}V + 8.51 \times 10^{-3}V^2 - 1.5 \times 10^{-5}V^3 \quad \textbf{20.15}$$

(8) Incorporate the vegetation factors into the erosion estimate.

$$E_5(\text{Mg ha}^{-1}\text{y}^{-1}) = a\,E_4^b \quad \textbf{20.16}$$

When working with these equations, it will become apparent that for high values of field length the effect of field length becomes minimal. This is because these equations predict erosion mainly by saltation and creep processes, and these processes are limited by the transport capacity of the site. To include suspended material, the reader should use the WEPS model, which allows for increased erosion at longer lengths as detached material becomes suspended at greater heights in the atmosphere (ARS, 2003) and is not limited by surface wind alone to detach and transport sediment.

Example 20.1

A field is 800 m long in central Kansas (from Figure 20-1 $C = 0.8$). From a sieve analysis, it is determined that the soil has 25 percent nonerodible clods (>0.84 mm). Several knolls with 3 percent slopes are in the field. A crop of forage sorghum was grown in 750-mm rows, and 500 kg/ha of 150-mm-tall stubble remains standing in the field. The ridge roughness is 100 mm. What is the estimated annual soil loss due to wind erosion on this field?

Solution. The soil loss is calculated following the steps described above.

(1) Calculate I from the clod content (Equation 20.2).

$$I = 525 \times e^{(-0.04\times2.5)} = 193 \text{ Mg ha}^{-1}\text{ y}^{-1}$$

(2) Calculate the effect of the 3 percent knolls from Table 20-2.

$$I = 193 \times 1.3 = 251 \text{ Mg ha}^{-1}\text{ y}^{-1}$$

(3) Calculate the roughness factor for a ridge roughness of 100 mm (Equation 20.4).

$$K = 0.34 + \frac{12}{100 + 18} + 6.2 \times 10^{-6} \times 100^2 = 0.5$$

(4) Calculate E_2 from Equation 20.9.

$$E_2 = 251 \times 0.5 = 126 \text{ Mg ha}^{-1}\text{ y}^{-1}$$

(5) Calculate E_3 from Equation 20.10 with $C = 0.80$.

$$E_3 = 193 \times 0.5 \times 0.8 = 100 \text{ Mg ha}^{-1}\text{ y}^{-1}$$

(6) Calculate L_o from Equation 20.11.

$$L_o = 1.56 \times 10^6(126)^{-1.26}e^{(-.00156\times 126)} = 2877 \text{ m}$$

(7) Calculate WF from Equation 20.12.

$$WF = 126 \times \left(1.0 - 1.2\left[\frac{800}{2877}\right]^{-0.383}\right)e^{800/2877} = 102$$

(8) Calculate E_4 from Equation 20.13.

$$E_4 = (102^{0.348} + 100^{0.348} - 126^{0.348})^{2.87} = 79.4 \text{ Mg ha}^{-1}\text{ y}^{-1}$$

(9) Calculate the small-grain equivalent from Equation 20.5. From Table 20-3, $a = 0.353$ and $b = 1.124$.

$$SG = 0.353 \times 500^{1.124} = 381 \text{ kg/ha}$$

(10) Calculate the vegetative cover factor from Equation 20.7.

$$V = 2.533 \times 10^{-4}(381)^{1.363} = 0.836$$

(11) Calculate vegetation factors from Equation 20.14 and 20.15.

$$a = e^{(-7.59\times .836 - 4.47\times 10^{-1}\times 0.836^2 + 2.95\times 10^{-4}\times 0.836^3)} = 0.513$$

$b = 1.0 + 8.93 \times 10^{-2} \times 0.836 + 8.51 \times 10^{-3} \times 0.836^2 - 1.5 \times 10^{-5} \times 0.836^3 = 1.08$

(12) Incorporate vegetation factors into erosion prediction with Equation 20.16.

$$E_5 = 0.513 \times 79.4^{1.08} = 57.9 \text{ Mg ha}^{-1}\text{ y}^{-1}$$

Thus, the estimated soil loss is 57.9 Mg ha^{-1} y^{-1}.

If the above loss is unacceptable, the loss can be reduced by reducing the length of the field with respect to the prevailing wind direction, by increasing the residue cover, or possibly by increasing the clod content of the soil surface through tillage.

Control Practices

20.6 Cultivated Crops

In general, close-growing crops are more effective for erosion control than are row crops. The effectiveness of crops is dependent on stage of growth, density of cover, row direction, width of rows, kind of crop, and climatic conditions. Pasture or meadow may accumulate soil from neighboring cultivated fields if there are good grazing management practices. Good management grazing practices such as rotation grazing are important to minimize erosion, because sparse rangelands are also susceptible to wind erosion, as can be seen from the vegetation coefficients for rangeland in Table 20-3.

Tillage and planting normal to the prevailing winds will reduce the risk of wind erosion. A crop rotation that will maintain soil structure and conserve water should be followed. Crops adapted to soil and climatic conditions and providing as much protection against erosion as practical are recommended. For instance, in the Great Plains region forage sorghum and Sudan grass are resistant to drought and are effective in preventing wind erosion. In more humid regions, stubble mulch farming and cover crops between row crops can reduce wind erosion between cropping seasons. In some dry regions, emergency crops with low water requirements may be established on summer fallow land before seasons of high-intensity winds. In muck soils where vegetable crops are grown, miniature windbreaks consisting of rows of small grain are sometimes planted.

Sand dunes can be stabilized by first planting drought-resistant grasses to provide protection until appropriate shrubs or trees can be established. The vegetation should have the ability to grow in the open on sandy soil, be wind resistant, and have a long life. Vegetation should also provide a dense cover during critical seasons, provide as uniform an obstruction to the wind as possible, reduce the surface wind velocity, and form an abundance of crop residue.

20.7 Windbreaks and Shelterbelts

Windbreaks are generally associated with mechanical or vegetative barriers for buildings, gardens, orchards, and feed lots (Figure 20-3). A shelterbelt is a longer barrier than a windbreak, usually consisting of shrubs and trees, and is intended for the conservation of soil and water and for the protection of field crops. About 450 000 km of windbreaks and shelterbelts have been planted in the United States since the middle of the 1800s. Windbreaks and shelterbelts are valuable for wind erosion control, reduce heating and cooling costs, increase livestock gains, reduce evaporation, reduce crop damage from hot winds, catch snow during the winter months, provide better fruiting in orchards, and make spraying of orchards for insect control more effective. Windbreaks may also improve offsite water quality, or provide wildlife habitat.

The relative wind velocity near a windbreak is shown in Figure 20-4. Depending on the effectiveness of the windbreak, reduction in wind velocity can occur for a distance up to 20 times the height of the windbreak. Shelterbelts should be moderately dense from ground level to treetops if they are to be effective in filtering the wind and lifting it from the surface. Long shelterbelts are more effective than short ones. An opening or break in an otherwise continuous belt will reduce the effectiveness. Roads through shelterbelts should be avoided, and, when essential, they should cross the belt at an angle or should be curved. In establishing the direction of shelterbelts, records of wind direction and velocity, particularly during vulnerable seasons, should be considered, and the barrier should be oriented as nearly as possible at right angles to the prevailing direction of winds. Such information for many locations in the United States is available from NRCS (2003).

(a)

(b)

Figure 20–3

Typical shelterbelts. (a) A windbreak of evergreens, deciduous trees, and shrubs protects a Kansas farmstead (from http://photogallery.nrcs.usda.gov). (b) Evergreen and broadleaf tree windbreak (From http://www.forestry.iastate.edu/ext/).

Generally, the distance of full protection from a windbreak is

$$d = 17h\left(\frac{v_m}{v}\right)\cos(\theta) \qquad \textbf{20.17}$$

where d = distance of full protection (L),

h = height of the barrier in the same units as d (L),

v_m = minimum wind velocity at 15-m height required to move the most erodible soil fraction (L/T),

v = actual wind velocity at 15-m height (L/T),

θ = the angle of deviation of prevailing wind direction from the perpendicular to the windbreak.

Chepil (1959) reported that v_m for a smooth bare surface after erosion has begun but before wetting by rainfall and subsequent surface crusting, was about 9.6 m/s.

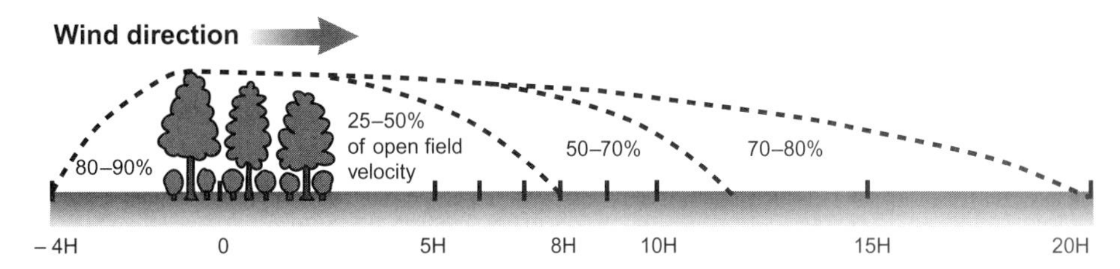

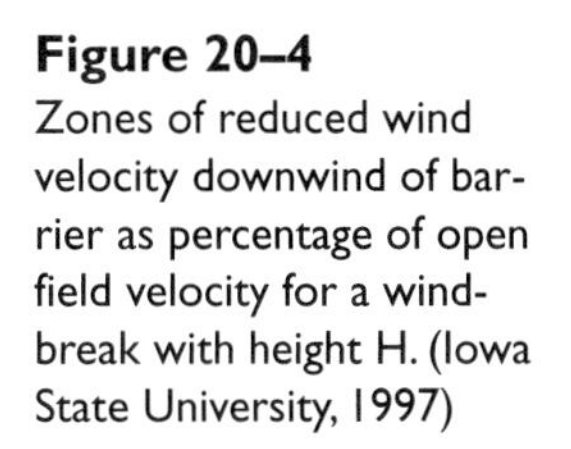

Figure 20–4
Zones of reduced wind velocity downwind of barrier as percentage of open field velocity for a windbreak with height H. (Iowa State University, 1997)

Equation 20.16 is valid only for wind velocities below 18 m/s. It may also be adapted for estimating the width of strips by using the crop height in the adjoining strip in the equation.

In shelterbelts, a tight row of shrubs on the windward side is desirable. When combined with conifers and low, medium, and tall deciduous trees, the shelterbelt provides a compact and rather dense barrier. Such an extensive shelterbelt may not always be required. Single-row belts are preferred in many areas because fewer trees and less land are needed, and the shelterbelt is easier to cultivate and maintain. Local recommendations should be followed for varieties, spacings, and other practices.

20.8 Strip Cropping

Strip cropping consists of growing alternate strips of clean-cultivated and close-growing crops in the same field (Figure 20-5). Field strip cropping is laid out parallel to a field boundary or other guideline. In some of the plains states, strips of fallow and grain crops are alternated. The chief advantages of strip cropping are (1) physical protection against blowing, provided by the vegetation; (2) soil erosion reduced within and between the vegetated strips; (3) greater conservation of water, particularly from snowfall; and (4) the possibility of earlier harvest. The chief disadvantages are machine problems in farming narrow strips and greater number of edges to protect in case of insect infestation.

The strips should be of sufficient width to be convenient to farm, yet not so wide as to permit excessive erosion. Strip width depends on the wind speed and direction,

Figure 20–5
Parallel strip cropping to reduce wind erosion. (ARS, 2003)

row direction, standing crop and stubble height, and erodibility of the soil. The strip width can be estimated with Equation 20.16, or is sometimes set at about 10 times the crop or barrier height. Widths may be adjusted to match the width of tillage, planting, or harvesting equipment.

20.9 Tillage

The objective of tillage for wind erosion control is to produce a rough, cloddy surface with some plant residue exposed on the surface. To obtain maximum roughness, the land should be cultivated as soon after a rain as possible to obtain large aggregates.

Ridges from tillage should be normal to the direction of prevailing wind for erosion control. The decrease in wind velocity and change in direction between the ridges cause soil deposition (Figure 20-6). In some areas, ridge tillage systems have completely eliminated wind erosion (Walker & Peterson, 1985).

Tillage is sometimes used as an emergency control measure. Soil blowing usually starts in a small area where the soil is less stable or is more exposed than in other parts of the field. If the entire field starts to blow, it is sometimes recommended that the surface be made rough and cloddy as soon as practicable. This tillage should begin at the windward side of the field and continue by making widely spaced trips across the field. When the field has been stripped, the areas between the strips may then be cultivated.

Crop residues on the surface are an effective means of erosion control, especially when combined with a rough soil surface. This practice is usually called stubble mulch tillage. Crop residues reduce wind velocity and trap eroding soil. Short stubble is generally less effective than long stubble. A mixture of straw and stubble on the surface provides more protection against erosion than equivalent amounts of straw or stubble alone. The higher the wind velocity, the greater the quantity of crop residue required. The effectiveness of surface residue on reducing wind erosion is shown by the importance of the vegetation cover terms in soil erosion prediction, as was demonstrated in the final step of Example 20.1.

20.10 Mechanical Methods

Mechanical barriers such as windbreaks are of limited importance for field crops, but they are frequently employed for the protection of farmsteads, areas of high-value vegetable production, and beach restoration. Mechanical control methods include slat or brush fences, board walls, and vertical fabrics, as well as the surface protection, such as brush matting, rock, and gravel. These techniques are sometimes used for the protection of vegetable crops in organic soils as well.

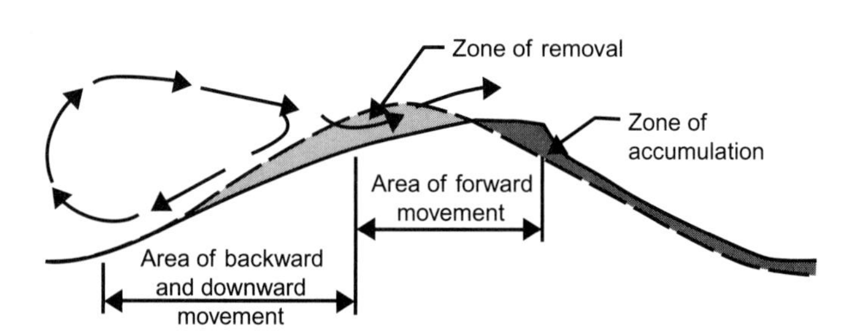

Figure 20–6
Soil movement across ridges as impacted by wind turbulence. (NRCS, 2002)

Terraces have some effect on wind erosion. In the Texas Panhandle, terraces lost less soil than the interterrace area and, in some instances, gained soil. Most of the soil that was lost from the interterrace area was collected in the terraces. Terraces can increase trapping of eroded sediment, and conserve water (see next section).

20.11 Managing Soil Water

The conservation of soil water, particularly in arid and semiarid regions, is important for wind erosion control and for crop production. Water conservation methods fall into three categories: increasing infiltration, reducing evaporation, and preventing unnecessary plant growth. Water conservation management practices include level terracing, contouring, mulching, and selection of suitable crops. Tillage of fallow lands immediately following precipitation will reduce weed growth to conserve water and will also tend to form clods, reducing water movement to the surface, and increasing the resistance of the surface to detachment.

Level or conservation bench terraces are often used to retain water to reduce wind erosion. They are suitable on slopes under about 4 percent so that the water can be spread over a relatively large area (Chapter 8). Such practices as contouring, strip cropping, and mulching are effective in increasing the total infiltration and thereby the total soil water available to crops. Field strip cropping generally does not conserve as much water as contour strip cropping, but it is somewhat more effective in reducing surface wind velocities.

Organic soils do not blow appreciably if the soil is moist. If the subsoil is wet, rolling the soil with a heavy roller will increase capillary movement and moisten the surface layer. Controlled drainage where the water level is maintained at a specified depth may also reduce blowing. In irrigated areas overhead sprinklers can be used to increase surface water contents during critical times of the year.

20.12 Conditioning Topsoil

Since wind erosion is influenced to a large extent by the size and apparent density of aggregates. An effective method of conditioning the soil against wind erosion is to use practices that produce nonerosive aggregates (greater than 1 mm in diameter). During the periods of the year when the soil is bare or has a limited amount of crop residue, control of erosion may depend on the degree and stability of soil aggregation.

Tillage may or may not be beneficial to soil structure, depending on the water content of the soil, type of tillage, and number of operations. For optimum resistance to wind erosion in semiarid regions, it is desirable to perform primary tillage as soon as practical after a rain. The number of operations should be kept to a minimum, because tillage has a tendency to reduce soil aggregate size. Secondary tillage for seedbed preparation should be delayed as long as practical.

Soil structure is affected by the climatic influences such as rainfall and temperature distributions. Freezing and thawing generally have a beneficial effect in improving soil structure where sufficient water is present; however, in dry regions the soil is more susceptible to erosion because of rapid breakdown of the clods into smaller aggregates. Resistance to erosion can also be increased with applications of chemicals such as polyacrylamide to the surface (Armbrust, 1997).

Internet Sites

Environmental Protection Agency Office of Air and Radiation
http://www.epa.gov/oar/

USDA Agricutural Research Service Wind Erosion Research Unit
http://www.weru.ksu.edu. Key search term: wind erosion

USDA NRCS, New Mexico Wind Erosion Equation Spreadsheet
http://www.nm.nrcs.usda.gov/technical/tech-notes/agro.html

Links to windbreak design and management resources
http://ilvirtualforest.nres.uiuc.edu. Key search term: windbreak

References

Armbrust, D. V. (1997). Effectiveness of polyacrylamide (PAM) for wind erosion control. *Proceedings of the International Symposium on Wind Erosion,* 3–5 June 1997. Online at http://www.weru.ksu.edu [Key search term: wind erosion] Accessed December 2003.

Chepil, W. S. (1959). Wind erodibility of farm fields. *Journal of Soil and Water Conservation, 14*(5), 214–219.

Chepil, W. S., F. H. Siddoway, & D. V. Armbrust. (1962). Climate factor for estimating wind erodibility of farm fields. *Journal of Soil and Water Conservation, 17*(4), 162–165.

Hagen, L. J. (1991). A wind erosion prediction system to meet user needs. *Journal of Soil and Water Conservation, 46*(2), 106–111.

Iowa State University. (1997). *Farmstead Windbreaks: Planning.* PM-1716. Online at http://www.exnet.iastate.edu/Publications/PM1716.pdf. Accessed July 2003.

Natural Resource Conservation Service (NRCS). (2002). *National Agronomy Manual.* Online at http://www.nrcs.usda.gov/technical/ECS/agronomy/. Accessed September 2004.

———. (2003). *Field Office Technical Guide: D. Erosion Prediction.* Online at http://www.nm.nrcs.usda.gov/technical/. Accessed December 2003.

Saxton, K., D. Chandler, L. Stetler, B. Lamb, C. Claiborn, & B.-H. Lee. (2000). Wind erosion and fugitive dust fluxes on agricultural lands in the Pacific Northwest. *Transactions of the ASAE, 43*(3), 623–630.

Skidmore, E. L. (1994). Wind erosion. In R. Lal, ed., *Soil Erosion Research Methods,* 2nd ed., pp. 264–293. Ankeny, IA: Soil and Water Conservation Society.

U.S. Department of Agriculture, Agricultural Research Service (ARS). (2003). *Wind Erosion Research Unit (WERU).* Online at http://www.weru.ksu.edu. Accessed September 2004.

U.S. Environmental Protection Agency (EPA). (2003). *Office of Air and Radiation.* Online at http://www.epa.gov/oar/. Accessed December 2003.

Walker, R., & D. Peterson. (1985). *A Plan for the Land Erosion-Control Alternatives.* Online at http://hermes.ecn.purdue.edu [Key search term: erosion control]. Accessed December 2003.

Woodruff, N. P., & F. H. Siddoway. (1965). A wind erosion equation. *Journal of the Soil Science Society of America, Proceedings, 29,* 602–608.

Problems

20.1 During a wind erosion event, the velocity of the wind is 10 m/s, the estimated erosion rate is 0.5 Mg/ha/h from a 40-ha field (200 m long and 2000 m wide), and the soil particulate matter less than 10 μm represents 1 percent of the eroded sediment. This material is carried in suspension within 100 m of the surface from the field to a nearby town. What is the PM-10 during this event?

20.2 During a wind erosion event, if the wind velocity is doubled and all other factors remain the same, what will be the percent increase in soil erosion?

20.3 Calculate the estimated annual erosion for a 500-m-long field located in north central Kansas that has a soil with 30 percent nonerodible clods (>0.84 mm), several knolls with 4 percent slopes in the field, a crop of flat wheat residue estimated to be 300 kg/ha, and a ridge roughness of 50 mm.

20.4 Calculate the percentage change in predicted erosion rates if a field that initially has a flat wheat residue cover of 500 kg/ha and a roughness of 30 mm is cultivated to give a cover of 100 kg/ha, but a ridge roughness of 125 mm. Comment on the effect of cultivation on erosion.

20.5 Determine the spacing between windbreaks that are 15 m high if the 5-year return period wind velocity at a 15-m height is 16 m/s and the wind direction deviates 10 degrees from the perpendicular to the field strip. Assume a smooth, bare soil surface and a fully protected field.

20.6 Determine the full protection strip width for field strip cropping if the crop in the adjacent strip is wheat 0.9 m tall and the wind velocity at 15-m height is 9 m/s at 90 degrees with the field strip.

20.7 Obtain a set of local weather records, and determine your local climate factor from methods presented in NRCS (2002).

20.8 Using the information in Example 20.1, vary the length of the field from 500 to 5000 m, and graph the erosion rate versus the field length. Comment on the shape of the curve.

APPENDIX

A Conversion Constants

TABLE A–1 English–SI Length Conversion Constants

Length	*in.*	*ft*	*yd*	*mi*	*m*	*km*
1 in.	1	0.083	0.027	-	0.0254	-
1 ft	12	1	0.333	-	0.305	
1 yd	36	3	1	-	0.914	-
1 mi (statute)	-	5280	1760	1	1609	1.609
1 m	39.37	3.281	1.094	-	1	0.001
1 km	-	3281	1094	0.621	1000	1

TABLE A–2 English–SI Area Conversion Constants

Area	*in.*^{2}	*ft*^{2}	*yd*^{2}	*ac*	*m*^{2}	*ha*
1 in.^{2}	1	0.007	-	-	0.00064	-
1 ft^2	144	1	0.1111	-	0.0929	
1 yd^2	1296	9	1	-	0.836	-
1 ac	-	43 560	4840	1	4047	0.405
1 m^2	1550	10.76	1.20	-	1	0.0001
1 ha	-	107 639	11 960	2.47	10 000	1

TABLE A–3 English–SI Volume Conversion Constants

Volume	*in.*^{3}	*ft*^{3}	*yd*^{3}	*U.S. gal*	*L*	*m*^{3}	*ac-ft*	*ha-m*
1 in.^{3}	1	-	-	0.0043	0.0164	-	-	-
1 ft^3	1728	1	0.037	7.481	28.32	0.0283	-	-
1 yd^3	-	27	1	201.97	-	0.765	-	-
1 U.S. gal	231	0.134	-	1	3.785	0.0038	-	-
1 L	61.02	0.0353	-	0.2642	1	0.001	-	-
1 m^3	61 024	35.31	1.308	264.2	1000	1	0.00081	0.0001
1 ac-ft	-	43 560	-	325 872	-	1233.4	1	0.1233
1 ha-m	-	353 146	-	-	1×10^7	10 000	8.108	1

TABLE A–4 Miscellaneous Conversion Constants

Pressure and Force

1 in. Hg = 3386.4 Pa	1 Pa = 1 N/m^2
= 0.491 psi	1 atm = 1013 mb
1 mm Hg = 133.3 Pa	= 101.3 kPa
1 mm water = 9.8 Pa	= 760 mm. Hg
1 psi = 51.7 mm. Hg	= 33.93 ft water
1 psi = 6.895 kPa	1 mb = 100 Pa
1 lb f = 4.45 N	1 mb = 10.2 mm water
1 lb m = 0.454 kg	1 ft water = 2.985 kPa
1 lb f/ft2 = 47.88 Pa	1 kPa = 0.335 ft water

Power and Energy

1 Btu = 1055 J	1 kW = 1.341 hp
1 cal = 4.19 J	1 kWh = 3.6 × 10^6 J
1 hp = 550 ft-lb/s	
= 746 W	

Volume, Weight, and Mass

1 U.S. gal = 8.34 lb	1 oz = 28.35 g
1 Imp. gal = 10.02 lb	1 lb = 453.6 g
1 ft^3 water = 62.4 lb	1 lb/ft^3 = 16.02 kg/m^3
	1 short ton (t) = 907.18 kg = 2000 lb
	1 metric tonne =1000 kg (Mg)
	= 2205 lb
	1 t/ac = 2.24 Mg/ha (metric tonne)
	1 kg = 2.205 lb mass

Velocity

1 ft/s = 0.305 m/s	1 mph = 1.61 km/h = 0.447 m/s
= 1097 m/h	1 km/h = 0.621 mph
	= 0.278 m/s

APPENDIX B

Manning Roughness Coefficients n

TABLE B–1 Roughness Coefficient *n* for the Manning Formula

Line No.	*Type and Description of Conduits*	*n Value*[a]		
		Minimum	*Design*	*Maximum*
Channels, Lined				
1	Asphaltic concrete, machine placed		0.014	
2	Asphalt, exposed prefabricated		0.015	
3	Concrete	0.012	0.015	0.018
4	Concrete, rubble	0.017		0.030
5	Metal, smooth (flumes)	0.011		0.015
6	Metal, corrugated	0.021	0.024	0.026
7	Plastic	0.012		0.014
8	Shotcrete	0.016		0.017
9	Wood, planed (flumes)	0.010	0.012	0.015
10	Wood, unplaned (flumes)	0.011	0.013	0.015
Channels, Earth				
11	Earth bottom, rubble sides	0.028	0.032	0.035
	Drainage ditches, large, no vegetation			
12	(a) <0.8 m, hydraulic radius	0.040		0.045
13	(b) 0.8–1.2 m, hydraulic radius	0.035		0.040
14	(c) 1.2–1.5 m, hydraulic radius	0.030		0.035
15	(d) >1.5 m, hydraulic radius	0.025		0.030
16	Irrigation border strip, no vegetation		0.04	
17	Irrigation furrow, no vegetation		0.04	
18	Small drainage ditches	0.035	0.040	0.040
19	Stony bed, weeds on bank	0.025	0.035	0.040
20	Straight and uniform	0.017	0.0225	0.025
21	Winding, sluggish	0.0225	0.025	0.030
Channels, Vegetated				
	(grassed waterways) (Chapter 8) Dense, uniform stands of green vegetation more than 250 mm. long			

22	(a) Bermudagrass	0.04		0.20
23	(b) Kudzu	0.07		0.23
24	(c) Lespedeza, common	0.047		0.095
	Dense, uniform stands of green vegetation cut to a length less than 60 mm			
25	(a) Bermudagrass, short	0.034		0.11
26	(b) Kudzu	0.045		0.16
27	(c) Lespedeza	0.023		0.05
28	Irrigation border strip, low density vegetation (wheat drilled parallel to flow)		0.15	
29	Irrigation border, dense vegetation (grass)		0.25	
30	Irrigation furrow with vegetation		0.15	0.25
31	Sorghum, 1-m rows	0.04		0.15
32	Wheat, mature poor	0.08		0.15
Natural Streams				
33	(a) Clean, straight bank, full stage, no rifts or deep pools	0.025		0.033
34	(b) Same as (a) but some weeds and stones	0.030		0.040
35	(c) Winding, some pools and shoals, clean	0.035		0.050
36	(d) Same as (c), lower stages, more ineffective slopes and sections	0.040		0.055
37	(e) Same as (c), some weeds and stones	0.033		0.045
38	(f) Same as (d), stony sections	0.045		0.060
39	(g) Sluggish river reaches, rather weedy or with very deep pools	0.050		0.080
40	(h) Very weedy reaches	0.075		0.150
Pipe				
41	Cast iron, coated or uncoated	0.011	0.013	0.015
42	Clay or concrete drain tile (76–760-mm dia.)	0.011	0.013	0.020
43	Concrete or clay vitrified sewer pipe	0.01	0.014	0.017
44	CPT, Corrugated plastic tubing, 76–203-mm. dia.		0.015	
45	CPT, 254–305-mm dia.		0.017	
46	CPT, >305-mm dia.		0.02	
47	CPT, smooth wall inside		0.009	
48	Metal, corrugated, ring	0.021	0.025	0.026
49	Metal, corrugated, helical	0.013	0.015	
50	PVC and PE		0.01	
52	Steel, riveted and spiral	0.013	0.016	0.017
53	Wood stave	0.010	0.013	
54	Wrought iron, black	0.012		0.015
55	Wrought iron, galvanized	0.013	0.016	0.017

[a]Selected from numerous references.

APPENDIX

C Pipe and Conduit Flow

The area of a segment of a circle, or the cross section of flow of a partially full circular conduit is given by

$$A = r^2 \cos^{-1}\left(\frac{r-d}{r}\right) - (r-d)\sqrt{2rd - d^2} \qquad \text{C.1}$$

where r is the radius of the circular conduit and d is the depth of flow in the conduit.

TABLE C–1 Friction Loss Coefficients for Circular and Square Pipe at Bends

$\frac{R}{D} = \frac{\text{Bend Radius to Pipe Centerline}}{\text{Pipe Diameter}}$	Bend Coefficient, K_b	
	45° Bend	90° Bend
0.5	0.7	1.0
1	0.4	0.6
2	0.3	0.4
5	0.2	0.3

Source: U.S. Soil Conservation Service. (1951). *National Engineering Handbook Hydraulics,* Section 5, (Washington, DC: Author).

TABLE C–2 Head Loss Coefficients for Circular Pipe Flowing Full (SI units)

$$K_c = \frac{1\ 244\ 522 n^2}{d^{4/3}} \quad \text{where } d = \text{diameter (mm)}$$

Pipe Inside Diameter, mm (in.)	*Flow Area, mm²*	Manning roughness Coefficient, n: 0.010	0.013	0.016	0.020	0.025
13 (0.5)	133	4.071	6.881	10.423	16.286	25.447
25 (1)	491	1.702	2.877	4.358	6.810	10.641
51 (2)	2043	0.658	1.112	1.685	2.632	4.113
76 (3)	4536	0.387	0.653	0.990	1.546	2.416
102 (4)	8171	0.261	0.441	0.669	1.045	1.632
127 (5)	12 668	0.195	0.329	0.499	0.780	1.218
152 (6)	18 146	0.153	0.259	0.393	0.614	0.959
203 (8)	32 365	0.104	0.176	0.267	0.417	0.652
254 (10)	50 671	0.0774	0.131	0.198	0.309	0.484
305 (12)	73 062	0.0606	0.102	0.155	0.242	0.379
381 (15)	114 009	0.0451	0.0761	0.115	0.180	0.282
457 (18)	164 030	0.0354	0.0598	0.0905	0.141	0.221
533 (21)	223 123	0.0288	0.0487	0.0737	0.115	0.180
610 (24)	292 247	0.0241	0.0407	0.0616	0.0962	0.150
762 (30)	456 037	0.0179	0.0302	0.0458	0.0715	0.112
914 (36)	656 119	0.0140	0.0237	0.0359	0.0561	0.0877
1219 (48)	1 167 071	0.00956	0.0162	0.0245	0.0382	0.0597
1524 (60)	1 824 147	0.00710	0.0120	0.0182	0.0284	0.0444

Note: K_c (English units) = K_c (SI units)/3.28.

TABLE C–3 Head Loss Coefficients for Square Conduits Flowing Full

$$K_c = \frac{19.6 n^2}{R^{4/3}} \quad \text{where } R = \text{hydraulic radius (m)}$$

Conduit Size m × m (ft × ft)	*Flow Area (m²)*	Manning Coefficient of Roughness, n: *0.012*	*0.014*	*0.016*	*0.020*
0.61 × 0.61 (2 × 2)	0.372	0.0347	0.0472	0.0616	0.0963
0.91 × 0.91 (3 × 3)	0.828	0.0203	0.0277	0.0361	0.0564
1.22 × 1.22 (4 × 4)	1.488	0.0138	0.0187	0.0245	0.0382
1.52 × 1.52 (5 × 5)	2.310	0.0103	0.0140	0.0182	0.0285
1.83 × 1.83 (6 × 6)	3.349	0.00800	0.0109	0.0142	0.0222
2.13 × 2.13 (7 × 7)	4.537	0.00653	0.00889	0.0116	0.0181
2.44 × 2.44 (8 × 8)	5.954	0.00545	0.00742	0.00970	0.0152
2.74 × 2.74 (9 × 9)	7.508	0.00467	0.00636	0.00831	0.0130
3.05 × 3.05 (10 × 10)	9.303	0.00405	0.00551	0.00720	0.0113

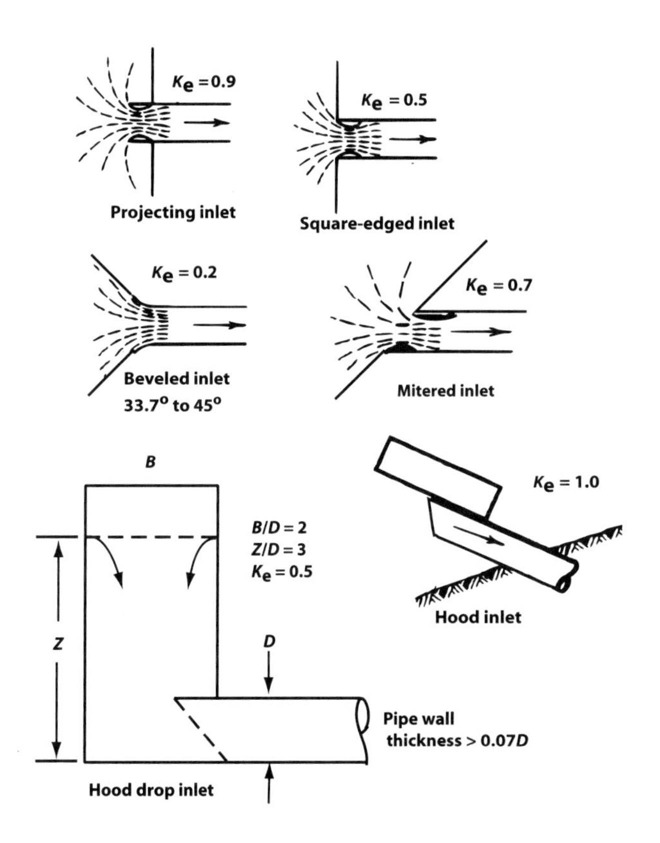

Figure C–1

Entrance loss coefficient for pipe conduits. (U.S. Soil Conservation Service, (1951), *National Engineering Handbook Hydraulics,* Sect. 5, Washington, DC: U.S. Government Printing Office. F.W. Blaisdell & C.A. Donnelly (1956), Hood inlet for closed conduit spillways, Agricultural Engineering, 37, 670–672; and K. Yalamanchile & F.W. Blaisdell (1975), *Hydraulics of Closed Conduit Spillways, and the Hood Inlet,* ARS-NC-23, Washington, DC: USDA. See Chapter 9)

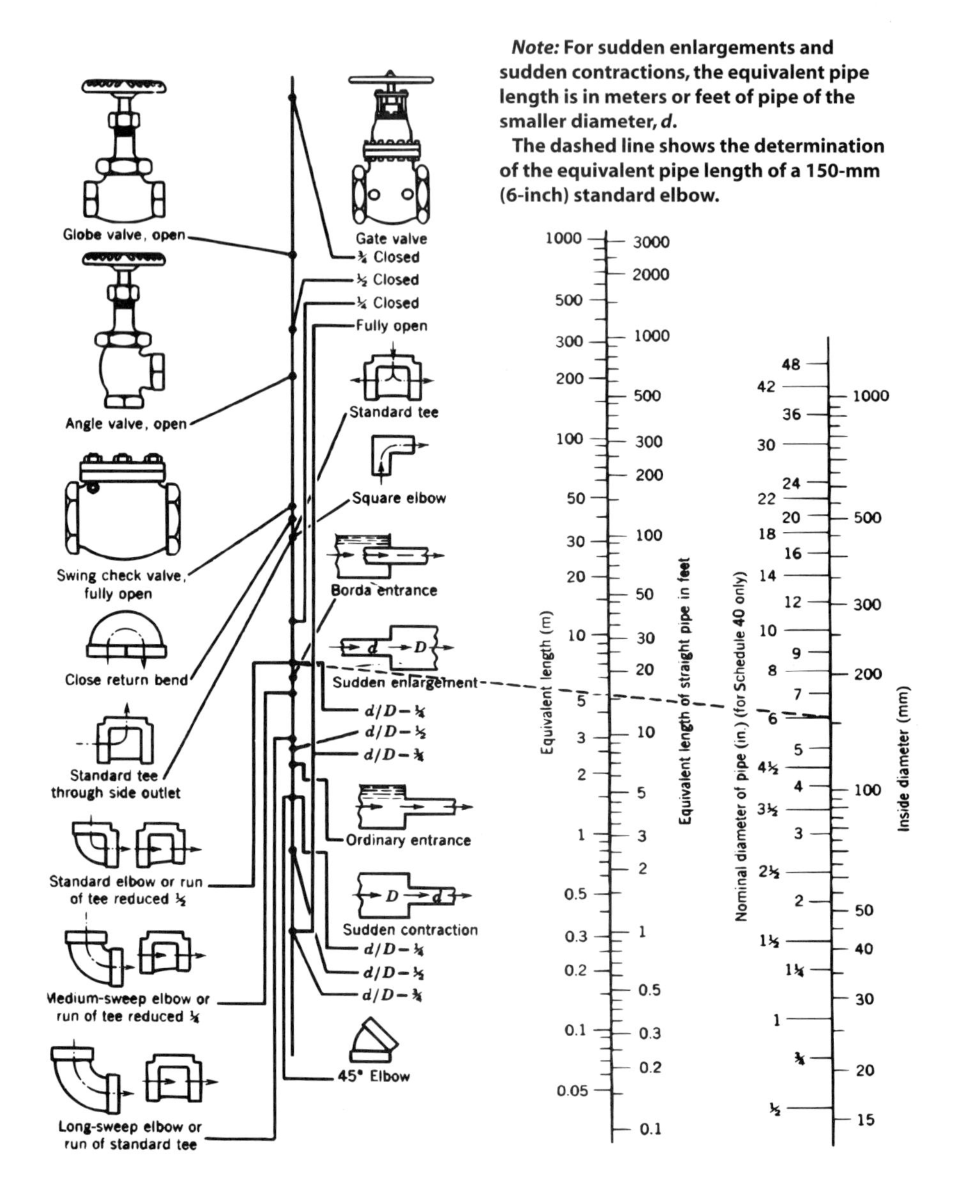

Figure C–2

Equivalent lengths of straight pipe for valves and fittings. (Revised from J. W. Wolfe (1950), *Friction Lossses for pipe and Fittings*, Oregon Agricultural Experimental Station Bulletin No. 181)

APPENDIX D

Pipe and Drain Tile Specifications

Pipe Specifications

The following section contains dimensions for pipe commonly used in soil and water conservation engineering applications.

D.1 PVC Schedule 40 IPS Plastic Pipe and Schedule 40 Standard Steel Pipe

Because Schedule 40 PVC and Schedule 40 Standard steel pipe have the same sizes the dimensions are included in one table (Table D-1). The outside diameter is fixed so that fittings are interchangeable. The wall thickness increases with pipe diameter to maintain the pressure rating.

D.2 Class 160 PVC Pipe

Class 160 PVC pipe has a pressure rating of 160 psi or 1100 kPa and is often used in low-pressure irrigation systems because it is less costly. The dimensions are given in

TABLE D–1 Dimensions of Schedule 40 PVC and Steel Pipe

Nominal size (mm)	*O.D. (mm)*	*I.D. (mm)*	*Wall thickness (mm)*	*O.D. (in.)*	*I.D. (in.)*	*Wall thickness (in.)*	*Nominal size (in.)*
15	21.336	15.799	2.769	0.840	0.622	0.109	1/2
20	26.670	20.930	2.870	1.050	0.824	0.113	3/4
25	33.401	26.645	3.378	1.315	1.049	0.133	1
35	42.164	35.052	3.556	1.660	1.380	0.140	1 1/4
40	48.260	40.894	3.683	1.900	1.610	0.145	1 1/2
50	60.325	52.502	3.912	2.375	2.067	0.154	2
60	73.025	62.713	5.156	2.875	2.469	0.203	2 1/2
75	88.900	77.927	5.486	3.500	3.068	0.216	3
100	114.300	102.260	6.020	4.500	4.026	0.237	4
150	168.275	154.051	7.112	6.625	6.065	0.280	6

Table D-2. The wall is thinner for Class 160 pipe that for Schedule 40 pipe because it has a lower pressure rating.

D.3 Aluminum Irrigation Pipe

Aluminum irrigation pipe was standardized based on the outside diameter in inches (Table D-3). For higher pressure or strength requirements—for example, side-roll laterals—the wall thickness is increased, which decreases the inside diameter.

For side-roll irrigation systems where the pipe acts as the axle, the wall thickness should be 1.8 mm or greater. Pipe diameters of 100 or 125 mm are commonly used for side-roll systems.

D.4 Polyethylene Plastic Pipe, PE, for Microirrigation Laterals

Polyethylene plastic pipe for microirrigation laterals is available in three sizes (Table D-4). Not all manufacturers make all sizes. These pipes contain additives to reduce cracking and UV damage. Couplings may be barbed inserts with stainless steel clamps placed over the barbed portion of the fitting to hold it in the pipe,

TABLE D-2 Dimensions of Class 160 PVC Pipe

Nominal size (mm)	*O.D.* (mm)	*I.D.* (mm)	*Wall thickness* (mm)	*O.D.* (in.)	*I.D.* (in.)	*Wall thickness* (in.)	*Nominal size* (in.)
25	33.401	30.353	1.524	1.315	1.195	0.060	1
35	42.164	38.913	1.626	1.660	1.532	0.064	1 1/4
40	48.260	44.552	1.854	1.900	1.754	0.073	1 1/2
50	60.325	55.702	2.311	2.375	2.193	0.091	2
60	73.025	67.437	2.794	2.875	2.655	0.110	2 1/2
75	88.900	82.042	3.429	3.500	3.230	0.135	3
100	114.300	105.512	4.394	4.500	4.154	0.173	4
150	168.275	155.321	6.477	6.625	6.115	0.255	6

TABLE D-3 Dimensions of Aluminum Irrigation Pipe.

Nominal size (mm)	*O.D.* (mm)	*I.D.* (mm)	*Wall thickness* (mm)	*Pipe use*[a]	*O.D.* (in.)	*I.D.* (in.)	*Wall thickness* (in.)	*Nominal size* (in.)
75	76.2	73.660	1.270	L	3.0	2.900	0.050	3
100	101.6	99.060	1.270	L, M	4.0	3.900	0.050	4
100	101.6	97.942	1.829	S	4.0	3.856	0.072	4
125	127.0	124.409	1.295	M	5.0	4.898	0.051	5
125	127.0	123.038	1.981	S	5.0	4.844	0.078	5
150	152.4	149.454	1.473	M	6.0	5.884	0.058	6
200	203.2	199.949	1.626	M	8.0	7.872	0.064	8
250	254.0	250.749	1.626	M	10.0	9.872	0.064	10
300	304.8	301.549	1.626	M	12.0	11.872	0.064	12

[a] L = lateral line; M = main line; S = side-roll.

TABLE D–4 Dimensions of Polyethylene Pipe for Microirrigation Laterals

Nominal size (mm)	O.D. (mm)	I.D. (mm)	O.D. (in.)	I.D. (in.)
16	16.26	13.97	0.64	0.55
18	18.03	15.75	0.71	0.62
20	20.32	18.03	0.80	0.71

Note: These dimensions may vary slightly depending on the manufacturer.

compression fittings that fit over the outside of the pipe, or locking fittings that have an insert and a threaded lock portion over the outside of the pipe. The lock portion clamps the pipe to the fitting. The design engineer should determine what sizes and fittings are locally available.

Standard polyethylene pipe is available having inside diameters identical with Schedule 40 IPS dimensions. High-density polyethylene HDPE is sometimes used for irrigation mains or submains. HDPE has special fittings or can be fused at connections.

Drain Tile Specifications

The following sections contain dimensions and strength information on tile and piping used for drainage applications.

D.5 Clay Drain Tile

The test requirements given in Tables D-5 and D-6 are condensed from ASTM C4-62, *Tentative Specifications for Clay Drain Tile*, but the most current standard should be checked for possible changes.

Drain tile subject to these specifications may be made from clay, shale, fire clay, or mixtures thereof, and burned. The quality of tile selected should be such that the strength exceeds the soil load by a suitable margin. Where tile are subjected to extreme freezing and thawing, extra-quality or heavy-duty tile are recommended.

Size and Minimum Lengths. The nominal sizes of clay drain tile shall be designated by their inside diameter. Tile less than 305 mm in diameter shall be not less than 0.3 m in length; 305- to 760-mm tile, not less than their diameter; and tile larger than 760 mm, not less than 0.76 m in length.

Other Physical Properties. Some of the general physical requirements for the three classes of clay drain tile are given in Table D-6. Drain tile, while dry, shall give a clear ring when stood on end and tapped with a light hammer. They shall also be reasonably straight and smooth on the inside. Drain tile shall be free from cracks and checks extending into the tile that would decrease its strength appreciably. They shall be neither chipped nor broken so as to decrease their strength materially or to admit soil into the drain.

D.6 Concrete Drain Tile

The specifications given in Table D-7 are condensed from ASTM C412-83. Standard and extra-quality concrete tile are intended for ordinary soils; special-quality tile are

TABLE D–5 Physical Test Requirements for Clay Drain Tile

	Standard		*Extra-Quality*		*Heavy-Duty*	
Internal Diameter (in.)	*Average Minimum Strength[a] (lb/ft)*	*Average Maximum Absorption[b] (%)*	*Average Minimum Strength[a] (lb/ft)*	*Average Maximum Absorption[b] (%)*	*Average Minimum Strength[a] (lb/ft)*	*Average Maximum Absorption[b] (%)*
4,5,6	800	13	1100	11	1400	11
8	800	13	1100	11	1500	11
10	800	13	1100	11	1550	11
12	800	13	1100	11	1700	11
15	870	13	1150	11	1980	11
18			1300	11	2340	11
21			1450	11	2680	11
24			1600	11	3000	11

[a]Average of five tiles using the three-edge bearing method.

[b]Average of five tiles using the 5-h boiling test.

Note: Tile diameters 14, 16, 27, and 30 are omitted for brevity. Diameter in mm = diameter in in. × 25.4; N/m = lb/ft × 14.6.

TABLE D–6 Distinctive General Physical Properties of Clay Drain Tile

Physical Properties Specified	*Standard*	*Extra-Quality and Heavy-Duty*
Number of freezings and thawings (reversals)	36	48
Permissible variation of average diameter below specified diameter (%)	3	3
Permissible variation between maximum and minimum diameters of same tile (% of thickness of wall)	75	65
Permissible variation of average length below specified length (%)	3	3
Permissible variation from straightness (% of length)	3	3
Permissible thickness of exterior blisters, lumps, and flakes that do not weaken tile and are few in number (% of thickness of wall)	20	15
Permissible diameters of above blisters, lumps, and flakes (% of inside diameter)	15	10
General inspection	Rigid	Very rigid

for soils or drainage waters that are markedly acid (pH of 6.0 or lower) or that contain unusual quantities of soil sulfates, chiefly sodium and magnesium, singly or in combination (assumed to be 3000 ppm or more).

Special-quality tile should have the same wall thickness and strength as extra-quality tile. Average maximum absorption is 8 percent, and closer tolerances are required for wall thicknesses than for extra-quality tile. The 10-min, room-temperature maximum soaking absorption shall be 3 percent for individual tile. The hydrostatic pressure test may be made in lieu of the above 10-min test. For sulfate exposures, sulfate-resistant cement shall be specified.

TABLE D-7 Physical Test Requirements for Concrete Drain Tile

Internal Diameter (in.)	*Average Minimum Strength[a] (lb/ft)*	*Average Maximum Absorption[b] (%)*	*Wall Thickness[c] (in.)*	*Average Minimum Strength[a] (lb/ft)*	*Average Maximum Absorption[b] (%)*
4	800	10	1/2	1100	9
5	800	10	9/16	1100	9
6	800	10	5/8	1100	9
8	800	10	3/4	1100	9
10	800	10	7/8	1100	9
12	800	10	1	1100	9
15			1 1/4	1100	9
18			1 1/2	1200	9
21			1 3/4	1400	9
24			2	1600	9

[a]Average of five tiles using the three-edge bearing method.

[b]Average of five tiles using the 5-h boiling test.

[c]Minimum diameters shall not be less than the nominal diameters by more than 1/4 in. for 4- and 5-in. tile, 3/8 in. for 6- and 8-in. tile, 1/2 in. for 10- to 14-in. tile, 5/8 in. for 15- to 18-in. tile, and 3/4 in. for 20- to 24-in. tile. No wall thickness is specified for standard quality.

Note: Tile diameters 14, 16, and 20 are omitted for brevity. Diameter in mm = diameter in in. × 25.4; N/m = lb/ft × 14.6.

Size and Minimum Lengths. Concrete tile less than 305 mm in diameter shall not be less than 0.3 m in length, and 305- to 610-mm diameter tile shall have nominal lengths not less than their diameter.

D.7 Corrugated Plastic Tubing

Specifications for tubing and fittings are given in ASTM F405, F667, and D2412. Pipe stiffness is the slope of the load–deflection curve (kPa) at a specified percentage deflection based on the original inside diameter. The load is applied between two parallel plates at a constant deflection rate of 12.7 mm/min, at a test temperature of 23°C. The test specimen shall be 305 mm long. Minimum tubing stiffness and maximum elongation are given in Table D-8. Heavy-duty tubing is required for leach beds.

Tubing Size and Perforations. Nominal diameters range from 76 to 203 mm. in 25.4-mm increments. Perforations shall be cleanly cut and uniformly spaced along the length and circumference of the tubing in a size, shape, and pattern to suit the needs of the user.

Elongation. A 1.27-m-long specimen shall be tested with the axis vertical using a test load of 5D lb, where D is the nominal inside tubing diameter in inches.

D.8 Technical Specifications

ASTM and ASAE standards provide technical specifications for various types of pipe and tubing, and their installation procedures. A list of these standards is given in Table D.9.

TABLE D–8 Physical Test Requirements for Corrugated Plastic Tubing (76 to 203 mm in Diameter) ASTM F405-76b

Physical Property	*Standard [MPa (psi)[b]]*	*Heavy Duty[a] [MPa (psi)[b]]*
Pipe stiffness at 5% deflection, minimum	0.17(24)	0.21 (30)
Pipe stiffness at 10% deflection, minimum	0.13(19)	0.175(25)
Elongation, maximum %	10	5

[a]Pipe stiffness for 254-, 305-, and 381-mm diameters as per ASTM F667-80.

[b]lb/in. per in.

TABLE D–9 Specifications for Tile, Pipe, Hose, and Tubing and Installation

No.	*Type and Specification*	*Specification No.*
1	Clay drain tile	ASTM[a] C4-Yr[b]
2	Clay drain tile, perforated	ASTM C498
3	Clay pipe, vitrified, perforated	ASTM C700
4	Concrete drain tile	ASTM C412
5	Concrete pipe, perforated	ASTM C444M
6	Concrete, nonreinforced, irrigation pipe systems, design and installation	ASAE S261.7[c]
7	Concrete sewer pipe	ASTM C14M
8	Concrete pipe for irrigation or drainage	ASTM C118M
9	Concrete pipe, reinforced, culvert, storm drain, and sewer	ASTM C76M
10	Concrete tile or pipe, method of testing	ASTM C497M
11	Drain and sewer pipe, bituminized fiber	ASTM D1861-2
12	Drain and sewer pipe, plastic	Commercial standard CS-228[d]
13	Hose and couplings, underground irrigation, thermoplastic pipelines	ASAE 376.1
14	Hose and couplings, self-propelled hose-drag irrigation systems	ASAE S394
15	Irrigation wells, design and construction	ASAE EP400.2T
16	Pipe, polyethylene for drip/trickle irrigation laterals	ASAE S435
17	Pipe, polyvinyl chloride, drain, waste, vent, and fittings	ASTM D2665
18	Pipe, plastic, properties by parallel-plate loading	ASTM D2412
19	Tubing, corrugated polyethylene and fittings	ASTM F405, F667
20	Tubing, corrugated polyvinyl chloride and fittings	ASTM F800
21	Tubing, corrugated thermoplastic installation for agricultural drainage or water table control	ASTM F449
22	Tubing, thermoplastic pipe and corrugated tubing installation in septic tank leach fields	ASTM F481

[a]American Society for Testing Materials, ASTM International, 100 Barr Harbor Drive, PO Box C700, West Conshohocken, PA, 19428-2959.

[b]Specifications include last two digits of calendar year (Yr) when approved. T for tentative and M for metric where appropriate, i.e., C4-90TM.

[c]ASAE Standards, ASAE, 2950 Niles Rd., St. Joseph, MI 49085.

[d]Available from GPO, Washington, DC 20250.

Index

D

E

F

G

H

M

N

O

P

Q

R

S

W